CAMBRIDGE MONOGRAPHS
ON MATHEMATICAL PHYSICS

General Editors: P.V. Landshoff, W.H. McCrea, D.W. Sciama, S. Weinberg

GRAVITATIONAL PHYSICS OF STELLAR AND GALACTIC SYSTEMS

FRONTISPIECE: A cluster of galaxies in the southern constellation of Pavo about 300 million light years away. This illustrates several types of galaxies: spiral, elliptical, barred and box-shaped. The giant galaxy in the center may have formed by the merging of smaller galaxies, or it may have formed first and attracted other galaxies to cluster around it. (Original negative by U.K. Schmidt Telescope Unit. Photography by Photolabs, Royal Observatory, Edinburgh, with permission.)

GRAVITATIONAL PHYSICS

OF

STELLAR

AND

GALACTIC SYSTEMS

WILLIAM C. SASLAW

University of Cambridge
University of Virginia
National Radio Astronomy Observatory, USA

CAMBRIDGE UNIVERSITY PRESS

CAMBRIDGE

LONDON   NEW YORK   NEW ROCHELLE

MELBOURNE   SYDNEY

Published by the Press Syndicate of the University of Cambridge
The Pitt building, Trumpington Street, Cambridge CB2 1RP
32 East 57th Street, New York, NY 10022, USA
10 Stamford Road, Oakleigh, Melbourne 3166, Australia

First published 1985

Printed in Great Britain at the University Press, Cambridge

Library of Congress catalogue card number:
84–12127

British Library cataloguing in publication data
Saslaw, William C.
Gravitational physics of stellar and
galactic systems. – (Cambridge monographs
on mathematical physics)
1. Gravitation
I. Title
531′.14   QC178

ISBN 0 521 23431 X

TM

to my parents

Contents

Preface

The plan of this monograph is divided into four main parts. These parts develop in order of decreasing symmetry, from idealized infinite homogeneous systems to finite flattened irregular systems. Along this sequence, the ratio of model applications to fundamental physical ideas and techniques increases. Even so, I have tended to emphasize the basic physics over detailed applications. Specific astronomical models wax and wane as data and fashions change, but the principles on which they are built have much longer lifetimes. Thus the degree to which various topics are discussed does not always reflect their popularity in today's, or yesterday's, literature.

Nearly all the theory described in this book is based on classical Newtonian gravity. Relativistic generalizations of almost every aspect are possible, and there was a flurry of these generalizations in the 1960s and early 1970s. It was greatly stimulated by possible applications to quasars. Although quasars still are unsolved, no evidence has developed that relativistic star clusters are needed to explain them. That, plus the fact that there are enough fascinating things to say about observed non-relativistic systems, persuaded me to restrict this book to classical gravity.

The book is reasonably self-contained in that most of its results can be obtained directly from preceding ones, sometimes with two or three intermediate algebraic steps to be added by the reader. Usually these steps are straightforward and they are outlined in the text. So the book is meant to be read with a notepad and pencil. If, having bought the book, you can no longer afford a notepad, feel free to write on the pages. It is a copy to be used, not a sacred relic.

Although this is a theoretical treatise, I have also tried to describe just enough related observations to set the scene and provide some motivation for the physical problems. In some cases, observations are described after a conceptual framework for them is developed. Computer simulations of N-body systems have reached the stage where they can almost be regarded as quasi-observations, or at least as experiments, and I have included some representative examples.

I have not attempted to put results in an historical context, or trace them back to their origins. Some acquaintance with the early literature has convinced me that this is a major task, interesting in its own right but inappropriate here. There are many cases, incidentally, where the original ideas go back much earlier than generally recognized. Although the book contains almost 300 references, the bibliographies are not meant to be complete, but just to be small gates into a vast literature.

Some of these gates are in sections called 'Problems and extensions' where the reader can try out his skill and pursue non-trivial problems into the literature. Results given in these problems are occasionally used later in the text. Other, more straightforward problems, are lightly sprinkled throughout the text. There are also a number of suggestions for research topics, and I have not hesitated to point out loose ends of the theory.

Gravitational physics has developed through many applications, and, inevitably, conflicts of notation arise. There are disadvantages to imposing a superficial unity on the subject by requiring each symbol to stand for a unique physical quantity. It would require too many symbols, many of which would be unfamiliar. Rather, I have generally kept to the usual notation in each branch of the subject, trusting the reader to keep his wits about him and not confuse the mass of a galaxy with the index of a tensor. Where serious confusion might occur, or I have had to use a variant notation, I have normally remarked on it in the text. Most symbols are defined within sight of the equations where they are first used in each section. For some applications it is convenient (or traditional) to think of the gravitational potential as positive; for other applications as negative. In either case, the sign in Poisson's equation is compensated by the sign in the equation of motion, where the direction of the force is usually obvious.

Many colleagues have been helpful and influential with regard to this book, and I would not have been able to complete a project this large without their assistance. For their careful reading of the manuscript, numerous helpful comments and friendly criticism, I especially thank Douglas Heggie, Ray White and Andrew Hamilton. I am also glad to thank Dennis Sciama for encouraging both the writing of this book and my early work in astrophysics. Richard Epstein and Chris Pethick invited me to give a series of lectures on gravitational physics at NORDITA, which was the catalyst that started my writing. Subsequently while writing, I enjoyed the hospitality of NORDITA, the Niels Bohr Institute, the Tata Institute and the Peking Observatory. I owe a particular debt to Larry Frederick, Dave Heeschen and Morton Roberts in Charlottesville, and Fred Hoyle, Donald Lynden-Bell, Martin Rees, the late Sir Denys Page, Sir Alan Cottrell, and the Fellows of Jesus College in Cambridge, England. They have encouraged the somewhat unusual administrative arrangements which enabled me to spend half of each year at the University of Virginia and the National Radio Astronomy Observatory in Charlottesville, and the other half at the Institute of Astronomy and Jesus College, Cambridge University.

At this stage in the acknowledgements, authors often thank colleagues 'too numerous to mention' for reading parts of the manuscript and for helpful discussions. But I shall try to mention all who, having been helpful in these respects over the years, come to mind at the moment, with double apologies to those inadvertently left out: S. Aarseth, G.B. van Albada, G.D. van Albada, T.S. van Albada, P. Anderson, J. Aoudouze, H. Ardavan, D. Arnett, J.M. Bardeen, J. Barnes, J. Barrow, J. Bartlett, M. Begelman, J. Binney, R. Bond, S. Bonometto, G. Burbidge, M. Burbidge, S. van den Burgh, W. Burke, A.G.W. Cameron, R. Cannon, B. Carr, S. Chandrasekhar, S. Colgate, G. Contopoulos, A. Cottrell, P. Crane, R. Davies, M. Davis, D. De Young, M. Disney, G. Efstathiou, P. Eggleton, R. Epstein, M. Fall, J. Faulkner, J. Felten, G. Field, W. Fowler, R. Fox, J. Franck, K. Freeman, C. Frenk, K. Fricke, M. Geller, T. Gold, P. Goldreich, J.R. Gott, J. Gunn, D. Ter Haar, E.R. Harrison, R. Havlen, M. Hénon, J. Hindmarsh, S. von Hoerner, H. van Horn, F. Hoyle, V. Icke, S. Inagaki, M. Iye, B. Jones, A. Kalnajs, J. Katz, T. Kiang, I. King, G. Lake, M. Lecar, A. Lightman, D.N. Limber, C.C. Lin, D. Lin, D. Lynden-Bell, W. Mathews, D. Merritt, L. Mestel, R. Miller, J. Monahagn, P. Morrison, R. Nakatsuka, C. Norman, J. Oort, J. Ostriker, P.J.E. Peebles, C. Pethick, K. Prenderghast, W. Press, J. Primack, M. Rees, M. Roberts, W. Roberts, M. Ruderman, E.E. Salpeter, R. Sanders, W. Sargent, B. Schutz, M. Schwarzschild, J. Sellwood, F. Shu, J. Silk, E.A. Spiegel, L. Spitzer, G. Steigman, P. Strittmatter, B. Stromgren, P. Sturrock, K. Thorne, T. Thuan, A. Toomre, J. Toomre, S. Tremain, A. Tubbs, J.A. Tyson, M. Valtonen, P. Vandervoort, G. de Vaucouleurs, J. Wadiack, J. Wheeler, S. White, M. Whittle, A. Whitworth, A.M. Wolfe, P. Woodward, A. Zee.

Saundra Mason typed a difficult manuscript with truly outstanding ability and patience. She said she was only doing her job, but she did it extraordinarily well. George Kessler, also at NRAO, drew the diagrams with a steady and skilful hand. I am very grateful to them both. It has also been a pleasure, as I expected it would be, to work with Simon Mitton, Rufus Neal, and the staff of the Cambridge University Press.

W. Saslaw

Introduction

Glendower: I can call spirits from the vasty deep.
Hotspur: Why, so can I, or so can any man; but will
they come when you do call for them?

Shakespeare

The spirit beneath the surface of nearly any astronomical phenomenon is gravitation. The reason why gravity is the motive force for much of the Universe is not hard to see. What primarily interests us about the Universe is its structure, including ourselves. And the physical reason for the existence of this structure is gravity. Even in the case of ourselves, it is the force of gravity in massive stars which drives their nuclear reactions to produce heavy elements, then eventually causes the star to explode and spew these elements throughout the galaxy. Some of them collect into new stars and planets, partly through the more gentle ministrations of gravity – and here we are! Of course, the mass of humanity, though important to ourselves, is only about 10^{-65} of the mass of the Universe. To put it another way, we contribute about 10^{-31} km s^{-1} Mpc^{-1} to the Hubble constant.

As our own origin, through star formation and evolution, was driven by gravity, so even more directly does gravity govern the dynamics of other astronomical structure: stellar clusters, the shapes and evolution of galaxies, and the motions of the entire system of galaxies. Richard Hooker, writing in the seventeenth century, thought the relations between human and gravitational phenomena were even closer:

> If the frame of that heavenly arch erected over our heads should loosen and dissolve itself; if celestial spheres should forget their wonted motions, and by irregular volubility turn themselves any way as it might happen; if the prime of the lights of heaven, which now as a giant doth run his unwearied course, should as it were through a languishing faintness begin to stand and to rest himself; if the moon wander from her beaten way, the times and seasons of the year blend themselves by disordered and confused mixture, the winds breathe out their last gasp, the clouds yield no rain, the earth be defeated of heavenly influence . . . : what would become of man himself, whom these things now do all serve? See we not plainly that obedience of creatures unto the law of nature is the stay of the whole world?

So you see, there are dire consequences of not understanding this subject!

The scope of gravitational many-body physics and its applications to astronomical systems is very broad. To gain a bit of perspective, consider two major ways in which the gravitational problem differs from other branches of many-body physics such as superfluids, condensed nuclear matter, or amorphous solids. First, we have had our Newton, so we start with a known microscopic description of the physics. The gravitational Hamiltonian was written down more than 150 years ago – by Hamilton. Thus our problem is often to find a suitable macroscopic description of the physics. This development is historically inverted compared to other branches of physics where macroscopic descriptions were followed by microscopic ones. For example: thermodynamics led to statistical mechanics, fluid mechanics was followed by kinetic theory, and a description of the bulk properties of solids preceded our understanding of their molecular and atomic structure.

Because we understand the microscopic interaction of point Newtonian particles completely (even when the 'particles' represent galaxies with diameters of 100 000 parsecs!) it is possible to do numerical experiments with them on large computers. So we can both simulate natural astronomical systems (the great analog computer in the sky) and test our macroscopic ideas of many-body properties. This has been one of the recent growth industries of astronomy and some of its results will be incorporated into our discussion.

We notice the second difference between gravitational and other many-body systems immediately on writing down the Hamiltonian. For a system of gravitationally interacting particles with momenta p_i and constant masses m_i, the Hamiltonian is the total energy of the system:

$$H = \sum_i \frac{p_i^2}{2m_i} - G \sum_{\substack{pairs \\ i \neq j}} \frac{m_i m_j}{r_{ij}}.$$

This is a simple but highly non-linear function of the separations r_{ij} of the particles. The equations of motion are also non-linear and, since all pairs are attractive, they do not saturate, unlike plasmas with overall neutrality. Already there is a hint that gravitating systems like to evolve toward more and more condensed phases – toward more negative gravitational energy. However, it does not follow that the entire system develops into a singular state with $r_{ij} \to 0$ for all i and j. Since H is constant, small values of r_{ij} must be balanced by large values of momenta. So some particles acquire high velocities and, if the system is finite, they escape it.

Thus part of the system can condense, and the rest expand to conserve total energy; so that the particles' harmonic mean separation proportional to

$$1 \left/ \sum_{\substack{pairs \\ i \neq j}} r_{ij}^{-1} \right.$$

becomes very small. This tendency toward subclustering will lurk behind much of our discussion of gravitating systems.

Although the gravitational Hamiltonian was written down in 1834, for more than

70 years its applications were mainly to the orbits of two or three bodies, with the perturbations of celestial mechanics. Observations of the motions of large numbers of stars in our galaxy had led to interest in the many-body problem by the first decade of this century. By the end of its second decade, several classic papers of Charlier, Eddington, Jeans and others treated the problem as an analog of the Maxwell–Boltzmann kinetic theory of gases. A major difference between the two approaches was the lack of short-range collisions among stars, so they used a collisionless Boltzmann approximation. In the 1930s and 40s, much effort went into using the collisionless Boltzmann equation to try to determine the gravitational potential and structure of our galaxy from observations of its stellar distribution and assumptions that its velocity distribution could be parametrized by a Gaussian ellipsoid. This was the so-called inverse Jeans problem. It did not turn out to be very fruitful for understanding the structure of our galaxy, and was superseded in this respect by the 1420 MHz measurements of the distribution of neutral hydrogen. In the 1940s also there were calculations of the evaporation of stars from clusters and in the 1950s and 1960s many astronomers began applying variants of techniques developed for plasma physics to the gravitational problem.

In the last two decades, there has been a resurgence of many aspects of this problem. The revival has been motivated mainly by observations of *quasars* and violent activity in *galactic nuclei*, consisting of highly concentrated stellar systems, new analyses of *galaxy clustering*, increased understanding of *spiral structure* of galaxies as gravitational density waves, observations of *triaxial elliptical* galaxies, and the discovery of X-ray emission from *globular clusters* of stars in our galaxy, previously thought to be very stable and boring astronomical objects but now quite lively.

PART I

Idealized homogeneous systems – basic ideas and gentle relaxation

In reality we apprehend nothing exactly, but only
as it changes according to ... the things that
impinge on or offer resistance to it.

Democritus

To understand some of the most important properties of gravitating systems we temporarily put aside the effects of their density and velocity gradients, their components of different masses and sizes, and any external forces which may act upon them. Imagine an idealized, isolated, homogeneous gravitating system of particles. Usually in discussing the physics of these systems we will call their particles stars for brevity, although when discussing many astronomical contexts they will often be galaxies, or even clusters of galaxies. These idealized systems will sometimes be finite, sometimes infinite. In later sections we will find that both sizes of homogeneous clusters turn out to be unstable. But never mind that for now; there are more basic properties to consider.

1

The average and fluctuating gravitational fields

Little by little does the trick.
Aesop

The self-consistent motions of all the stars or galaxies in an isolated cluster are completely determined by their mutual gravitational forces, so long as each object can be treated as a point mass. For a system of objects, each with radius d, this criterion requires that the volume occupied by the objects be small compared to the volume of the system, so that bodily collisions are infrequent. Two spherical objects of radius d have an effective radius for grazing collision of $2d$, and thus an effective cross section $\sigma = 4\pi d^2$. Randomly moving objects with number density n will have a mean free path λ_G to geometric collisions given by

$$\lambda_G \approx 1/n\sigma = \frac{R^3}{3Nd^3}d \tag{1.1}$$

in a spherical system of radius R containing N uniformly distributed members. The number of times an object can move through its own diameter before colliding is essentially the ratio of the cluster's volume to that occupied by objects. Moreover, the number of cluster radii the object could traverse before colliding is essentially the ratio of the projected area of the cluster to that of the objects.

For galaxies in clusters: $N \approx 10^3$, $R \approx 3$ Mpc, and $d \approx 0.01$ Mpc, so $\lambda_G/d \approx 10^4$ and $\lambda_G/R \approx 30$. For stars in globular clusters: $N \approx 10^5$, $R \approx 10$ pc, and $d \approx 3 \times 10^{-8}$ pc, so $\lambda_G/d \approx 10^{20}$ and $\lambda_G/R \approx 3 \times 10^{11}$. For stars in moderately dense galactic nuclei: $N \approx 10^6$, $R \approx 0.1$ pc, and $d \approx 3 \times 10^{-8}$ pc, so $\lambda_G/d \approx 10^{13}$ and $\lambda_G/R \approx 3 \times 10^6$. Therefore in all these cases direct collisions rarely affect the motions of most galaxies or stars. This is not to say that collisions are always unimportant for the overall evolution of the clusters, and we shall later examine the effects when they dominate.

The criterion that a system be isolated means that forces (potential gradients) from the stars within the system are much greater than forces from outside sources. An external mass M_e at distance l typically produces a force per unit mass $\sim GM_e/l^2$, if it is sufficiently spherical that its multipole contributions are unimportant. Internal forces on a star arise from two contributions: an average force produced by many distant stars and a fluctuating force produced by a few nearby stars. To make a rough estimate (which will be sharpened later), consider a star at distance r from the centre of a uniform spherical system. Within this distance the total mass is $M(r) = (\frac{4}{3})\pi mnr^3$ and it produces a mean field force $\sim GM(r)/r^2 \approx 4Gmnr$.

Thus the average internal force dominates an external force if

$$\frac{r}{R} \gtrsim \frac{R^2}{l^2}\frac{M_e}{M},\tag{1.2}$$

where $M = M(R)$. The contribution from the few nearest neighbors, which cannot be averaged because its time fluctuations are of the same order as its instantaneous value, is approximately the mass m of the nearest star times G divided by the square of the interparticle distance, i.e., $Gmn^{2/3} \approx GmN^{2/3}R^{-2}$. This fluctuating force exceeds the mean force for $r/R \lesssim N^{-1/3}$. The internal fluctuating force dominates an external force if

$$N^{-1/3} \gtrsim \frac{R^2}{l^2}\frac{M_e}{M}.\tag{1.3}$$

As a first example, consider the forces on stars within our previous globular cluster (supposing each star to be one solar mass) at the edge of a spherical galaxy 10 kpc in radius containing $10^{11}\,M_\odot$. From inequality (1.2) we find that in the central region of the cluster the external force dominates the average internal force by about an order of magnitude, and the forces are about the same near the cluster's edge. Equation (1.3) shows that the internal fluctuating forces are much less than the external force. Next, consider stars within a galaxy influenced by a companion galaxy of the same mass at $l \approx 10\,R$. The situation is different. Now the internal average force dominates, except for very small r/R. A caveat arises because the mass density in the center is idealized to be constant, giving a simple harmonic restoring force. In real galaxies this is most unlikely to be true.

It might seem, in the previous globular cluster example, that we cannot isolate the cluster from the galaxy when analyzing its internal dynamics. This is only true, however, in a restricted sense. Insofar as the external force is the same for the entire cluster, it only determines the motion of the cluster's center of mass, i.e., its orbit around the galaxy. It is the differential force on the cluster – the tidal force – which affects its internal dynamics. The effect of the external mass on the outer stars in a nearby cluster differs from the effect at the center by an amount

$$F(R) - F(0) \approx \frac{\partial F}{\partial r}\Delta r \approx \frac{F}{l}R.\tag{1.4}$$

For the globular cluster example this reduces the tidal force relative to the external gravitational force by a factor $R/l \approx 10^{-3}$. So the galaxy does not determine the cluster's internal dynamics, except near the tidal radius.

There is an interesting case when a galaxy can influence a globular cluster's internal dynamics strongly, even if the tidal forces are small. If the external field varies rather sharply in space, it can give the cluster a shock or impulse every time its orbit comes around to the irregularity. Although each shock may be small, their cumulative result over many orbits can add enough energy to the cluster to shake it apart. We return to this in more detail in Part IV; for the present we consider isolated systems.

One of the most important ideas which arose in the previous discussion is that we can separate the internal field into the irregularly fluctuating field of near neighbors plus the calmer averaged field of more distant stars. Many many-body approximations use this idea, so we now examine it further.

We start with a static case, imagining a star in a uniform random distribution. The force on this star from N stars in a small cell at distance r is proportional to N/r^2, so cells at different distances which produce the same force must contain numbers of stars $N \propto r^2$. In a uniform random distribution N will be the average number of stars in the cell, but there will also be a typical relative fluctuation $\Delta N/N \approx N^{-1/2} \propto r^{-1}$ from cell to cell. Thus, the relative importance of fluctuations in cells producing the same magnitude of gravitational force decreases inversely as their distance.

Let us look at it another way. Suppose the star of interest is at the center of a spherical cluster. If the matter were distributed in a continuous uniform way, the central star would feel no force. But because the matter is condensed into stars, its distribution is grainy. Graininess imposes a net force. A spherical shell surrounding the star will contain $N_{\text{shell}} = 4\pi r^2 n \, \Delta r$ stars, on average. Thus, the fluctuating component of the force due to the shell would be

$$\Delta F_{\text{shell}} \approx -\frac{2\pi^{1/2} G m^2 n^{1/2} (\Delta r)^{1/2}}{r} \qquad (1.5)$$

for, say, a positive number fluctuation in the shell. This force would arise, for example, if one-half the shell contained about $\frac{1}{2}(N + \sqrt{N})$ stars. No longer would the shell be completely spherical, so the central star would be deflected. The net effect on the star would result from all the shells, there being $R/\Delta r$ of them. However, the deflection of each shell is also in a random direction. The net residual effect is approximately proportional to the square root of the number of shells, $R^{1/2}/(\Delta r)^{1/2}$. Thus, the total fluctuating force is approximately

$$\Delta F_{\text{tot}} \approx \Delta F_{\text{shell}} R^{1/2}(\Delta R)^{-1/2} \approx -2\pi^{1/2} G m^2 n^{1/2} R^{1/2} \langle r^{-1} \rangle \approx -\frac{3\sqrt{3}}{2} \frac{G m^2 N^{1/2}}{R^2},$$
$$(1.6)$$

where $\langle r^{-1} \rangle$ is a convenient average reciprocal radial distance,

$$\langle r^{-1} \rangle = \frac{3}{4\pi R^3} \int_0^R \frac{1}{r} 4\pi r^2 dr = \frac{3}{2R}, \qquad (1.7)$$

which would arise from integrating the fluctuating force over the system.

Notice that the shell thickness actually cancels in the calculation, so its particular value is irrelevant. In fact, the system behaves essentially as though $N^{1/2}$ of its stars were placed in a lump roughly half-way out from the center, and this lump provides a typical fluctuation in the gravitational force. As the number of stars increases, the medium becomes more continuous and the ratio of the fluctuating to the average force decreases as $N^{-1/2}$.

These fluctuations will give the star a random component of velocity, in addition to the velocity it gains from the average field. The simple static picture we have used so far is not adequate to calculate this random velocity. As the star moves, the fluctuations in the surrounding stars also change their position and strength. Moreover, the motion of the star through the system can lead to collective effects in which the positions and velocities of nearby stars are not random, but correlated. So we must turn from static to dynamic calculations.

2
Gentle relaxation: timescales

I have moved this way and that: gradually, this
way and that, but mostly this way.

Mervyn Peake

Since gravity is a binary interaction, occurring between pairs of stars, the deflection
which results from all the interactions in a system is essentially the sum of all pair
interactions. The dynamical effect of each interaction is to deflect the star's previous
orbit. It is easy to estimate this deflection for weak scattering. Consider, as in
Figure 1, a massive star m_1 deflecting a much less massive star m_2, moving initially
with velocity v perpendicular to the impact parameter (distance of closest approach
of the undeflected orbit) b. The gravitational acceleration Gm_1/b^2 acts for an effective
time $2b/v$ to produce a component of velocity

$$\Delta v \approx \frac{2Gm_1}{bv} \tag{2.1}$$

approximately perpendicular to the initial velocity. Since $\Delta v \ll v$, the effects are
linear and they give a scattering angle $\psi \approx \Delta v/v$. Although this is the physical essence
of the problem, an exact treatment which follows the orbits in detail (available from
most standard texts in classical mechanics) shows that, more generally,

$$\tan \frac{\psi}{2} = \frac{G(m_1 + m_2)}{v^2 b} \tag{2.2}$$

and

$$\frac{\Delta v}{v} = 2 \sin \frac{\psi}{2}. \tag{2.3}$$

Equations (2.2) and (2.3) reduce to (2.1) for $m_2 \ll m_1$ and $\psi \ll 1$. They were originally
applied to stellar systems by Jeans (1913) working in analogy with Maxwell's kinetic
theory of gases. Throughout its development, there have been many productive
exchanges of techniques and results between kinetic theory – and later plasma
physics – and gravitational many-body physics.

These results show that large velocity changes, and large scattering angles, occur
when two objects are so close that their potential energy is about equal to the kinetic
energy. Are these close encounters typical? Characterizing typical encounters by an
impact parameter about equal to the mean interparticle separation $b \approx RN^{-1/3}$, an
average mass $m_1 \approx M/N$, and an initial velocity of about the equilibrium velocity

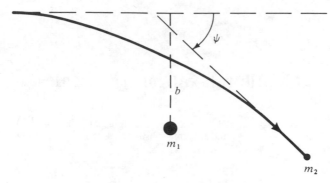

Fig. 1. The orbit of a deflected star.

dispersion $v^2 \approx GM/R$ (see Equation (9.30)), then shows that

$$\frac{\Delta v}{v} \approx \psi \approx N^{-2/3} \tag{2.4}$$

for $N \gg 1$. So the typical neighborly interaction of stars in a system near equilibrium involves very little energy or momentum exchange. Of course, there will always be a few stars in any cluster which have low relative velocities. Usually, these are stars near turning points of their orbits in the averaged potential, in the outer regions where they linger longer. In these positions they are particularly susceptible to large deflections. However, the probability that they encounter a similarly slow star nearby is small enough to make these large deflections rare. Nevertheless, there is one case where they are important. In inhomogeneous clusters with objects of different masses, the most massive members accumulate at the center and strongly alter each other's orbits. We will return to this aspect in Part III.

Although individual deflections are small, their cumulative effect is not. We may see this in several ways. The simplest approximation is to suppose that each small deflection in a typical encounter alters the orbit deflection angle ψ and velocity v in a completely random direction, by an amount given by (2.2) and (2.3). The magnitudes of v and ψ then undergo a random walk and their root mean square values increase in proportion to the square root of the number of scatterings. The average number of scatterings of any given star by surrounding stars having velocity between v and $v + \mathrm{d}v$, and with an impact parameter between b and $b + \mathrm{d}b$ in time $\mathrm{d}t$ is

$$2\pi b f(v, t) v \, \mathrm{d}t \, \mathrm{d}b \, \mathrm{d}v \tag{2.5}$$

where $f(v, t)\mathrm{d}v$ is the number of stars with velocity between v and $v + \mathrm{d}v$ at time t, normalized so that its integral over all velocities is the total number density of stars n. In the steady state, $f(v)$ is independent of t; moreover the fact that it depends only on the magnitude of v implies that we are taking the velocity distribution to be isotropic around any given point.

To estimate the mean square velocity change, we multiply the number of these scatterings by the square of the velocity change $(\Delta v)^2$ in each scattering described by

(2.2) and (2.3) for the case $\Delta v \ll v, m_1 = m_2$, and $f(v,t) = f(v)$. Then we integrate over dt, db, and dv. This procedure overestimates the effects of the last few scatterings. But since each scattering is small, it does not seriously affect the result.

The results of each scattering add up in this linear manner for $(\Delta v)^2$ because every deflection is small and uncorrelated with any other. In a steady state, the integral over time is trivial. The integral

$$\int_0^\infty v^{-1} f(v) \mathrm{d}v = n \langle v^{-1} \rangle \tag{2.6}$$

is just the average inverse velocity. The integral over impact parameter,

$$\int_{b_1}^{b_2} \frac{\mathrm{d}b}{b} = \ln \frac{b_2}{b_1}, \tag{2.7}$$

introduces a new feature of the problem. Its limits diverge at both extremes, as either b_2 becomes very large or b_1 very small, and so we must introduce a cutoff. The upper cutoff is fairly straightforward. Physically, it arises from the long-range nature of the gravitational potential and represents the largest distance over which any disturbances in the system can interact. In a finite system, this is approximately the system's radius R. In infinite systems, it is the distance over which disturbances can be correlated, and we shall find in Part II that this is closely related to the 'Jeans length' at which the kinetic and potential energies of perturbations are about equal. Its plasma analog is the Debye length.

The lower cutoff arises because very close scatterings disturb v in a violently non-linear way. These large-angle deflections are rare, and are eliminated by equating b_1 to the impact parameter for, say, $90°$ scattering. From (2.2) we then get $b_1 = 2Gm/v^2$, the distance at which the kinetic and potential energies are equal. Thus $\ln(b_2/b_1) \approx \ln(Rv^2/2Gm)$. For a finite system close to equilibrium, the virial theorem (described in Section 9) shows that $v^2 \approx GmN/R$, i.e., that the average kinetic energy of a star is about one-half its potential energy in the cluster. Thus the logarithmic term is approximately $\ln(N/2)$. An alternative approach is to take the lower cutoff to be the distance at which a nearby star would exert as much gravitational force as the average force of the cluster: $m/b_1^2 \approx M/R^2$. In this case $\ln(b_2/b_1) \approx 1/2 \ln N$. A third alternative would be the nearest neighbor separation: $b_1 \approx (4\pi R^3/3N)^{1/3}$. Although some interesting physical problems surround this ambiguity in b_1, its very weak effect on the result shows that it is not a great barrier to our understanding of gentle relaxation.

Putting the three integrations together gives a cumulative mean square velocity perturbation

$$\left(\frac{\Delta v}{v}\right)^2 \approx 32\pi G^2 m^2 n v^{-3} \ln(N/2) t. \tag{2.8}$$

Apart from the ambiguity of the logarithmic factor, the main uncertainty in this formula comes from the identification of $\langle v^{-1} \rangle$ with the inverse of the velocity dispersion in the system. Thus (2.8) should be multiplied by a numerical factor, usually between 0.1 and 1.0, which depends on the velocity distribution function. For example, the more detailed discussion of Chandrasekhar (1960) shows that with a Maxwellian distribution, this factor is about 0.2 for test stars having speeds close to the velocity dispersion. In practice, these uncertainties are usually much less than the uncertainties in the astronomically observed values of m, n, v, and t. Therefore numerical experiments are more useful than observations for checking the theory of stellar deflections.

Lecar & Cruz-Gonzáles (1972) designed a numerical experiment to check the independence and linearity of individual deflections. One hundred field stars were placed uniformly at random throughout a sphere and told to remain at rest. A massless test star was then shot into the system with an initial velocity along a diameter, and its orbit computed. Encounters, both near and far, with the motionless field stars turned its velocity vector through an angle ψ in time t. The results were averaged over an ensemble of cases with different random distributions of test stars. This gave an experimental value for $\langle t^{-1} \sin^2 \psi \rangle$. Next, for each case the deflections ψ_i caused by each field star were calculated separately, as though no other field stars were present, and added up to find $\langle t^{-1} \Sigma \sin^2 \psi_i \rangle$. The whole process was repeated for several initial velocities, ranging over a factor of six.

For each initial velocity, the two computations of average deflection agreed with each other to within one or two standard deviations (as measured by the dispersion of the ensemble results). The values of $\langle t^{-1} \Sigma \sin^2 \psi_i \rangle$ also agreed well with the theoretical results of Equation (2.8). However, values of $\langle t^{-1} \sin^2 \psi \rangle$ systematically fell somewhat lower. Moreover, the vectorial sum of the individual deflections was not generally the same as the combined deflection of all the field stars acting at once. This suggests that the theory is indeed a good description of cumulative, independent linear deflections. However, in the real situation there are also a few non-linear scatterings which do not add independently and they can make an important contribution. It is not understood how to treat them theoretically in a systematic way.

Since deflections increase the expectation value of the root mean square velocity, we can easily define a timescale τ_R, using (2.8), for this many-body relaxation effect to become significant for a typical star. We just set $(\Delta v/v) = 1$, whence

$$\tau_R = \frac{v^3}{32\pi G^2 m^2 n \ln (N/2)}$$

$$= \frac{v^3 R^3}{24 G^2 m M \ln (N/2)} \tag{2.9}$$

where M is the total mass. In systems near equilibrium, $GM \approx Rv^2$. We can define a characteristic 'crossing time'

$$\tau_c = R/v \tag{2.10}$$

and write

$$\tau_R = \frac{0.04 \, N}{\ln (N/2)} \tau_c. \tag{2.11}$$

Since for globular clusters, galaxies, and rich clusters of galaxies $N \gtrsim 10^3$, the relaxation time is much longer than the dynamical crossing time. So cumulative small deflections do not amount to much over any one orbital period, but their secular effect can dominate after many orbits. Only for small globular clusters and dense galactic nuclei, however, will gentle relaxation be important on cosmological timescales.

It is amusing to compare these relations with the analogous calculation for a system of point masses confined to a disk. Consider just the orders of magnitude. Noting that the number of encounters during time t with impact parameters between b and $b + db$ is now $2vst\,db$ with $s \approx N/R^2$ the surface density, we obtain the disk analog of (2.8)

$$\left(\frac{\Delta v}{v}\right)^2 \approx \frac{G^2 m^2 st}{v^3} \int_{b_1}^{b_2} \frac{db}{b^2}. \tag{2.12}$$

The first interesting difference is that the impact parameter integral no longer diverges for large distances. This is because the volume filled by distant particles is less by one spatial dimension when everything is confined to a two-dimensional disk. Comparing the results for $b_2 = 2b_1$ and $b_2 = \infty$ in (2.12), shows that both close and distant encounters now contribute about equally to the cumulative deflection. Distant encounters dominated the three-dimensional case. Using the approximate velocity relation $v^2 \approx GNm/R$ again from the virial theorem, and taking again $b_1 \approx Gm/v^2 \approx R/N$, we now find

$$\tau_R \approx \tau_c \approx R/v \tag{2.13}$$

instead of the result (2.11). Thus the second interesting difference, implied by the first, is that the crossing and gentle relaxation timescales are not much different. In this sense, stars strictly confined to a disk interact more strongly with each other than stars allowed to wander through all three dimensions.

Returning to the three-dimensional case, we notice that (2.8) seems to imply that the mean square velocity can increase indefinitely, even though the rate of increase will slow as v^{-1}. Of course, this cannot be since the star would eventually reach escape velocity and leave the cluster. We will describe this process in detail in Part III. But we can already make a rough estimate of the timescale for stars to evaporate from the cluster. The escape velocity, $v_{esc} \approx (4GM/R)^{1/2}$, is typically a factor two greater than the velocity dispersion in a cluster near equilibrium (see Section 45). So an average star evaporates when $\Delta v \approx v$, and from (2.9) this just takes a time τ_R. Thus τ_R can also be interpreted as the timescale for a cluster to change its structure significantly by losing its members.

How many 'effective encounters' does a star undergo before it reaches escape velocity? This is essentially the problem of a random walk through velocity space with an absorbing barrier at the escape velocity. Figure 2 illustrates this random walk schematically. If the velocity increments have a Gaussian distribution with mean value zero, so that there is no net velocity drift, and dispersion σ, then it is physically plausible (and can be shown rigorously (e.g., Cox & Miller, 1965)), that

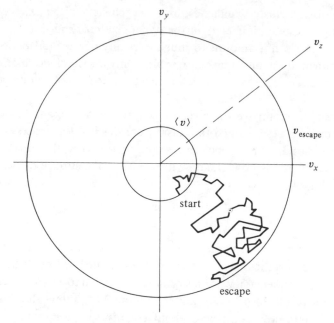

Fig. 2. Random walk of a star in velocity space from $\langle v \rangle$ to v_{esc}.

the expected number I of increments necessary to reach escape velocity is given by

$$v_{\text{esc}} \approx \sigma \sqrt{I}. \tag{2.14}$$

Now in a Gaussian distribution, the velocity dispersion σ is also the most probable value of a velocity increment Δv. We can estimate it from (2.2) and (2.3) for $m_1 = m_2$ by using the volume weighted average of $1/b$ as in (1.7), $\langle b^{-1} \rangle = \frac{3}{2} R$. Thus $\sigma = \langle \Delta v \rangle = 6Gm/Rv$ and with $v_{\text{esc}} = 2v$ we find that a star escapes after undergoing approximately

$$I = \frac{v_{\text{esc}}^2}{\sigma^2} = \frac{R^2 v^4}{9G^2 m^2} = \frac{N^2}{9} \tag{2.15}$$

encounters. Here N is again the number of stars in the cluster and the last equality applies for virial equilibrium.

   Since most encounters are very distant ones, occurring among stars separated by about a cluster radius, they are never really completed. The long-range nature of gravitational forces gives many-body systems a long 'memory' of their previous history. Therefore (2.15) gives only a rough sense of the number of encounters because most encounters do not have a well-defined beginning or end. Nevertheless, the result indicates again how very large numbers of very weak long-range interactions accumulate to change the nature of the cluster.

   While the simple approaches to gravitating systems we have used so far help to establish some basic notions and results, they obviously have their limitations. For example, they do not take account of correlations of nearby orbits, nor do they include more distant collective effects. In both these cases, the velocity increments do not have a Gaussian distribution. Moreover, these models do not consider that when stars are near the edge of a system they feel different fluctuating forces than when they are embedded within the system. But most objectionable is that these models are very difficult to generalize in a self-consistent way for more complicated systems. They cannot handle realistic inhomogeneous clusters, for example.

We therefore need a more powerful description of many-body systems. To establish some of the relevant ideas we start with simple generalizations of the orbit relaxation model and develop some of their properties. Then we will switch to the most complete description of the gravitational many-body problem which is possible. This description will be so powerful that we will have to specialize it again in various ways to obtain useful information from it. So these next few sections will establish the basic mathematical descriptions of the problem.

3

The dynamics of random impulsive forces

The slings and arrows of outrageous fortune.
Shakespeare

Previous sections have made it plausible that an object in a gravitating system near equilibrium can be considered to be immersed in a bath of fluctuating forces, along with an average mean field force. We now consider a simple mathematical model for the time evolution of orbits. We use this intuitive physical picture to try to capture the essence of the problem in a fairly simple way. An advantage of this procedure is that it readily suggests modifications of the description for an improved physical picture. The results can always be checked against N-body computer experiments, and we will discuss their more exact derivation in Section 10.

At first sight, the simplest model might seem to represent the motion of each star by Newton's equation of motion with a stochastic force $\boldsymbol{\beta}(t)$ which fluctuates in time, i.e., $m\dot{\boldsymbol{v}} = \boldsymbol{\beta}(t)$. But this turns out to be too simple. It makes the velocity undergo Brownian motion (for a Gaussian distribution of fluctuations) with an ever-increasing root mean square value $v_{\mathrm{rms}} \propto t^{1/2}$. Correspondingly, the root mean square position of an average star also departs monotonically from its initial value. These two properties are inconsistent with conservation of total energy, for the increase in kinetic energy $\propto v_{\mathrm{rms}}^2$ must be compensated by a contraction of the system to decrease the potential energy. But the Brownian increase of every star's root mean square position from its initial value prevents the system from becoming very small. Moreover, it is clear that this model has no possibility of a stationary state. Now it is not obvious, at this stage of our discussion, that gravitating systems have rigorous stationary states and, as we shall eventually show, they do not. Nonetheless, numerical simulations show that stars in these systems do not have a tendency to increase both their root mean square velocities and positions indefinitely. Furthermore, systems such as globular clusters and some elliptical galaxies seem to have existed in nearly their present state for about 10^{10} yr. This suggests they have reached at least a quasi-stationary state of some sort, or at least one in which global diffusion is slow compared to the fluctuation timescale of order $N/\sqrt{(G\rho)}$.

Of course, what this simple model has left out is the effect of the average drag force. A star moving faster than its neighbors deflects the surrounding orbits so that the average density behind the star is slightly greater than in its direction of motion.

The resulting excess gravitational force behind the star tends to slow it down. This process is known as dynamical friction (see Sections 4, 5, 13, 14.3, 17.2 and 40.3). Since we want to inhibit large departures from the star's average motions, the simplest phenomenological description would have a restoring force proportional to the star's velocity. So we might examine a model equation of the form

$$m\frac{\mathrm{d}\mathbf{v}}{\mathrm{d}t} = -\alpha\mathbf{v} + \boldsymbol{\beta}(t). \tag{3.1}$$

This approach was adopted by Langevin as early as 1902 to describe Brownian motion, and (3.1) is usually called the Langevin equation. It is one of the first examples of finding new physical results by looking for the simplest adequate mathematical description of a phenomenon – a technique which Einstein brought to a peak during the following two decades.

What are the consequences of the Langevin equation and how do they apply to gravitating systems? There is a naive way and a sophisticated way to solve (3.1). We shall use the naive method, but also remark on the more sophisticated method which resolves some of the mathematical inconsistencies – but yields the same results. Then we describe a more sophisticated version of the naive method which resolves these problems in a simple manner, and also permits an interesting generalization.

If we pretend that (3.1) is an ordinary linear differential equation for each vector component, its solution is formally

$$\mathbf{v}(t) = \mathbf{v}(0)\exp\left(-\frac{\alpha}{m}t\right) + \frac{1}{m}\int_0^t \exp\left[-\frac{\alpha}{m}(t-\tau)\right]\boldsymbol{\beta}(\tau)\mathrm{d}\tau \tag{3.2}$$

for any star at a given time. To determine the average velocity of a group of stars at an arbitrary time we need to average $v(t)$. So we must specify the fluctuating force $\boldsymbol{\beta}(\tau)$. This could be done by solving the entire stellar dynamical problem self-consistently, but we are not at that stage yet. Instead we make an educated guess about the major properties of $\boldsymbol{\beta}(\tau)$, namely its first and second moments. Since the fluctuating field is the sum of a large number of small perturbations, we can suppose that these perturbations are independent. A positive force being as probable as a negative one, the time average value of $\boldsymbol{\beta}(\tau)$ is zero.

$$\langle \boldsymbol{\beta}(\tau) \rangle = 0. \tag{3.3}$$

Since perturbations at different times and different vector components are independent

$$\langle \beta_i(\alpha\tau/m)\beta_j(\alpha\tau'/m) \rangle = \alpha m\sigma^2 \delta[\alpha(\tau-\tau')/m]\delta_{ij} \tag{3.4}$$

where σ^2 represents an amplitude (chosen to have the dimensions of (velocity)^2 time^{-1}) and the standard delta functions are zero unless their arguments are identical

$$\delta_{ij} = \begin{cases} 0 & i \neq j \\ 1 & i = j \end{cases}$$

$$\delta(x-x') = 0, \quad x \neq x',$$

$$\int \delta(x - x') dx = 1 \quad \text{if } x' \text{ is included in the integral}$$

$$= 0 \quad \text{if } x' \text{ is not included,}$$

with $x \equiv \alpha\tau/m$. These are the properties of a Gaussian stochastic process. They pervade many statistical problems in physics because they are simple and, as texts on probability theory show, they follow from the central limit theorem as the natural outcome of most processes which respond to many independent inputs.

The physical meaning of our assumption that the perturbations in (3.2) are independent is that they are uncorrelated after times longer than any other timescale in the problem. Only one other timescale appears in (3.2): the relaxation time m/α for an initial velocity to decay. This relaxation time also contributes to the exponential weighting of the fluctuating force in the integral on the righthand side of (3.2). But since $\boldsymbol{\beta}(\tau)$ is uncorrelated at different times, $\mathbf{v}(t)$ will not depend on the detailed history of fluctuations, in this approximation.

How can we estimate α/m? If the fluctuating force in (3.1) were zero, the only timescale characterizing the restoring force would be m/α. On the other hand, if there were no restoring force, the fluctuating force would be characterized by the diffusion timescale τ_R of (2.9). If both these forces are present, and they are to lead to a balanced, nearly stationary state, their characteristic timescales must be approximately equal. This phenomenological argument suggests that

$$\frac{m}{\alpha} \approx \tau_R. \tag{3.5}$$

The situation for gravitating systems differs from the Brownian motion of small particles, to which Langevin's equation is often applied. For small particles, viscosity is the restoring force and α is given by Stokes law (e.g., Einstein, 1955). In gravitating systems, momentum is not transferred by physical collisions so viscosity loses its ordinary interpretation. This is easily seen by noting that the ratio of the mean free path for large-angle scattering, $\lambda = 2/n\sigma$, where the cross section $\sigma \approx b_1^2 \approx G^2m^2/v^4 \approx R^2/N^2$, to the size of the system is $O(N) \gg 1$. It is possible to redefine an 'effective viscosity' η for gravitational scattering by requiring it to satisfy the analog of Maxwell's relation $\tau_R = 2\eta v^2/\rho$ for the relaxation of a gas. However, this would amount to little more than putting old wine in new bottles.

Now we can derive some basic properties of (3.1), which also serve to illustrate more general concepts. First integrate (3.2) over t for $t > 0$ and $t > \tau$ to find

$$\mathbf{x}(t) = \mathbf{x}_0 + \dot{\mathbf{x}}_0 m\alpha^{-1} \left[1 - \exp\left(-\frac{\alpha}{m}t \right) \right] + \alpha^{-1} \int_0^t$$

$$\cdot \left\{ 1 - \exp\left[-\frac{\alpha}{m}(t - \tau) \right] \right\} \boldsymbol{\beta}(\tau) d\tau. \tag{3.6}$$

This holds for any individual star. To determine the position of a typical star, we average over many stars. Recalling (3.3) gives

$$\langle \mathbf{x}(t) \rangle = \mathbf{x}_0 + \dot{\mathbf{x}}_0 m\alpha^{-1}(1 - e^{-\alpha t/m}). \tag{3.7}$$

Similarly, from (3.2),

$$\langle \mathbf{v}(t) \rangle = \mathbf{v}_0 e^{-\alpha t/m}. \tag{3.8}$$

Thus the average velocity of a group of stars decays from its initial value toward zero on a timescale $m/\alpha \sim \tau_R$ as a result of the random forces acting on them. For short times, the average position starts increasing proportionally to $\dot{\mathbf{x}}_0 t = \mathbf{v}_0 t$, but at long times the average star's position reaches an asymptotic departure from where it started.

However, the average position is not the same as the root mean square position – a familiar property of Brownian motion. Since positive and negative values both contribute to the average position, it can be much less than the root mean square position, and its time development will be different. This will also be true for the root mean square velocity, which is calculated from (3.2)

$$\langle v^2(t) \rangle = v_0^2 e^{-2\alpha t/m} + \frac{1}{m^2} \int_0^t \int_0^t \langle \boldsymbol{\beta}(\tau) \cdot \boldsymbol{\beta}(\tau') \rangle \exp{-\frac{\alpha}{m}(t - \tau + t - \tau')} \mathrm{d}\tau \mathrm{d}\tau'$$

$$= v_0^2 e^{-2\alpha t/m} + \frac{\sigma^2 m}{2\alpha}(1 - e^{(-2\alpha/m)t}). \tag{3.9}$$

The cross terms which would arise in (3.9) vanish in the averaging by (3.3), and we have used (3.4) to evaluate the integral, after changing to the dimensionless variable $x = \alpha\tau/m$. Thus after long times, $t \gg t_R$, the root mean square velocity $\langle v^2(t) \rangle^{1/2}$ or velocity dispersion relaxes to an equilibrium value $\sigma(m/2\alpha)^{1/2}$, even though the average velocity is zero. A similar calculation of the mean square position with $x_0 = 0$ gives

$$\langle x^2(t) \rangle = v_0^2 m^2 \alpha^{-2}(1 - 2e^{-\alpha t/m} + e^{-2\alpha t/m})$$

$$+ \frac{m^3 \sigma^2}{2\alpha^3}(2\alpha m^{-1} t - 3 + 4e^{-\alpha t/m} - e^{-2\alpha t/m}). \tag{3.10}$$

Expanding the exponentials for short times shows that when $t \ll m/\alpha$, $\langle x^2 \rangle^{1/2} \approx v_0 t$, and the motion resembles that of ordinary reversible mechanics. For much longer times

$$\langle x^2 \rangle^{1/2} \approx \tau_R^{3/2}\left(\frac{\alpha t}{m}\right)^{1/2} \sigma \tag{3.11}$$

and irreversibility sets into our description as a result of averaging the fluctuating forces.

This process describes the Brownian motion of an average star. The motion has several amusing properties described in more detail in most texts on stochastic processes. For example, if you follow the motion of a Brownian particle long enough, you will find that it has written your name in an arbitrarily good forgery. Not only will it be written on one particular scale, but if you look at the trajectory at any magnification, you will find your name on all scales. And not only your name, but everything that has ever been written, as well as everything that will be written in the future. The only problem (and what a problem!) is distinguishing the message from the noise. These results follow directly from the uncorrelated nature of Brownian motion, and the basic principle is similar to that of many monkeys typing forever at random. There are also more subtle properties, such as the space-filling nature of Brownian trajectories in one and two dimension, but not in three.

At the beginning of this analysis of the Langevin equation, I mentioned that although it gives the correct results, it is mathematically inconsistent. We can now illustrate why this is so. The Langevin equation involves a velocity derivative and its solution assumes that the average value of this derivative exists and is equal to the derivative of the average value (since the averages are taken over an ensemble, not over time). But this derivative does not exist! For, from (3.9) we see, setting $v_0 = 0$, that

$$\lim_{\Delta t \to 0} \frac{\Delta \langle v^2(t) \rangle^{1/2}}{\Delta t} \sim \frac{(\Delta t)^{1/2}}{\Delta t} \sim \frac{1}{(\Delta t)^{1/2}}, \qquad (3.12)$$

which is infinite. The basic physical problem is that for very short times the motions at successive instants are correlated and the assumption of (3.4) breaks down. One could replace this assumption by a more general correlation with a memory. Alternatively, one could regard this assumption as posing a definite mathematical problem which can be solved consistently by redefining the notions of stochastic derivatives and integrals. Both these approaches are discussed in detail in an excellent review by Fox (1978). Either one can make these essential physical results more rigorous.

Here, however, we shall take a different tack. It will introduce a basic approach of stellar dynamics – the Fokker–Planck equation. Later we will rederive this from fundamental physical principles. Now we see how it follows naturally from the Langevin equation and provides an equivalent solution.

To illustrate this in the simplest way, we rewrite the content of the Langevin equation (3.1) in the one-dimensional case as

$$\mathrm{d}v = a(v, t)\mathrm{d}t + Z(t)\sigma(v, t)\sqrt{(\mathrm{d}t)}. \qquad (3.13)$$

This is a stochastic differential equation in the form discussed by Ito (1951). It says that the change in velocity $\mathrm{d}v$ (or, more precisely, the change in a stochastic function $v(t)$ which takes values $v(t) = v$) in a short time interval $\mathrm{d}t$ is caused by two effects. The first is a continuous process which has mean value $a(v, t)\mathrm{d}t$; this is the 'friction' term $-\alpha v$ (for a unit mass). The second process is a stochastic force $Z(t)$ producing a displacement proportional to $\sigma t^{1/2}$ with a variance σ^2, as in (3.11).

Now, instead of trying to solve (3.13) directly for v, we will solve for the distribution of v resulting from the random forces. Quantitatively, this distribution is described by a time dependent density function $f(v, t)$ such that $f(v, t)\mathrm{d}v$ is the probability of finding the velocity in a small interval between v and $v + \mathrm{d}v$. This probability of course has values between 0 and 1, has, at most, a finite number of step discontinuities, and is normalized such that its integral over all velocities is unity. We may also specify it at some initial time. Imagine we have found $f(v, t)$. Then at any time we can use it to compute the expectation (mean) value of v (or any other variable)

$$E(v) = \bar{v}(t) = \int_{-\infty}^{\infty} vf(v, t)\mathrm{d}v \qquad (3.14)$$

and its variance

$$E((v - \bar{v})^2) = \int_{-\infty}^{\infty} (v - \bar{v})^2 f(v, t)\mathrm{d}v \qquad (3.15)$$

These expectation values are essentially moments of v with respect to f, and there is an easier way to calculate them than by doing lots of integrals. For, if we once calculate the two-sided Laplace transform of f

$$f^*(\Theta, t) = \int_{-\infty}^{\infty} \mathrm{e}^{-\Theta v} f(v, t)\mathrm{d}v, \qquad (3.16)$$

then all the moments of v follow just by taking derivatives of f^* evaluated at $\Theta = 0$

$$\left. \frac{\partial^n f^*(\Theta, t)}{\partial \Theta^n} \right|_{\Theta=0} = (-1)^n \int_{-\infty}^{\infty} v^n f(v, t)\mathrm{d}v = (-1)^n E(v^n). \qquad (3.17)$$

So the expectation value of $\mathrm{e}^{-\Theta v}$ is the moment generating function of v.

The change of the distribution function with time can be written in two ways, and equating the results yields the Fokker–Planck equation (not named after its discoverer, Lord Rayleigh). First, from (3.16) we have directly

$$\frac{\partial f^*(\Theta, t)}{\partial t} = \int_{-\infty}^{\infty} \mathrm{e}^{-\Theta v} \frac{\partial f(v, t)}{\partial t}\mathrm{d}v. \qquad (3.18)$$

Second, we go back to the basic definition of this derivative

$$\frac{\partial f^*}{\partial t} = \lim_{\Delta t \to 0} \frac{\Delta f^*}{\Delta t} = \lim_{\Delta t \to 0} \frac{1}{\Delta t}[f^*(\Theta, t + \Delta t) - f^*(\Theta, t)]$$

$$= \lim_{\Delta t \to 0} \frac{1}{\Delta t} E[\mathrm{e}^{-\Theta\{v(t) + \Delta v(t)\}} - \mathrm{e}^{-\Theta v(t)}]. \qquad (3.19)$$

The stochastic Langevin equation (3.13) describes Δv in the limit $\Delta t \to 0$ for a given value of v, i.e., Δv is conditional on v. Moreover, since the expectation value of an expectation value is just the original expectation value, we can rewrite the expectation value of (3.19) as

$$E[(\mathrm{e}^{-\Theta \Delta v(t)} - 1)\mathrm{e}^{-\Theta v(t)}] = \int_{-\infty}^{\infty} E[(\mathrm{e}^{-\Theta \Delta v(t)} - 1)\mathrm{e}^{-\Theta v(t)}]f(v, t)\mathrm{d}v$$

$$= \int_{-\infty}^{\infty} E[\mathrm{e}^{-\Theta \Delta v(t)} - 1]\mathrm{e}^{-\Theta v(t)}f(v, t)\mathrm{d}v. \qquad (3.20)$$

To calculate the expectation value on the right hand side of this equation we use

(3.13), noting that $E[Z(t)] = 0$, $E[Z^2(t)] = 1$, and $E[Z^3(t)] = O(1)$. Substituting (3.13) for Δv into (3.20), expanding the exponential $e^{-\Theta \Delta v}$ and integrating then shows that to terms of order Δt

$$E[e^{-\Theta v(t)} - 1] = -\Theta a(v, t)\Delta t + \frac{\Theta^2}{2}\sigma^2(v, t)\Delta t. \tag{3.21}$$

Inserting (3.21) and (3.20) back into (3.19) and taking the limit $\Delta t \to 0$ gives

$$\frac{\partial f^*}{\partial t} = \int_{-\infty}^{\infty}\left[-\Theta a(v, t) + \frac{\Theta^2}{2}\sigma^2(v, t) \right]e^{-\Theta v}f(v, t)\mathrm{d}v. \tag{3.22}$$

Integrating this by parts twice to obtain a form similar to (3.18),

$$\frac{\partial f^*}{\partial t} = \int_{-\infty}^{\infty} e^{-\Theta v}\left[\frac{1}{2}\frac{\partial^2}{\partial v^2}(\sigma^2 f) - \frac{\partial}{\partial v}(af)\right]\mathrm{d}v \tag{3.23}$$

since the distribution function decreases sufficiently fast at the limits. Then equating the integrands of Equations (3.18) and (3.23) yields

$$\frac{\partial f(v, t)}{\partial t} = -\frac{\partial}{\partial v}[a(v, t)f(v, t)] + \frac{1}{2}\frac{\partial^2}{\partial v^2}[\sigma^2(v, t)f(v, t)] \tag{3.24}$$

which is the Fokker–Planck equation. It describes the time change of the velocity distribution function which solves the Langevin equation. It is easy to generalize this approach for the full six-dimensional velocity and position distributions. Other stochastic equations also lead to modified versions of the Fokker–Planck equation. We will see how to apply the Fokker–Planck equation to spherical stellar systems in Section 40.3.

4

General properties of Fokker–Planck evolution

Before specializing to the most interesting cases for gravitational systems, it helps to develop a feeling for the general nature of the simple one-dimensional Fokker–Planck description. First of all, notice that if the frictional force $a(v, t)$ in (3.24) is zero, and the dispersion σ is independent of velocity, the result

$$\frac{\partial f}{\partial t} = \frac{\sigma^2}{2} \frac{\partial^2 f}{\partial v^2},$$ (4.1)

is just the normal diffusion equation, albeit for the density in velocity space. Phenomenologically, diffusion equations result from supposing that a quantity, such as mass density ρ, has a flux $\rho v = \mathbf{J}$ which is proportional to its gradient

$$\mathbf{J} = -D\nabla\rho.$$ (4.2)

Here D is the 'diffusion constant' and it depends on the particular properties of the interacting particles. The particles satisfy a continuity equation

$$\frac{\partial \rho}{\partial t} = -\nabla \cdot \mathbf{J}$$ (4.3)

since they are neither created nor destroyed, but merely flow throughout the system. Taking the divergence of (4.2) and substituting (4.3) for $\nabla \cdot \mathbf{J}$ immediately gives a result of the form of (4.1).

Since the diffusion equation involves only a first order time derivative, its evolution is not time reversible. If this process were used to describe a stellar system, the stars would not conserve energy, as mentioned at the beginning of Section 3. Nor would the entropy of the system reach a constant equilibrium value; instead it would increase forever. To see this, and also to introduce a powerful technique for extracting information from partial differential equations, we multiply both sides of (4.1) by v^2 and integrate over all velocities (the infinite limits are really symbolic, but suffice for most cases when the distribution function decreases sufficiently fast). This is a simple example of taking moments of the equation, and since the second order velocity moment of $f(v, t)$ is twice the average energy (per unit mass), it is a moment to remember

$$\int_{-\infty}^{\infty} v^2 \frac{\partial f}{\partial t} dv = \frac{\partial}{\partial t} \langle v^2 \rangle = \frac{\sigma^2}{2} \int_{-\infty}^{\infty} v^2 \frac{\partial^2 f}{\partial v^2} dv = \sigma^2. \qquad (4.4)$$

The right hand side of (4.1) has been integrated by parts twice, recalling the normalization $\int_{-\infty}^{\infty} f(v,t)dv = 1$. Integrating Equation (4.4), we see that if we introduce a distribution of stars into the system, initially with zero velocity, they will diffuse with an average energy per unit mass $\frac{1}{2}\langle v^2 \rangle = \sigma^2 t/2$, the familiar result of Brownian motion. Indeed, we can readily derive this same result from the Langevin approach by taking the limit $\alpha \to 0$ for any fixed time in (3.9).

The continuous production of entropy by diffusion is also easy to demonstrate. Standard texts on statistical mechanics show that for a weakly interacting classical system (remember this does not describe gravitating systems yet) the entropy S is related to the distribution function by

$$S = -\int_{-\infty}^{\infty} f(v,t)\ln f(v,t)dv. \qquad (4.5)$$

(For further properties of entropy, see Sections 10 and 29.) Taking the time derivative of S, substituting (4.1), and integrating once by parts presuming that the gradient of f is zero at its limits gives

$$\frac{dS}{dt} = \frac{\sigma^2}{2} \int_{-\infty}^{\infty} f^{-1} \left(\frac{\partial f}{\partial v} \right)^2 dv. \qquad (4.6)$$

Since $f > 0$ and $\partial f/\partial v \neq 0$ over a finite interval of velocities, we have $dS/dt > 0$. A similar proof holds for the Fokker–Planck equation. These are simple cases of Boltzmann's H-theorem, which can be proved with great rigor and generality.

Next, consider (3.24) for the steady state $\partial f/\partial t = 0$. In any small volume element, the outward diffusion is being balanced by particles which have diffused in and lost velocity through dynamical friction. Thus, the particular distribution function which describes this case has reached equilibrium and no longer generates entropy. With no dependence of f, σ, or a on time, (3.24) is easy to solve once the velocity dependence of $\sigma(v)$ and $a(v)$ are specified. The simplest form of σ would be constant throughout the system, independent of velocity. For the right form of $a(v)$ such an equilibrium would represent an isothermal distribution which we know applies to a weakly coupled gas, for example. However it will not generally apply to gravitating systems, except as a first approximation when they evolve very slowly (e.g., by evaporating stars) compared to their dynamical crossing time. If we assume $\sigma =$ constant, we will need a consistent form for the steady state friction term $a(v)$. If this were independent of velocity, then from (3.13) each object would change its velocity by the same increment in a short time interval. For constant $a > 0$, say, objects of either positive or negative velocities would be accelerated to large positive velocities. This would clearly be the case for any additional dependence of $a(v)$ on even powers of v. However, if a were to depend on only odd powers of v with negative coefficients, the negative velocities would become less negative and the positive velocities less positive. Thus if we expand $a(v)$ in odd powers of v, the phenomenological form $a = -\alpha v$ (for a unit mass) can be regarded as the first term of such an expansion, providing the simplest form of dynamical friction. This was the intuition behind Langevin's analysis.

With these forms for the coefficients of $f(v)$, the normalized solution of the steady state Fokker–Planck equation is

$$f(v) = \sigma^{-1}(\alpha/\pi)^{1/2}e^{(-\alpha/\sigma^2)v^2}.$$ (4.7)

This result should not be surprising, since it is just the Maxwell–Boltzmann distribution. Its derivation from the Fokker–Planck equation, however, exhibits how detailed assumptions about the kinetic motion influence the equilibrium state.

Of course, the main virtue of this approach is that it enables us to examine departures from equilibrium. The time dependent form of (3.24) with the present constant values of the coefficients

$$\frac{\partial f(v,t)}{\partial t} = \alpha \frac{\partial}{\partial v}(vf) + \frac{\sigma^2}{2}\frac{\partial^2 f}{\partial v^2}$$ (4.8)

will illustrate this for linear departures from equilibrium. Non-linear departures would entail higher order terms in the Fokker–Planck expansion (3.24) and, in some cases, completely different relaxation mechanisms. Equation (4.8) is a second order partial differential equation which is most easily solved by the following rather powerful technique. By taking its Fourier transform

$$f^*(\xi,t) = (2\pi)^{-1/2}\int_{-\infty}^{\infty} e^{-i\xi v}f(v,t)dv$$ (4.9)

we decrease the order of the equation by one. Then we solve the differential equation for the transform of the distribution function and invert the transform to get

$$f(v,t) = (2\pi)^{-1/2}\int_{-\infty}^{\infty} e^{i\xi v}f^*(\xi,t)d\tau.$$ (4.10)

So, multiplying (4.8) by $(2\pi)^{-1/2}e^{-i\xi v}$ and integrating over all velocities, using one integration by parts for the first term on the right hand side and two for the second, shows that $f^*(\xi,t)$ satisfies the first order equation

$$\frac{\partial f^*}{\partial t} + \alpha\xi\frac{\partial f^*}{\partial \xi} = -\tfrac{1}{2}\sigma^2\xi^2 f^*.$$ (4.11)

This is the simplest basic form of partial differential equation solved in many standard texts. Briefly, the function $f^*(t,\xi)$ defines a surface over the t–ξ plane. Since

$$0 = \frac{\partial f^*}{\partial t}dt + \frac{\partial f^*}{\partial \xi}d\xi - df^*$$

$$\equiv f_t^* dt + f_\xi^* d\xi - df^*,$$ (4.12)

this surface has a normal vector $(f_t^*, f_\xi^*, -1)$ at any point on the surface. Moreover, the scalar product

$$(1, \alpha\xi, -\tfrac{1}{2}\sigma^2\xi^2 f^*) \cdot (f_t^*, f_\xi^*, -1) = 0 \text{ by (4.11) so } (1, \alpha\xi, -\tfrac{1}{2}\sigma^2\xi^2 f^*)$$

is perpendicular to the normal and tangent to the plane. Being tangent means it is parallel to a small surface increment $(dt, d\xi, df*)$ so we can write

$$\frac{dt}{1} = \frac{d\xi}{\alpha\xi} = \frac{df*}{-\frac{1}{2}\sigma^2\xi^2 f*}. \tag{4.13}$$

Thus the solutions of (4.13), which are known as characteristic curves, sweep out a surface $f*(t, \xi)$ which satisfies (4.11). Solving the two trivial independent ordinary differential equations in (4.13) leads to two constants of integration, say, η and ϕ, and the result is

$$\xi = \eta e^{\alpha t}, \tag{4.14a}$$

$$f* = \phi e^{-\sigma^2\xi^2/4\alpha}. \tag{4.14b}$$

On a two-dimensional surface, however, it must be possible to move from one characteristic curve to another by changing just one parameter (say the distance along the surface between selected points on different characteristics). Therefore the solution (4.14) can involve at most one free parameter, and one of the integration constants must be a function of the other. For the general solution of (4.11) we then have

$$f*(\xi, t) = \phi(\eta)e^{-\sigma^2\xi^2/4\alpha} = \phi(\xi e^{-\alpha t})e^{-\sigma^2\xi^2/4\alpha}, \tag{4.15}$$

where ϕ is now an arbitrary function (corresponding to the arbitrary constant of integration for ordinary differential equations).

In our case, the arbitrary function is determined by initial conditions. An exemplary initial condition is to suppose that all the objects start with the same velocity v_0

$$f(v, 0) = \delta(v - v_0), \tag{4.16}$$

so from (4.9),

$$f*(\xi, 0) = (2\pi)^{-1/2}e^{-iv_0\xi} \tag{4.17}$$

and from (4.15) at $t = 0$

$$\phi(\xi) = (2\pi)^{-1/2}e^{-iv_0\xi + \sigma^2\xi^2/4\alpha}. \tag{4.18}$$

Hence

$$\phi(\xi e^{-\alpha t}) = (2\pi)^{-1/2}e^{-iv_0\xi\exp(-\alpha t) + \sigma^2\xi^2\exp(-2\alpha t)/4\alpha} \tag{4.19}$$

and from (4.15) at any time

$$f*(\xi, t) = (2\pi)^{-1/2}e^{-iv_0\xi\exp(-\alpha t) - \sigma^2\xi^2[1 - \exp(-2\alpha t)]/4\alpha}. \tag{4.20}$$

Already we can learn much from $f*(\xi, t)$ without having to invert it to get $f(v, t)$. Notice that the Fourier transform (4.9), like the Laplace transform (3.16), is also a moment generating function. Proceeding analogously to the earlier discussion shows that the first order moment, the average velocity, decreases with time according to

$$E(v) = \langle v(t) \rangle = v_0 e^{-\alpha t} \tag{4.21}$$

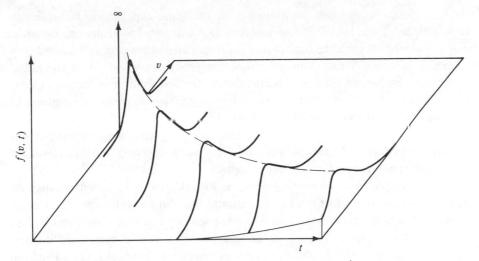

Fig. 3. Schematic illustration of the time evolution of the distribution function
$f(v, t)$.

and eventually becomes arbitrarily small. But the velocity dispersion increases until it relaxes to a non-zero value

$$\langle (v - \langle v \rangle)^2 \rangle = \frac{\sigma^2}{2\alpha}(1 - e^{-2\alpha t}), \qquad (4.22)$$

which is what we have come to expect for Brownian motion (see Equations (3.8) and (3.9) for a unit mass).

At this stage it is straightforward to perform the inversion of (4.10) by completing the square of the exponential in (4.20) involving ξ and integrating. Out steps the complete solution

$$f(v, t) = \left[\frac{\alpha}{\pi \sigma^2 (1 - e^{-2\alpha t})} \right]^{1/2} \exp \left[-\frac{\alpha (v - v_0 e^{-\alpha t})^2}{\sigma^2 (1 - e^{-2\alpha t})} \right] \qquad (4.23)$$

properly normalized. It is none other than the time dependent Gaussian. After times long compared to the relaxation timescale α^{-1} the distribution function has evolved into the Maxwell–Boltzmann form (4.7). During the course of this evolution, $f(v, t)$ starts as a very narrow Gaussian spike around $v = v_0$. Gradually the peak moves toward $v = 0$ and broadens as it moves, always retaining its Gaussian shape. Figure 3 shows this schematically.

In a sense, this type of time evolution entails the minimum departure from the final equilibrium since its basic form remains the same. More complicated initial conditions can be approximated by a sum of delta functions, each of which will shift toward $v = 0$ and decay according to (4.23) since the Fokker–Planck equation is linear in f. Different modes do not interact in this description. Indeed, recalling the central limit theorem of statistics, we can see that this leads to a more general result. The central limit theorem essentially states that if a set of mutually independent random variables are each distributed with the same expectation but different variances, and if the sum of their variances is large compared with any one variance, then the sum of the variables tends toward a Gaussian distribution as the number of variables increases. Thus, as many δ function groups of velocities in the initial distribution evolve toward $v = 0$ independently, the distribution of their sum – the total velocity distribution – will be a single evolving Gaussian eventually centered at $v = 0$ (and not a superposition of Gaussians). This is the state of ultimate relaxation. Although gravitating systems never reach this state, they strive mightily toward it. It is useful to understand this goal because the *process* of trying to relax toward it determines many properties of these systems.

Another helpful way of viewing Fokker–Planck evolution is with transition elements. Consider again the one-dimensional distribution function $f(v, t)$ as an illustration. It is the probability that a star has velocity v at time t. As a result of all the interactions among stars, after an interval δt a new distribution function $f_*(v', t + \delta t)$ forms. A star originally of velocity v may now have velocity v'. The form of the new distribution function may also be different. It seems reasonable to suppose that there is a transition probability $W(v, v')$ for the system to change its velocity distribution from $f(v, t)$ to $f_*(v', t + \delta t)$ after this interval. This assumption, however, is not as innocuous as it may first appear. For it presumes that the transition from v to v' depends only on v at any given time, and not on earlier values of v during the system's previous history. (In statistics this is called the Markov property.) Microscopically this means that transitions occur only through instantaneous collisions, and the system has no memory. Although the discussion following (2.16) leads us to expect that this condition gives only a crude approximation to gravitating systems, it does provide useful insight and has been the basis of many calculations for globular clusters and galactic nuclei. We shall see later how it can be modified.

With the concept of a transition probability $W(v, v')$, we can describe the change of the velocity distribution during a time δt as caused by two processes. First there is the probability that a star with arbitrary velocity v' at $t - \delta t$ will have velocity v at time t. This is the probability that the star had velocity v' times the probability that it made the transition, or $f_*(v', t - \delta t)W(v', v)\delta t$. Second, there are those stars with velocity v at time t which will have velocity v' at $t + \delta t$, and the chance of this occurring is $f(v, t)W(v, v')\delta t$. Figure 4 illustrates these scattering processes. Integrating these two processes over all possible velocities v' gives the total gains and losses for the number of stars of velocity v. Subtracting the losses from the gains, in the limit $\delta t \to 0$, then gives the net instantaneous rate of change of the distribution function

$$\frac{\partial f(v, t)}{\partial t} = \int_{-\infty}^{\infty} [f_*(v', t)W(v', v) - f(v, t)W(v, v')]\mathrm{d}v'. \qquad (4.24)$$

This is known as the 'master equation' and, so far, is just an empty formalism without any physics apart from the Markov assumption.

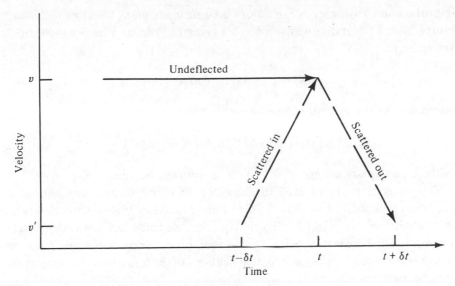

Fig. 4. The scattering processes contributing to the master equation.

To build on this skeleton we first give W some general features. Near equilibrium the velocity of a star rarely changes by a large amount in a short time. Thus $W(v', v)$ may be regarded as a sharply peaked function of $\Delta v = v' - v$ and either v or v'. At equilibrium, $f_*(v, t) - f(v, t)$ and $W(v', v) = W(v, v')$ giving $\partial f / \partial t = 0$, as expected. So we start by supposing that very close to equilibrium $f_* \approx f$ and that the flow of stars between different velocities is so nearly reversible that $W(v', v) \approx W(v, v')$.

Since f_* does not differ very much from f, we expand it in a Taylor series around f

$$f_*(v', t) = f(v, t) - \left.\frac{\partial f(v', t)}{\partial v'}\right|_v \Lambda + \frac{1}{2}\left.\frac{\partial^2 f}{\partial v'^2}\right|_v \Lambda^2 + \ldots, \tag{4.25}$$

where

$$\Lambda = v' - v = \Delta v. \tag{4.26}$$

We consider the velocity change in (4.25) to be negative in order to obtain the usual 'forward' Fokker–Planck equation. More rigorous derivations (found in texts on probability theory or stochastic processes) show this is because a negative velocity change implies differentiation with respect to the final time of the interval δt. A positive velocity change would give the corresponding 'backward' (Kolmogorov) equation, useful for finding the probability distribution of initial conditions which would lead to a given final state.

With W, we can go a step further. Its expansion may be written

$$W(v', \Lambda) = W(v, \Lambda) - \frac{\partial W(v, \Lambda)}{\partial v} \Lambda + \frac{1}{2}\frac{\partial^2 W}{\partial v^2} \Lambda^2 + \ldots. \tag{4.27}$$

The truncation of these expansions with the second derivative is also a statement of proximity to equilibrium. Far from equilibrium, large-angle scattering becomes important and Taylor series expansions become inadequate. Inserting (4.25)–(4.27) into (4.24) and integrating with respect to Λ gives the Fokker–Planck equation (3.24) yet again with

$$a(v) = \int_{-\infty}^{\infty} \Lambda W(v, \Lambda) d\Lambda = \langle \Delta v \rangle \qquad (4.28)$$

interpreted as the average change of velocity *per unit time* and

$$\sigma^2(v) = \int_{-\infty}^{\infty} \Lambda^2 W(v, \Lambda) d\Lambda = \langle (\Delta v)^2 \rangle \qquad (4.29)$$

as the mean square change of velocity in a *unit time interval*.

This derivation shows explicitly how the first and second moments of the transition probability determine the distribution function. It also shows clearly how the validity of the Fokker–Planck equation depends on both the Markov assumption and the system being close to equilibrium. Small departures from these conditions can easily be fitted into modifications of the formalism. But violent non-linear departures cannot.

5

Fokker–Planck description of gravitating systems

So far we have examined just an illustrative one-dimensional example of the Fokker–Planck equation. To apply it to gravitating systems we need to generalize it to higher dimensions and make the average velocity changes represent gravitational scattering.

Generalization to higher dimensions is necessary because the distribution function depends just on the modulus v of the velocity only for a completely isotropic system. Once anisotropies are present $f = f(v_1, v_2, v_3, t)$. Similarly, the transition probability is $W = W(v_1, v_2, v_3, v_1', v_2', v_3')$, still assuming it to be stationary. It is straightforward to extend the derivation of the last section using Taylor series expansions to generalize (3.24) for anisotropic or N-dimensional distributions

$$\frac{\partial f}{\partial t} = -\sum_{i=1}^{N} \frac{\partial}{\partial v_i}[a_i(v_1, \ldots, v_N)f] + \frac{1}{2}\sum_{j,k}\frac{\partial^2}{\partial v_j \partial v_k}[\sigma_{jk}^2(v_1, \ldots, v_N)f]. \tag{5.1}$$

The coefficients a_i and σ_{jk}^2 are defined analogously to (4.28) and (4.29).

Next we see how a_i and σ_{jk}^2 can describe the effects of many small-angle gravitational scatterings. Each two-body interaction scatters the relative velocity magnitude $u = |v_i' - v_i|$ through an angle θ in the center-of-mass frame. With an inverse square force, the effective area, per unit solid angle, for this change is the famous Rutherford differential cross section for two masses m_a and m_b with reduced mass $m_{ab} = m_a m_b/(m_a + m_b)$

$$S(\theta, u) = \left(\frac{Gm_a m_b}{2m_{ab}u^2}\right)^2 \sin^4\frac{\theta}{2} \tag{5.2}$$

(see, e.g., Landau & Lifshitz, 1976). An analysis of the kinematics of the deflection relates the scattering angle θ to the change of velocity Δv_i. To find the average velocity change a_i for all deflections, Δv_i is first integrated over all scattering angles and then over all velocities v_i of the scattering star. Each of these velocities occurs with a probability $f(v_i)$, so $a_i = \langle \Delta v_i \rangle$ will itself depend on an integral over the distribution function occurring in the Fokker–Planck equation (5.1). The result for stars of mass m_a encountering stars of mass m_b, which may represent an average mass for the field stars, is (Rosenbluth, MacDonald & Judd, 1957)

$$a_i = \langle \Delta v_i \rangle = \Gamma \frac{\partial h}{\partial v_i}, \tag{5.3}$$

$$\sigma_{jk}^2 = \langle \Delta v_j \Delta v_k \rangle = \Gamma \frac{\partial^2 g}{\partial v_j \partial v_k}, \tag{5.4}$$

where

$$\Gamma = 4\pi m_b^2 G^2 \ln \left[\frac{D \langle v^2 \rangle}{2m_b G} \right], \tag{5.5}$$

with D the upper limit to the impact parameter of the deflections and

$$h = \frac{m_a + m_b}{m_b} \int f(\mathbf{v}') |\mathbf{v} - \mathbf{v}'|^{-1} d\mathbf{v}', \tag{5.6}$$

$$g = \int f(\mathbf{v}') |\mathbf{v} - \mathbf{v}'| d\mathbf{v}'. \tag{5.7}$$

Next, we specialize h and g to gravitating systems.

It looks as though to solve the Fokker–Planck equation we must already know its solution in order to use the proper coefficients. We break this deadlock in the usual way for differentio-integral equations by assuming a distribution function $f_0(v')$ for the coefficients, calculating $f_1 = f_0 + \Delta_1 f$ from the equation, substituting f_1 for f_0 in the coefficients, calculating $f_2 = f_1 + \Delta_2 f$, etc. The hope is that such an iterative procedure will converge. In practice no one has had the stamina, or found it necessary, to do more than one iteration analytically. This is because in a uniform, or even a fairly uniform, system, it is reasonable to suppose that near equilibrium the background distribution of field star velocities is essentially Gaussian with a constant normalization A up to some maximum velocity V_{max}

$$f = Ae^{-\beta^2 v^2}, \quad v \le v_{max},$$
$$= 0 \quad , \quad v > v_{max}. \tag{5.8}$$

Thus the velocities are isotropic, $\partial h / \partial v_1 = 3^{-1/2} \partial h / \partial v$, etc., and only the derivatives of h and g with respect to the speed v appear. Moreover, these are no cross correlations of orthogonal velocity changes so $\sigma_{jk} = 0$ when $j \ne k$. In this case the derivatives are (Michie, 1963)

$$\frac{\partial h}{\partial v} = \frac{2n_0}{\sqrt{\pi}} \left[1 + \frac{m_a}{m_b} \right] \left[\frac{\beta}{v} e^{-\beta^2 v^2} - \frac{1}{v^2} F(\beta v) \right], \tag{5.9}$$

$$\frac{\partial g}{\partial v} = \frac{n_0}{\sqrt{\pi}} \left[-\tfrac{4}{3} \beta v e^{-\beta^2 v_{max}^2} + F(\beta v) \left(2 - \frac{1}{\beta^2 v^2} \right) + \frac{1}{\beta v} e^{-\beta^2 v^2} \right], \tag{5.10}$$

$$\frac{\partial^2 g}{\partial v^2} = \frac{n_0}{\sqrt{\pi}} \left[-\tfrac{4}{3} \beta e^{-\beta^2 v_{max}^2} - \frac{2}{\beta v^2} e^{-\beta^2 v^2} + \frac{2F(\beta v)}{\beta^2 v^3} \right], \tag{5.11}$$

where

$$F(x) = \int_0^x e^{-z^2} dz = \frac{\sqrt{\pi}}{2} \mathrm{erf}(x) = \sum_{n=0}^{\infty} \frac{(-1)^n x^{2n+1}}{n!(2n+1)} \tag{5.12}$$

and n_0 is the average number density.

Expanding $F(\beta v)$ and the exponentials (for $\beta^2 v_{\max}^2 \gg 1$), using (5.3)–(5.5) and noting $\langle v^2 \rangle = 3/2\beta^2$, shows that for $\beta v < 1$

$$a \equiv \sqrt{3}\, a_i = \langle \Delta v \rangle = -\frac{16\sqrt{\pi}}{3} n_0 m_b^2 G^2 \left[1 + \frac{m_a}{m_b} \right] \ln \left[\frac{3D}{4m_b G \beta^2} \right] \beta^3 v [1 - \tfrac{3}{5}\beta^2 v^2 + \cdots] \tag{5.13}$$

and

$$\sigma^2 \equiv 3\sigma_{jk}^2 = \langle (\Delta v)^2 \rangle = \frac{16\sqrt{\pi}}{3} n_0 m_b^2 G^2 \ln \left[\frac{3D}{4m_b G \beta^2} \right] \beta [1 - \tfrac{3}{5}\beta^2 v^2 + \ldots]. \tag{5.14}$$

Thus we now find more rigorously the result $a \propto -v$, or the coefficient of dynamical friction $\alpha = -a/v$ is independent of velocity for small velocities – a relation which we earlier derived heuristically. Moreover (5.14) is essentially our earlier result (2.8), now reduced by a factor of about 0.2 from the velocity integrations (remember also that σ^2 here is the mean square velocity change per unit time). Furthermore, comparison between Equations (5.13) and (2.9) shows that $a \approx v/\tau_R$, which might also have been found from heuristic considerations (i.e., an educated guess).

For large velocities, $\beta v \gg 1$, the dynamical friction is different. Since $F(x) \to \frac{1}{2}\sqrt{\pi}$ as $x \to \infty$, we now have

$$a \approx -\frac{1}{\tau_R} \frac{\langle v^2 \rangle^{3/2}}{v^2} \tag{5.15}$$

and

$$\sigma^2 \propto \frac{\langle v^2 \rangle}{v^3}. \tag{5.16}$$

Again, this could have been guessed. Since relatively rapidly moving stars do not have time to perturb their neighbors significantly (see (5.15)), the dynamical friction on them is less and they take longer to slow down. Nor do their velocities diffuse as fast as those of stars with $v \approx v_{\mathrm{rms}}$. Note that molecular or atomic systems in which short-range, hard-core collisions dominate will behave oppositely. In microscopic systems, the faster moving particles make more collisions each second, so are slowed down more quickly. This is another example where the analogy between molecular viscosity and gravitational friction breaks down. The form of α_i and σ_{jk}^2, both normalized to 1 at $\beta v = 0$, from Equations (5.3)–(5.12) is shown in Figure 5.

To gain some preliminary insight into the importance of anisotropy in the velocity distribution, consider an imaginary, artificial system. Let it be spatially homogeneous with three types of stars – red, white and blue, say. Suppose the red stars move in the x direction, white stars in y and blue stars in z throughout the system.

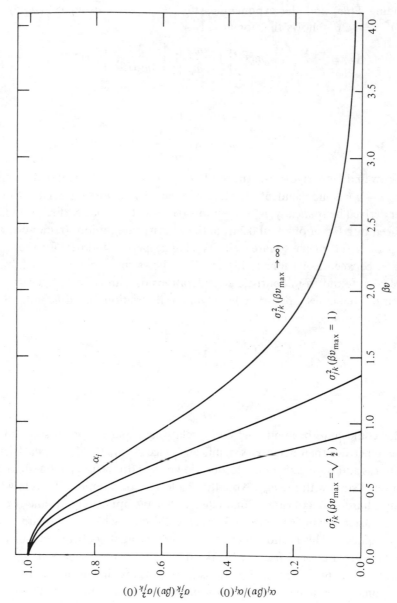

Fig. 5. The coefficient of dynamical friction α_i and the fluctuating force σ_{jk}^2 for a gravitating system described by the Fokker–Planck equation.

Each class of stars has a Gaussian velocity distribution but with different dispersions represented by β_x, β_y and β_z. Of course, in reality this would never work since streaming instabilities would also produce relaxation on the same timescale $\sim \tau_R$ as dynamical friction. So we must either imagine that stars of different colors do not interact with each other, or, alternatively, examine such a system just for times shorter than τ_R after these initial conditions are set up. The motivation for this system is that each coefficient α_i of dynamical friction will have the same form as in our previous discussion, with β_i replacing β. Thus from (5.13) and (5.15) we see that the friction is a strong function of β_i, typically varying as β_i^3. Insert a green star of arbitrary low velocity, less than any velocity dispersion, into such a system. It tends to increase its velocity more rapidly in the directions of smaller velocity dispersion. Encounters with stars of smaller velocity dispersion generally last longer and give greater deflections. On the other hand, a green star of arbitrary high velocity tends to decrease its velocity less rapidly in directions of smaller velocity dispersion since those stars usually have less time to deflect it. A green star whose initial velocity is less than the dispersion in some directions and greater than it in others will tend toward the average velocity in each direction. But the greater its initial difference, the longer it will take to come to equilibrium.

Superimposed on these effects is the ever-present tendency to increase the velocity by diffusion. From (5.14) and (5.16), a low velocity green star increases its velocity by diffusion faster in directions of low dispersion. A high velocity star increases its velocity faster in directions of high dispersion. Again the coupling is stronger when the velocities of test stars and field stars are closer. In real systems, anisotropic velocity dispersions are most often produced by rotation. But for these cases, the density is also anisotropic and leads to further interesting behavior.

Behind our dynamical analysis so far lie three basic assumptions. First, the development of the system at any given time depends only on the state of the system at that time, and not on its previous history. Second, the field stars are uncorrelated in position or velocity. In particular, the test star does not change the distribution of field stars as it passes through them. Third, the distribution evolves through a series of stellar encounters, each of which makes only a slight fractional change in the stars' velocities. Naturally we want to know what happens when these assumptions do not hold, since they are rather special. So we will describe non-Markovian effects in the next section, collective effects starting in Section 13, and violent relaxation in Part III.

6

Dynamics with a memory: non-Markovian evolution

Those who do not remember the past are
condemned to relive it.

George Santayana

Why should the evolution of a gravitating system depend on its previous history, and when is this important? In discussing the properties of systems of colliding objects, two timescales are especially critical. One is the time *of* a collision τ_* during which the objects affect each other significantly. The other is the time *between* collisions τ_0. Ordinary systems of atoms or molecules interact with a clear separation of these two timescales. Consider a room-temperature gas of hydrogen atoms with cross section $\sigma \approx 10^{-16} \text{cm}^2$ so that $\tau_* \approx (m\sigma/12\pi kT)^{1/2} \approx 10^{-14}\text{s}$ but $\tau_0 \approx (m/3kT)^{1/2}(n\sigma)^{-1} \approx 3 \times 10^{10}n^{-1}\text{s}$. Then for $n \leq 10^{24}\text{cm}^{-3}$, we have $\tau_0 > \tau_*$. When there is a clear separation of these timescales, the particles undergo a series of buffetings, each of which can be idealized as a random impulse to its orbit. Each buffet is over well before the next one begins. One can start to describe such a system's evolution at any time with initial conditions just at that time. It is not necessary to know what happened earlier, for the system has no memory. This is why the Fokker–Planck equation is a differential equation and not a differentio-integral equation; it operates locally in time.

Now consider gravitating systems. The analog of short-range collisions for point objects is encounters which deflect their orbits by $\sim 90°$. (Short-range physical collisions of finite objects, like stars, will be analyzed quite differently in Part III.) From (2.1) these have a cross section $\sigma \approx 4\pi b^2 \approx 16\pi G^2 m^2/v^4$. Therefore, using the subscript s to denote short range,

$$\tau_{*\text{s}} = \frac{b}{v} \approx \frac{2GM}{v^3} \approx 2N^{-3/2}\tau_\text{G}, \tag{6.1}$$

and

$$\tau_{0\text{s}} = \frac{1}{n\sigma v} \approx \frac{v^3}{16\pi G^2 m^2 n} \approx \frac{1}{12}N^{1/2}\tau_\text{G}. \tag{6.2}$$

In the last term of both these equalities, we have used the virial relation $v^2 = GmN/R$ and set

$$\left(\frac{Gm}{R^3}\right)^{-1/2} \equiv \tau_\text{G}. \tag{6.3}$$

The length R is the radius for a finite spherical system, and for an infinite system it represents a scale over which orbits can be correlated. For an infinite system, this correlation length may result from a density enhancement $\Delta n \approx n$ giving rise to enhancements of order v in the peculiar (i.e., non-streaming) velocities of the objects. The time τ_G is basically the contribution that one object of mass m would make to the gravitational response time if it were smoothed over the whole system. From Equations (6.1) and (6.2) we see that the time between collisions is much longer than the time of a short-range collision, by a factor of about $0.04\,N^2$.

The opposite result holds for long-range collisions (subscript L), as might be expected from their name. For these, the time of a collision is

$$\tau_{*\mathrm{L}} = \frac{R}{v} = \frac{GM}{v^3} \approx N^{-1/2}\tau_\mathrm{G}, \tag{6.4}$$

while the time between collisions is

$$\tau_{0\mathrm{L}} = \frac{1}{\pi R^2 n v} \approx N^{-3/2}\tau_\mathrm{G}. \tag{6.5}$$

Indeed the time of a long-range collision is greater than the time between collisions by a factor $\sim N$, and as a result a long-range collision is never complete!

So it would seem that long-range collisions, by their very nature, must be non-Markovian. But this is not the whole story. Since long-range collisions are so gentle, it takes about $10^{-2}\,N$ times as many of them to produce the same effect, $\Delta v \approx v$, as one short-range collision (see (2.10), (2.11) and (6.4)). Comparing $10^{-2}N\,\tau_{*\mathrm{L}}$ with $\tau_{0\mathrm{s}}$ shows that the time $\tau_{*\mathrm{L}}$ for long-range collisions to have a major effect on the orbits is about one-tenth the time between short-range collisions. In a realistic system with density variations and non-Maxwellian distributions, at least roughly one-tenth the long-range collisions are quenched by a short-range nearly random scattering. The time between short-range collisions as a function of velocity change varies as $\tau_{0\mathrm{s}} \propto b^{-2} \propto (\Delta v/v)^2 v^3$. The long-range relaxation time τ_R has the same velocity dependence (see (2.8)). So the quenching is about the same at any level of velocity perturbation.

Other phenomena also tend to quench long-range deflections in realistic systems. Any density or velocity inhomogeneities will scatter individual objects strongly. If these inhomogeneities, representing collective motions, change rapidly with time – as in a system far from equilibrium – they can be much more important than single long-range collisions. This state, called violent relaxation, is discussed in Section 38. Even without violent relaxation, we should not forget that an object responds to the mean gravitation field, as well as to the fluctuations. Global asymmetries of scale length D in the average field will affect the orbit on times of order D/v (the crossing time for finite systems). If the amplitude of the asymmetry is large, as in a contracting ellipsoidal system of stars, it can dominate the fluctuations. Finally, the tidal effects of external systems may also influence internal orbits.

Nevertheless, there may be cases where non-Markovian effects are important and we should at least construct a simple theory to see how they can alter a system's evolution. To do this we return to the Langevin analysis of (3.1) and ask how it is modified by perturbing forces which still have zero average value (3.3) but are no longer unrelated at different times. What form will $\langle \beta(\tau)\beta(\tau')\rangle$ now have? In other words, how long will a star travel under the influence of a long-range encounter before a new interaction destroys this memory?

Let $P(x)$ be the probability that a given star has not had a new memory-destroying interaction after moving a distance x. The probability that it will have such an interaction between x and $x + dx$ is $P(x)$ times the memory-destroying interaction cross section Ω per field star times the number density of field stars n times the distance dx. This reduces the probability of not having had an interaction by the amount $dP(x) = -\Omega n P(x)dx$ so

$$P(x) = e^{-\Omega n(x - x_0)} \tag{6.6}$$

since $P(x_0) = 1$. Between strong interactions, $x = x_0 + v(t - t_0)$ to a good approximation, and

$$P(t - t_0) = e^{-(t - t_0)/\tau_m}, \tag{6.7}$$

where

$$\tau_m \equiv (\Omega n v)^{-1} \tag{6.8}$$

is the memory-decay timescale. It measures the duration of a typical long-range encounter, which is the interval between strong deflections. This suggests that in a non-Markovian Langevin model the time correlation of perturbations should have a form similar to (6.7). Thus we replace (3.4) by

$$\langle \beta(\tau/\tau_m)\beta(\tau'/\tau_m)\rangle = \frac{m^2\sigma^2}{2\tau_m} e^{-|\tau - \tau'|/\tau_m}, \tag{6.9}$$

with σ^2 again having dimensions of (velocity)^2 (time)$^{-1}$. The whole analysis of Section 3 can now be repeated with this modification.

As an illustration of the results, consider the velocity dispersion which follows from (6.9) and the first part of (3.9)

$$\langle v^2(t)\rangle = v^2(0)e^{-(2\alpha/m)t} + \frac{\sigma^2 m}{2\alpha}\left[\frac{1}{1+z} - \frac{1}{1-z}e^{-(2\alpha/m)t} + \frac{2z}{1-z^2}e^{-(\alpha/m)(1+z^{-1})t}\right]$$

$$\equiv v^2(0)e^{-(2\alpha/m)t} + \frac{\sigma^2 m}{2\alpha}M(z), \tag{6.10}$$

where (using 3.5)

$$z \equiv \frac{\alpha\tau_m}{m} \approx \frac{\tau_m}{\tau_R}, \tag{6.11}$$

and $M(z)$ is a memory function.

In the limit of very short memory, $\tau_m \to 0$, Equation (6.9) approaches a δ function and (6.10) reduces to the previous result of (3.9). The main effect of increasing τ_m is to decrease the equilibrium velocity dispersion

$$\lim_{t \to \infty} \langle v^2(t)\rangle \propto (1 + z)^{-1}.$$

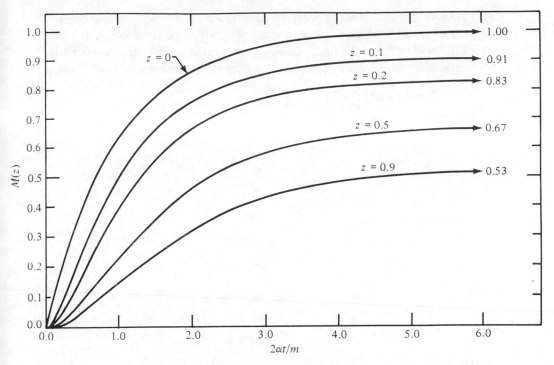

Fig. 6. The memory function $M(z)$ in Equation (6.10).

Systems with longer dynamical memories have less entropy than those with short memory. If τ_m is increased, the velocity dispersion is also lower at a given time, as Figure 6 illustrates.

Dynamical memory resolves the problem of non-differentiable solutions of the Langevin equation. We expect this on physical grounds since the lack of differentiability when $t \to 0$ arose from the completely uncorrelated nature of successive perturbations. Now that perturbations are correlated over time τ_m, the orbits are 'smooth' and differentiable. This is readily seen by expanding (6.10) for small times (again for $v(0) = 0$) to find

$$\lim_{\Delta t \to 0} \frac{\Delta \langle v^2(t) \rangle^{1/2}}{\Delta t} = \frac{\sigma}{\sqrt{(2\tau_m)}}, \tag{6.12}$$

which is finite, in contrast to (3.12). Moreover, this result also shows that at a given small value of t, the mean square velocity dispersion is proportional to τ_m^{-1}, so for larger τ_m the system takes longer to reach a given velocity dispersion – correlations hold back the trend to equilibrium. In the non-Markovian analog of the Fokker–Planck description, this effect shows up through the time dependence of the diffusion coefficients (an exercise for the reader).

For gravitating systems, the exact value of τ_m is subject to uncertainties in the quenching processes described earlier in this section. A very generous upper limit is the time between short-range collisions, and from (6.2), (6.4) and (2.11) this would give values of $z \approx 10$, reducing $\langle v^2 \rangle^{1/2}$ to a factor ~ 3 below its Markovian equilibrium value. However, a better estimate of τ_m would be close to τ_{*L}, in which case $z \approx 10^2 N^{-1} \ll 1$ and the final velocity dispersion is not affected significantly. This is especially the case for systems forming from an inhomogeneous distribution of stars.

7

The Boltzmann equation

When all forces except the mean field can be ignored, there is a relatively simple description of slowly evolving gravitating systems. It is sufficiently simple that we can relax the condition of homogeneity and suppose that the distribution function $f(\mathbf{r}, \mathbf{v}, t)$ depends on position as well as on velocity and time. Imagine a differential volume element in six-dimensional position – velocity phase space (Figure 7). Let it be large compared to the separation of galaxies, but small compared to the size of the Universe, so that it contains $f(\mathbf{r}, \mathbf{v}, t)\, \mathrm{d}\mathbf{r}\, \mathrm{d}\mathbf{v}$ galaxies at any time t. If galaxies do not jump quickly from one volume element to another, but flow smoothly, we expect $f(\mathbf{r}, \mathbf{v}, t)$ to satisfy the equation of continuity

$$\frac{\partial f}{\partial t} = -\nabla \cdot (f\mathbf{v}) - \nabla_{\mathbf{v}} \cdot (f\dot{\mathbf{v}}). \tag{7.1}$$

This says that the time change of f within the volume $\mathrm{d}\mathbf{r}\,\mathrm{d}\mathbf{v}$ is just due to the galaxies entering or leaving through the boundaries. Here ∇ denotes the positional vector gradient and $\nabla_{\mathbf{v}} = \hat{\mathbf{i}}(\partial/\partial v_1) + \hat{\mathbf{j}}(\partial/\partial v_2) + \hat{\mathbf{k}}(\partial/\partial v_3)$ is the velocity space gradient; notice they are both treated the same way in phase space. Using the vector identity $\nabla \cdot (f\mathbf{w}) = f\nabla \cdot \mathbf{w} + \mathbf{w} \cdot \nabla f$ and recalling that $\mathbf{v}$ and $\mathbf{r}$ are considered here as independent variables, and that the acceleration is independent of velocity, simplifies (7.1) to

$$\frac{\partial f}{\partial t} + \mathbf{v} \cdot \nabla f + \dot{\mathbf{v}} \cdot \nabla_{\mathbf{v}} f = 0. \tag{7.2}$$

Evidently the total time derivative of f which follows the motions of the galaxies through phase space is zero. This is the collisonless Boltzmann equation – one of the most useful descriptions of stellar dynamics.

Now we derive (7.2) more carefully to clarify the assumptions on which it rests. By definition, the number of galaxies in the phase space volume $\mathrm{d}\mathbf{r}\,\mathrm{d}\mathbf{v}$ at $\mathbf{r}$ and $\mathbf{v}$ is

$$\mathrm{d}N = f(\mathbf{r}, \mathbf{v}, t)\,\mathrm{d}\mathbf{r}\,\mathrm{d}\mathbf{v} \tag{7.3}$$

at time t. At a slightly later time, $t + \Delta t$, each galaxy has moved to a new position

$$\mathbf{r}' = \mathbf{r} + \mathbf{v}\Delta t \tag{7.4}$$

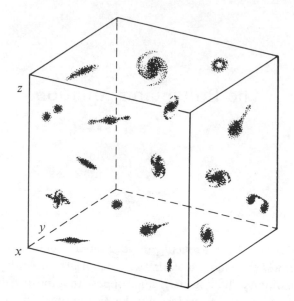

Fig. 7. Three-dimensional positional projection of six-dimensional position –
velocity phase space.

and has a new velocity

$$\mathbf{v}' = \mathbf{v} + \dot{\mathbf{v}}\Delta t = \mathbf{v} + \frac{\mathbf{F}}{m}\Delta t, \qquad (7.5)$$

where $\mathbf{F}/m$ is the gravitational force per unit mass. The volume element $d\mathbf{r}\,d\mathbf{v}$
containing the original bunch of galaxies has also changed to a new volume element
$d\mathbf{r}'\,d\mathbf{v}'$ containing these same galaxies, so that

$$dN' = dN = f(\mathbf{r}', \mathbf{v}', t + \Delta t)\,d\mathbf{r}'\,d\mathbf{v}'. \qquad (7.6)$$

If galaxies had jumped into or out of the volume element, say by collisions or large-
angle scatterings which modify (7.4) and (7.5), we would need to write $dN' = dN +
\Delta N$ and represent the detailed scattering by the form of ΔN (as is done when
applying the Boltzmann equation to atomic gas dynamics). Equation (7.5) does not
mean that $\dot{\mathbf{v}}$ or $\mathbf{F}$ cannot change with time. It just requires that any time change in $\mathbf{F}$
be slower than the change of $\mathbf{v}$ caused by $\mathbf{F}$ during the interval Δt. Here the alert
reader will notice a subtlety. After all, is not the change in the mean field $\mathbf{F}$ caused by
the motion of the galaxies and therefore are not these two timescales roughly equal?
The point is that since a great many galaxies contribute to the mean field, the
motions of some galaxies will usually be compensated by other galaxies, and the net
field from many galaxies will change more slowly than the contribution of any small
subset of galaxies.

From (7.3) and (7.6),

$$f(\mathbf{r}, \mathbf{v}, t)\,d\mathbf{r}\,d\mathbf{v} = f(\mathbf{r}', \mathbf{v}', t + \Delta t)\,d\mathbf{r}'\,d\mathbf{v}'. \qquad (7.7)$$

Texts on integral vector calculus show that two volume elements connected by a transformation such as (7.4)–(7.5) are related by the Jacobian of the transformation

$$d\mathbf{r}'d\mathbf{v}' = \left| \frac{\partial(\mathbf{r}',\mathbf{v}')}{\partial(\mathbf{r},\mathbf{v})} \right| d\mathbf{r}\,d\mathbf{v} \tag{7.8}$$

where the Jacobian is the absolute value of the 6×6 determinant

$$J(\mathbf{r}',\mathbf{v}'/\mathbf{r},\mathbf{v}) = \left| \frac{\partial(\mathbf{r}',\mathbf{v}')}{\partial(\mathbf{r},\mathbf{v})} \right| = \begin{vmatrix} \dfrac{\partial r'_1}{\partial r_1} & \dfrac{\partial r'_1}{\partial r_2} & \dfrac{\partial r'_1}{\partial r_3} & \dfrac{\partial r'_1}{\partial v_1} & \dfrac{\partial r'_1}{\partial v_2} & \dfrac{\partial r'_1}{\partial v_3} \\ \dfrac{\partial r'_2}{\partial r_1} & & \cdots & & & \dfrac{\partial r'_2}{\partial v_3} \\ \vdots & & & & & \vdots \\ \dfrac{\partial v'_3}{\partial r_1} & & & & & \dfrac{\partial v'_3}{\partial v_3} \end{vmatrix}. \tag{7.9}$$

Substituting (7.4) and (7.5) into the Jacobian readily reveals that to first order in Δt, which is all we need, the Jacobian is just unity. A more physical interpretation of this result follows from the relation between the six-dimensional divergence of velocity $\mathbf{u}$ in position – velocity phase space and the Jacobian

$$\nabla_6 \cdot \mathbf{u} = \frac{1}{J} \frac{\partial J}{\partial t}. \tag{7.10}$$

The right hand side is the fractional change of six-dimensional phase space volume occupied by the galaxies as they move along. To first order in Δt, this is zero, showing that the motion of the particles representing the galaxies in phase space is divergence-free. This is the analog of the equation for mass conservation of an ordinary fluid in ordinary space

$$\nabla_3 \cdot \mathbf{v} = -\frac{1}{\rho} \frac{D\rho}{Dt}, \tag{7.11}$$

where ρ is the fluid density and

$$\frac{D}{Dt} = \frac{\partial}{\partial t} + \mathbf{v} \cdot \nabla \tag{7.12}$$

is the usual convective derivative following the motion. We therefore see how the motion of the phase space particles resembles an incompressible flow. We will meet this property in a more violent context in Part III.

Returning to Equation (7.7), we can now cancel the phase space volume elements from both sides, expand the right hand side in a Taylor series around $f(\mathbf{r},\mathbf{v},t)$ for a short interval Δt, let $\Delta t \to 0$ and obtain the collisionless Boltzmann equation (7.2).

The structure of this equation should come as no surprise. It is essentially the Fokker–Planck equation (5.1), without the diffusion term, for a distribution function which depends on position as well as velocity coordinates. There would then be coefficients a_i which represent the average change in position per unit time (the velocity), as well as those representing the acceleration. If these coefficients, arising from dynamical friction, were calculated self-consistently in the mean field of all the other galaxies, with no diffusion, one would indeed obtain the collisionless Boltzmann equation. The inclusion of the diffusion term adds much more information to the Fokker–Planck description, which then becomes equivalent to a Boltzmann equation with a collision term that describes scattering caused by small mean field fluctuations. This is quite different from the hard-core collisional terms which appear in atomic gas kinetics.

In its compact form (7.2), the collisionless Boltzmann equation looks transparently, but misleadingly, simple. It looks linear in f, but this is only for systems in which the mean field force, proportional to $\dot{\mathbf{v}}$, is externally imposed. Gravitating systems, however, determine the mean field internally in a self-consistent way depending on f. This makes the problem a more complicated one with greater potential for interesting behavior. The distribution function enters in a non-linear way through Poisson's equation

$$\nabla^2 \phi(\mathbf{r}, t) = -4\pi G\rho = -4\pi G\bar{m} \int f(\mathbf{r}, \mathbf{v}, t)\, d\mathbf{v} \qquad (7.13)$$

for the gravitational potential ϕ whose gradient gives the acceleration

$$\dot{\mathbf{v}} = \nabla \phi. \qquad (7.14)$$

With these equations we can find, for example, the internal distribution of stationary spherical systems as an approximation to clusters of galaxies, globular clusters, or galactic nuclei. Moreover, they are useful for describing the spiral structures of rotating galaxies (Section 62).

The Boltzmann equation also has another interpretation which is quite useful. Its characteristic curves (see Section 4) are given by

$$\frac{dt}{1} = \frac{dx}{v_x} = \frac{dy}{v_y} = \dots = \frac{dv_z}{\dot{v}_z}, \qquad (7.15)$$

using Cartesian coordinates as the simplest example. Equating different combinations of (7.15) gives $v_x = dx/dt$, $\dot{v}_x = dv_x/dt = F_x/m$, etc. But these trivial-looking results are actually the equations of motion of the stars. Since the characteristics sweep out solutions of the Boltzmann equation, we see that these solutions are equivalent to knowing the orbits of the stars. This is not surprising because if we know their initial positions and velocities, and their motion in the mean field, we can always add up the number within any small range of positions and velocities at any given time (or solve a collisionless Boltzmann equation) to get the distribution function. Section 63 describes an important application of this result.

In a Hamiltonian formulation, the Boltzmann equation looks especially elegant and suggestive. Now we consider $f(\mathbf{q}, \mathbf{p}, t)$ to depend on the Hamiltonian coordinates q_i and momenta p_i which satisfy the canonical equations of motion

$$\frac{dq_i}{dt} = \frac{\partial H}{\partial p_i}, \quad \frac{dp_i}{dt} = -\frac{\partial H}{\partial q_i}, \qquad (7.16)$$

where the Hamiltonian $H(q_i, p_i)$ determines the evolution of the system. For freely moving stars $H = (p_1^2 + p_2^2 + p_3^2)/2m$ with $p_1 = p_x = mv_x$ (in Cartesian coordinates), etc. A star moving in a gravitational field has its potential energy, which depends just on the q_i, added to the Hamiltonian. It is a simple matter to use (7.16) and rewrite (7.2) in terms of canonical coordinates to obtain

$$\frac{\partial f}{\partial t} = \sum_{i=1}^{3} \left(\frac{\partial H}{\partial q_i} \frac{\partial f}{\partial p_i} - \frac{\partial H}{\partial p_i} \frac{\partial f}{\partial q_i} \right) \equiv \{H, f\}. \tag{7.17}$$

The quantity on the right hand side is known as the Poisson bracket. In general the total time derivative of a dynamical function is its partial time derivative minus its Poisson bracket with the Hamiltonian

$$\frac{\mathrm{d}f}{\mathrm{d}t} = \frac{\partial f}{\partial t} - \{H, f\}. \tag{7.18}$$

For example, substituting $H(q_i, p_i)$ for f in (7.18) shows that energy is conserved since $\dot{H} = 0$ as an identity.

If we multiply (7.17) by $i = \sqrt{-1}$ and define the operator operating on f as the Boltzmann operator B, then

$$Bf = i\{H, f\} \tag{7.19}$$

and the evolution equation is

$$i\frac{\partial f}{\partial t} = Bf. \tag{7.20}$$

This is very similar in form to Schrödinger's equation

$$i\hbar \frac{\partial \psi}{\partial t} = H\psi \tag{7.21}$$

where ψ is the wave function, $2\pi\hbar$ is Planck's constant and H is the Hamiltonian operator. In Section 14, we will see how to exploit this similarity.

8

Some properties of the Boltzmann equation

Without much effort, we can draw some interesting conclusions about simple systems which satisfy the collisionless Boltzmann equation. First we see that a uniform, equilibrium, isothermal gravitating system cannot exist (at least in this description). The requirements of equilibrium and homogeneity mean that the first two terms of (7.2) vanish. An isothermal distribution depends only on the energy of its particles which have a single uniform velocity dispersion. Thus $\mathbf{V}_v f \neq 0$ and so by (7.2) the acceleration $\dot{\mathbf{v}} = 0$. This might also have been concluded directly from the symmetry of the mean field. Now (7.14) implies $\mathbf{V}\phi = 0$ and (7.13) in turn implies $\nabla^2 \phi = 0$ and so $f(\mathbf{v}) = 0$. The only system satisfying our simple conditions is one which is empty, everywhere and forever.

This is quite unlike the case of a perfect non-gravitating gas, which can easily satisfy these equilibrium conditions. The difference between the two cases is caused by the different nature of the forces which try to maintain equilibrium. In a perfect gas model, the atoms hit one another like billiard balls. These short-range forces add a collision term to the right hand side of (7.2) whose properties inhibit departures from the uniform, isothermal equilibrium distribution. In fact, from the property that the collisions are isotropic and the equilibrium distribution is homogeneous, we can, following Maxwell, deduce the form of the distribution function. It will depend only on velocity and we need just assume that in every Cartesian coordinate system the three orthogonal random velocity components are independent. Considering the distribution of two velocity components, for example, this says that their joint probability distribution is the product of their individual distributions, i.e.,

$$f_{12}(v_1, v_2) = f_1(v_1)f_2(v_2). \tag{8.1}$$

Now the velocities change just by collisions, and if the collisions are isotropic the independence of $f_1(v_1)$ and $f_2(v_2)$ does not change when we rotate the coordinate system through an angle θ to obtain new values for the velocities

$$v_1' = v_1 \cos \theta + v_2 \sin \theta, \tag{8.2a}$$

$$v_2' = - v_1 \sin \theta + v_2 \cos \theta, \tag{8.2b}$$

as in Figure 8.

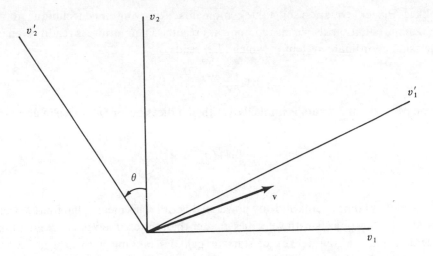

Fig. 8. Rotation of the coordinate system which leaves the velocity distribution invariant.

Rotation of the coordinate system leaves the velocity distribution invariant. This invariance of independence means that

$$f'_1(v'_1)f'_2(v'_2) = f_1(v_1)f_2(v_2),\qquad(8.3)$$

or

$$\ln f'_1 + \ln f'_2 = \ln f_1 + \ln f_2.\qquad(8.4)$$

Differentiating first with respect to v_1 and then with respect to v_2 gives

$$\frac{d^2 \ln f'_1(v'_1)}{dv'^2_1} - \frac{d^2 \ln f'_2(v'_2)}{dv'^2_2} = 0.\qquad(8.5)$$

Varying v_1 and v_2 so that v'_2 remains constant, which is possible for any given θ, gives a constant for the second term of (8.5). Integrating the remaining term shows that

$$f'_1(v'_1) = e^{a + bv'_1 + cv'^2_1},\qquad(8.6)$$

where a, b, and c are constants, determined by isotropy and homogeneity since (8.6) holds for each velocity component. Thus the distribution function of a homogeneous, isotropic, equilibrium, billiard ball gas must have the Maxwell–Boltzmann form.

For an infinite gravitating system, we could of course start off with a Maxwell–Boltzmann distribution and regard it as the simplest initial condition in the sense of requiring the least information to specify it since there are no correlations. Because gravitational scattering is attractive and anisotropic, the first scattering changes the gravitational field which influences the second scattering, which in turn alters the third scattering, etc., and such a system would not remain homogeneous and uncorrelated for very long. This theme is developed for galaxy clustering in Part II.

In Section 4 we found we could extract some useful information from the

Fokker–Planck equation by taking moments. Applying this technique to the Boltzmann equation also leads to important results. Their nature is readily seen in a Cartesian coordinate system in which (7.2) reads

$$\frac{\partial f}{\partial t} + v_i \frac{\partial f}{\partial x_i} + \dot{v}_i \frac{\partial f}{\partial v_i} = 0, \tag{8.7}$$

where $i = 1, 2, 3$ and a sum is implied over the i indices occurring twice in any term. At large velocities, the distribution function is assumed to decrease so rapidly that

$$\left. \begin{array}{c} \lim v_i^n f = 0, \\[1em] |v_i| \to \infty, \end{array} \right\} \tag{8.8}$$

as

for all velocity components and any power n. Basically this means that finite systems contain very few objects with velocities greater than the escape velocity, and infinite systems contain a low density of stars or galaxies moving with arbitrarily high relativistic velocities. Next, define the average velocity moments of order $l + k$

$$\int_{-\infty}^{\infty} v_i^l v_j^k f(\mathbf{x}, \mathbf{v}, t) dv_1 \, dv_2 \, dv_3 = n(\mathbf{x}, t) \overline{v_i^l v_j^k}, \tag{8.9}$$

where $n(\mathbf{x}, t)$ is the number density and the bar denotes the velocity average. Multiply (8.7) by v_j^k and integrate all over velocity space. The first term is just

$$\int_{-\infty}^{\infty} v_j^k \frac{\partial f}{\partial t} dv_1 \, dv_2 \, dv_3 = \frac{\partial}{\partial t} (n \overline{v_j^k}). \tag{8.10}$$

The second set of terms are

$$\int_{-\infty}^{\infty} v_1 v_j^k \frac{\partial f}{\partial x_i} dv_1 \, dv_2 \, dv_3 = \frac{\partial}{\partial x_i} (n \overline{v_i v_j^k}). \tag{8.11}$$

The third set of terms reduces to

$$\dot{v}_i \int_{-\infty}^{\infty} v_j^k \frac{\partial f}{\partial v_i} dv_1 \, dv_2 \, dv_3 = \begin{cases} 0 \text{ when } i \neq j \\ -k \dot{v}_j n(\mathbf{x}, t) \overline{v_j^{m-1}} \text{ when } i = j \text{ (no summation on } j) \end{cases} \tag{8.12}$$

since gravitational acceleration does not depend on velocity and we integrate by parts using (8.8). Combining these results

$$\frac{\partial}{\partial t} (n \overline{v_j^k}) + \frac{\partial}{\partial x_i} (n \overline{v_i v_j^k}) = k \dot{v}_j n(\mathbf{x}, t) \overline{v_j^{k-1}} \text{ (no summation over } j). \tag{8.13}$$

There are three equations, one for each value of j.

It takes the merest moment to interpret what this technique produces. Letting $k = 0$, the three equations (8.13) are all the same

$$\frac{\partial n}{\partial t} + \nabla \cdot (n \bar{\mathbf{v}}) = 0. \tag{8.14}$$

This is just the equation of continuity (4.3) for number (mass) conservation of the average flow in the system. Letting $k = 1$ gives three equations, each of the form

$$\frac{\partial(n\bar{v}_1)}{\partial t} + \frac{\partial}{\partial x_1}(n\overline{v_1^2}) + \frac{\partial}{\partial x_2}(n\overline{v_1 v_2}) + \frac{\partial}{\partial x_3}(n\overline{v_1 v_3}) = \dot{v}_1 n(\mathbf{x}, t), \tag{8.15}$$

from which the remaining two also follow by cyclic permutation, i.e., changing the indices $1 \to 2$, $2 \to 3$, and $3 \to 1$, then repeating this transformation. Equations (8.15) describe the transport of momentum. In this case it is momentum per unit mass for equal mass objects, but it can easily be generalized for a distribution function which depends on mass as well as on position, velocity and time (an exercise for the reader). The results (8.14)–(8.15) were first derived by Jeans (1915, 1919, 1922) and are often called the equations of stellar hydrodynamics.

The transport equations can be made to look more like ordinary fluid hydrodynamics by rewriting the nine terms of the form $\partial(n\overline{v_1 v_2})/\partial x_2$. These terms represent, for example, the gradient in the x_2 direction of the average flux of $n v_1$ momentum along x_2. This generally differs from $n\bar{v}_1 \bar{v}_2$, which is the flux of the average momentum $n\bar{v}_1$ along x_2. The reason is essentially that the fluctuations of opposite sign in the mean flow make it unequal to the root mean square flow, as also occurs in the Langevin and Fokker–Planck descriptions. Supposing that each velocity component is the sum of an average and a fluctuating term,

$$v_i = \bar{v}_i + v_i', \tag{8.16}$$

and that the fluctuations are not autocorrelated,

$$\bar{v}_i' = 0, \tag{8.17}$$

gives

$$\overline{v_i v_j} = \bar{v}_i \bar{v}_j + \overline{v_i' v_j'} \equiv \bar{v}_i \bar{v}_j + \frac{P_{ij}}{mn}, \tag{8.18}$$

where $P_{ij} = P_{ji}$ is a pressure tensor and m is the mass of a star in the system. The lack of autocorrelation is a very strong assumption, much less realistic for gravitating systems than for atomic gases. But it is consistent with the collisionless Boltzmann approximation and, as we will see, simplifies the situation so much that it is often used to give approximate results which provide some insight into particular astronomical problems. Most of these problems are so difficult that even small insights become significant.

Substituting (8.18) into (8.15), recalling from (8.9) that $\bar{v}_i$ is generally a function of x and t even though v_i is not, and using the continuity equation (8.14) transforms the momentum transport equation into its more familiar form

$$\frac{\partial \bar{v}_1}{\partial t} + \bar{v}_1 \frac{\partial \bar{v}_1}{\partial x_1} + \bar{v}_2 \frac{\partial \bar{v}_1}{\partial x_2} + \bar{v}_3 \frac{\partial \bar{v}_1}{\partial x_3} + \frac{1}{mn}\left(\frac{\partial P_{11}}{\partial x_1} + \frac{\partial P_{12}}{\partial x_2} + \frac{\partial P_{13}}{\partial x_3}\right) = \frac{F_1}{m}, \tag{8.19}$$

with $F_1 = m\dot{v}_1$ the 'external' force due to the mean field. Similar equations obtained by cyclic permutation hold for the v_2 and v_3 components. When the pressure is isotropic so $P_{ij} = P\delta_{ij}$, all these equations can be combined in the vector form

$$\frac{\partial \mathbf{v}}{\partial t} + \mathbf{v} \cdot \nabla \mathbf{v} + \frac{1}{mn}\nabla P = \mathbf{F}m^{-1}. \tag{8.20}$$

Poisson's equation (7.13) determines $\mathbf{F}$ from n, and mass conservation (8.14) relates changes in n and $\mathbf{v}$.

But something is lacking. Even in the simplest case of isotropic pressure, we have eight variables, v_1, v_2, v_3, n, P, F_1, F_2, F_3, and only seven equations to determine them. There is not enough information in these moments to solve them self-consistently. If only there were another equation for, say the pressure $\overline{v'^2}$. Aha! If we go back to the moment Equation (8.13) and take the next higher moment with $k = 2$ we get a transport equation for energy. And after applying the separation of (8.16) we would have our equation for the pressure. The catch, however, is that the pressure equation contains terms involving third order moments arising from $\partial(nv_iv_j^k)/\partial x_i$. The trend is now clear: equations for second order moments contain third order moments, those for third order moments contain fourth order moments and so on to infinity. It seems we cannot break this chain. What started as a promising technique turns out disappointing.

All is not lost, however. The problem just calls for a new approximation; there is more than one way to proceed. We could calculate many moments until we tired and then set the highest moment equal to zero to close the set of equations. This would have the advantage that differences in the solutions found by closing the equations at different levels would give an idea of their accuracy. The more moments used, the greater the constraints placed on the solution to behave like the solution of the original Boltzmann equation. Infinite knowledge of all the solution's moments would be equivalent to knowing the solution itself. (This follows from an extension of the arguments after (3.14) along the lines that all the moments of a function give its transform which can then be inverted to obtain the function. The functions need to have standard properties of continuity and good behavior.)

Without infinite knowledge, we can take a more physical approach. If the motion of the stars or galaxies is close to Maxwellian, then $P_{ij} = P_{ii} = v_i'^2$ and this average square random velocity is approximately the velocity dispersion of the distribution function. In a finite spherical system, the kinetic and potential energies approximately balance, so $\overline{v_i'^2} \approx GM/R$. At a stroke we can then truncate the coupled moment equations. This only works, however, for the equilibrium case where we half know the answer anyway. Nevertheless, an important series of questions involves perturbations around equilibrium states, for which these equations are ideally suited. We will discuss them several times, in different contexts and coordinate systems, in later sections.

9

The virial theorem

One by one the moments fall; some are coming,
some are going; do not strive to grasp them all.

Adelaide Proctor

This result is used so often throughout so many branches of dynamical astronomy
that it is worth discussing separately, even though it is just a special moment of the
equations of motion. For a first look at the virial theorem, we start with the equation
of motion (8.20) written in component form as

$$\rho \frac{dv_i}{dt} = -\frac{\partial P}{\partial x_i} + nF_i. \tag{9.1}$$

Here $\rho = mn$ is the mass density and the total derivative

$$\frac{dv_i}{dt} = \frac{\partial v_i}{\partial t} + v_j \frac{\partial v_i}{\partial x_j} \tag{9.2}$$

follows the motion. In a discrete distribution the ith component of gravitational
force on the αth star is

$$F_i^{(\alpha)} = m^{(\alpha)} \frac{\partial \phi}{\partial x_i}, \tag{9.3}$$

where the potential ϕ results from all the other stars

$$\phi = G \sum_{\beta \neq \alpha} \frac{m^{(\beta)}}{|\mathbf{x}^{(\alpha)} - \mathbf{x}^{(\beta)}|}. \tag{9.4}$$

Since we are dealing here with a continuous distribution, this sum is replaced by the
volume integral

$$\phi(\mathbf{x}) = G \int \frac{\rho(\mathbf{x}')}{|\mathbf{x} - \mathbf{x}'|} d\mathbf{x}', \tag{9.5}$$

which is the solution to Poisson's equation (7.13). We have changed the notation
slightly because it is now more convenient to think explicitly in terms of Cartesian
coordinates.

Take the first positional moment of (9.1) by multiplying each term with x_j and
integrating over the system's volume. The left hand term is

$$\int \rho(\mathbf{x}) x_j \frac{dv_i}{dt} d\mathbf{x} = \frac{d}{dt} \int \rho(\mathbf{x}) x_j v_i d\mathbf{x} - \int \rho(\mathbf{x}) v_i v_j d\mathbf{x} \tag{9.6}$$

since mass is conserved. The pressure integral is

$$\Pi\delta_{ij} = \int P\,d\mathbf{x}\,\delta_{ij} \tag{9.7}$$

since the pressure vanishes on the boundary (integrating by parts) for a finite system, or is symmetric for a homogeneous infinite system. The gravitational force term (for all masses equal to $m^{(\alpha)}$) is

$$\int \rho(\mathbf{x})x_j\frac{\partial\phi}{\partial x_i}d\mathbf{x} = -G\int\int\rho(\mathbf{x})\rho(\mathbf{x}')x_j\frac{(x_i-x_i')}{|\mathbf{x}-\mathbf{x}'|^3}d\mathbf{x}'d\mathbf{x}$$

$$= +G\int\int\rho(\mathbf{x})\rho(\mathbf{x}')x_j'\frac{(x_i-x_i')}{|\mathbf{x}-\mathbf{x}'|^3}d\mathbf{x}'d\mathbf{x}, \tag{9.8}$$

where the primed and unprimed variables are interchanged in the second form. To obtain a symmetric version of this term we add the two forms of (9.8) and take half their sum. This gives

$$W_{ij} = -\frac{G}{2}\int\int\rho(\mathbf{x})\rho(\mathbf{x}')\frac{(x_i-x_i')(x_j-x_j')}{|\mathbf{x}-\mathbf{x}'|^3}d\mathbf{x}'d\mathbf{x}, \tag{9.9}$$

which is the symmetric potential energy tensor. Contracting it by setting $i=j$ and summing over the repeated index yields the gravitational potential energy of the system. Writing

$$T_{ij} = \tfrac{1}{2}\int\rho(\mathbf{x})v_iv_j\,d\mathbf{x} \tag{9.10}$$

for the symmetric energy momentum tensor due to the average velocities, the contribution from random velocities being separated off as the pressure term, and combining (9.6)–(9.10)

$$\frac{d}{dt}\int\rho(\mathbf{x})x_jv_i\,d\mathbf{x} = 2T_{ij} + W_{ij} + \Pi\delta_{ij}. \tag{9.11}$$

Although this is the basic moment equation, it can readily be made more informative. The symmetry of the right hand side suggests that we write the left hand side as the sum of an antisymmetric and a symmetric component

$$\frac{d}{dt}\int\rho(\mathbf{x})x_jv_i\,d\mathbf{x} = \frac{1}{2}\frac{d}{dt}\int\rho(\mathbf{x})(x_jv_i - x_iv_j)\,d\mathbf{x}$$

$$+ \frac{1}{2}\frac{d}{dt}\int\rho(\mathbf{x})(x_jv_i + x_iv_j)\,d\mathbf{x}. \tag{9.12}$$

Since the symmetry of both sides of the equation must be the same, the antisymmetric part must vanish

$$\frac{d}{dt}\int\rho(\mathbf{x})(x_jv_i - x_iv_j)\,d\mathbf{x} = 0, \tag{9.13}$$

showing how angular momentum is conserved. The symmetric part of (9.11) remains

$$\frac{1}{2}\frac{d^2 I_{ij}}{dt^2} = 2T_{ij} + W_{ij} + \Pi\delta_{ij}, \tag{9.14}$$

where

$$I_{ij} = \int \rho(\mathbf{x})x_i x_j d\mathbf{x} \tag{9.15}$$

is the inertia tensor of the system. This is generally known as the tensor virial theorem. Its contracted form, setting $i = j$ and summing,

$$\frac{1}{2}\frac{d^2 I}{dt^2} = 2T + W + 3\Pi \tag{9.16}$$

is the usual scalar version.

These results are nothing more than concatenated moments of the collisionless Boltzmann equation. The first velocity moment of the distribution function gave the Eulerian equations of motion. Now the first position moment of the Euler equations gives the virial theorem. Higher order moments would constrain the system's behavior still further. Indeed, we see again how these higher moments enter to produce a coupled hierarchy: the inertia tensor is a second order moment in position. To close the moment equations at this stage, we must assume a new physical constraint on the system's moment of inertia.

Several simplifying conditions are possible. If the system is in a sufficiently steady state that $d^2 I_{ij}/dt^2 = 0$, then the pressure, gravity, and bulk motions just balance. If, further, it is in hydrostatic equilibrium between pressure and gravity, with no bulk motions of stars, then $-W_{ij} = \Pi\delta_{ij}$. Alternatively, we may be able to examine a system long enough to obtain a time average defined for an arbitrary quantity, say α, by

$$\langle \alpha \rangle = \lim_{t \to \infty} \frac{1}{\tau} \int_0^\tau \alpha(t)dt. \tag{9.17}$$

For the moment of inertia

$$\left\langle \frac{d^2 I}{dt^2} \right\rangle = \lim_{\tau \to \infty} \frac{1}{\tau} \int_0^\tau \frac{d\dot{I}}{dt}dt = \lim_{\tau \to \infty} \frac{1}{\tau}[\dot{I}(\tau) - \dot{I}(0)]. \tag{9.18}$$

This time average can be zero either if the system is localized in position and space so that $\dot{I}(\tau)$ has an upper bound for all τ, or if the orbits are periodic so that $\dot{I}(\tau) = \dot{I}(0)$ as $\tau \to \infty$. In either of these cases, (9.16) becomes

$$2\langle T \rangle + \langle W \rangle + 3\langle \Pi \rangle = 0. \tag{9.19}$$

In this form of the virial theorem, we must remember that $\langle T \rangle$ does not refer to the total kinetic energy, but just to the portion of the kinetic energy which involves the average motion. If we observe a stellar system, we have no way of knowing how to separate these components. Nor, as a practical matter, can we average a system's internal motions over times which are much longer than its dynamical relaxation time. The first problem can be solved easily by recasting the virial theorem, and we will do that straightaway. The second difficulty is often assumed away by supposing that the configuration we observe actually is the time average. The validity of this assumption will be discussed in Section 41.

Whereas the previous derivation of the virial theorem treated the system as a continuum, this one starts with the basic equations of motion for discrete objects. This will avoid the need to separate the kinetic energy into mean flow and 'thermal' terms. Moreover, we shall generalize the situation by allowing the mass of each object to change with time. This represents mass loss from stars or galaxies, accretion and merging of galaxies, or, more speculatively, cosmologies in which mass varies. The equations of motion for each object are

$$\frac{\mathrm{d}}{\mathrm{d}t}(m^{(\alpha)}v_i^{(\alpha)}) = F_i^{(\alpha)} = m^{(\alpha)}\frac{\partial \phi}{\partial x_i^{(\alpha)}} = - Gm^{(\alpha)} \sum_{\beta \neq \alpha} \frac{m^{(\beta)}(x_i^{(\alpha)} - x_i^{(\beta)})}{|\mathbf{x}^{(\alpha)} - \mathbf{x}^{(\beta)}|^3}, \qquad (9.20)$$

using (9.4). The (α) and (β) indices serve the purpose of the primed and unprimed variables in the continuous case. The masses now are functions of time. For simplicity the objects lose mass isotropically, although the anisotropic case can easily be described by denoting the mass loss in each direction by $\dot{m}^{(\alpha i)}$ in the momentum derivative in (9.20).

Multiplying equation (9.20) by $x_j^{(\alpha)}$ and summing over α

$$\sum_{\alpha} x_j^{(\alpha)} \frac{\mathrm{d}}{\mathrm{d}t}(m^{(\alpha)}v_i^{(\alpha)}) = - G \sum_{\alpha} \sum_{\beta \neq \alpha} m^{(\alpha)}m^{(\beta)}\frac{x_j^{(\alpha)}(x_i^{(\alpha)} - x_i^{(\beta)})}{|\mathbf{x}^{(\alpha)} - \mathbf{x}^{(\beta)}|^3}. \qquad (9.21)$$

Symmetrizing the right hand side in the manner of Equation (9.8) gives the potential energy tensor

$$W_{ij} = - \frac{G}{2} \sum_{\alpha} \sum_{\beta \neq \alpha} m^{(\alpha)}m^{(\beta)}\frac{(x_i^{(\alpha)} - x_i^{(\beta)})(x_j^{(\alpha)} - x_j^{(\beta)})}{|\mathbf{x}^{(\alpha)} - \mathbf{x}^{(\beta)}|^3}. \qquad (9.22)$$

Rewriting the left hand side

$$\sum_{\alpha} x_j^{(\alpha)}\frac{\mathrm{d}}{\mathrm{d}t}(m^{(\alpha)}v_i^{(\alpha)}) = \frac{\mathrm{d}}{\mathrm{d}t}\sum_{\alpha} m^{(\alpha)}x_j^{(\alpha)}v_i^{(\alpha)} - \sum_{\alpha} m^{(\alpha)}v_i^{(\alpha)}v_j^{(\alpha)}$$

$$= \frac{\mathrm{d}}{\mathrm{d}t}\sum_{\alpha} \tfrac{1}{2}m^{(\alpha)}\{(x_j^{(\alpha)}v_i^{(\alpha)} + x_i^{(\alpha)}v_j^{(\alpha)}) + (x_j^{(\alpha)}v_i^{(\alpha)} - x_i^{(\alpha)}v_j^{(\alpha)})\} - \sum_{\alpha} m^{(\alpha)}v_i^{(\alpha)}v_j^{(\alpha)}.$$

$$(9.23)$$

The antisymmetric contribution to (9.23) is the only such term in (9.21), so it must be zero, again demonstrating conservation of angular momentum. The symmetric contribution to (9.23) is

$$\frac{1}{2}\frac{\mathrm{d}}{\mathrm{d}t}\sum_{\alpha} m^{(\alpha)}(x_j^{(\alpha)}v_i^{(\alpha)} + x_i^{(\alpha)}v_j^{(\alpha)}) - \sum_{\alpha} m^{(\alpha)}v_i^{(\alpha)}v_j^{(\alpha)}$$

$$= \frac{1}{2}\frac{\mathrm{d}^2}{\mathrm{d}t^2}\sum_{\alpha} m^{(\alpha)}x_i^{(\alpha)}x_j^{(\alpha)} - \frac{1}{2}\frac{\mathrm{d}}{\mathrm{d}t}\sum_{\alpha} \dot{m}^{(\alpha)}x_i^{(\alpha)}x_j^{(\alpha)} - \sum_{\alpha} m^{(\alpha)}v_i^{(\alpha)}v_j^{(\alpha)}, \qquad (9.24)$$

which is motivated by differentiating the inertia tensor, given the form of (9.14) as a clue.

For the kinetic energy tensor we now have

$$T_{ij} = \frac{1}{2} \sum_{\alpha} m^{(\alpha)} v_i^{(\alpha)} v_j^{(\alpha)}, \tag{9.25}$$

for the inertia tensor

$$I_{ij} = \sum_{\alpha} m^{(\alpha)} x_i^{(\alpha)} x_j^{(\alpha)}, \tag{9.26}$$

and for the mass variation tensor

$$J_{ij} = \sum_{\alpha} \dot{m}^{(\alpha)} x_i^{(\alpha)} x_j^{(\alpha)}. \tag{9.27}$$

Thus the symmetric part of (9.21) is

$$\frac{1}{2} \frac{d^2 I_{ij}}{dt^2} - \frac{1}{2} \frac{d}{dt} J_{ij} = 2T_{ij} + W_{ij}. \tag{9.28}$$

Equation (9.28) is the tensor virial theorem for a discrete collection of gravitationally interacting objects. Its kinetic energy term includes the entire motion of the objects without separating it into an average bulk motion plus a pressure term for fluctuations. A special rate of mass loss proportional to the mass, $\dot{m}^{(\alpha)} = f(t) m^{(\alpha)}(t)$, simplifies the result slightly since then $J_{ij} = f(t) I_{ij}$. Taking the time average of (9.28) shows that if x_i and v_i remain bounded, as before, and $\dot{m}(t)$ does not increase as fast as t for $t \to \infty$, then

$$2\langle T_{ij} \rangle + \langle W_{ij} \rangle = 0. \tag{9.29}$$

The time averages will not, except fortuitously, be equal to the instantaneous average here, since T_{ij} and W_{ij} are explicitly time dependent. However, this time dependence does not change the form of the virial theorem. Contracting and summing over the indices gives

$$2\langle T \rangle + \langle W \rangle = 0, \tag{9.30}$$

which is the most quoted version.

Any number can play in the virial theorem. It applies to a satellite going around a planet as well as to a cluster of galaxies. In the simplest case, a two-body circular orbit, for example, we know from first principles that centrifugal and gravitational forces balance: $m^{(\alpha)} v^2/r = G m^{(\alpha)} m^{(\beta)}/r^2$. Multiplying through by r gives the virial theorem. In this case the time average is equal to the instantaneous value since the orbit is periodic and symmetric. When applied to clusters of many objects, the virial theorem gives the order-of-magnitude relation used in Section 2 between the size of the cluster and its velocity dispersion: $v^2 \approx GmN/R$. About a half-century ago this relation was used to estimate the mass of several clusters of galaxies from their observed radii and velocity dispersions. An alternative estimate from the calibrated luminosity–mass relation of the galaxies gave a cluster mass about an order of magnitude less than the dynamical mass. The disagreement of these two estimates was called 'the mystery of the missing mass' and is discussed further in Section 42.1.

10

The grand description – Liouville's equation and entropy

All for one, one for all, that is our device.
Alexandre Dumas, Elder

Langevin's equation, the Fokker–Planck equation, the master equation, and Boltzmann's equation are all just partial descriptions of gravitating systems. Each is based on different assumptions, suited to different conditions. They all arise from physical, rather intuitive, approaches to the problem. But there is also a more general description from which our previous ones emerge as special cases. We know this must be true because Newton's equations of motion provide a complete description of all the orbits. The trouble with Newton's equations is that they are not very compact: N objects generate $6N$ equations. True, the total angular and linear momenta, and energy, are conserved, at least for isolated systems, but this is not usually a great simplification.

By extending our imagination, we can cope with the problem. We previously imagined a six-dimensional phase space for the collisionless Boltzmann equation. Each point in this phase space represented the three position and three velocity (or momentum) coordinates of a single particle. It was a slight generalization of the two-dimensional phase plane whose coordinates are values of a quantity and its first derivative resulting from a second order differential equation for that quantity. The terminology probably arose from the case of the harmonic oscillator where this plane gave the particular stage or phase in the recurring sequence of movement of the oscillator.

Now consider a bigger 'phase space' having $6N$ dimensions. Each point represents not a single object, but the entire system of N objects. As the system evolves, the trajectory of its phase point traces out this evolution. So far, nothing new has been added except a pictorial representation of the dynamics. The next step is a great piece of intuitive insight due to Gibbs. Although only one system with a particular set of properties may really exist in nature, suppose there were many such systems, an ensemble. If all the objects in each of these systems had *exactly* the same positions and velocities at some time, the systems would all share the same phase point. More interestingly, consider such a Gibbs ensemble in which each system has a different internal distribution of positions and velocities (for the same number of objects). This ensemble is represented by a cloud of points in $6N$-dimensional phase space. At any time, the probability density for finding a system in the ensemble within a particular range of $6N$ coordinates will be denoted by

$$f^{(N)}(\mathbf{x}^{(1)}, \ldots, \mathbf{x}^{(N)}, \mathbf{v}^{(1)}, \ldots, \mathbf{v}^{(N)}, t) \mathrm{d}\mathbf{x}^{(1)} \ldots \mathrm{d}\mathbf{v}^{(N)}. \tag{10.1}$$

This shorthand notation avoids the boredom of writing millions of coordinates for, say, a globular cluster. The value of $f^{(N)}$ is the fraction of systems in the ensemble with the desired range of velocities and positions. Thus the integral of $f^{(N)}$ over all phase space is unity

$$\int_{-\infty}^{\infty} f^{(N)} \mathrm{d}\mathbf{x}^{(1)} \ldots \mathrm{d}\mathbf{v}^{(N)} = 1. \tag{10.2}$$

The next conceptual step is to assume, following Gibbs, that the probability distribution of all members of the ensemble is the same as the probability of finding a given set of coordinates in any one member of the ensemble. For ordinary statistical mechanics this is justified by supposing that all members of the ensemble are fairly similar and represent different microscopic realizations of systems with the same macroscopic (average) properties such as temperature and density. Then one appeals to the ergodicity of the ensemble. Some physicists find this intuitively obvious and note that it leads to many experimentally verified predictions. Others find it intuitively implausible, and thus all the more remarkable for seeming to be true. They have therefore sought rigorous proofs in statistical mechanics and generated a considerable industry. In gravitational (and other explicitly Hamiltonian) systems, the situation is perhaps more straightforward: the Gibbs concept implies the exact equations of motion of the system.

To see this, we first determine how $f^{(N)}$, considered as the probability of finding a given system in the ensemble, changes with time. Initially the probability that a given system has coordinates $(\mathbf{x}_0^{(1)}, \ldots, \mathbf{v}_0^{(N)}, t_0)$ lying within a small $6N$-dimensional volume, with boundary S_0, of phase space is

$$A(t_0) = \int_{S_0} f_0^{(N)}(\mathbf{x}_0^{(1)}, \ldots, \mathbf{v}_0^{(N)}, t_0) \mathrm{d}\mathbf{x}_0^{(1)} \ldots \mathrm{d}\mathbf{v}_0^{(N)}. \tag{10.3}$$

At some later time, the coordinates $\mathbf{x}_0, \mathbf{v}_0$ in the system will evolve dynamically into $\mathbf{x}$ and $\mathbf{v}$, the distribution function will become $f^{(N)}(\mathbf{x}^{(1)}, \ldots, \mathbf{v}^{(N)}, t)$ and the boundary S_0 will change to S_t. The probability that the evolved system now lies within S_t is

$$A(t) = \int_{S_t} f^{(N)}(\mathbf{x}^{(1)}, \ldots, \mathbf{v}^{(N)}, t) \mathrm{d}\mathbf{x}^{(1)} \ldots \mathrm{d}\mathbf{v}^{(N)}. \tag{10.4}$$

However, since the dynamical evolution is continuous, i.e., the system does not suddenly jump into a new state, the probability that it is in the transformed region is the same as its probability of being in the original region

$$A(t) = A(t_0) = \text{constant}. \tag{10.5}$$

This is similar to the argument leading to Equation (7.7) and the collisionless Boltzmann equation, with one essential difference. Whereas the collisionless Boltzmann equation follows only if the objects' motions result just from a slowly

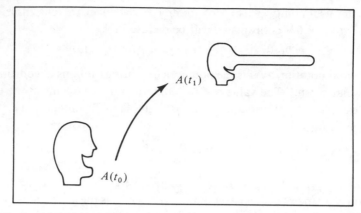

Fig. 9. Schematic evolution of volume in $6N$-dimensional phase space.

varying mean field, Equation (10.5) merely requires the system's equations of motion to be continuous. The invariant A may be regarded as the fractional volume of phase space occupied by systems having a particular evolving range of coordinates. The volume is conserved during the evolution, as shown schematically in Figure 9.

Consider in more detail how this volume stays the same. Its value at time $t_0 + \Delta t$ is related to its value at an arbitrary earlier time t_0 by

$$A(t_0 + \Delta t) = \int_{S_{t_0 + \Delta t}} f^{(N)}(\mathbf{x}^{(1)}, \ldots, \mathbf{v}^{(N)}, t_0 + \Delta t) \mathrm{d}\mathbf{x}^{(1)} \ldots \mathrm{d}\mathbf{v}^{(N)}$$

$$= \int_{S_{t_0}} f^{(N)}(\mathbf{x}^{(1)}, \ldots, \mathbf{v}^{(N)}, t_0 + \Delta t) \frac{\partial(\mathbf{x}^{(1)}, \ldots, \mathbf{v}^{(N)})}{\partial(\mathbf{x}_0^{(1)}, \ldots, \mathbf{v}_0^{(N)})} \mathrm{d}\mathbf{x}_0^{(1)} \ldots \mathrm{d}\mathbf{v}_0^{(N)}. \quad (10.6)$$

So far we have just changed the region of integration. We next relate the distribution function at the two times by a Taylor series expansion

$$f^{(N)}(\mathbf{x}^{(1)}, \ldots, \mathbf{v}^{(N)}, t_0 + \Delta t) = f_0^{(N)}(\mathbf{x}_0^{(1)}, \ldots, \mathbf{v}_0^{(N)}, t_0) + \left(\frac{\partial f_0^{(N)}}{\partial \mathbf{x}_0^{(1)}} \cdot \frac{\mathrm{d}\mathbf{x}_0^{(1)}}{\mathrm{d}t} + \cdots \right.$$

$$\left. + \frac{\partial f_0^{(N)}}{\partial \mathbf{v}_0^{(N)}} \cdot \frac{\mathrm{d}\mathbf{v}_0^{(N)}}{\mathrm{d}t} + \frac{\partial f_0^{(N)}}{\partial t} \right) \Delta t + O[(\Delta t)^2] + \cdots. \quad (10.7)$$

The coordinates and velocities at the two times are related by

$$\mathbf{x}^{(n)} = \mathbf{x}_0^{(n)} + \frac{\mathrm{d}\mathbf{x}_0^{(n)}}{\mathrm{d}t} \Delta t + \cdots \quad (10.8)$$

and

$$\mathbf{v}^{(n)} = \mathbf{v}_0^{(n)} + \frac{\mathrm{d}\mathbf{v}_0^{(n)}}{\mathrm{d}t} \Delta t + \cdots, \quad (10.9)$$

which give the Jacobian

$$\frac{\partial(\mathbf{x}^{(1)}, \ldots, \mathbf{v}^{(N)})}{\partial(\mathbf{x}_0^{(1)}, \ldots, \mathbf{v}_0^{(N)})} = 1 + \left(\frac{\partial \dot{x}_0^{(1)}}{\partial x_0^{(1)}} + \frac{\partial \dot{y}_0^{(1)}}{\partial y_0^{(1)}} + \frac{\partial \dot{z}_0^{(1)}}{\partial z_0^{(1)}} + \cdots + \frac{\partial \dot{v}_{0x}^{(N)}}{\partial v_{0x}^{(N)}} + \frac{\partial \dot{v}_{0y}^{(N)}}{\partial v_{0y}^{(N)}} + \frac{\partial \dot{v}_{0z}^{(N)}}{\partial v_{0z}^{(N)}} \right) \Delta t$$

$$+ O[(\Delta t)^2] + \cdots = 1 + \sum_{\alpha=1}^{N} (\nabla_{\mathbf{x}_0} \cdot \dot{\mathbf{x}}_0^{(\alpha)} + \nabla_{\mathbf{v}_0} \cdot \dot{\mathbf{v}}_0^{(\alpha)}) \Delta t + \cdots,$$

$$(10.10)$$

using Cartesian component notation. Multiplying (10.7) and (10.10),

$$f^{(N)} \frac{\partial(\mathbf{x}^{(1)}, \dots, \mathbf{v}^{(N)})}{\partial(\mathbf{x}_0^{(1)}, \dots, \mathbf{v}_0^{(N)})} = f_0^{(N)} + \sum_{\alpha=1}^{N} [\mathbf{V}_{\mathbf{x}_0} \cdot (f_0^{(N)} \dot{\mathbf{x}}_0^{(\alpha)}) + \mathbf{V}_{\mathbf{v}_0} \cdot (f_0^{(N)} \dot{\mathbf{v}}_0^{(\alpha)})] \Delta t$$

$$+ \frac{\partial f_0^{(N)}}{\partial t} \Delta t + \cdots. \tag{10.11}$$

Note that the position and velocity coordinates are treated symmetrically. Substituting (10.11) into (10.6), recalling (10.3) and taking the limit as $\Delta t \to 0$, gives

$$\frac{\mathrm{d}A(t)}{\mathrm{d}t} = \int_{S_{t_0}} \left(\frac{\partial f_0^{(N)}}{\partial t} + \sum_{\alpha=1}^{N} [\mathbf{V}_{\mathbf{x}_0} \cdot (f_0^{(N)} \dot{\mathbf{x}}_0^{(\alpha)}) + \mathbf{V}_{\mathbf{v}_0} \cdot (f_0^{(N)} \dot{\mathbf{v}}_0^{(\alpha)})] \right) \mathrm{d}\mathbf{x}_0^{(1)} \dots \mathrm{d}\mathbf{v}_0^{(N)}. \tag{10.12}$$

Since $\mathrm{d}A/\mathrm{d}t = 0$ for any volume S_{t_0}, from (10.5), the integrand of (10.12) must be zero. Moreover, one initial time is as good as another, so we may drop the subscript in the integrand, to obtain

$$\frac{\partial f^{(N)}}{\partial t} + \sum_{\alpha=1}^{N} [\mathbf{V}_{\mathbf{x}} \cdot (f^{(N)} \dot{\mathbf{x}}^{(\alpha)}) + \mathbf{V}_{\mathbf{v}} \cdot (f^{(N)} \dot{\mathbf{v}}^{(\alpha)})] = 0. \tag{10.13}$$

This is Liouville's equation. It applies more generally to the phase space representing any set of continuous first order differential equations, not just to dynamics. Immediately, we notice its close resemblance to the equation of continuity (7.1) in six-dimensional phase space. Liouville's equation is indeed no more than a generalized equation of continuity for $6N$-dimensional phase space, and so could have been derived from physical arguments in the manner of (7.1). It says there are neither sources nor sinks as the probability ebbs and flows throughout phase space.

Regarding $f^{(N)}(q_i, p_i)$ as a function of $6N$ coordinates and momenta, rather than velocities, and substituting Hamilton's equations (7.16) into (10.13) gives

$$\frac{\partial f^{(N)}}{\partial t} + \{f^{(N)}, H\} = 0, \tag{10.14}$$

where the Poisson bracket is now summed over all objects

$$\{f^{(N)}, H\} = \sum_{i=1}^{3N} \left(\frac{\partial f^{(N)}}{\partial q_i} \frac{\partial H}{\partial p_i} - \frac{\partial f^{(N)}}{\partial p_i} \frac{\partial H}{\partial q_i} \right). \tag{10.15}$$

In other words, the Jacobian (10.10) of the dynamical transformation was really just unity, as the alert reader may have noticed. Therefore density is conserved in phase space.

The result (10.14) can thus be derived directly by requiring the total time derivative of $f^{(N)}$ to be zero,

$$\frac{\mathrm{d}f^{(N)}}{\mathrm{d}t} = \frac{\partial f^{(N)}}{\partial t} + \sum_{i=1}^{3N} \left(\frac{\partial f^{(N)}}{\partial q_i} \frac{\mathrm{d}q_i}{\mathrm{d}t} + \frac{\partial f^{(N)}}{\partial p_i} \frac{\mathrm{d}p_i}{\mathrm{d}t} \right) = 0, \tag{10.16}$$

and substituting Hamilton's equations. This also shows that the total time derivative of any dynamical quantity is its partial time derivative plus its Poisson bracket with the Hamiltonian. If the Hamiltonian is explicitly independent of time, for example, then the total energy of the system is constant since $\{H, H\} = 0$.

Physically, the reason for constant phase space density is that all the positions and velocities are determined completely and forever by their values at any one time. Since the Hamiltonian equations (7.16) are first order in time, they determine a unique trajectory through any point in six-dimensional phase space for each object in a system, and thus a unique trajectory in $6N$-dimensional phase space for each point representing an entire system in the Gibbs ensemble. Phase points cannot cross the changing boundary of any initial region in phase space, and the volume of this region is invariant as the ensemble evolves, so the phase space density is constant.

Although Liouville's equation (10.14) has a close formal resemblance to the collisionless Boltzmann equation (7.17), it is really much more powerful. This is because it contains complete information about all the orbits in the system, with no assumptions or approximations. Indeed the characteristics of (10.14)

$$\frac{dt}{1} = \frac{dq_1}{\partial H/\partial p_1} = -\frac{dp_1}{\partial H/\partial q_1} = \frac{dq_2}{\partial H/\partial p_2} = -\frac{dp_2}{\partial H/\partial q_2} = \cdots = -\frac{dp_{3N}}{\partial H/\partial q_{3N}} \quad (10.17)$$

are the exact equations of motion for all N objects. We now have seen how the distribution function for an ensemble describes the orbits within a member system. This is the analog of (7.15) for the collisionless Boltzmann equation. Again, the result is somewhat formal because H involves a sum or integral over the distribution function, which is not known until the orbits are all solved.

Despite this difficulty, Liouville's equation is very useful, not only as one compact representation of all the orbits. We illustrate this from three points of view.

If $f^{(N)}$ really does contain all the information about the system, then the entropy in $6N$-dimensional phase space should remain constant since no information is lost as the system evolves. This is in contrast with the six-dimensional Boltzmann entropy which is constant only in equilibrium and otherwise increases. Define the $6N$-dimensional entropy $S^{(N)}$ analogously to (4.5), i.e., proportional to the average value of the logarithm of the probability $f^{(N)}$

$$S^{(N)} = -\int f^{(N)} \ln f^{(N)} dV^{(N)} \quad (10.18)$$

(where Boltzmann's constant of proportionality $k = 1$ and the integral is over $6N$ phase space). Its derivative is

$$\frac{dS^{(N)}}{dt} = -\int \frac{\partial f^{(N)}}{\partial t} (\ln f^{(N)} + 1) dV^{(N)}. \quad (10.19)$$

Now we develop the first term on the right hand side using Liouville's equation (10.14) and an integration by parts assuming that $f^{(N)}$ vanishes on the boundary

$$\int \frac{\partial f^{(N)}}{\partial t} \ln f^{(N)} dV^{(N)} = -\int \{f^{(N)}, H\} \ln f^{(N)} dV^{(N)}$$

$$= \int f^{(N)} \{\ln f^{(N)}, H\} dV^{(N)}$$

$$= \int \{f^{(N)}, H\} \, \mathrm{d}V^{(N)}$$

$$= - \int \frac{\partial f^{(N)}}{\partial t} \, \mathrm{d}V^{(N)}. \tag{10.20}$$

Hence the two right hand terms of (10.19) cancel and

$$\frac{\mathrm{d}S^{(N)}}{\mathrm{d}t} = 0. \tag{10.21}$$

In order to obtain an entropy which increases with time, we must relax the exact description of the system. There are two ways to do this. Following Gibbs, we could coarse grain the distribution function, replacing $f^{(N)}$ by its average over a finite volume element of phase space

$$\bar{f}^{(N)} = \frac{1}{\Delta V^{(N)}} \int_{\Delta V} f^{(N)} \mathrm{d}V^{(N)}. \tag{10.22}$$

As an initial set of points representing systems in phase space – an element of 'phase fluid' – evolves, its boundary will generally become very convoluted even though its volume is constant. This is because different systems in the ensemble evolve at different rates and their representative points move off in different directions. After a while, a fixed volume of phase space which was initially filled may have some gaps in it, so $\bar{f}^{(N)} \leq f^{(N)}$. Consequently

$$\bar{S}^{(N)} = - \int \bar{f}^{(N)} \ln \bar{f}^{(N)} \mathrm{d}V^{(N)} \geq S^{(N)}. \tag{10.23}$$

The temporal increase of coarse grained entropy

$$\bar{S}^{(N)}(t_1) \leq S^{(N)}(t_2 > t_1) \tag{10.24}$$

is a result of the tendency of most regions of the phase fluid to become more and more filamentary as systems evolve. Normally there are only a few small regions of phase space, representing objects locked into resonances and periodic orbits, which do not participate. Eventually systems may evolve so that their phase fluid is so finely interwoven that a sort of 'detailed balance' applies. As one filament moves out of the averaging volume $\Delta V^{(N)}$, another moves in to take its place, so that for any two times $\bar{f}^{(N)}(t_1) = \bar{f}^{(N)}(t_2)$. This is the sign that the system has reached equilibrium. It does not happen, however, for gravitating systems.

A problem with Gibbs' coarse graining is that its detailed result will clearly depend on the size of $\Delta V^{(N)}$. If the averaging volume is extremely fine, the orbits of individual objects will show up in the motion of the representative point, and in the limit we are back to $S^{(N)}$ rather than $\bar{S}^{(N)}$. Using very large averaging elements, on the other hand, means that the representative point will spend a long time in one element, then quickly move to another, possibly distant, element. Obviously in the extreme case, averaging over the entire phase space, nothing ever happens. Sudden jumps between large averaging elements will therefore depend on the system's evolution over a finite segment of its previous history, during which it actually moved around within the element, but where this motion was unnoticed in the averaging. In this case the description of the phase space evolution becomes non-Markovian.

The standard statistical mechanical description of entropy depends upon there being an intermediate case. It must be possible to find a scale of averaging on which the representative points appear to move about in a random Markovian manner. The broader the range of scales to which this applies, the better the description. On these scales particles appear to diffuse through phase space, and this diffusion generates entropy in a manner analogous to the discussion in Section 4.

There is another way to define entropy which avoids this 'eye of the beholder' aspect (see Grad, 1958). In the next section we will find that it also provides a most useful way to extract information from a system. Basically the idea is to form reduced distribution functions $f^{(n)}$ by integrating out all the $6(N-n)$ phase space coordinates

$$ f^{(n)} = \int \cdots \int f^{(N)} \mathrm{d} V^{(N-n)}. \tag{10.25} $$

Thus $f^{(2)}(\mathbf{x}_1, \mathbf{v}_1, \mathbf{x}_2, \mathbf{v}_2, t) \equiv f^{(2)}(1,2)$, for example, is the probability of finding an object with position $\mathbf{x}_1$ and velocity $\mathbf{v}_1$ in $\mathrm{d} V^{(1)}(1) = \mathrm{d}\mathbf{x}_1 \mathrm{d}\mathbf{v}_1$, as well as an object with position $\mathbf{x}_2$ and velocity $\mathbf{v}_2$ in $\mathrm{d} V^{(1)}(2) = \mathrm{d}\mathbf{x}_2 \mathrm{d}\mathbf{v}_2$. In this notation, the argument or subscript refers to the particle label; the superscript refers to the number of particles. Consider, for the present, just the one- and two-particle distribution functions. Since the objects must be somewhere in phase space, the simplest normalizations are

$$ \int f^{(1)}(1)\mathrm{d} V^{(1)}(1) = \int f^{(1)}(2)\mathrm{d} V^{(1)}(2) = \int f^{(2)}(1,2)\mathrm{d} V^{(2)}(1,2) = 1 \tag{10.26} $$

and

$$ \int f^{(2)}\mathrm{d} V^{(1)}(1) = f^{(1)}(2); \quad \int f^{(2)}\mathrm{d} V^{(1)}(2) = f^{(1)}(1), \tag{10.27} $$

where $\mathrm{d} V^{(2)}(1,2) \equiv \mathrm{d} V^{(1)}(1)\mathrm{d} V^{(1)}(2)$. Now if the objects did not interact at all their distributions would be independent. Since independent probabilities multiply, we would have

$$ f^{(2)}(1,2) = f^{(1)}(1)f^{(1)}(2). \tag{10.28} $$

Any interactions, however, will introduce correlations between the objects. The probability of finding two objects in a given element of phase space will no longer be the probability of finding one object times the probability of finding another object. The presence of one object may increase or decrease the probability of finding another object nearby. To allow for this, we modify (10.28) by introducing a pair correlation function $P(\mathbf{x}_1, \mathbf{v}_1, \mathbf{x}_2, \mathbf{v}_2, t)$

$$ f^{(2)}(1,2) = f^{(1)}(1)f^{(1)}(2) + P(1,2). \tag{10.29} $$

In the limit as the interaction vanishes, $P(1,2)$ vanishes also. If the interaction is weak we might also expect $P(1,2) \ll f^{(1)}(1)f^{(1)}(2)$. However, this is not always the case in gravitating systems since the gravitational energy is non-linear with mass and does not saturate. Therefore a large enough number of objects, each interacting weakly, can produce large pair correlations. From (10.29) and the normalizations (10.26) we see that

$$\int P(1,2)\mathrm{d}V^{(2)}(1,2) = 0, \tag{10.30}$$

which is just a sophisticated way of saying that mass is conserved. Positive correlations in one region must be balanced by negative correlations elsewhere.

The entropy of the single-particle distribution function

$$S^{(1)}(1) = -\int f^{(1)}(1)\ln f^{(1)}(1)\mathrm{d}V^{(1)}(1) \tag{10.31}$$

is the form most similar to the entropy we have discussed before (see Equations (4.5), (4.6)). It increases with time (unless there is equilibrium) because it is not a complete description of the system. But we can now also define a two-particle entropy

$$S^{(2)}(1,2) = -\int f^{(2)}(1,2)\ln f^{(2)}(1,2)\mathrm{d}V^{(2)}(1,2). \tag{10.32}$$

If there were no interactions, substituting (10.29) and (10.27) shows that $S^{(2)}(1,2)$ would just be the sum of the one-particle entropies. Moreover, a similar result would hold for $S^{(3)}(1,2,3)$ and so on up the line to $S^{(N)}$. Thus when the single-particle entropies increase with time, so do the multiparticle entropies up to $S^{(N-1)}$. The result ceases to hold for $S^{(N)}$ because $f^{(N)}$, and only $f^{(N)}$, satisfies Liouville's equation and is equivalent to complete orbital information.

How do interactions change the entropy? Looking at $S^{(2)}$, we see from (10.32) and (10.29) that it will contain the somewhat awkward factor $\ln[f^{(1)}(1)f^{(1)}(2) + P(1,2)]$ in which both terms may be any size. We can cast this into a neater form using the Taylor series expansion for any analytic function around a value x

$$f(x+y) = \sum_{j=0}^{m} \frac{y^j}{j!}\frac{\mathrm{d}^j f(x)}{\mathrm{d}x^j} + \frac{y^{m+1}}{(m+1)!}\frac{\mathrm{d}^{m+1}f(x')}{\mathrm{d}x'^{m+1}}\bigg|_{x'=x+\alpha y}, \tag{10.33}$$

where the number α is bounded by $0 \leq \alpha \leq 1$. The last term is called the remainder; for normal use $y \ll x$ and m is large so that the remainder is as small as desired. Here, however, we take the opposite extreme with $m = 0$, so the result may be nearly 'all remainder' for large P. This gives

$$\ln[f^{(2)}(1,2)] = \ln[f^{(1)}(1)f^{(1)}(2) + P(1,2)]$$

$$= \ln[f^{(1)}(1)f^{(1)}(2)] + \frac{P(1,2)}{f^{(1)}(1)f^{(1)}(2) + \alpha P(1,2)} \tag{10.34}$$

From (10.26)–(10.34) we next calculate the difference,

$$S^{(2)}(1,2) - [S^{(1)}(1) + S^{(1)}(2)] = -\int P(1,2)\left[1 + \frac{(1-\alpha)P(1,2)}{f^{(1)}(1)f^{(1)}(2) + \alpha P(1,2)}\right]\mathrm{d}V^{(2)}(1,2)$$

$$= -\int \frac{(1-\alpha)P^2(1,2)}{f^{(1)}(1)f^{(1)}(2) + \alpha P(1,2)}\mathrm{d}V^{(2)}(1,2), \tag{10.35}$$

using sophisticated mass conservation (10.30) for the last step. With no interaction or correlation we clearly recover our previous result that the single-particle entropies add to produce the two-particle entropy. If the pair correlation is positive, then the integrand in (10.35) is also positive. If $P(1, 2)$ is negative, then

$$f^{(1)}(1)f^{(1)}(2) + \alpha P \geq f^{(1)}(1)f^{(1)}(2) + P = f^{(2)}(1, 2) \geq 0 \qquad (10.36)$$

and the integrand is still positive. Therefore in both cases

$$S^{(2)}(1, 2) \leq S^{(1)}(1) + S^{(1)}(2). \qquad (10.37)$$

Correlations always decrease the two-particle entropy.

In Part II we shall see how this result provides insight into the tendency of gravitating systems to cluster. The gravitational contribution to entropy is negative and the correlations of clustering decrease this entropy. If we retain the notion that systems evolve in he direction of an entropy extreme (a maximum negative value in the gravitational case), then we should expect infinite systems of galaxies to form tighter and tighter clumps over larger and larger scales. The expansion of the Universe modifies the details of clustering somewhat, as we shall find in Part II, but the qualitative features remain. This theme also occurs in stellar dynamics. Spherical systems of stars evolve toward maximum negative gravitational entropy. Clustering within the constraint of spherical symmetry leads the core of the system to become denser and denser, until finally stars collide bodily. Then they explode.

We are now in a position to see how correlations modify the virial theorem. Just as we derived a virial theorem by taking moments of the Boltzmann equation, we can process Liouville's equation in a similar way. Starting with (10.13) and using (9.20) for the acceleration, Liouville's equation in Cartesian components reads

$$\frac{\partial f^{(N)}}{\partial t} + \sum_\alpha v_i^{(\alpha)} \frac{\partial f^{(N)}}{\partial x_i^{(\alpha)}} + Gm \sum_\alpha \sum_{\beta \neq \alpha} \frac{x_i^{(\beta)} - x_i^{(\alpha)}}{|\mathbf{x}^{(\beta)} - \mathbf{x}^{(\alpha)}|^3} \frac{\partial f^{(N)}}{\partial v_i^{(\alpha)}} = 0. \qquad (10.38)$$

Here all the masses have a constant value for simplicity, and terms are summed for Cartesian components over the repeated i index. Since the distribution function cannot distinguish among objects of identical mass (or any other property), it must be symmetric in the Greek indices. For objects of different properties, this symmetry can be broken in interesting ways.

If we integrate (10.38) over the positions and velocities of all the objects except one, then

$$\frac{\partial f^{(1)}}{\partial t} + v_i^{(1)} \frac{\partial f^{(1)}}{\partial x_i^{(1)}} + (N-1)Gm \int \int \frac{x_i^{(2)} - x_i^{(1)}}{|\mathbf{x}^{(2)} - \mathbf{x}^{(1)}|^3} \frac{\partial f^{(2)}}{\partial v_i^{(1)}} d\mathbf{x}^{(2)} d\mathbf{v}^{(2)} = 0. \quad (10.39)$$

The summation over $\beta \neq \alpha$ gives a factor $(N-1)$ through the symmetry of $f^{(N)}$. Except for the complexity of the last term, this is rather similar to the Boltzmann equation (8.7). Once again, we are clearly led to a hierarchy of equations since the evolution of $f^{(1)}$ depends on $f^{(2)}$. Nevertheless, we can proceed analogously to Sections 8 and 9, but now averaging over ensembles rather than over a particular system. The resulting virial theorem has the form (Chandrasekhar & Lee 1968)

$$\frac{1}{2}\frac{d^2 I_{ij}}{dt^2} = 2T_{ij} + \Pi_{ij} + W_{ij}. \tag{10.40}$$

This is very similar to (9.14) except that P_{ij}, which defines Π_{ij} through the equation analogous to (9.7), is an ensemble average rather than the system average of (8.18), and now

$$W_{ij} = \tfrac{1}{2}(N-1)G \int \int \rho^{(2)}(\mathbf{x}, \mathbf{x}', t) \frac{(x_i - x_i')(x_j - x_j')}{|\mathbf{x} - \mathbf{x}'|^3} d\mathbf{x}' d\mathbf{x} \tag{10.41}$$

instead of (9.9).

The main difference between this and the collisionless Boltzmann virial theorem lies in the two-particle density function

$$\rho^{(2)}(\mathbf{x}, \mathbf{x}', t) = \rho^{(2)}(\mathbf{x}', \mathbf{x}, t) = \int \int f^{(2)}(\mathbf{x}, \mathbf{v}, \mathbf{x}', \mathbf{v}', t) d\mathbf{v} d\mathbf{v}'. \tag{10.42}$$

This is the probability of finding an object at $\mathbf{x}$ as well as one at $\mathbf{x}'$ with any velocity. If the two probabilities are independent, we would have $\rho^{(2)}(\mathbf{x}, \mathbf{x}') = \rho^{(1)}(\mathbf{x})\rho^{(1)}(\mathbf{x}')$, similar to (10.28). In this case (10.41) would reduce to (9.9), renormalized. However, if the positions of the objects are correlated, there is an additional term in W_{ij} since $\rho^{(2)}$ is replaced by $\rho^{(2)}(\mathbf{x}, \mathbf{x}') = \rho^{(1)}(\mathbf{x})\rho^{(1)}(\mathbf{x}') + P^{(2)}(\mathbf{x}, \mathbf{x}')$. When the correlation term involving $P^{(2)}$ is contracted, W_{ii} gives the gravitational correlation energy of the system. This correlation energy plays an important role in understanding galaxy clustering.

11

Extracting knowledge: the BBGKY hierarchy

Where is the knowledge we have lost in
information?

T.S. Eliot

BBGK and Y stand for the names of the physicists who independently, or in different contexts, developed a rigorous way to extract knowledge from the information in many-variable partial differential equations: Born, Bogoliubov, Green, Kirkwood and Yvon. The basic idea is to turn one equation with a large number of variables into many equations, each with a small number of variables. The many equations cannot be the orbit equations since each orbit equation contains the positions and velocities of all the other objects. Virial and higher order moment equations are useful for answering some questions, but for others they average out too much of the information. However, an hierarchial set of coupled equations involving the reduced distribution functions (10.25) may provide just the right amount of detail.

To place various approximations of the many-body problem in some perspective, Figure 10 shows the relations among their information content. Both the master equation and Liouville's equation provide independent starting points for a many-body description. Liouville's equation is completely consistent and self-contained (represented by the box around it) whereas the master equation requires further physical input before it is useful. The Langevin equation is so phenomenological as to be in a class of its own.

Arrows in Figure 10 show the direction in which information is sacrificed to obtain solvability. Information, here, is represented by the detail of the many-body description. It may be decreased by reducing the number of variables, or by truncating a Taylor series expansion, or by averaging over moments. The art of many-body physics is to find the simplest level of description which can give enough information to solve the problem. An aspect of this art is the development of specific models to bypass the formalism on which we are about to embark. This detailed formalism, in any event, is needed to provide a foil against which the accuracy of models can sometimes be tested.

From Liouville's equation in component form (10.38) we develop the hierarchy of equations for the reduced distribution functions. Changing the normalization of $f^{(n)}$ from (10.25) to

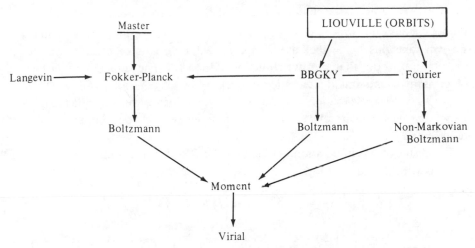

Fig. 10. Relations between different descriptions of the many-body problem.

$$f^{(n)}(\mathbf{x}_1, \mathbf{v}_1, \ldots, \mathbf{x}_n, \mathbf{v}_n, t) = f^{(n)}(1, \ldots, n) = \frac{(N/\rho)^N}{(N-n)!} \int f^{(N)}(1, \ldots, N) d\mathbf{x}_{n+1} d\mathbf{v}_{n+1} \ldots d\mathbf{v}_N$$

(11.1)

and

$$\int f^{(N)} dV^{(N)} = (\rho/N)^N N!$$

(11.2)

will make the resulting equations appear less encumbered with coefficients. Here ρ denotes the average number density of objects. The factorials in these normalizations arise from the number of ways of interchanging labels among the identical objects in the symmetric distribution functions $f^{(n)}$ (see (10.39)).
Thus

$$f^{(n)}(1, \ldots, n) d\mathbf{x}_1 d\mathbf{v}_1 \ldots d\mathbf{x}_n d\mathbf{v}_n \equiv f^{(n)} dV^{(n)}$$

(11.3)

gives the probability of finding any n objects simultaneously in the different volume elements $d\mathbf{x}_1, \ldots, d\mathbf{x}_n$ at positions $\mathbf{x}_1, \ldots, \mathbf{x}_n$ and in the different velocity elements $d\mathbf{v}_1, \ldots, d\mathbf{v}_n$ at velocities $\mathbf{v}_1, \ldots, \mathbf{v}_n$.

Multiply (10.38) by $(N/\rho)^N/(N-n)!$ and integrate over $dV^{(N-n)} = d\mathbf{x}_{n+1} \ldots d\mathbf{v}_N$ for an arbitrary value of n. The first term is just $\partial f^{(n)}/\partial t$. In the second set of terms, those terms for which $\alpha < n$ give

$$\sum_{i=1}^{n} \mathbf{v}_i \cdot \frac{\partial f^{(n)}}{\partial \mathbf{x}_i},$$

where the notation is simplified by replacing the i index of (10.38) with the vector dot product, and the (α) index by the dummy summation variable i over the particles. The remaining terms for $\alpha > n$ can be integrated by parts. In each of them the integral is zero since the v_i and x_i are independent variables. The contribution at the limits

$$v_i f^{(N)}(\mathbf{x}_{n+1})|_{\mathbf{x}_{n+1} \to -\infty}^{\mathbf{x}_{n+1} \to \infty}$$

will vanish if the system is finite. It also vanishes in infinite systems if they remain homogeneous on a large scale so the value of $f^{(N)}$ at the two limits is the same, and we will take this to be the case. A special example is a spherically symmetric distribution, or one that is an even function of the x_{n+1}. The reduction of the last set of terms involves a similar integration by parts. Now $f^{(n)}$ depends on all the $f^{(n')}$ with $n' > n$ since the acceleration depends on the positions of all the other particles. However, the symmetry of $f^{(N)}$ allows all these higher distribution functions to be represented by $f^{(n+1)}$.

The result of this integration of (10.38) is the **BBGKY** hierarchy for gravitating systems with no external force

$$\left[\frac{\partial}{\partial t} + \sum_{i=1}^{n} \mathbf{v}_i \cdot \frac{\partial}{\partial \mathbf{x}_i} - \sum_{i \neq j}^{n} \mathscr{V}(i,j) \right] f^{(n)}(1, \ldots, n)$$

$$= \sum_{i=1}^{n} \int \mathscr{V}(i, n+1) f^{(n+1)}(1, \ldots, n+1) \mathrm{d}\mathbf{x}_{n+1} \mathrm{d}\mathbf{v}_{n+1}. \tag{11.4}$$

Here we have written

$$\mathscr{V}(i,j) \equiv \frac{\partial \phi(i,j)}{\partial \mathbf{x}_i} \cdot \frac{\partial}{\partial \mathbf{v}_i} \tag{11.5}$$

for the interaction of the (i,j) pair whose gravitational potential is

$$\phi_{(i,j)} = - \frac{Gm}{|\mathbf{x}_i - \mathbf{x}_j|}. \tag{11.6}$$

The result (11.4) is actually a set of N equations, one for each value of n. The equation for each value of n is coupled to that for $n + 1$, until $n = N$ whereupon (11.4) becomes Liouville's equation. The case $n = 1$ would be the collisionless Boltzmann equation if the integral term involving $f^{(2)}$ were zero. Explicitly, this shows how the collisional term of the Boltzmann equation represents the combined effects of 'scattering' by all the other objects when not subject to the usual simplifications of billiard ball fluids. Each equation of the hierarchy may also be viewed as the 'projection' of the system's complete dynamics onto the evolution of its partial distribution functions.

Nothing has been gained or lost; the **BBGKY** hierarchy contains as many equations as there are stars. But by recasting the problem into this form we hope to find a suitable physical approximation which enables us to truncate the hierarchy at a solvable level. Originally (11.4), or its equivalent, was devised to treat problems involving the distribution of molecules in fluids and of ions in a plasma. These are systems close to equilibrium. Gravitating systems act differently, and to compare them we examine three timescales characterizing their kinetics (see Section 6).

First, there is the timescale during which a collision occurs, $\tau_* \approx r_0/v$, where r_0 is a typical interaction distance for large-angle scattering of two objects, and v is the relative velocity of these objects. Next, there is the time between collisions, $\tau_0 \approx \lambda/v$, where λ is a mean free path. Finally, there is a macroscopic relaxation time, $\theta_0 \approx L/v_0$, where L is the length scale over which a macroscopic parameter (such as density, temperature, or pressure) changes, propagating with velocity v_0. Often, but not always, v_0 is the same order of magnitude as the velocity dispersion.

Some kinetic systems, such as nearly homogeneous plasmas and molecular gases, appear to satisfy a hypothesis suggested by Bogoliubov. His idea was that if $\tau_* \ll \tau_0 \ll \theta_0$, then there are three distinct stages in the dynamical evolution of an N-body system:

(a) Initially all the $f^{(N)}$ distribution functions vary rapidly and chaotically for a time $\sim \tau_*$, except for $f^{(1)}$ which changes slowly because it does not involve the individual particle interactions directly.

(b) A kinetic stage occurs after a time $\sim \tau_*$. There is a first smoothing after which the detailed initial information is lost and the entire time dependence is contained in $f^{(1)}$ which satisfies a functional equation of the form

$$\frac{\partial f^{(1)}}{\partial t} = \mathscr{F}[\mathbf{x}, \mathbf{v}, f^{(1)}].\tag{11.7}$$

The Boltzmann and Fokker–Planck equations are special cases of (11.7).

(c) A hydrodynamic stage occurs after several collision times τ_0. There is a second smoothing after which the velocities are in equilibrium and the entire time dependence is contained in macroscopic variables (e.g., density, average velocity, temperature) which are velocity moments of $f^{(1)}$.

Gravitating systems are different. To determine their relevant timescales, we use the interaction distance $r_0 \approx b \approx Gm/v^2$ from (2.1) for large-angle scattering. This gives the kinetic timescales of Table 1 where the gravitational response time (6.3),

$$\tau_G \equiv (R^3/Gm)^{1/2}.\tag{11.8}$$

This is the response time a system would have if it contained only one object whose mass was smoothly distributed throughout the entire volume.

If the gravitating system is close to equilibrium, then $f^{(1)}$ is close to a Maxwell–Boltzmann distribution and the virial theorem applies approximately. Under these conditions, from Table 1

$$\tau_*/\tau_0 \approx N^{-2} \text{ and } \tau_0/\theta_0 \approx N.\tag{11.9}$$

In a globular cluster, for example, $N \approx 10^5\text{–}10^6$ and these systems are now stuck in a nearly 'collisionless' kinetic stage. For distant weak encounters in such a system, we

Table 1. *Kinetic timescales in gravitating N-body systems*

	General N-body system	General gravitating system	System close to virial equilibrium	System with 'Hubble expansion'
Time *of* collision	$\tau_* \approx r_0/v$	Gm/v^3	$\tau_G N^{-3/2}$	$\tau_G N^{-1/2}$
Time *between* collisions	$\tau_0 \approx \lambda/v$	$v^3 R^3/G^2 m^2 N$	$\tau_G N^{1/2}$	$\tau_G N^{-1/2}$
Macroscopic relaxation time	$\theta_0 \approx L/v_0$	R/v_0	$\tau_G N^{-1/2}$	$\tau_G N^{-1/2}$

have seen in Section 6 that $\tau_{*L} \approx N\tau_{0L}$ without quenching. Maximum quenching would imply that these long-range encounters lose their correlations on the macroscopic crossing time, so $\tau_{*L} \approx \tau_{0L} \approx \theta_0 \approx \tau_G N^{-1/2}$. Neither short- nor long-range collisions satisfy Bogoliubov's assumption, so the BBGKY approach cannot be used in direct analogy with fluids or plasmas. At best, it is good for calculating small departures from a pre-existing state of low initial correlation.

When the initial velocity distribution satisfies a 'Hubble law', two galaxies have relative velocity $v \approx rH$, where r is their separation and $H \approx (GmN/R^3)^{1/2} \approx \tau_G^{-1}N^{1/2}$. The 'Hubble constant' H may be thought of as the inverse of a characteristic macroscopic relaxation time or expansion time for the system. In these cosmic systems the interaction distance has the scale of large irregularities in the gravitational field, which is approximately the average distance between the nearest neighbors. So for these systems, all three timescales are about equal and different stages of kinetic evolution cannot be separated. In this case, however, an important problem can be solved with the BBGKY approach: the early growth of clustering in an initially uncorrelated infinite distribution of galaxies. We will explore it in Part II.

12

Extracting knowledge: the Fourier development

> Knowledge without thought is labor lost, thought
> without knowledge is perilous.
>
> *Confucius*

Prigogine (1962) and his associates have developed an alternative formalism to the BBGKY hierarchy for extracting important results from Liouville's equation. Here we describe the general method and some of the conclusions it has led to for gravitating systems, without going into detail. Section 14 will apply the essential idea behind this approach to the Boltzmann equation, and we will see how it can describe the effect of sending a massive probe into a cluster of stars or galaxies.

Start again with Liouville's equation (10.14) in the form resembling Schrödinger's equation (see (7.20))

$$i\frac{\partial f^{(N)}}{\partial t} = Lf^{(N)}, \tag{12.1}$$

with the Liouville operator L given by

$$Lf^{(N)} = i\{H, f^{(N)}\} = i\sum_{j=1}^{3N}\left(\frac{\partial H}{\partial q_j}\frac{\partial}{\partial p_j} - \frac{\partial H}{\partial p_j}\frac{\partial}{\partial q_j}\right)f^{(N)}. \tag{12.2}$$

The Hamiltonian is

$$H = \sum_{k=1}^{N}\frac{p_k^2}{2m} - \frac{Gm^2}{2}\sum_{j\neq k}^{N}|\mathbf{r}_j - \mathbf{r}_k|^{-1}. \tag{12.3}$$

(Where the limit on the summations is N, indices refer to objects; where the limit is $3N$, indices refer to the three vector components for each object in the system.) Since L does not itself depend on $f^{(N)}$, Equation (12.1) is linear in $f^{(N)}$ and solutions for different eigenfunctions can be superposed. The reason for multiplying by $i = \sqrt{-1}$ is to make the Liouville operator Hermitian, with the convenience of having real eigenvalues. To see this, consider two arbitrary functions, $f_1(p_1, \ldots, q_N)$ and $f_2(p_1, \ldots, q_N)$, which are integrable over phase space. The condition for Hermiticity, which holds for each term in the summation in (12.2), is that (for a typical term)

$$\int f_1^*(Lf_2)dp_j dq_j = i\int f_1^*\left[\frac{\partial H}{\partial q_j}\frac{\partial f_2}{\partial p_j} - \frac{\partial H}{\partial p_j}\frac{\partial f_2}{\partial q_j}\right]dp_j dq_j$$

$$= -i \int f_2 \left[\frac{\partial H}{\partial q_j} \frac{\partial f_1^*}{\partial p_j} - \frac{\partial H}{\partial p_j} \frac{\partial f_1^*}{\partial q_j} \right] dp_j dq_j$$

$$= \int f_2 (L^* f_1^*) dp_j dq_j, \tag{12.4}$$

integrating by parts and making the usual assumption that f_1 and f_2 vanish at the phase space limits.

To the eigenfunctions ψ_k of L, correspond eigenvalues λ_k given by

$$L\psi_k = \lambda_k \psi_k. \tag{12.5}$$

Multiplying (12.5) by ψ_k^*, integrating over phase space and noting as a special case of (12.4) that $\int \psi_k^* (L\psi_k) dp dq$ is real, shows that the eigenvalues λ_k are also real.

These eigenfunctions are assumed to form a complete orthonormal set so they can be superposed to give the distribution function

$$f^{(N)}(\mathbf{p}_1, \ldots, \mathbf{q}_N, t) = \sum_k a_k(t) \psi_k(\mathbf{p}_1, \ldots, \mathbf{q}_N). \tag{12.6}$$

They can also represent any other phase space function.

Inserting the expansion (12.6) into Liouville's equation (12.1) and using (12.5) shows that

$$i \sum_k \psi_k \frac{da_k}{dt} = \sum_k \lambda_k a_k \psi_k. \tag{12.7}$$

Multiplying by ψ_L^* and using the orthogonality relation

$$\int \psi_k \psi_L^* d\mathbf{p} d\mathbf{q} = \delta_{kL} \tag{12.8}$$

(which follows from (12.5), Hermiticity, and an assumed normalization of the eigenfunctions), then gives for each eigenvalue

$$i \frac{da_k}{dt} = \lambda_k a_k, \tag{12.9}$$

whence

$$a_k(t) = c_k e^{-i\lambda_k t}, \tag{12.10}$$

with c_k a constant. Therefore the solution of Liouville's equation can formally be written as

$$f^{(N)}(\mathbf{p}_1, \ldots, \mathbf{q}_N, t) = \sum_k c_k e^{-i\lambda_k t} \psi_k(\mathbf{p}_1, \ldots, \mathbf{q}_N). \tag{12.11}$$

Of course, finding the c_k, λ_k, and ψ_k which are suitable for the case in hand is what constitutes the hard work. Knowing that the system is time reversible provides an important insight. From (12.11), we see that $f^{(N)}(-t) = f^{(N)}(t)$ implies that for each eigenfunction with eigenvalue λ_k there is also an eigenfunction with value $-\lambda_k$. Moreover, the eigenfunctions are complex since $L^* = -L$.

If the objects in the system were to move freely, without interacting gravitationally, their Hamiltonian would be

$$H = \sum_{\alpha=1}^{N} \frac{p_\alpha^2}{2m} \tag{12.12}$$

giving a Liouville operator

$$L = -i \sum_{\alpha=1}^{N} \frac{\mathbf{p}_\alpha}{m} \cdot \frac{\partial}{\partial \mathbf{q}_\alpha}. \tag{12.13}$$

Its eigenfunctions are the products

$$\psi(\mathbf{k}_1, \ldots, \mathbf{k}_N) = D^{-(3N/2)} e^{i \Sigma \mathbf{k}_j \cdot \mathbf{q}_j} \tag{12.14}$$

and its eigenvalues are the sums

$$\lambda(\mathbf{k}_1, \ldots, \mathbf{k}_N) = \sum_j \mathbf{k}_j \cdot \frac{\mathbf{p}_j}{m}. \tag{12.15}$$

Here $\mathbf{k}_j$ is a real vector which depends on the boundary conditions and D is the
length of a cubic box which contains the galaxies. The theory is applied in the limit
$D \to \infty$, whereupon the boundary conditions become irrelevant and may, for
convenience, be replaced by the usual periodic boundary condition that

$$\mathbf{k}_j = \frac{2\pi}{D} \mathbf{n}_j, \tag{12.16}$$

where $\mathbf{n}_j$ is a vector having integer components. Substituting (12.14) and (12.15) into
(12.11) shows that the distribution function $f^{(N)}$ for non-interacting objects can be
expanded as a multiple Fourier series. We will examine a simpler application of this
technique in Section 14 in detail.

The general procedure for gravitating systems is to introduce the gravitational
force as a perturbation in the Hamiltonian. From the perturbed Liouville operator,
new perturbed eigenfunctions are calculated using the non-interacting Fourier
eigenfunctions as a basis set. The gravitational force introduces complicated
couplings between the Fourier components, which evolve with time. This leads to
kinetic equations for the Fourier components and, since gravitating systems have
memories, these kinetic equations are generally non-Markovian. Moreover, the
Fourier components are also closely related to density fluctuations and correlations
in the system. Thus the kinetic equations describe the evolving ebb and flow of
correlations. Non-Markovian integrals over time in the Fourier description are
related to correlation integrals over space in the BBGKY description. Both types of
integrals represent departures from randomness.

To solve these kinetic equations, it is necessary to make a number of physical
assumptions and approximations. For atomic gas in a box, it is easy to perturb away
from a stable equilibrium state. Assuming a weak coupling between atoms, this
kinetic theory successfully describes the gentle non-equilibrium evolution back
toward the stable state. But treating gravitating systems in this way is more difficult.

Gravitational terms control the Hamiltonian of finite systems near virial equilibrium. As we have seen in Section 1, these terms can be divided into a mean field and a fluctuating component. One approximation is to consider what happens in regions small compared to the size of the system. Then the mean field is fairly constant and does not significantly affect the local evolution. If there are local inhomogeneities of small scale and amplitude, their correlations can be followed by perturbing the Liouville operator. The small amplitude requirement also restricts the scale to be greater than the separation $\approx RN^{-1}$, which produces large-angle scattering between two objects (Equation (2.1)). When very close separations produce bound binary subsystems, they can be treated simply as single objects of greater mass. Triple and higher order subsystems are trickier since they are generally unstable; these have not yet been analyzed in the theory.

Infinite homogeneous systems present a different set of problems for the Fourier technique. Here the mean field is zero and one can treat linear departures from the uniform state. These introduce weak gravitational coupling among the objects. Approximating the non-Markovian evolution equations gives solutions for the growth of correlation kinetic energy in the system. However, the description only includes those processes which grow on timescales long compared with $(G\rho)^{-1/2}$. Infinite inhomogeneous systems include aspects of both finite systems and infinite homogeneous systems. In particular, the presence of non-linear clusters with crossing time $\tau_c \approx (G\rho)^{-1/2}$ as well as the long-range relaxation time τ_R (Equation (2.11)) introduces processes described by the hybrid timescale $(\tau_c \tau_R)^{1/2}$. These systems are particularly difficult to treat with the Fourier technique. Haggerty & Severne (1976) have given a detailed review of how this method may be applied to gravitating systems.

13

Collective effects – grexons

'All together now!'
Rowboat crew cox

A collapsing stock market, astronomers jumping aboard the bandwagon of the latest 'hot topic', rowboat crews, and galaxy formation – all are examples of collective effects. If the strokes of all the rowers in a boat have randomly distributed phases, some pulling upstream while others pull downstream, the boat at best performs a linear random walk. But if all the rowers row in phase, their strengths add nearly linearly (there is some turbulence) and away glides the boat. If buyers and sellers of stocks are about evenly balanced, the price of an average share records little change. A mild selling imbalance somewhat decreases prices. If the decrease deepens and spreads, people begin to worry that they may lose their fortunes. 'Sell out fast !,' the word goes 'round. Panic strikes. Everyone sells. The market collapses. In this case, not only do all the phases act together, but the action reinforces itself. It is highly non-linear and the result is catastrophe. Astrophysical bandwagons illustrate still other aspects of collective phenomena (left as an exercise for the reader).

Gravitating systems are prone to a variety of collective effects. Previously, whenever we examined the motion of a star in a system, we have always asked what the system can do to the star, not what the star can do to the system. Obviously a heavyweight star will do more to a system than a lightweight one will, and this change in the system in turn will react back on the motion of the star. To see how this affects the fundamental relaxation of the stellar orbits in Section 2, consider the following approximate physical description of a star moving through a cluster.

We found in Section 2 that if the moving star does not affect the cluster, then its motion is deflected from its original path by a small angle whose mean square is approximately (Equations (2.2), (2.3))

$$\langle \psi^2 \rangle \approx N\psi^2 \approx \frac{4G^2 m_*^2 N}{v^4 b^2} \text{ radians} \tag{13.1}$$

for $m_1 = m_2 \equiv m_*$. More exactly, using (2.8) and substituting the velocity dispersion of the cluster,

$$v_* = \left(\frac{Gm_* N}{R}\right)^{1/2}, \tag{13.2}$$

gives

$$\langle \psi^2 \rangle = \frac{24}{N} \left(\frac{v_*}{v} \right)^4 \ln \left(\frac{b_2}{b_1} \right) \tag{13.3}$$

after one crossing time $t = Rv^{-1}$. For a globular cluster with $N = 10^5$ and $v \gtrsim v_*$, this angle is very small. But suppose that for some unspecified reason the stars in the cluster were to clump into a few coherent subclusters. Then N would be small and if a star were injected into this system with $v \approx 2v_*$, say, its scattering angle would be quite large. In the limiting case, the star is deflected by the mass of the entire cluster acting coherently so $N \approx 1$ and $b \approx R$. Equation (13.1) then gives the maximum scattering angle

$$\langle \psi^2 \rangle_{max} \approx (v_*/v)^4. \tag{13.4}$$

Of course such clumps would not be in equilibrium (although they could arise from initial conditions, as in the merging of several clusters) and the large-angle collective scattering would rapidly redistribute the orbits of all stars, including those within clumps. This is the basic idea behind 'violent relaxation', discussed further in Section 38. Applying these results of Section 2 directly to collective effects pushes the formulae far beyond their derived range of validity, and we should not expect them to give a very precise description of collective scattering. But they do contain some of the essential physics of the problem, and provide a prelude to examining the more general case in detail.

In this more general case, we shoot a massive star M_* with velocity $v > v_*$ into the stellar system. Moving through the system, it deflects the orbits of all the other stars, as well as being deflected by them. These other stars are attracted by the more massive star, which tends to leave a wake of greater-than-average density behind it. Orbits become correlated and produce local concentrations of gravitational energy within the system. Some of the random kinetic motion of the stars sweeping by the intruder is converted into streaming motion. The lifetime of the concentrations, and their exact shapes, depend on detailed exchanges of energy among the stars in the wake. We may think of these concentrations of energy as gravitational excitations produced by the fast massive star, and call such regions grexons. The word comes from the Latin 'grex' ('herd' or 'flock' (of stars or galaxies)), and is also an acronym for 'gravitational excitation'. So it has a double etymology.

The forming of grexons is somewhat analogous to the excitation of plasmons by an electron moving through a plasma. Unlike plasmons, however, grexons may become unstable and contract, rapidly amplifying the initial fluctuations. This is particularly prone to happen if the stellar system has a (temporary) velocity dispersion which is colder than the equilibrium dispersion given by (13.2).

We now consider a very simple model which describes the back-reaction of grexons on the orbit of the massive star. In order for a grexon to form, the massive star must accrete surrounding stars and produce highly correlated orbits. This happens on the local dynamical timescale

$$\tau_{\text{form}} \approx (G\rho)^{-1/2}, \tag{13.5}$$

where the local density is

$$\rho = \bar{\rho} + \frac{M_*}{\frac{4}{3}\pi r_g^3} = \bar{\rho}\left(1 + \frac{M_* R^3}{M r_g^3}\right). \tag{13.6}$$

Here $\bar{\rho}$ is the background density of stars, r_g is the radius of the grexon and M is the total mass of stars in the cluster. To estimate the number of grexons that can form in a distance R, roughly the cluster's radius, we consider two requirements for grexon formation. First, they must form in a time less than the time for the massive star to move away from the region of accreted stars

$$\tau_{form} \lesssim r_g v^{-1}. \tag{13.7}$$

Second, in order for a well-defined (non-linear) grexon to form, the massive star must contribute significantly to the local density

$$\frac{M_*}{r_g^3} \gtrsim \frac{M}{R^3}. \tag{13.8}$$

The number of grexons that can form as the massive star moves through the system is then roughly

$$N_g \approx \frac{R}{r_g} \gtrsim \frac{M}{M_*}\left(\frac{v}{v_*}\right)^2 \tag{13.9}$$

from (13.5)–(13.8). Thus as the velocity of the disturbing star increases, or its mass decreases, more grexons are formed, but they have greatly reduced mass.

Next consider two limiting cases which illustrate the scattering of a massive star by its grexons. First, suppose the stellar velocities are close to equilibrium and the grexons form the most symmetric wake consistent with statistical fluctuations behind the massive star. Moreover, since the stellar system is so nearly stable, grexons do not last long and do not induce second order clustering transverse to the wake. Then the grexon accelerates the disturbing star by an amount approximately $Gm_g b^{-2} \sin \phi$, where m_g is the grexon's mass and $\sin \phi (\approx \phi$ since ϕ is small) is the angle between the orbit of the massive star and the center of mass of the grexon. If there are N_{*g} stars in the grexon, random uncorrelated statistical fluctuations will displace its center of mass from the initial orbit of the disturbing star by an amount $\approx r_g N_{*g}^{-1/2}$. Since these grexons are only somewhat non-linear, the number of stars they contain will be given to order of magnitude by $N_{*g} \approx N(r_g/R)^3$. Finally, the impact parameter $b \approx r_g$. Thus Equation (13.1) becomes

$$\langle \psi^2 \rangle \approx \frac{G^2 m_g^2 N_g \phi^2}{v^4 r_g^2} \approx \frac{1}{N}\left(\frac{v_*}{v}\right)^4, \tag{13.10}$$

which is essentially (13.3). The logarithmic factor does not occur here because only grexons in the narrow wake are assumed to scatter so there is no integration over the complete solid angle. Thus, in this extreme case of no correlation, the grexon picture brings us back to the original analysis. It provides another way of viewing the results of Section 2.

On the other hand, we can also extend it to the highly correlated case. If the stellar system is inherently 'cool' or otherwise unstable, a few massive grexons forming in the wake of the heavy star will cause the motions of more distant stars to become correlated. Secondary grexons may form throughout the system, at least temporarily. When the disturbing star moves slowly enough it can be affected by these higher collective modes distributed over a large solid angle. Now we would expect the scattering angle to be given approximately by (13.3) with N_g from (13.9) replacing the number of scattering centers N

$$\langle \psi^2 \rangle \approx \frac{M_*}{M} \left(\frac{v_*}{v} \right)^6 \ln \left(\frac{b_2}{b_1} \right). \tag{13.11}$$

When $M_* \approx m_*$, this result is similar to the uncorrelated case of (13.3) since $M = Nm_*$, except for two powers of (v_*/v). For the special cases of $M_* \approx m_*$ and large v/v_*, this shows that incoherent contributions dominate the scattering.

These estimates hold for small total scattering angles, roughly less than one radian. Moreover, the approximations we have used so far break down for $v \approx v_*$, since the response of the system to a nearly stationary massive star is much stronger. It then resembles accretion onto the massive star, rather than the formation of a wake of grexons (see Section 17). Slow motion can produce local non-linear density enhancements which alter the massive star's orbit on the local dynamical timescale $\approx \tau_{form}$ given by Equations (13.5)–(13.6). This speeds up the relaxation processes by a factor of roughly N, which can be considerable.

In the next section, we will derive the results (13.10) and (13.11) more rigorously as special cases of a detailed examination of how a stellar system responds to objects sent in to probe its reactions. Our technique applies a Fourier development to Boltzmann's equation. This is an analog to the solution of Schrödinger's equation for quantum mechanical scattering.

14

Collective scattering

14.1. The scattering probability

Shoot a massive star through a stellar system. Even if the system is symmetric and homogeneous, we saw in the last section that the probe star will be deflected by the fluctuations it itself induces in the stellar distribution. These fluctuations involve a transfer of energy and momentum between the system and the probe star. Since the total energy and momentum are conserved, we can find the deflection of the probe by calculating the induced change in the system.

Initially, the massive star has mass M_0, momentum $\mathbf{P}_0$ and kinetic energy $E_0 = P_0^2/2M_0$. Interaction with the field stars changes its momentum by an amount $-\mathbf{P}_1$, as shown in Figure 11. This momentum change can be divided into two parts: P_{1a} measures the change in the magnitude of $\mathbf{P}_0$, and is projected in the same direction as the new momentum $\mathbf{P}_0 - \mathbf{P}_1$. It therefore gives the energy change

$$\Delta E = \frac{1}{2M_0}[P_0^2 - (\mathbf{P}_0 - \mathbf{P}_1)^2] = \frac{P_{1a}P_0}{M_0} - \frac{P_{1a}^2}{2M_0} \tag{14.1}$$

since $(\mathbf{P}_0 - \mathbf{P}_1)^2 \equiv (|\mathbf{P}_0| - |\mathbf{P}_{1a}|)^2$. The second component $\mathbf{P}_{1b}$ measures the angle of deflection

$$\theta = 2\sin^{-1}\frac{|\mathbf{P}_{1b}|}{2|\mathbf{P}_0|}. \tag{14.2}$$

If the velocity of the intruder star is sufficiently large compared with the velocity dispersion of the system, the stars will not have time to respond much to its passage and the angle of deflection will be small. This is the case we will treat here since it is simplest. But we shall see, as a pleasant surprise and a bonus, that it is also a good approximation even at slow velocities. For small deflections the geometry of Figure 11 and (14.1)–(14.2) give

$$|\mathbf{P}_{1a}| = P_1^{\parallel} = M_0\Delta E/P_0, \tag{14.3}$$

$$|\mathbf{P}_{1b}| = P_1^{\perp} = P_0\theta \tag{14.4}$$

$$P_1 = P_0[\theta^2 + (\Delta E/2E)^2]^{1/2}, \tag{14.5}$$

to first order.

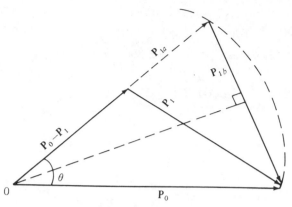

Fig. 11. Change in the initial momentum $\mathbf{P}_0$ of a massive star after it has moved through the stellar system.

Next we characterize the evolution of the stellar distribution. It will start out uncorrelated, so the simplest description assuming that correlations do not become important is the collisionless Boltzmann equation (7.20)

$$i\frac{\partial f}{\partial t} = Bf. \tag{14.6}$$

Initially the stellar distribution and therefore the Hamiltonian in the Boltzmann operator B are given, making it possible to treat (14.6) as a linear equation for short times – until major changes in the structure of the system grow to influence the probe star's orbit. The Hamiltonian of an arbitrary star in the system may be divided into three parts

$$H = \frac{P^2}{2m} + \mathscr{V}_1(\mathbf{r}) + \mathscr{V}_2(\mathbf{r}), \tag{14.7}$$

where

$$\mathscr{V}_1(\mathbf{r}) = \sum_k \mathscr{V}(|\mathbf{r} - \mathbf{r}_k|) \tag{14.8}$$

is the gravitational potential energy of a star at $\mathbf{r}$ in the field of all the other stars except for the massive probe star which contributes the potential energy

$$\mathscr{V}_2(\mathbf{r}) = \mathscr{V}(|\mathbf{r} - \mathbf{R}|) \tag{14.9}$$

at a distance $|\mathbf{r} - \mathbf{R}|$.

Although the distribution function is a real quantity, it will be algebraically neater to write it as the product of a complex amplitude, and its conjugate, $f = \psi\psi^*$, and to work in terms of the modulus of this amplitude, $\psi = \phi e^{i\theta}$. Inserting these relations into (14.6) readily shows that the modulus ϕ satisfies the same evolution equation as f itself

$$i\frac{\partial\phi}{\partial t} = B\phi. \tag{14.10}$$

There are several ways to solve (14.10). Perhaps the simplest, and one which does not obscure too much of the underlying physics, is by Fourier analysis. It starts analogously to the analysis of Liouville's equation in Section 12, and the formalism is similar to quantum mechanical scattering, but with a different interpretation. If the stellar system were time independent and sufficiently homogeneous that $\mathscr{V}_1$ and $\mathscr{V}_2$ were negligible contributions to the Hamiltonian, then the solution of (14.10) could be written in terms of its Fourier eigenfunctions $\sim e^{i\mathbf{k}\cdot\mathbf{r}}$ obtained from

$$B_0\phi_0 = -i\frac{\mathbf{p}}{m}\cdot\frac{\partial\phi_0}{\partial\mathbf{r}} = \omega_k\phi_0. \tag{14.11}$$

The distribution function, and therefore its amplitude, will generally depend on both momentum (velocity) and position. The momentum dependence can be represented by the amplitude of each Fourier coefficient, leading us to expand this time independent case as

$$\phi_0(\mathbf{p},\mathbf{r}) = D^{-3/2}\sum_k a_{\mathbf{k}}(\mathbf{p})e^{(i\mathbf{k}\cdot\mathbf{r})} \tag{14.12}$$

and

$$a_{\mathbf{k}}(\mathbf{p}) = D^{-3/2}\int\phi_0(\mathbf{p},\mathbf{r})e^{(-i\mathbf{k}\cdot\mathbf{r})}d^3\mathbf{r} \tag{14.13}$$

with the wavenumber $\mathbf{k}$ related to the eigenvalue by

$$\omega_{\mathbf{k}} = \frac{\mathbf{k}\cdot\mathbf{p}}{m} \tag{14.14}$$

similarly to (12.12)–(12.15). Again, we have imagined the stellar system to be in a cubical box whose sides have length D. This normalizes the distribution function. When the box becomes very large, in particular as $D \to \infty$, the effects of scattering at the boundaries become negligible compared to scattering in the bulk of the volume. Under these conditions, the boundary conditions are not physically relevant and can be chosen for mathematical convenience. Customarily, they are taken to be periodic so that $\phi_0(\mathbf{p},\mathbf{r}) = \phi_0(\mathbf{p},\mathbf{r}+\mathbf{D})$ and

$$\mathbf{k} = \frac{2\pi}{D}\mathbf{n}, \tag{14.15}$$

with $\mathbf{n}$ a vector whose components are integers, as in (12.16). The difference from the earlier discussion for Liouville's equation is that here the system is described by a single three-dimensional wave vector, rather than N such wave vectors, one for each star.

How may we interpret this Fourier representation of the stellar system? From (14.13) we see that $|a_0(\mathbf{p})|^2$ represents the initial probability of a given momentum, i.e., the initial momentum distribution function for the system. Values of $|a_{\mathbf{k}}(\mathbf{p})|^2$ for $\mathbf{k} \neq 0$ represent the contributions of the different spatial Fourier components of the total distribution function to the momentum distribution. In this representation, each spatial component is discrete. However, as $D \to \infty$, the difference between values of $\mathbf{k}$ for different values of $\mathbf{n}$ becomes arbitrarily small and the states $a_\mathbf{k}$ of the system become continuous. Since k^{-1} is the wavelength of the Fourier decomposition, $|a_\mathbf{k}|^2$ is the contribution of spatial fluctuations of a given wavelength to the momentum distribution. A spatial perturbation of a homogeneous system is produced by a velocity perturbation of the stellar motions, and the $a_\mathbf{k}(\mathbf{p})$ relate these two perturbations.

It is important to bear in mind that any given Fourier wavelength in the decomposition is non-local in the sense that it occurs equally throughout the system. Thus, a Fourier component with a given wavelength does not itself represent a general density perturbation, say, having that scale length. To find the actual perturbation, we need to sum all the Fourier wavelength components which contribute to it. This is just the difference between a plane wave and a wave packet representing a particle in quantum mechanics.

The initial states of a real system evolve with time according to (14.10). Having found the time independent eigenfunctions, we can use them to build up the form of the time dependent solution. Since the collisionless Boltzmann equation (14.10) is separable in time, we may write $\phi(\mathbf{p},\mathbf{r},t) = \phi_0(\mathbf{p},\mathbf{r})\phi_1(t)$, substitute this in (14.10) and using (14.11) solve for $\phi_1(t)$. The result is

$$\phi(\mathbf{p},\mathbf{r},t) = D^{-3/2} \sum_k a_\mathbf{k}(\mathbf{p}) e^{i(\mathbf{k}\cdot\mathbf{r} - \omega_k t)}. \tag{14.16}$$

But is this adequate to describe the effects of the massive perturbing star? No. Even if the initial system were to stay sufficiently homogeneous so that $\mathscr{V}_1$ could be neglected (Section 8 shows it must have a slight inhomogeneity), we cannot neglect $\mathscr{V}_2$. As the massive star moves rapidly through the system, it perturbs the density and velocity distributions of the field stars. We saw that these distributions are related by the Fourier coefficients $a_\mathbf{k}(\mathbf{p})$, so these coefficients must also become functions of time. Thus, the general form of the solution is

$$\phi(\mathbf{p},\mathbf{r},t) = D^{-3/2} \sum_\mathbf{k} a_\mathbf{k}(\mathbf{p},t) e^{i(\mathbf{k}\cdot\mathbf{r} - \omega_k t)}. \tag{14.17}$$

We next solve for the changes in these Fourier coefficients as the intruding star moves along. They will tell us how energy and momentum are transferred between the probe star and the system, how the probe star is scattered and slowed down. From (7.17), (14.9) and (14.10) we see that when the system is perturbed by the probe star it evolves according to

$$i\frac{\partial \phi}{\partial t} = (B_0 + \delta B)\phi, \tag{14.18}$$

where

$$\delta B = i\frac{\partial \mathscr{V}_2}{\partial \mathbf{r}} \cdot \frac{\partial}{\partial \mathbf{p}}. \tag{14.19}$$

Substituting (14.17) into (14.18) and using (14.11) gives

$$i\sum_{\mathbf{k}}\frac{da_{\mathbf{k}}}{dt}e^{i(\mathbf{k}\cdot\mathbf{r}-\omega_{k}t)} = \sum_{\mathbf{k}}\delta B a_{\mathbf{k}}e^{i(\mathbf{k}\cdot\mathbf{r}-\omega_{k}t)}. \tag{14.20}$$

To isolate the evolution of a single coefficient, multiply both sides by $D^{-3/2}$ exp $(-i\mathbf{k}'\cdot\mathbf{r})$ and integrate over the box using the normality and orthogonality relations of the Fourier eigenfunctions. Interchanging the dummy $\mathbf{k}$ and $\mathbf{k}'$ variables, the result is

$$i\frac{da_{\mathbf{k}}(\mathbf{p},t)}{dt} = D^{-3}\sum_{\mathbf{k}'}e^{i\omega_{k}t}\langle\mathbf{k}|\delta B|\mathbf{k}'\rangle e^{-i\omega_{k'}t}a_{\mathbf{k}'}\cdot(\mathbf{p},t), \tag{14.21}$$

where the operator matrix element

$$\langle\mathbf{k}|\delta B|\mathbf{k}'\rangle = \int e^{-i\mathbf{k}\cdot\mathbf{r}}\delta B e^{i\mathbf{k}'\cdot\mathbf{r}}d^{3}\mathbf{r} \tag{14.22}$$

operates on the quantities to the right of it.

The matrix element is proportional to the probability that the system changes its energy from the state $\omega_{\mathbf{k}}$ to another state $\omega_{\mathbf{k}'}$. To see this, suppose the system is initially in one particular 'base state' given by $a_{\mathbf{k}'}(\mathbf{p},0)$ so that $a_{\mathbf{k}}(\mathbf{p},0)=0$ for $k \neq k'$. Then integrating (14.21) from 0 to t and squaring the result gives the transition probability from state $\mathbf{k}'$ to state $\mathbf{k}$ to first order

$$|a_{\mathbf{k}}(\mathbf{p},t)|^{2} = 4D^{-6}|\langle\mathbf{k}|\delta B|\mathbf{k}'\rangle a_{\mathbf{k}'}(\mathbf{p},0)|^{2}\frac{\sin^{2}[(\omega_{\mathbf{k}}-\omega_{\mathbf{k}'})t/2]}{(\omega_{\mathbf{k}}-\omega_{\mathbf{k}'})^{2}}. \tag{14.23}$$

For the complete solution, we must remember that all the Fourier coefficients $a_{\mathbf{k}'}$ will change with time and enter into the summation in (14.21) which governs the growth of any one coefficient. Treating the solution as a series of increasingly accurate approximations, we would substitute each $a_{\mathbf{k}}$ from the first order integration of (14.21) back into the right hand side of (14.21) and obtain the second approximation from integrals over terms involving the products $\langle\mathbf{k}|\delta B|\mathbf{k}'\rangle \langle\mathbf{k}'|\delta B|\mathbf{k}''\rangle$ of transition matrices. The third approximation would involve triple products and so on.

This state of affairs, whose analogs occur in many branches of physics involving wave interactions – from fluid mechanical perturbation theory to quantum electrodynamics – is often represented schematically by diagrams such as Figure 12. In Figure 12(a) the system is in state $\mathbf{k}'$ until it interacts with the potential δB of the probe star. Then it changes to state $\mathbf{k}$. The second approximation is the product of two such interactions. If each interaction caused by $\mathscr{V}_{2}$ is weak, then their products will be still weaker and the higher order effects will not contribute much to the total energy exchange. Use of the first order result (14.23) is usually known as the Born approximation, originally employed extensively for quantum mechanics. When extended to second order interactions as in Figure 12(b), it is called (in some circles) the born again approximation.

Physically the Born approximation is designed for the case when each scattering is small. Thus it applies to a probe star moving faster than the random motions of the

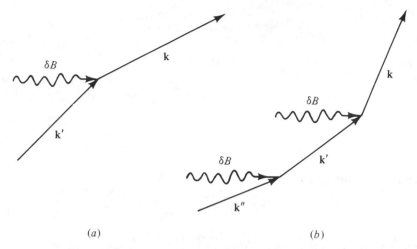

Fig. 12. Schematic diagram representation of (*a*) the first and (*b*) second order transitions between different momentum states of a stellar system.

field stars – basically an impulsive approximation. It takes into account the effect of the probe star on the initial system – and the perturbed initial system's effect on the probe star. But it does not describe how the growing modifications produced by the perturbations themselves interact with the probe star. By then, the probe star is gone.

From the transition probability (14.23) we can calculate the transfer of energy and momentum between the probe star and the system. Eventually, this will tell us the drag and scattering angle of the probe. There are two equivalent ways to proceed. We could calculate the transfer of energy and momentum into a particular final state from a range of initial states or we could calculate the transfer out of a particular initial state into a range of final states. Since we are interested mainly in the differential scattering per unit time of a given star into a given solid angle, the second approach is more direct.

First we integrate (14.23) over all the Fourier frequencies $\omega_{\mathbf{k}}$ (representing the energies) which contribute to the desired final range. Now the factor of the form $(\Delta\omega)^{-2}\sin^2(t\,\Delta\omega/2)$ in the integrand is a strongly peaked function of $\Delta\omega = \omega_{\mathbf{k}} - \omega_{\mathbf{k}'}$ for large t, as may be seen in Figure 13. Setting the derivative of this function equal to zero shows that its peaks occur at values of $t\,\Delta\omega/2 = \tan(t\,\Delta\omega/2)$ or $t\,\Delta\omega/2 = 0, 4.49,$ 7.73, 10.9, etc., and its zeros occur at $t\,\Delta\omega/2 = n\pi$ with $n = 0, 1, 2, \ldots$, etc. The height of the central peak is $t^2/4$, the next peak's height is reduced from this by a factor $\approx$ $\sin^2(4.49)/(4.49)^2 = 0.0472$, etc. For times long compared with the period $\sim (\Delta\omega)^{-1}$ of the frequency difference between initial and final states, this factor behaves as a delta function $\delta(\omega - \Delta\omega)$ described in Section 3. Because this factor is so strongly peaked, we can replace the integral over a finite range of final states $\omega_{\mathbf{k}}$ by the infinite integral $-\infty \leq \omega_{\mathbf{k}} \leq \infty$ in the limit $t \to \infty$. Using the result

$$\int_{-\infty}^{\infty} \frac{\sin^2[(\omega_{\mathbf{k}} - \omega_{\mathbf{k}'})t/2]}{(\omega_{\mathbf{k}} - \omega_{\mathbf{k}'})^2}\,\mathrm{d}\omega_{\mathbf{k}} = \frac{\pi}{2}t, \qquad (14.24)$$

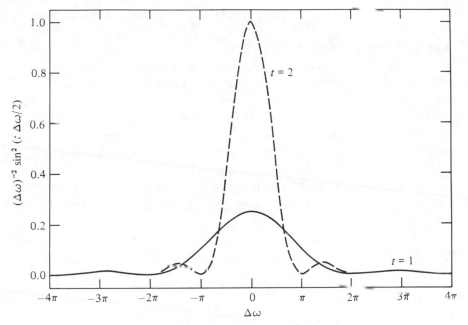

Fig. 13. Graphs of $(\Delta\omega)^{-2}\sin^2(t\,\Delta\omega/2)$ for $t = 1$ (solid line) and $t = 2$ (dashed line).

which confirms the normalization of the δ function, and summing (14.23) over all final momentum states **k** in the solid angle $d\Omega$ which are related to the initial state by the matrix element δB, gives the probability $\Delta\omega$ per unit time and solid angle for the momentum and energy transfer

$$\frac{\Delta W}{t\,d\Omega} = \frac{2\pi}{D^6} \sum_{\mathbf{k}} (|\langle \mathbf{k}|\delta B|\mathbf{k}'\rangle| |a_{\mathbf{k}'}(\mathbf{p},0)|)^2 \delta(\omega - \Delta\omega_{\mathbf{kk}'}). \qquad (14.25)$$

Here we have added the subscripts to $\Delta\omega_{\mathbf{kk}'} = \omega_{\mathbf{k}} - \omega_{\mathbf{k}'}$ as a reminder that ω depends on **k** through the basic scattering equations ((14.1) and (14.14)), which also give the connection relating $\Delta\omega$ to ΔE.

The physical interpretation of (14.25) involves an interesting subtlety. It is the analog of the 'golden rule' in quantum mechanics, but here it applies to celestial objects (a connection perhaps less surprising in Chinese). The factor $\delta(\omega - \Delta\omega_{\mathbf{kk}'})$ implies that transitions conserve energy, so that the frequency of the final state is related to the frequency of the initial state by just the difference of the initial and final frequencies. But to obtain this result we had to go to the limit of long times, during which we would expect higher order corrections to the solution of (14.21) to become important. Now there is also a countervailing tendency. As the corrections to the $a_{\mathbf{k}}$ grow they also change the basic eigenfrequencies of the system. The resulting frequencies $\omega_{\mathbf{k}'} + \omega_{\mathbf{k}'}^{(1)}$ no longer resonate well with the $\omega_{\mathbf{k}}$, thus diminishing their later contribution to the final energy state. As time advances, the final state absorbs energy from a frequency range $|\omega_{\mathbf{k}} - \omega_{\mathbf{k}'}| \approx \pi/t$ which becomes even smaller. So a small modification to $\omega_{\mathbf{k}'}$ can reduce the resonance absorption more and more with increasing t.

Pictorially, the massive probe star is scattered by a grexon for a short time, then the grexon is modified and usually dispersed by surrounding stars. New grexons may form in the wake of the probe star. Each grexon scatters the probe by a small amount during its lifetime. Then the probe moves on. If it moves into an essentially homogeneous medium, it can restart the collective scattering process from essentially the same initial conditions further along its trajectory. Thus the cumulative scattering over long times can be described quite well by the sum of brief uncorrelated energy and momentum exchanges. This leads to the experimental result, described in Section 14.3, that the Born approximation is good even for quite slowly moving massive probes. It is another example where a calculation rises above its origin.

Clearly, the next step then is to calculate the matrix element in (14.25). From its definition (14.22) and the perturbed Boltzmann operator (14.19), we have

$$\langle \mathbf{k} | \delta B | \mathbf{k}' \rangle = i \int e^{-i\mathbf{k}\cdot\mathbf{r}} \frac{\partial \mathscr{V}_2}{\partial \mathbf{R}} \cdot \frac{\partial}{\partial \mathbf{p}} e^{i\mathbf{k}'\cdot\mathbf{r}} d^3\mathbf{r}. \tag{14.26}$$

Here the perturbing potential $\mathscr{V}_2$ is the sum over all the stars which transfer energy between the system and the probe star at $\mathbf{R}$

$$\mathscr{V}_2(\mathbf{R}) = \sum_j \mathscr{V}(\mathbf{r}_j - \mathbf{R}). \tag{14.27}$$

This is essentially what gives rise to the collective effects on the probe. We can rewrite $\mathscr{V}_2$ in a way which separates out its coordinate dependence by using the Fourier decomposition

$$\mathscr{V}_2 = \sum_j \sum_{\mathbf{k}} \mathscr{V}_{\mathbf{k}} e^{i\mathbf{k}\cdot(\mathbf{r}_j - \mathbf{R})}. \tag{14.28}$$

Here $\mathscr{V}_{\mathbf{k}}$ (which we shall soon calculate explicitly) is the Fourier transform of $\mathscr{V}(\mathbf{r})$. Interchanging the order of the summations, we can then rewrite the inner summation as

$$\sum_j e^{i\mathbf{k}\cdot\mathbf{r}_j} = \sum_j \int e^{i\mathbf{k}\cdot\mathbf{r}} \delta(\mathbf{r} - \mathbf{r}_j) d^3\mathbf{r}$$

$$= \int n(\mathbf{r}) e^{i\mathbf{k}\cdot\mathbf{r}} d^3\mathbf{r} \equiv n_{\mathbf{k}}^+. \tag{14.29}$$

Thus the summation of $\delta(\mathbf{r} - \mathbf{r}_j)$ over j is a counting function which adds up the number of stars at $\mathbf{r}$ to give the density, and $n_{\mathbf{k}}^+$ is the Fourier transform of this density using a positive exponential. When the wavenumber $\mathbf{k} = 0$, n_0^+ is the total number of stars N. This can be normalized by the volume to give the average density. When $\mathbf{k} \neq 0$, the $n_{\mathbf{k}}^+$ give the fluctuations of wavelength $\lambda = 2\pi/k$ of the stellar density around this average. Since the wavenumbers do not depend on position in the system, they are, again, a non-local description applying equally throughout the system.

Thus Fourier components of the density can be used to describe the collective modes excited by the probe. From the last two equations we have

$$\mathscr{V}_2 = \sum_{\mathbf{k}} \mathscr{V}_{\mathbf{k}} n_{\mathbf{k}}^+ e^{-i\mathbf{k}\cdot\mathbf{R}}, \tag{14.30}$$

which separates the coordinate and momentum dependence. Substituting this into (14.26) gives the matrix element for a transition from $\mathbf{k}'$ into a particular state $\mathbf{k}$ (which requires dropping the summation of (14.29) over all states) as

$$\langle \mathbf{k}|\delta B|\mathbf{k}'\rangle = \mathscr{V}_{\mathbf{k}}(n_{\mathbf{k}}^+)_{\mathbf{kk}'} e^{-i\mathbf{k}\cdot\mathbf{R}} \mathbf{k}\cdot\frac{\partial}{\partial\mathbf{p}}. \tag{14.31}$$

These are the matrix elements summed over the relevant $\mathbf{k}$ states in (14.25). The factor $(n_{\mathbf{k}}^+)_{\mathbf{kk}'}$ is the matrix element of the number density fluctuations between the states $\mathbf{k}$ and $\mathbf{k}'$. We retain it in this general form until we find the spectrum of fluctuations.

First we find $\mathscr{V}_{\mathbf{k}}$. An integration shows that

$$\mathscr{V}_{\mathbf{k}} = -GmM_0 \int \frac{1}{|\mathbf{r}-\mathbf{R}|} e^{-i\mathbf{k}\cdot(\mathbf{r}-\mathbf{R})} d^3(\mathbf{r}-\mathbf{R})$$

$$= -GmM_0 \int \frac{1}{r} e^{-i\mathbf{k}\cdot\mathbf{r}} d^3\mathbf{r}$$

$$= -2\pi GmM_0 \int_0^\infty \int_{-1}^1 \frac{1}{r} e^{-ikr\cos\theta} r^2 d(-\cos\theta)dr$$

$$= -\frac{4\pi GmM_0}{k^2} \tag{14.32}$$

(since the oscillating term of the integral averages to zero for $k \neq 0$ as $r \to \infty$). Note the conceptual convenience – for thinking of transitions to particular states – of doing one Fourier transform in (14.28) as an integral and its dual as a discrete sum, even though a continuous limit for the sum is implied. Next, combining (14.31) and (14.32) with (14.25) gives

$$\frac{\Delta W(\mathbf{p},\omega)}{td\Omega} = \frac{32\pi^3 G^2 m^2 M_0^2}{D^6 k^4} \sum_{\mathbf{k}} \left[(n_{\mathbf{k}}^+)_{\mathbf{kk}'} \mathbf{k}\cdot\frac{\partial}{\partial\mathbf{p}}|a_{\mathbf{k}'}(\mathbf{p},0)|\right]^2 \delta(\omega-\Delta\omega_{\mathbf{kk}'}). \tag{14.33}$$

We have been able to remove the factor k^{-4} from the sum since it must have the same value for each term of the transition from the homogeneous initial state, i.e., the value which conserves energy as implied by $\delta(\omega-\omega_{\mathbf{kk}'})$. The sum over all the final momentum states which have this energy remains. The result is general for scattering in a collisionless gravitating system using the first order Born approximation. In order to apply it to specific systems we turn to a deeper interpretation of the summation and a means of determining its value.

14.2. Fluctuations, correlations, form factors and the f sum rule for stellar systems

Let us look again at the unperturbed state of a stellar system. Its density fluctuates with time as stars move about. In previous sections we employed equations of motion for orbits or distribution functions, using exact forces or diffusion approximations, to calculate these fluctuations. We can, however, reverse the procedure and regard the fluctuations themselves as fundamental. From the detailed phases and correlations of these fluctuations at different times and places, we could recover the exact equations of motion. It would be a hard way to rediscover Newton's law. But we will follow this path a short distance to place useful constraints on the fluctuation transitions $(n_{\mathbf{k}}^+)_{\mathbf{kk'}}$.

Since the density $n(\mathbf{r}, t)$ is just the integral of the distribution function over its momenta, and the distribution function has the same evolution (14.6) as does the modulus ϕ in (14.10), it changes with time in the unperturbed ($\mathscr{V}_2 = 0$) state according to

$$n(\mathbf{r}, t) = n(\mathbf{r}, 0) e^{-i\omega_{\mathbf{k}} t}. \tag{14.34}$$

Taking the spatial Fourier transform of both sides of (14.34) just replaces n by $n_{\mathbf{k}}$, giving the same evolution for density fluctuations of a particular wavenumber as for the density itself. This is because linear amplitude fluctuations ($\mathscr{V}_1$ is small) are uncoupled. So we may consider just a representative term of the Fourier series. Now we can see how these fluctuations correlate throughout the system between an initial time $t = 0$ and a later time t. To measure temporal correlation, we take the product $n_{\mathbf{k}}(\mathbf{r}, t) n_{\mathbf{k}}^+(\mathbf{r}, 0)$ and average it over the entire system. This is equivalent to taking the transition matrix element from the state $\mathbf{k'}$ to that same state at a later time

$$
\begin{aligned}
F(\mathbf{k}, t) &\equiv \langle \mathbf{k'} | n_{\mathbf{k}}(t) n_{\mathbf{k}}^+(0) | \mathbf{k'} \rangle \\
&= \sum_{\mathbf{k''}} \langle \mathbf{k'} | n_{\mathbf{k}}(t) | \mathbf{k''} \rangle \langle \mathbf{k''} | n_{\mathbf{k}}^+(0) | \mathbf{k'} \rangle \\
&= \sum_{\mathbf{k''}} |(n_{\mathbf{k}}^+)_{\mathbf{k''k'}}|^2 e^{-i\omega_{\mathbf{k''k'}} t}.
\end{aligned}
\tag{14.35}
$$

The second expression is a sum over all the eigenstates of the unperturbed system and the third expression includes the time dependence explicitly. This sum over eigenstates has the physical interpretation of a sum over all 'virtual states' into which the system could make a transition and then return to its original state.

Thus the quantity $F(\mathbf{k}, t)$ also determines the spectral density of states into which the system can make a transition. Taking its Fourier transform with respect to time by multiplying (14.35) by $e^{i\omega t}$ and integrating over time gives

$$F(\mathbf{k}, \omega) = \sum_{\mathbf{k''}} |(n_{\mathbf{k}}^+)_{\mathbf{k''k'}}|^2 \delta(\omega - \omega_{\mathbf{k''k'}}), \tag{14.36}$$

which is known as the dynamic form factor. It measures the excitation spectrum of density fluctuations directly, and their correlations through its Fourier transform.

This relation between fluctuations and correlations is a special example of the Wiener–Khintchine theorem of Fourier analysis.

Now the dynamic form factor resembles the sum in (14.33) except for the modification by the probe star. To calculate this modification we suppose the initial distribution is fairly uniform so that $a_{\mathbf{k}'} = 0$ unless $\mathbf{k}' = 0$, i.e., very long wavelength fluctuations dominate. An isotropic Maxwellian is then a good estimate for the initial stellar momentum distribution

$$a_0^2(\mathbf{p},0) = a_0^2(p^2,0) = (3/2\pi)^{3/2} \langle p^2 \rangle^{-3/2} e^{-3\mathbf{p}\cdot\mathbf{p}/2\langle p^2 \rangle}. \tag{14.37}$$

It is normalized to unity and has root mean square momentum $\langle p^2 \rangle^{1/2}$. Then, using (14.14),

$$\mathbf{k} \cdot \frac{\partial a_0}{\partial \mathbf{p}} = -\frac{3}{2} \frac{a_0(\mathbf{p},0)}{\langle p^2 \rangle} \mathbf{k} \cdot \mathbf{p} = -\frac{3m\omega_k a_0(\mathbf{p},0)}{2\langle p^2 \rangle}. \tag{14.38}$$

Inserting this into (14.33) gives the scattering contribution from any one field star with momentum $\mathbf{p}$. Integrating over the momentum distribution gives the total contribution

$$\frac{\Delta W(\mathbf{k}, \omega)}{t d\Omega} = \frac{9(2\pi)^3 G^2 M_0^2 m^4 \delta(\omega - \omega_{\mathbf{k}\mathbf{k}'})}{D^6 \langle p^2 \rangle^2 k^4} \sum_{\mathbf{k}} \omega_{\mathbf{k}}^2 (n_{\mathbf{k}})_{\mathbf{k}\mathbf{k}'}^2. \tag{14.39}$$

Thus the modification by the probe star leads to a second order moment of the dynamic form factor with respect to the transition frequency.

To calculate this moment, we develop the f sum rule for classical mechanics. It is the analog of the f sum rule for oscillator strengths in atomic transitions. From (14.29) and the Fourier transform of (14.34), we notice that the product of a density fluctuation with its time derivative looks rather like a term of the summation in (14.39). Moreover, (7.17) shows that the sum over all terms is related to the Poisson bracket of $n_{\mathbf{k}}^+$ with its time derivative. But the time derivative is itself the Poisson bracket with the Hamiltonian. This suggests we examine the nature of the quantity

$$\left\{ -\frac{\partial n_{\mathbf{k}}}{\partial t}, n_{\mathbf{k}}^+ \right\} = \{\{n_{\mathbf{k}}, H\}, n_{\mathbf{k}}^+\}. \tag{14.40}$$

Using (14.29) enables us to work out the inner bracket

$$-\frac{\partial n_{\mathbf{k}}}{\partial t} = \{n_{\mathbf{k}}, H\} = \left\{ \sum_j e^{-i\mathbf{k}\cdot\mathbf{r}_j}, \sum_j \frac{\mathbf{p}_j \cdot \mathbf{p}_j}{2m} + \frac{1}{2} \sum_{i \neq j} \mathscr{V}(\mathbf{r}_i - \mathbf{r}_j) \right\}$$

$$= -\frac{i}{m} \mathbf{k} \cdot \sum_j \mathbf{p}_j e^{-i\mathbf{k}\cdot\mathbf{r}_j} \equiv -i\mathbf{k} \cdot \mathbf{J}_{\mathbf{k}}. \tag{14.41}$$

This result has an important physical interpretation. The flow of stars in the system is described by its current density

$$\mathbf{J}(\mathbf{r}) = \sum_j \frac{\mathbf{p}_j}{m} \delta(\mathbf{r} - \mathbf{r}_j) \tag{14.42}$$

and $\mathbf{J_k}$ is just its Fourier transform. So (14.41) is the Fourier transform of the equation of continuity, representing mass conservation.

The double Poisson bracket then becomes

$$\{\{n_\mathbf{k}, H\}, n_\mathbf{k}^+\} = \left\{ -\frac{i}{m} \mathbf{k} \cdot \sum_j \mathbf{p}_j e^{-i\mathbf{k}\cdot\mathbf{r}_j}, \sum_j e^{i\mathbf{k}\cdot\mathbf{r}_j} \right\}$$

$$= -\frac{k^2 N}{m} \qquad (14.43)$$

since the Poisson bracket, being linear, is a distributive operator and the cross products $(\partial/\partial p_j)(\partial/\partial r_i) = 0$ leaving N terms in the summation.

There is an alternative way to calculate the double Poisson bracket, and comparing the two results will yield the sum rule we are after. From (14.29) and (14.34) we have that

$$\frac{\partial n_\mathbf{k}}{\partial t} = i\omega_\mathbf{k} n_\mathbf{k}(t) \qquad (14.44)$$

for the complex conjugate of $n_\mathbf{k}^+$. Next, just as we can expand a function describing the perturbed $\mathbf{k}$ state of the system in terms of the $\mathbf{k}'$ base state eigenfunctions, so we can expand the base state in terms of the perturbed state eigenfunctions, or any other complete set. In this manner, taking the expectation value of the double Poisson bracket in the base state and expanding it in terms of the complete $\mathbf{k}''$ gives

$$\langle \mathbf{k}' | \{\{n_\mathbf{k}, H\}, n_\mathbf{k}^+\} | \mathbf{k}' \rangle = \langle \mathbf{k}' | \{ -i\omega_\mathbf{k} n_\mathbf{k}, n_\mathbf{k}^+\} | \mathbf{k}' \rangle$$

$$= 2\sum_{\mathbf{k}''} \{ \langle \mathbf{k}' | -i\omega_\mathbf{k} n_\mathbf{k} | \mathbf{k}'' \rangle, \langle \mathbf{k}'' | n_\mathbf{k}^+ | \mathbf{k}' \rangle \}$$

$$= 2\sum_{\mathbf{k}''} \{ -i\omega_{\mathbf{k}''\mathbf{k}'}(n_\mathbf{k})_{\mathbf{k}''\mathbf{k}'}, \quad (n_\mathbf{k}^+)_{\mathbf{k}''\mathbf{k}'} \}$$

$$= 2\sum_{\mathbf{k}''} \left\{ \frac{-i\mathbf{k}''\cdot\mathbf{p}}{m}(n_\mathbf{k})_{\mathbf{k}''\mathbf{k}'}, \quad (n_\mathbf{k}^+)_{\mathbf{k}''\mathbf{k}'} \right\}$$

$$= -2\sum_{\mathbf{k}''} \frac{k''^2}{m}(n_\mathbf{k})_{\mathbf{k}''\mathbf{k}'}^2 . \qquad (14.45)$$

The factor two in the second step arises because every state $\mathbf{k}''$ is degenerate with its time reversed state. In the fourth step we used (14.14). Since this result is equivalent to (14.43) we now have the f sum rule

$$\sum_{\mathbf{k}''} k''^2 (n_\mathbf{k})_{\mathbf{k}''\mathbf{k}'}^2 = \frac{Nk^2}{2}. \qquad (14.46)$$

In the atomic physics version, usually cast in terms of the transition frequency, each term of the left hand side multiplied by $2k^{-2}$ is called the oscillator strength $f_{\mathbf{k}''\mathbf{k}'}$ of the transition.

There are two main ways in which sum rules can be useful. First, if the spectrum of fluctuations is known for nearly all frequencies (or wavenumbers), the sum rule gives an estimate of the spectrum for the remaining frequencies. Second, if the spectrum is dominated by one frequency, the sum rule gives an estimate of the amplitude of the fluctuations for that frequency. It is this second application we use here.

The relationship between frequency and wavenumber is provided by the plane-wave expansion (14.17). This represents wavefronts moving with phase velocity $\mathbf{V}_p$ determined by requiring the phase of the wave to be a constant (which may as well be zero): $\mathbf{k} \cdot \mathbf{r} - \omega_k t = 0$. The phase velocity for waves moving parallel to the wavenumber vector $\mathbf{k}$ is

$$\mathbf{V}_p = \frac{\mathbf{r}}{t} = \omega k^{-2} \mathbf{k}, \tag{14.47}$$

and the magnitude of $\mathbf{k}$ is $|\mathbf{k}| = \omega / V_p$.

Although a plane wave is not a star, the motion of a star can be represented as a superposition of plane waves having different phases. The plane waves into which the distribution function is decomposed can be recombined to give the changing probability $F(\mathbf{r}, t)$ of finding a star at a given position, $\mathbf{r}$

$$F(\mathbf{r}, t) = \int_{-\infty}^{\infty} F(\mathbf{k} - \mathbf{k}_0) e^{\mathrm{i}\mathbf{k} \cdot (\mathbf{r} - \mathbf{r}_0) - \mathrm{i}\omega(k)t} \mathrm{d}^3 \mathbf{k}, \tag{14.48}$$

Here $F(\mathbf{k} - \mathbf{k}_0)$ is a sharply peaked function around $\mathbf{k} = \mathbf{k}_0$ which defines the shape of the wave packet. Generally the frequency will be a function of wavenumber; this is called a dispersion relation since it originally referred to the dispersion of waves in a medium whose index of refraction n is different for different wavelengths. (In this case the angular frequency $\omega = ck/n(k)$.) Equation (14.48) shows that the waves of the packet interfere most constructively where the phase of the exponential is an extremum, and there the probability of finding the star is greatest. Thus at this place

$$\frac{\mathrm{d}}{\mathrm{d}\mathbf{k}} [\mathrm{i}\mathbf{k} \cdot (\mathbf{r} - \mathbf{r}_0) - \mathrm{i}\omega(\mathbf{k})t] = 0 \tag{14.49}$$

or

$$\frac{\mathbf{r} - \mathbf{r}_0}{t} = \frac{\mathrm{d}\omega}{\mathrm{d}\mathbf{k}} \tag{14.50}$$

and the maximum of the wave packet moves through space with the 'group velocity'

$$\mathbf{V}_g = \left(\frac{\mathrm{d}\omega}{\mathrm{d}\mathbf{k}} \right) \Bigg|_{\mathbf{k} = \mathbf{k}_0}. \tag{14.51}$$

For a simple isotropic linear dispersion relation, $\omega = ck$, the phase velocity is the same as the group velocity. We shall see in Section 15 that this applies – not exactly but to a good approximation – to our case of small-angle scattering. The reason is that the dispersion results from the non-equilibrium response of the field stars to the massive moving star. To the approximation that the collective modes are so weak they only induce small-angle scattering, their fundamental gravitational response frequency is also unchanged. Then the dominant response frequency of the system is

$$\omega = \omega_g = (4\pi G \rho_0)^{1/2} = \frac{\langle p^2 \rangle^{1/2}}{m} k = \frac{\langle p^2 \rangle^{1/2}}{m} k_g, \tag{14.52}$$

which might also be guessed on dimensional grounds. Combining (14.52) with (14.47) for an arbitrary momentum of the star then gives

$$\mathbf{p} = \frac{m\omega}{k^2}\mathbf{k} = \frac{m\omega_g}{k_g^2}\mathbf{k} = \frac{\langle p^2 \rangle^{1/2}}{k_g}\mathbf{k} = \frac{\langle p^2 \rangle}{m\omega_g}\mathbf{k}. \tag{14.53}$$

Substituting the previous two equations into (14.46) we find

$$\sum_{\mathbf{k}''}\omega_{\mathbf{k}''}^2(n_\mathbf{k})_{\mathbf{k}''\mathbf{k}'}^2 = \frac{N\omega_g^2 p^2}{2\langle p^2\rangle}. \tag{14.54}$$

This form of the sum rule provides a simple solution to the summation in the scattering probability (14.39). We can now complete its calculation.

14.3. The deflection angle, dynamical friction again, and a numerical test

The problem of finding the deflection angle is now almost solved. As a bonus along the way we also get a detailed expression for the dynamical friction. Substituting (14.52), (14.53) and (14.54) into (14.39) gives

$$\frac{\Delta W(\mathbf{p},\omega)}{t\,d\Omega} = \frac{(3\pi)^2 M_0^2 G\langle p^2\rangle\delta(\omega - \omega_g)}{mD^3p^2} \tag{14.55}$$

for a given momentum transfer $\mathbf{p}$. To integrate over all momentum transfers we need to know the density of states in momentum space. This is related to the density η of states in k space by (14.53) and (14.15). Since $d\eta = (D/2\pi)^3 dk_x dk_y dk_z$, where $k_x = 2\pi l/D$, etc., with l, etc., an integer, we find that in momentum space

$$d\eta = \left(\frac{Dm\omega_g}{2\pi\langle p^2\rangle}\right)^3 d^3\mathbf{p}. \tag{14.56}$$

Exploiting the symmetry in momentum space we use cylindrical coordinates oriented along the initial momentum of the massive star, so that from (14.4)

$$d^3\mathbf{p} = 2\pi p_\perp^1 dp_\perp^1 dp^{\|} = p_0^2 dp^{\|} d\Omega. \tag{14.57}$$

To obtain the angular scattering probability rate per unit energy exchange we integrate (14.55) over the momentum coordinates (except the small solid angle $d\Omega = 2\pi\theta d\theta$) for which the energy change is ΔE

$$\begin{aligned} T(\theta,\omega) &= \left(\frac{Dm\omega_g}{2\pi\langle p^2\rangle}\right)^3 p_0^2 \int \frac{\Delta W(\mathbf{p},\omega)}{t\,d\Omega}\delta(\Delta E - p_1^{\|}V_0)dp_1^{\|} \\ &= \frac{9 M_0^2 Gm^2\omega_g^3\delta(\omega - \omega_g)}{8\pi\langle p^2\rangle^2 V_0[\theta^2 + (\Delta E/2E)^2]}. \end{aligned} \tag{14.58}$$

Here we have made use of the relation

$$\delta(f(x) - f(x_0)) = \frac{1}{\left|\dfrac{df}{dx}\right|}\delta(x - x_0), \tag{14.59}$$

which follows from the properties of the δ function described after Equation (3.4). To derive this, use the fact that $\delta(f)df = \delta(x)dx$. Note the case for $f(x_0) = 0$.

Next we calculate the total rate of energy transfer between the massive star and the stellar system. An energy exchange $\Delta E = p\Delta p/m$ is related to the frequency by (14.47), $p = m\omega/k$. Employing our assumption that the exchange is dominated by the gravitational frequency, we have $\Delta E = m\omega_g\omega/k_g^2 = \langle p^2 \rangle \omega/m\omega_g$. Thus the rate of energy transfer into the solid angle Ω is

$$\frac{dE(\Omega)}{dt} = \int_0^\infty (\Delta E)T(\theta, \omega)d(\Delta E) = \frac{9M_0^2 G\omega_g^2}{8\pi V_0[\theta^2 + a^2]}, \qquad (14.60)$$

where

$$a \equiv \frac{\Delta E_g}{2E} \qquad (14.61)$$

and for an individual scattering $\Delta E_g = \langle p^2 \rangle/m$. Integrating (14.60) over solid angle $2\pi \sin\theta d\theta \approx 2\pi\theta d\theta$ gives the total rate of energy transfer between the massive star and the stellar system

$$\frac{dE}{dt} = \frac{9GM_0^2\omega_g^2}{8V_0} \ln\left(\frac{\theta_{max}^2}{a^2} + 1\right). \qquad (14.62)$$

Here θ_{max} is the largest angle through which an individual scattering transfers energy (not necessarily the largest angle through which the massive star can be scattered). The angle θ_{max} is related to k_{max}, the largest wavenumber into which energy is transferred by $p_{1max}^\perp = p_0\theta_{max} = \langle p^2 \rangle k_{max}/m\omega_g$ so that

$$\frac{\theta_{max}}{a} = \frac{k_{max} V_0}{\omega_g}. \qquad (14.63)$$

An energy relaxation time for collective scattering follows from (14.62)

$$\tau_{R\,coll} = \frac{E}{\dot{E}} = \frac{V_0^3}{9\pi G^2 M_0 mn \ln\left(\frac{\theta_{max}^2}{a^2} + 1\right)}. \qquad (14.64)$$

Comparison with (2.9) shows directly how important collective effects are. In the special case of an ordinary star with $M_0 = m$ and $V_0 = v$, recalling from the discussion following (2.8) that for the Maxwellian distribution used in deriving (14.64) the right hand side of (2.9) must be multiplied by a factor ~ 5, we see that $\tau_{R\,coll} \approx 0.7\,\tau_R$. Thus, the collective relaxation is as strong as the ordinary relaxation. In general

$$\tau_{R\,coll} \approx 0.7\left(\frac{V_0}{v_*}\right)^3 \frac{m}{M_0}\tau_R. \qquad (14.65)$$

For massive stars which do not move much faster than the velocity v_* of the field stars, collective relaxation will clearly dominate. Stars with undistinguished masses do not make dominant wakes.

To find the probability distribution of the scattering angle after the massive star has moved through a distance R, we return to (14.58) and integrate it over all frequencies. In this case we can estimate ΔE_g from $\Delta E_g \approx \dot{E}R/V_0$, which gives (using (14.61)–(14.63) and the virial relation)

$$a \approx \frac{27}{8} \frac{M_0}{mN} \left(\frac{v_*}{V_0}\right)^4 \ln\left[\left(\frac{k_{max} V_0}{\omega_g}\right)^2 + 1\right]. \tag{14.66}$$

Moreover, the integral can be normalized to unity between $0 \leq \theta \leq \theta_1$, where θ_1 is the largest scattering angle produced by this process. The result is

$$T(\theta) = \frac{a}{\tan^{-1}(\theta_1/a)} \frac{1}{\theta^2 + a^2}. \tag{14.67}$$

The distribution of deflection angles is strongly peaked at $\theta < a$ and falls off as θ^{-2} for larger θ (still satisfying $p_1 \ll p_0$). In statistics, this relation is known as a Cauchy distribution and is quite common. Here it has a cutoff at θ_1. This is the detailed distribution behind the simple results for mean square deflections described in Section 13.

To recover these simple results, we calculate the mean square deflection from

$$\langle \psi^2 \rangle = \int_0^{\theta_1} \theta^2 T(\theta) \mathrm{d}\theta = \frac{a\theta_1}{\tan^{-1}(\theta_1/a)} - a^2, \tag{14.68a}$$

$$\approx 0.28\,\theta_1^2 \quad \text{for } \theta_1 < a, \tag{14.68b}$$

$$\approx \frac{2}{\pi} a\theta_1 \quad \text{for } \theta_1 > a. \tag{14.68c}$$

The quantity $\langle \psi^2 \rangle a^{-2}$ from (14.68a) is plotted as a function of $\theta_1 a^{-1}$ in Figure 14.

When the scattering is uncorrelated and the massive star scatters against approximately $\sqrt{N}$ stars a distance $\sim R$ away, we obtain $\theta_1 \approx N^{-1/2}(v_*/V_0)^2 < a$ and (14.68b) gives $\langle \psi^2 \rangle \approx N^{-1}(v_*/V_0)^4$, so we have recovered the earlier result (13.10). On the other hand, when the massive star is scattered mainly by a small number of grexons, $\theta_1 \approx (v_*/V_0)^2$ from (13.4) and with (14.68c) we recover the result (13.11).

Choosing a value for k_{max} in (14.66) will depend on the nature of the system. Fortunately, as in (2.7), the results are not sensitive to its exact value. If the scattering is essentially uncorrelated, the maximum wavenumber is approximately the inverse interparticle distance, $N^{1/3}R^{-1}$, and $k_{max}^2 V_0^2 \omega_g^{-2} \approx N^{2/3}(V_0/v_*)^2 \gg 1$. For highly correlated scattering there are several possibilities, all giving roughly the same result. If grexons are distributed along a line in the wake of a very massive star and $k_{max} \approx N_g R^{-1}$, the intergrexon distance, then $k_{max}^2 V_0^2 \omega_g^{-2} \approx N_g^{2/3}(V_0/v_*)^2 \gtrsim 1$. But, if the distribution of grexons is more spherical, so $k_{max} \approx N_g^{1/3}R^{-1}$, then $k_{max}^2 V_0^2 \omega_g^{-2} \approx N_g^{2/3}(V_0/v_*)^2 \gtrsim 1$. On the other hand, if $k_{max} \approx r_{grexon}^{-1}$, the grexon scale length, then (13.9) gives $k_{max}^2 V_0^2 \omega_g^{-2} \gtrsim (Nm/M_0)^2(V_0/v_*)^3 > 1$. At this stage, the problem is not sufficiently well-defined to choose between these possibilities. However, in all cases the k_{max} term dominates the logarithm.

Equations (14.66) and (14.67) provide the probability distribution of the scattering angle for a a massive star moving through a stellar system. This calculation has

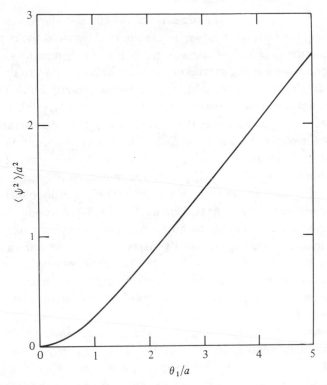

$\langle \psi^2 \rangle / a^2$

θ_1/a

Fig. 14. Graph of Equation (14.68a) for the mean deflection angle.

involved several simplifying assumptions. The main physical assumptions are that the initial system is homogeneous, and that the interaction between the massive star and the system is weak enough that the system's response is determined by its properties in the absence of the massive star. The system can even consist of clumps of stars if these clumps are homogeneously distributed scattering centers whose internal structure can be neglected. The matrix elements of the perturbed Boltzmann operator, which conserve wavenumber and, therefore, momentum in a homogeneous system, describe the transitions between different momentum states of the system. The changes of the massive star's orbit reflect these transitions in the system. This analysis can be generalized to more complicated situations in a relatively straightforward way. Perturbations of a stellar system by a time dependent external force, as in globular clusters moving around the galaxy, are just one example. Tidal encounters of two or more galaxies are another. Rotating and anisotropic systems are a third. Transfer of energy into a clump of stars from a massive star is a fourth.

Dynamical friction, in forms similar to (14.62), also plays a more general role in interacting astronomical systems. For example, colliding galaxies which interpenetrate experience dynamical friction. The detailed effects of friction depend on the relative velocity of the intruding stars and the galaxy, and on the original and tidally induced spin of the galaxies. Usually the two galaxies tend to merge.

Dynamical friction acts to equalize average velocities. A star moving faster than its surrounding stars will slow down in relation to them. And the converse is also true (see Equation (3.1) and its consequences). If a star is moving slower than its surroundings on average, the overtaking stars are deflected toward the laggard and their orbits become slightly focused into the region toward which the laggard is moving. The gravity of this enhanced density tugs at the laggard, speeding it up.

In many astronomical examples the relative velocity of a disturbing star and the field stars is not very large. For these cases, I have argued in Section 14.1 that the scattering angle is usually still small so (14.62) should still apply. These sorts of arguments should always be tested by N-body simulations in order to confirm their validity, in lieu of a laboratory experiment. White (1978) has done such a test. He let 250 stars, each of unit (scaled) mass, form a relaxed self-gravitating cluster. It had a density distribution $n(r)$ similar to an isothermal sphere with $n(r) \approx v^2/12\pi G r^2$, where v is the three-dimensional velocity dispersion. A heavy star, of mass 25, was put into the system on a roughly circular orbit. This orbit soon became more circular since a circular orbit has the minimum energy for a given angular momentum, and dynamical friction was removing its energy. On a longer term, continuing friction caused the massive star to spiral into the center.

Now consider the corresponding analytic model. Instantaneously, the massive star's gravitational energy is $E = E_0 - (2/3)M_0 v^2 \ln(r/r_0)$, where the scale is set at $E = E_0$ and $r = r_0$ (using the isothermal sphere with $n \propto r^{-2}$ and assuming the circular orbit has an average velocity equal to the one-dimensional velocity dispersion of the field stars). Then (14.62) takes the form $\dot{E} = 0.6\,GM_0^2(v/\sqrt{3})r^{-2} \cdot \ln[k_{max}^2 V_0^2 \omega_g^{-2} + 1]$, which gives $r = r_0(1 - t/T)^{1/2}$ and $E = E_0 - M_0(V_0^2/3) \cdot \ln(1 - t/T)$ with $T = v r_0^2/GM_0 \ln(k_{max}^2 V_0^2 \omega_g^{-2} + 1)$. White showed that these relations were good fits to the N-body simulation of dynamical friction until t nearly became equal to T. The best fit was for $k_{max} V_0 \omega_g^{-1} \approx 1.1$, indicating that in this case $k_{max} \sim R^{-1}$ and the effective number of grexons is about one. Of course, when the massive star is very close to the center, V_0 becomes very small and the whole basis of the analysis breaks down. Until then, however, small-angle scattering gives a good approximation even in this rather extreme case.

All the analyses so far can also be generalized to include correlations. Instead of perturbing the collisionless Boltzmann operator, one could perturb the Fokker–Planck and BBGKY operators. These would be necessary to describe the detailed evolution of an initially cold system, far from equilibrium. This is the case, for example, if a massive star moves through a cloud of gas in which stars have formed so recently that they have not had time to reach gravitational equilibrium. A massive star moving through a galactic cluster where the main motions of the stars are their orbits around the galaxy, rather than their self-gravitating velocities, would also need a more detailed analysis along these lines.

15

Linear response and dispersion relations

15.1. Basic result

In the last section, we showed how to determine the response of a system to a single star. In this section, we see how it responds to groups of stars. Again, we start with an idealized spatially homogeneous system. While not strictly consistent, it can be considered as a local region of a system which is initially inhomogeneous only on much larger scales. The analysis gives the essential properties of more realistic systems.

Now suppose we perturb the system by forming bunches of stars. Will these groups grow, or will they gradually disperse? Even without a detailed microscopic treatment of the stars' orbits and distribution function, we can make a rough estimate. The bunching of stars with total mass M creates a gravitational well of characteristic radius $\sim R$ whose negative gravitational energy is approximately GM^2R^{-1}. If these stars have roughly random motions characterized by a velocity dispersion V within the well, then their total positive 'thermal' energy is approximately MV^2 (not worrying about factors of 2). As the mass of a bunch having a fixed radius increases, its gravitational energy dominates its thermal energy and we would then expect the bunch to contract. Similarly this should also happen when the radius becomes sufficiently small for a fixed mass. If the perturbation's amplitude is small, then its density is not altered much, $\rho \approx MR^{-3}$ and equating the gravitational and thermal energies gives a critical radius

$$R_c \approx V(G\rho)^{-1/2}. \tag{15.1}$$

Bunches of larger radii (for fixed density) contract gravitationally; those of smaller radii disperse. To order of magnitude, we would also expect that R_c characterizes the size of isolated clusters whose gravity and thermal energies are in overall balance, e.g., in virial equilibrium.

Although this dimensional argument gives the basic result, it glides over several fundamental questions. First, it may not be obvious that the energy of the well should involve its total mass, and not just the perturbed mass. Second, there should be significant differences between the cases of a gas or continuum fluid, and of a discrete stellar system. In a discrete system, the graininess of the gravitational field should lead to a much richer and more interesting variety of interacting modes. Third, the instability should depend to some extent on how the bunch is formed. Fourth, what happens if the velocity distribution of the stars is not Maxwellian? After all, only some of the systems we find in the Universe are relaxed. We discuss the first two questions here, and take on the third and fourth in Section 16.

15.2. Gaseous systems

Some aspects of a gravitating gas are simpler than those of a stellar system and serve as a good introduction and comparison. The instability of an infinite, initially homogeneous gravitational continuum was first analyzed by Jeans in 1902 (see Jeans, 1928). These systems are described by the equation of continuity for mass conservation

$$\frac{\partial \rho}{\partial t} + \mathbf{V} \cdot (\rho \mathbf{v}) = 0, \tag{15.2}$$

Euler's force equation for momentum conservation

$$\rho \frac{\partial \mathbf{v}}{\partial t} + \rho(\mathbf{v} \cdot \mathbf{V})\mathbf{v} = -\mathbf{V}P + \rho \mathbf{V}\phi, \tag{15.3}$$

and Poisson's equation coupling mass and the gravitational potential

$$\nabla^2 \phi = -4\pi G \rho. \tag{15.4}$$

The first two of these equations are moments of the collisional Boltzmann equation for gaseous systems. Unlike collisionless gravitating systems, these moments form a closed set, so long as microscopic equilibrium holds and detailed balance restricts the symmetry properties of the collisions. Texts on the kinetic theory of gases discuss these matters.

A collective perturbation can be represented by small alterations in the density, pressure, gravitational potential and velocity

$$\rho = \rho_0 + \rho_1(\mathbf{r}, t), \tag{15.5}$$

$$P = P_0 + P_1(\mathbf{r}, t), \tag{15.6}$$

$$\phi = \phi_0 + \phi_1(\mathbf{r}, t), \tag{15.7}$$

$$v = v_0 + v_1(\mathbf{r}, t), \tag{15.8}$$

where $\rho_1 \ll \rho_0$, etc. Since the unperturbed gas is uniform, the zero order quantities are constants. For full consistency they are all zero. Taking $\rho_0 \neq 0$ is where the so-called 'Jeans swindle' enters. This problem is not important here since we can suppose the infinite medium to be inhomogeneous over scales large compared to regions of possible instability. Realistic finite systems are always inhomogeneous, however, and gradients must usually be taken into account in their perturbation analyses. Substituting (15.5)–(15.8) into (15.2)–(15.4) and retaining only the first order terms gives three linear equations for small variations in the four variables ρ_1, P_1, ϕ_1 and v_1. To obtain a necessary fourth relation among the variables, it is usual to specify an adiabatic or isothermal perturbation. Then the pressure and density fluctuations are connected by

$$P_1 = \frac{\partial P}{\partial \rho}\rho_1 = c^2\rho_1 = \frac{\gamma P}{\rho}\rho_1, \tag{15.9}$$

where c is the speed of sound, ignoring the effects of gravity. Since $P \propto \rho^\gamma$, isothermal perturbations are covered by $\gamma = 1$, from the perfect gas equation of state.

Substituting (15.9) into the linearized Euler equation, taking its divergence and using the linearized Poisson and continuity equations to eliminate ϕ_1 and v_1 gives

$$\left(\nabla^2 - \frac{1}{c^2}\frac{\partial^2}{\partial t^2}\right)\rho_1 + k_J^2\rho_1 = 0, \tag{15.10}$$

where we have written the Jeans wavenumber

$$k_J \equiv \frac{(4\pi G\rho_0)^{1/2}}{c} = \frac{2\pi}{\lambda_J}. \tag{15.11}$$

If there were no gravity, $k_J = 0$, this would describe normal sound waves in a perfect gas. To see how gravity rules the waves, we seek solutions propagating along the x axis with the form

$$\rho_1 = \text{constant } e^{i(kx - \omega t)} \tag{15.12}$$

Substituting into (15.10) shows that the requirement for such solutions is

$$\omega^2 = c^2(k^2 - k_J^2). \tag{15.13}$$

This is the dispersion relation relating frequency and wavenumber. From it, as in (14.47), we obtain the phase velocity of the waves

$$V_p = \frac{\omega}{k} = c\left(1 - \frac{k_J^2}{k^2}\right)^{1/2}. \tag{15.14}$$

This is plotted in Figure 15. Since the waves occur equally throughout the gas, rather than as a localized pulse, we are interested in their phase velocity here rather than their group velocity. Notice that the results do not depend on the wave's amplitude; this is a feature of all linear perturbation theory since the equations are homogeneous in the perturbation. The amplitude is therefore arbitrary, so long as it is small.

Several features produced by adding gravity become apparent. For small wavelength perturbations, the medium closely resembles a perfect gas with normal acoustic waves. As the wavelength increases, gravity's effects become more strongly felt. Waves become more sluggish until when $\lambda = \lambda_J$ they cease to propagate at all. Their self-gravity is now comparable to their thermal energy. Except for a factor $\sqrt{\pi}$ this is the same as the criterion (15.1). For longer wavelengths, gravity dominates and the perturbation's density increases with time. The dispersion relation $\omega(k)$ becomes imaginary and the growing mode of the density increases at a rate which

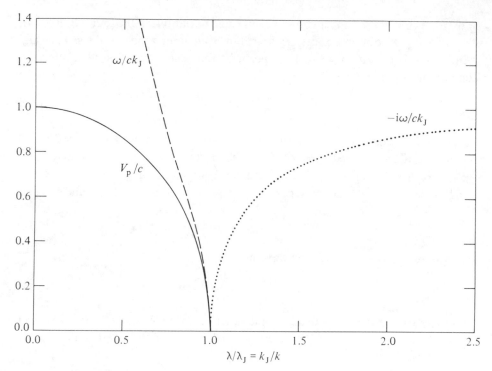

Fig. 15. The phase velocity of propagation (solid line) in a gravitating gas and the Jeans dispersion relation (dashed line for real ω, dotted line for imaginary ω).

asymptotically approaches

$$\rho_1 \propto e^{t/\tau} \qquad (15.15)$$

for $\lambda \gg \lambda_J$ with $\tau = (4\pi G\rho_0)^{-1/2}$. There is also a damped mode $\propto e^{-t/\tau}$, but this rapidly disappears. The dispersion relation $\omega = ick_J$ is essentially (14.52), as we will see in more detail when considering stellar systems.

Contraction cannot continue forever in the form (15.15), for when $\rho_1 \approx \rho_0$ the growth becomes non-linear. Although it is no longer described by this perturbation analysis, the basic equations (15.2)–(15.4) still apply. They, along with radiative transfer for energy flow, describe the further evolution of a contracting region as an isolated dense cloud. If the gravitational energy produced by contraction were instantly turned into heat and radiated, free fall would ensue and in about another period $\approx \tau$ the cloud would collapse to a point. The real situation is much more complicated and would take us too far afield here (for simple examples see von Hoerner & Saslaw, 1976).

Linear analysis therefore shows how gravity reduces the 'elasticity' of sound waves. And it has answered our first question, since the contraction timescale involves the total density $\rho \approx \rho_0$ of the fluctuation, rather than just the enhancement ρ_1. But one wavelength does not make a cloud. There are other forms of solutions to (15.10) and we next turn to these.

Instead of analyzing an arbitrary wave, we can use (15.10) to follow the evolution of a specific shape of perturbation. Thus we select an initial condition simple (i.e., artificial) enough to provide insight. Of course, any initial configuration could be decomposed into Fourier components and we could use (15.10) to watch the spectrum evolve and reconstitute the perturbation at any time. But that approach, as Section 14 showed, can sometimes be lengthy. To simplify the problem even more, we invert this procedure by specifying the growth of a perturbation and seeing what spatial form it takes.

As an example, consider a perturbation which grows exponentially but with an arbitrary value of τ. For a spherical perturbation, the density takes the form

$$\rho_1(r,\ t) = \rho_1(r)e^{t/\tau}, \tag{15.16}$$

which, substituted into (15.10), gives a Helmholtz equation

$$(\Delta^2 + \chi^2)\rho_1(r) = 0, \tag{15.17}$$

where

$$\chi^2 = k_J^2 \quad c^{-2}\tau^{-2}, \tag{15.18}$$

In spherical coordinates, it takes the form

$$\frac{d^2\rho_1}{dr^2} + \frac{2}{r}\frac{d\rho_1}{dr} + \chi^2\rho_1 = 0. \tag{15.19}$$

By the trivial change of variable,

$$\rho = ur^{-1}, \tag{15.20}$$

this is transformed into the simple harmonic oscillator equation for $u(r)$

$$\frac{d^2u}{dr^2} + \chi^2 u = 0. \tag{15.21}$$

There are three classes of solution, depending on the value of χ^2. For $\chi = 0$, the density would have the form $\rho = a + br^{-1}$. This would be finite at the center only if $b = 0$, giving a constant density consistent with the infinite wavelength required for $\chi = 0$, but not very interesting. For $\chi^2 < 0$, the inhomogeneity would collapse faster than free fall, and this is ruled out on physical grounds. For $\chi^2 > 0$, however, there are reasonable solutions and these have the form

$$\rho_1(r) = \rho_1(r = 0)\frac{\sin(\chi r)}{\chi r}, \tag{15.22}$$

with finite density at the center. If one were to lay out this density distribution initially, it would maintain its damped sinusoidal shape with an exponentially increasing amplitude. Its corresponding gravitational potential, obtained by inserting (15.22) into Poisson's equation, is

$$\phi_1(r) = c_1 + \frac{c_2}{\chi_r^2} + \frac{4\pi G\rho_1(r = 0)}{\chi^2}\frac{\sin(\chi r)}{}. \tag{15.23}$$

Thus ϕ_1 is a superposition of a constant term (which can be chosen to be zero), a term representing a point charge at the origin, and a damped shielding term which becomes negative for $\chi r > \pi$ and oscillates in sign as r increases.

The total perturbed mass of the inhomogeneous cloud is

$$M = 4\pi \int_0^R \rho_1(r) r^2 \mathrm{d}r = 4\pi \rho_1(0) \chi^{-3} [\sin(\chi R) - \chi R \cos(\chi R)]. \qquad (15.24)$$

The density does not decrease fast enough to prevent a volume divergence. At some finite radius R, the influence of surrounding inhomogeneities effectively cuts off the distribution (15.22). If the cloud is produced by adding a preshaped mass M to the system, (15.24) determines its radius. However, if the cloud is produced by regrouping gas already in the system, its net mass will be zero provided R satisfies

$$\chi R = \tan(\chi R). \qquad (15.25)$$

Then the shells of enhanced density just balance those of decreased density. Critical values of χR occur at approximately 4.5, 7.7, 10.9, 14.1, 17.2, 20.4, 23.5, 26.7, etc. This oscillating density is forced on the cloud by mass conservation – a lump in the middle means a deficit farther out – and by the inability of a cloud to react coherently over timescales shorter than the sound crossing time.

Notice that inhomogeneities which change as $\rho_1 \propto a_1 + a_2 t$ – or indeed which have any harmonic time dependence – will also take this shape.

These time harmonic inhomogeneties also serve to introduce the important notion of gravitational shielding. In a plasma, the electron cloud around a positive ion shields its distant electrostatic potential. At first sight it might seem that nothing like this could occur in a gravitating system since mass has only one sign, unlike electric charge. However, we can renormalize the density by substracting the average density of the medium. This leaves the net density zero, and regions of positive and negative density mimic positive and negative masses. The negative shells effectively shield outer particles from inner positive shells, producing a fluctuating gravitational force. The net force outside a cloud satisfying (15.25) is zero.

15.3. Stellar systems

Having established some basic principles using continuum systems, we now enter the much richer pastures of discrete dynamics. The gravitational graininess of stellar systems leads to small scale behavior which can accumulate until it has significant effects over much larger scales. Techniques to analyze these effects have often been borrowed from plasma physics over the last three decades. This is only fair, since many of the early plasma techniques, such as adiabatic invariants and the Boltzmann equation itself, were borrowed from stellar dynamics, which in turn adopted them from celestial mechanics and gas kinetic theories. There has been a strong interplay among all these related subjects. At present, plasma physics is in the lead, spurred on by the quest for controlled thermonuclear fusion.

To obtain the dispersion relation analogous to (15.13) for a stellar system, we start with the collisionless Boltzmann equation in the form

$$\frac{\partial f}{\partial t} + \mathbf{v} \cdot \frac{\partial f}{\partial \mathbf{r}} + \nabla \phi \cdot \frac{\partial f}{\partial \mathbf{v}} = 0 \tag{15.26}$$

and Poisson's equation

$$\nabla^2 \phi = -4\pi G m \int_{\mathbf{v}} f \, d\mathbf{v} \tag{15.27}$$

for the single-particle distribution function. Matters may be made more complicated by adding a mass distribution, or using the Fokker–Planck or BBGKY descriptions. Since the purpose here is to describe the basic technique, these elaborations are mostly left to the reader.

When the initial system is homogeneous, its equilibrium distribution function $f_0(\mathbf{v})$ depends on velocity only. Perturbing some of the stars away from equilibrium creates a small disturbance represented by a change in f

$$f = f_0(\mathbf{v}) + f_1(\mathbf{r}, \mathbf{v}, t). \tag{15.28}$$

The gravitational potential is similarly perturbed

$$\phi = \phi_0 + \phi_1(\mathbf{r}, t). \tag{15.29}$$

Substituting these last two equations into the previous pair and subtracting the zero order equilibrium equations, we are left with the first order linearized system

$$\frac{\partial f_1}{\partial t} + \mathbf{v} \cdot \frac{\partial f_1}{\partial \mathbf{r}} + \nabla \phi_1 \cdot \frac{\partial f_0}{\partial \mathbf{v}} = 0, \tag{15.30}$$

$$\nabla^2 \phi_1 = -4\pi G m \int_{\mathbf{v}} f_1 \, d\mathbf{v}, \tag{15.31}$$

for the collective perturbation.

Next we look for solutions in which f_1 and ϕ_1 have the forms (15.12) of traveling plane waves

$$f_1(\mathbf{r}, \mathbf{v}, t) = g(\mathbf{v}) e^{i(\mathbf{k} \cdot \mathbf{r} - \omega t)}, \tag{15.32a}$$

$$\phi_1(\mathbf{r}, t) = a e^{i(\mathbf{k} \cdot \mathbf{r} - \omega t)}. \tag{15.32b}$$

Substituting (15.32a) into (15.30), solving for f_1, integrating over velocity, and inserting ϕ_1 from (15.31) and (15.32b) shows that the condition for such solutions is that

$$1 = -\frac{4\pi G m}{k^2} \int \frac{\mathbf{k} \cdot \dfrac{\partial f_0}{\partial \mathbf{v}}}{\mathbf{k} \cdot \mathbf{v} - \omega} \, d^3\mathbf{v}. \tag{15.33}$$

This is the dispersion relation between k and ω analogous to (15.13). It should puzzle the reader.

Unlike the continuum case, (15.33) is available only in integral form, which makes it harder to relate ω to k explicitly. (The electron plasma has the same dispersion

relation except for a different constant and the opposite sign.) But the main problem
is that (15.33) relates ω to k only under special circumstances. The singularity in the
integral means that for a given value of k there can be many values of ω which satisfy
(15.32) as v changes. Physically these modes correspond to streams of stars, all
moving with the same velocity v which happens to be the phase velocity of the wave.
To set up these modes, it is necessary to find all the stars in the system with the same
velocity and form a spatial perturbation of just these stars, giving it a wavenumber $\mathbf{k}$
and frequency $\omega = \mathbf{k} \cdot \mathbf{v}$. The stream will then carry this perturbation along.
Superimposing different streams can produce a wave of arbitrary ω and k. As the
waves propagate, their amplitude may grow or decay.

These modes are innocuous, but artificial. Usually a perturbation arises among
stars of all velocities in a given region of space. We will treat this more realistic initial
condition in Section 16, but first there are several things to be learned from (15.33).
Moreover, we will find that this dispersion relation appears even in the more realistic
case, where it plays a more complex role.

When the distribution function is such that there are no stars moving at exactly
the phase velocity of a particular wave, and the distribution is an extremum there, so
$(\partial f_0/\partial v) = 0$ at $v = v_0$, then (15.33) can be integrated by parts to give

$$1 = -4\pi Gm \int \frac{f_0(\mathbf{v})}{(\mathbf{k} \cdot \mathbf{v} - \omega)^2} d^3\mathbf{v} \qquad (15.34)$$

if the surface term vanishes in the usual way as $f_0(\mathbf{v})$ decreases rapidly for large $\mathbf{v}$. To
simplify matters, we can align $\mathbf{k}$ with the waves' velocity and drop the vector signs.
For the artificial but instructive case of many cold constant velocity streams of stars,
each with number density n_β, moving in the $v = v_x$ direction,

$$f_0(\mathbf{v}) = \sum_\beta n_\beta \delta(v - v_\beta). \qquad (15.35)$$

The dispersion relation becomes

$$4\pi Gm \sum_\beta \frac{n_\beta}{(k\mathbf{v}_\beta - \omega)^2} = -1. \qquad (15.36)$$

Clearly this can only be satisfied if at least one of the ω is complex, characterizing
streams in which the amplitude of oscillation grows or decays with time.

If there is only one stream moving along, the solution of (15.33) is

$$\omega = kv \pm i(4\pi G\rho)^{1/2}. \qquad (15.37)$$

The waves borne along the stream are unstable (unlike the electron plasma case),
with the familiar gravitational timescale. What happens is that the regions of density
enhancement create gravitational wells which move along with the wave. These
perturb the velocities of other stars in the stream. Those stars whose perturbed
motion is in the same direction that a well is moving will gradually overtake it. Stars
which are perturbed in the direction opposite the stream velocity are gradually
overtaken by a well. These stars oscillate inside their gravitational wells and, by

increasing their well's effective mass, they make it deeper. It can then attract stars from farther away more forcefully, and so the instability grows. As the well becomes steeper it holds the stars inside it tighter, for longer, and this positive feedback contributes to the exponential nature of the growth. Since the unperturbed stream is cold, i.e., it has no velocity dispersion, all wavelengths are unstable on the free-fall timescale.

The next simplest case is two cold streams which interpenetrate. This even has astronomical applications such as a highly idealized description of a star cluster moving through a galaxy at a speed much faster than the internal velocity dispersion of either. Let the streams have number densities n_1 and $n_2 = \alpha n_1$. It simplifies matters a bit to move in a velocity frame such that $v_2 = -\alpha v_1$ (an inverse center-of-mass frame). Then the dispersion relation becomes

$$\frac{4\pi G m n_1 (1 + \alpha)(\omega^2 + \alpha k^2 v_1^2)}{[\alpha k^2 v_1^2 + k v_1 (1 - \alpha)\omega - \omega^2]^2} = -1. \tag{15.38}$$

The reader can spend a pleasant afternoon working through this to see how a changing density ratio α alters the growth of modes. Here we extract the main surprise by considering two streams of the same density. For $\alpha = 1$, Equation (15.38) is easily solved

$$\omega^2 = k^2 v_1^2 - 4\pi G m n_1 \left[1 \pm \left(1 - \frac{k^2 v_1^2}{\pi G m n_1} \right)^{1/2} \right]. \tag{15.39}$$

As $v_1 \to 0$, the positive square root gives the same unstable mode as (15.37) for a single stream (but now with twice the previous density). The condition for complex growing modes is that the square root be imaginary, or

$$\lambda < 2v_1 \left(\frac{\pi}{G m n_1} \right)^{1/2}. \tag{15.40}$$

You may not have expected the unstable modes to be the small ones. Normally we think of gravity as operating more effectively over the larger scales. Here, however, what happens is that the second stream sends fast stars through the gravitational wells at such a high rate that the wells are smoothed out. The higher the velocity v_1, the larger the scale which can be smoothed and made stable during the normal free-fall time.

When many streams interpenetrate, the dispersion relation becomes much more complicated and very many modes will correspond to any value of k. A realistic continuum distribution of velocities is represented by an infinite number of streams, so we might expect its mode analysis to be completely intractable. Fortunately this is not so, and in some ways it even becomes simpler.

A new simplicity arises because in smooth velocity distributions we are not so interested in the modes borne along each thin velocity stream. Rather, we ask about the behavior of modes encompassing all the velocities in a given spatial region. Among all these modes there is a very special one. It does not grow or decay. Neither does it propagate. It does nothing. Just sits there. It is the critical mode $\omega = 0$ at the threshold of Jeans instability. With a Maxwellian distribution for velocity in the direction of **k**,

$$f_0(v) = \frac{n}{\sqrt{(2\pi)}\langle v^2 \rangle^{1/2}} e^{-v^2/2\langle v^2 \rangle}, \qquad (15.41)$$

it is easy to solve the dispersion relation (15.33) with $\omega = 0$ to obtain

$$k_j^2 = \frac{4\pi G\rho}{\langle v^2 \rangle}. \qquad (15.42)$$

This is the dispersion relation (14.52) for the gravitational response frequency.

Thus the Jeans wavenumber in a stellar system is essentially the same as in a gas, (15.11). Only the relevant speed is slightly different in each case. For a perfect gas the sound speed and the velocity dispersion are related by a factor of order unity depending on the specific heat. In a discrete collisionless system there is nothing exactly like a sound speed since the moment equations cannot be truncated exactly. Here the velocity dispersion is the more fundamental quantity.

Another useful case of (15.33) can be examined readily. Suppose the distribution function has a cutoff at some maximum speed v_{max}. Then for perturbations with sufficiently high frequency or long wavelength that $\omega/k > v_{max}$ there is no problem with the singularity. Physically this is the case where the stars do not have time to respond strongly to a passing wave. Therefore, we can use (15.34) and expand the integrand in terms of the small parameter $\mathbf{k}\cdot\mathbf{v}/\omega$

$$1 = -\frac{4\pi G m}{\omega^2} \int \left[1 + 2\frac{\mathbf{k}\cdot\mathbf{v}}{\omega} + 3\left(\frac{\mathbf{k}\cdot\mathbf{v}}{\omega}\right)^2 + \ldots \right] f_0(\mathbf{v}) d^3\mathbf{v}. \qquad (15.43)$$

If the distribution is isotropic, the second integral vanishes leaving

$$\omega^2 = -\omega_g^2 \left[1 + 3\frac{\langle(\mathbf{k}\cdot\mathbf{v})^2\rangle}{\omega^2} \right] = -\omega_g^2 \left[1 + \frac{k^2\langle v^2 \rangle}{\omega^2} \right]$$

$$\approx -(\omega_g^2 + k^2\langle v^2 \rangle), \qquad (15.44)$$

where the last expression is correct to second order.

All these long wavelength modes are unstable. They could not occur in finite systems near equilibrium, since there they would correspond to wavelengths longer than the system itself! But, in an infinite system, or a small cold patch within a larger system, they could form and condense. Moreover, since these perturbations grow faster than the free-fall time, they would have to start off with an inward push. So the conditions under which they occur are still rather special.

More naturally occurring perturbations, alluded to earlier, can be examined by specifying the distribution function $f(\mathbf{r}, \mathbf{v}, t = 0)$ at some initial time, and following its evolution. This requires a slightly different treatment in which we integrate the transform of the Boltzmann equation starting at $t = 0$. It will also resolve the problem of the singularities in (15.33). As a bonus, we will also find new insight into the general nature of distribution functions which produce growing modes, and into what happens to modes which do not grow.

16

Damping and decay

16.1. Physical description

If you look for a particular type of response mode in analyzing a mechanical system you might find it. But, if you do not look for it you will never find it. Lack of attention to this truism has often held back progress by promoting the feeling that a system is understood when its Fourier modes have been examined. Thus, it came as a surprise in plasma physics when Landau (1946) discovered that waves in a collisionless plasma, described by the collisionless Boltzmann equation, can damp and decay. At the time it was especially unexpected because damping had been associated with viscosity, and there is no viscosity in such plasmas. Similar behavior occurs in gravitating systems.

The physical reason for collisionless damping arises from the detailed interaction of a wave with the orbits of background stars which are not part of the wave. Thus, this process would not show up in a continuum approximation where the waves could merrily propagate without growth or decay.

To see how this works, consider the idealized triangular (rather than sinusoidal) wave shown in Figure 16. We look at it in a co-moving frame, traveling along with the wave's phase velocity $v = \omega/k$. This ripple has been set up at $t = 0$ by suddenly imposing a periodic triangular perturbation in the density which gives a similar perturbation to the gravitational potential ϕ throughout the entire volume. Unlike our previous perturbations, this one involves stars of all velocities in the given volume. However, it is still a one-dimensional situation where the stars move in the direction of the wave along the x axis.

Some of the stars will move a bit faster than the wave, others a bit slower, even though their average background number density n is constant. First consider the stars that move a bit faster, with a small velocity v_1 relative to the wave at $t = 0$. Stars in region A feel a constant force F_A since the potential has a constant slope there. Those in region B feel the opposite force $F_B = - F_A$. After a short time t stars in A will have changed their original position relative to the wave by an amount

$$\Delta x_A = v_1 t + \frac{F_A}{m}\frac{t^2}{2},$$ (16.1)

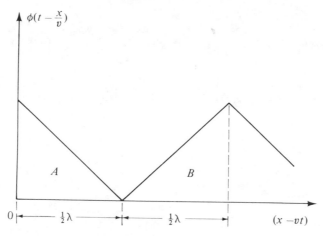

Fig. 16. The triangular wave in a co-moving frame.

and stars in B by an amount

$$\Delta x_B = v_1 t + \frac{F_B}{m} \frac{t^2}{2}. \tag{16.2}$$

If stars in A are speeded up, those in B are slowed down. The number of stars leaving A and moving into B is $n|\Delta x_A|$, while the number of stars slipping behind B into A is $n|\Delta x_B|$. Thus, A now contains

$$N_A = n\left(\frac{\lambda}{2} - |\Delta x_A| + |\Delta x_B|\right)$$

$$= n\left(\frac{\lambda}{2} + \frac{F_B - F_A}{m} \frac{t^2}{2}\right) = n\left(\frac{\lambda}{2} - \frac{F_A}{m} t^2\right) \tag{16.3}$$

stars, while B contains

$$N_B = n\left(\frac{\lambda}{2} - |\Delta x_B| + |\Delta x_A|\right) = n\left(\frac{\lambda}{2} + \frac{F_A}{m} t^2\right) \tag{16.4}$$

of them. Of course, what happens in A and B is representative of the slippage of stars along the entire wave since it is periodic.

Since there is a net flux of stars between A and B, this changing distribution exchanges energy with the wave. Equivalently, the wave does work on the stars. The rate at which this work is done is the total force exerted by the potential times the velocity of the wave relative to the bulk of the stars. Assuming $v = \omega/k > v_{\text{therm}} \gg v_1$, this rate is approximately

$$\dot{W} = (F_A N_A + F_B N_B)\left(\frac{\omega}{k}\right) = -2\frac{F_A^2}{m} n \frac{\omega}{k} t^2. \tag{16.5}$$

Thus, the work of the wave on the stars is negative, and the wave gains energy from those fast moving stars. The stars lose energy and tend to be captured by the waves. This amplifies the waves.

But, stars originally moving a bit slower than the waves compete in the opposite way. If v_1 is negative, then N_A and N_B are reversed. A similar analysis shows that slow stars gain the same absolute amount of energy from the wave as in (16.5). Thus, they will damp the wave. If there are more slow stars than fast ones with velocities near ω/k, the net effect is to damp the wave. So the sign of the result depends on the slope of the velocity distribution at $v = \omega/k$. Moreover, the strength of the damping (or growth) depends on the total net number of stars involved. This will usually be greater near the peak of the velocity distribution (where the slope will also generally be greater) than far out in the tail. So we would expect that for a roughly Maxwellian velocity distribution the damping would be greatest for waves whose phase velocity is near the velocity dispersion. And this brings us back, once again, to the Jeans criterion.

Waves longer than the Jeans length cannot propagate anyway since their 'pressure elasticity' fails to overcome their self-gravity. Only waves shorter than the Jeans length will, therefore, be affected by this Landau damping. A characteristic time for stars of different velocities having a dispersion $\langle v \rangle$ to move through the wave is $\lambda/\langle v \rangle$. As a rough estimate, we would expect this to be the timescale for damping of the wave.

16.2. Calculation of Landau damping rate

With this approximate physical picture in mind, we next sketch the more detailed analysis of linear wave–star interaction. It begins with the Boltzmann and Poisson equations (15.30) and (15.31). Instead of analyzing general plane-wave solutions, we now look for waves produced by realistic initial conditions. So we want to be able to specify $f_1(\mathbf{r}, \mathbf{v}, 0)$ and see how the final results depend on it. The technique is to use integral transforms of the basic equations. We Fourier transform it in space and Laplace transform it in time.

The spatial Fourier transform amounts to again decomposing $f_1(\mathbf{r}, \mathbf{v}, t)$ into terms representing periodic disturbances of a given wavenumber

$$f_1(\mathbf{r}, \mathbf{v}, t) = f_1(\mathbf{v}, t)e^{i\mathbf{k}\cdot\mathbf{r}}. \tag{16.6}$$

We could take a shaped wave, but this is simpler. The potential ϕ_1 has the same form. Directing $\mathbf{k}$ along the x axis, (15.30) and (15.31) become

$$\frac{\partial f_1}{\partial t} + ikv_x f_1 + ik\phi_1 \frac{\partial f_0(\mathbf{v})}{\partial v_x} = 0 \tag{16.7}$$

and

$$k^2 \phi_1 = 4\pi G m \int_{-\infty}^{\infty} f_1 \, d^3\mathbf{v}, \tag{16.8}$$

where $f_1(\mathbf{v}, t)$ and $\phi_1(t)$ contain just the velocity and time dependent parts of the perturbation. As a result, the spatial partial derivatives are reduced to an algebraic operator: multiplication by ik.

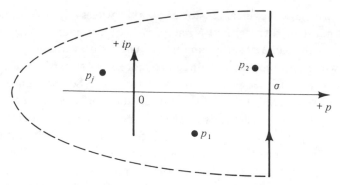

Fig. 17. Integration path for the inverse Laplace transform.

To reduce the time derivatives we use the Laplace transform

$$F_1(p,\mathbf{v}) = \int_0^\infty f_1(\mathbf{v},t)\mathrm{e}^{-pt}\mathrm{d}t. \tag{16.9}$$

(The standard parameter p should not be confused with momentum!) The expression for ϕ_1 is similar. This stores information about the initial distribution $f_1(\mathbf{v},0)$. The value of p must be large enough that the integral converges. In general, p will be complex, allowing for stable, unstable and oscillating behavior. Recovering this behavior requires the inverse Laplace transform over the complex plane

$$f_1(\mathbf{v},t) = \frac{1}{2\pi\mathrm{i}} \int_{-\mathrm{i}\infty+\sigma}^{+\mathrm{i}\infty+\sigma} F_1(p,\mathbf{v})\mathrm{e}^{pt}\mathrm{d}p. \tag{16.10}$$

The path of integration is shown in Figure 17. It goes along a line parallel to the imaginary p axis which keeps all the singularities of $F_1(p,\mathbf{v})$, located at $p_1, p_2, \ldots, p_j$, to the left. It is closed by the semicircle at very large p (assuming there are no branch points). This path enables us to determine $f_1(\mathbf{v},t)$ by the properties of the singularities of its transform, as long as these singularities are all poles. (Technically the transform is a meromorphic function; refer to elementary texts on complex analysis and integration.) It is basically the same process as determining the gravitational field throughout space once one knows the locations of the contributing point masses, which are singularities in the potential.

Multiplying (16.7) and (16.8) by e^{-pt} and integrating (the partial time derivative by parts) over t from 0 to ∞ gives algebraic expressions for the Laplace transforms of f_1 and ϕ_1

$$(p + \mathrm{i}kv_x)F_1(\mathbf{v},p) + \mathrm{i}k\phi_1\frac{\partial f_0}{\partial v_x} = f_1(\mathbf{v},0) \tag{16.11}$$

and

$$k^2\phi_1(p) = 4\pi Gm \int_{-\infty}^\infty F_1(\mathbf{v},p)\mathrm{d}^3\mathbf{v}. \tag{16.12}$$

Unlike the Fourier mode analysis, these results will clearly depend on the initial perturbation $f_1(\mathbf{v}, 0)$.

It is easier to work with $\phi_1(p)$, which depends on one variable, rather than $F_1(\mathbf{v}, p)$, which depends on two. Eliminating F_1 between these two equations, we therefore solve for ϕ_1

$$\phi_1(p) = \frac{4\pi Gm \displaystyle\int_{-\infty}^{\infty} \frac{f_1(\mathbf{v}, 0)}{p + ikv_x} d^3\mathbf{v}}{k^2 g(p, k)}, \tag{16.13}$$

where

$$g(p, k) = \left[1 - \frac{4\pi Gm}{k} \int_{-\infty}^{\infty} \frac{\partial f_0/\partial v_x}{ip - kv_x} d^3\mathbf{v} \right]. \tag{16.14}$$

The velocity space integrals over v_y and v_z are easily done formally. Setting $v_x = u$, this amounts to just replacing v_x by u, replacing $f_0(\mathbf{v})$ by

$$f_0(u) = \int f_0(\mathbf{v}) dv_y dv_z, \tag{16.15}$$

replacing $f_1(\mathbf{v}, 0)$ by

$$f_1(u, 0) = \int f_1(\mathbf{v}, 0) dv_y dv_z, \tag{16.16}$$

and replacing integrals over $d^3\mathbf{v}$ by integrals over du in (16.13) and (16.14). Consider it done.

Stability or instability of these perturbations shows up in both ϕ_1 and f_1, since a growing potential field requires a growing density perturbation. Thus inverting ϕ_1, similarly to (16.10)

$$\phi_1(t) = \frac{1}{2\pi i} \int_{-i\infty + \sigma}^{+i\infty + \sigma} \phi_1(p) e^{pt} dp \tag{16.17}$$

determines the growth or damping of waves. Clearly $g(p, k)$ plays an important role in the waves' developments. This quantity is analogous to the dielectric permeability of electrodynamic media. The value of $1/g(p, k)$ is a measure of the response ratio of the gravitational field to a small change in the distribution of stars. Notice that for very small wavelength perturbations of the distribution function, as k becomes large, the field responds only weakly. Obviously this is because the mass involved is small and the perturbations average out over smaller and smaller distances. Similarly, if the gravitational force were to become weak, for which we can imagine G becoming smaller, the field would be less affected by a perturbed distribution.

In an electrodynamic medium, the dielectric permeability (or dielectric constant) is the ratio of the induction D to the electric field E. It measures the polarization of the medium caused by a perturbed distribution function. A typical example is a moving electron or ion. The gravitational response of the system to a moving mass, described in Section 14, can also be interpreted analogously to an electric polarization. The energy which goes into producing the gravitational polarization comes from the moving mass. The mass slows down. This provides another way to determine the dynamical friction by inverting (16.13) with the form of f_1 which represents a single moving star. Then the gradient of this $\phi_1(\mathbf{r}, t)$ yields the perturbed force. The work this force does on the moving mass gives dynamical friction. Rather than examining dynamical friction yet again, we will use the inversion of (16.17) to continue our study of collective stability.

Inversion of transforms may be a subtle procedure. First one must find just the right contour. So we analyze ϕ in the p plane. From (16.13) and (16.14) we see that the singularities of ϕ_1 consist of poles of the numerator and zeros of the denominator $g(p, k)$. The zeros of g also give poles of ϕ_1. At each of these poles, $\phi_1(p)e^{pt}$ will have a residue, defined as the coefficient a_{-1} in the Laurant series expansion $\Sigma a_n(p-a)^n$ of $\phi_1(p)e^{pt}$ around the pole at $p_j = a$. To evaluate (16.17) we can use the theorem of residues. This states that if $f(z)$ is an analytic function, except for a finite number of poles, within a region of the complex plane bounded by a closed contour C, then the value of the contour integral $\oint f(z)dz$ around C is $2\pi i$ times the sum of the residues of $f(z)$ at the poles within the contour. Details of the contour's shape do not matter. Thus, in our case,

$$\phi_1(t) = \sum_{j=0}^{N} e^{p_j t} \text{Res } \phi_1(p_j). \tag{16.18}$$

Immediately this shows how the position of the poles determines the physical behavior of $\phi_1(t)$. Poles off the real axis lead to oscillations. Poles to the left of the imaginary axis give damping; those to the right give growth. Poles farther to the right of the imaginary axis dominate those closer to it, especially after a long time.

The contour in Figure 17 is chosen because it has the property of containing all the poles, while giving a contour integral which vanishes along the dashed part, so that it corresponds to the straight integration path of (16.17). By closing the original straight path, we can use the residue theorem. The vanishing act occurs because $\phi_1(p) \propto 1/p$ for large values of p and e^{pt} is small for the large negative values of p in the region where the contour crosses the real axis. There is a little catch in the last part of this statement. Only for non-zero values of t will e^{pt} be sufficiently small to make a negligible contribution near the crossing. This means that the analysis will not physically describe very short term, transient, time behavior just after the initial perturbations are set up. This involves just the initial acceleration of stars by the wave.

Next we search for the poles of $\phi_1(p)$. At first sight it looks like the numerator of (16.13) has poles at $p = -iku$, the phase velocity modes of Section 15 (let $p = -i\omega$). But now this is true only if $f_1(u, 0)du \neq 0$ as $du \to 0$ for these streams. Since we have explicitly taken initial perturbations in a region of space regardless of velocities, these modes do not now exist. (To make them exist we would have to give $f_1(u, 0)$ a δ function structure around them.) So we go on to search for zeros of the denominator.

The denominator, $g(p, k)$, contains a familiar expression – the dispersion relation

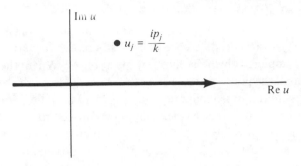

Fig. 18. Integration path for Re $p > 0$.

(15.33). When this relation is satisfied, $g(p_j, k) = 0$ and $\phi_1(p)$ has a pole at p_j

$$g(p_j, k) = 0 = 1 - \frac{4\pi Gm}{k} \int_{-\infty}^{\infty} \frac{\partial f_0/\partial u}{ip_j - ku} du. \qquad (16.19)$$

Originally we defined the Laplace transforms in (16.9) only for values of p which are positive and large enough that the integral converges. This clearly means that in the complex p plane, $g(p, k)$ is defined only to the right of the poles. To find the poles, we need to know $g(p, k)$ for all values of p. If we know an analytic function in one part of the complex plane we can extend it over the rest of the plane by the process of analytic continuation so long as it has no natural boundary. (There are no natural boundaries for functions with a finite number of poles.)

Since (16.19) involves an integral over velocity it is more direct to analytically continue $g(p, k)$ in the complex u plane. Figure 18 shows the path of integration of (16.19) in the u plane for the values of p whose real part is positive. These are the values, giving poles in the upper half of the u plane, which we have already dealt with in defining the Laplace transform. Now for the analytic continuation we must allow for poles corresponding to negative real parts of p which lie below the Re u axis. So we need to modify the contour of integration. Poles which were above the original contour have to stay there since the extension must not change the original integral. Moreover, $\phi_1(t)$ and $g(p, k)$ are continuous functions so poles cannot cross the line contour as we let Re p take on negative values. A pole crossing a contour causes a discontinuity in the integral. This can be seen from the residue theorem if $f_0(u)$ and its derivatives vanish for large imaginary velocities (e.g., for a Maxwellian with a cutoff) and the contour is closed in the upper half plane. Therefore, the only way to include poles in the lower half of the u plane is to deform the contour under them as shown in Figure 19. This applies even when, as for a Maxwellian without a cutoff, the contour cannot be closed by an infinite semicircle because this contribution would not vanish, indeed might even dominate.

At this point another subtlety intrudes. Anticipating the use of this contour for gravitational Landau damping, I have drawn it as a 'lollipop'. The contributions along the stem cancel and we are left with the contribution along the real velocity axis plus the integral around the off-axis negative pole. But the contour might have just been an arc or a semicircle under the pole (as in the plasma physics case for long wavelengths). What is the difference? The position of the negative pole controls the degree of damping, which is the real part of p_j. When the damping is infinitesimally small the pole is barely below the real axis. Since the behavior of the physical system is continuous, an infinitesimal change of the pole below the real axis must give an infinitesimal change of the damping compared to when the pole is on the axis. If the pole were actually on the axis, then to include its effect we would have to deform the contour below it. The resulting contribution would be πi times the residue there, just half the result from going completely around the pole. However, when the damping is finitely large, as we shall find in the gravitational case, the pole is far from the real axis and the contour encircles it completely.

When performing the integration of (16.19), we recall that perturbations with wavelengths longer than the critical Jeans length grow. So there is not much point in examining them for damping. In the continuum approximation shorter per-turbations propagate. Now we are equipped to calculate what happens to these in the discrete stellar description.

To simplify the analysis, let us look at very short wavelengths as $k \to \infty$. Separating the real and imaginary contributions to the perturbations' time development, we write

$$p = -i\omega - \gamma. \tag{16.20}$$

Waves damp if $\gamma > 0$. Earlier, the physical description of damping suggested that the damping timescale $\sim \lambda/\langle v \rangle$ so γ, which is the inverse timescale, is very large for small λ. We will anticipate the result that for short wavelengths $|\gamma| \gg \omega$ and $\omega/k \to 0$ while $|\gamma|/k \to \infty$. Then, in the complex u plane the pole $u = ip/k = -i\gamma/k + \omega/k$ will be far from the real axis but close to the imaginary one, as Figure 19 indicates.

Most physically reasonable equilibrium distribution functions $f_0(u)$ will be even functions of velocity, decreasing to zero for large real values of $|u|$. Like the Maxwellian (15.41), they generally increase without limit for large imaginary values of u. Then the pole dominates the contribution to the dispersion relation (16.19), and we can neglect the path along the real axis to obtain

$$1 = 2\pi i \left(\frac{4\pi Gm}{k^2} \right) \frac{\mathrm{d} f_0}{\mathrm{d} u} \bigg|_{ip/k}$$

$$= (2\pi)^{1/2} \frac{\omega_g^2 p}{\langle k^2 v^2 \rangle^{3/2}} e^{p^2/2\langle k^2 v^2 \rangle}, \tag{16.21}$$

where the second expression substitutes the Maxwellian. A dependence on the slope of the distribution function is expected from our earlier physical description. To see how γ and ω depend on k we substitute (16.20) into (16.21) and take the modulus of both sides of the result in the approximation $|\gamma| \gg \omega$

$$\gamma e^{\gamma^2/2\langle k^2 v^2 \rangle} = \frac{\langle k^2 v^2 \rangle^{3/2}}{(2\pi)^{1/2} \omega_g^2}. \tag{16.22}$$

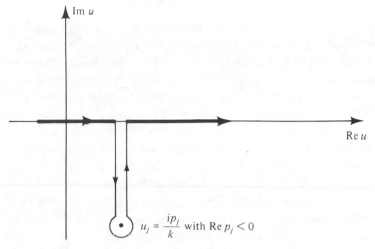

Fig. 19. Integration path for Re $p < 0$.

This shows that γ is a positive quantity and the wave does indeed decay with a timescale on the order of $\lambda/\langle v \rangle$, as we had expected.

To obtain ω, note that the phase of (16.21) is approximately $-e^{i\omega\gamma/k^2\langle v^2 \rangle}$ and, since this must have a real positive value,

$$\frac{\omega\gamma}{k^2\langle v^2 \rangle} = \pi. \tag{16.23}$$

There are higher order poles when the phase is equated to 3π, 5π, etc., but because these give smaller values of γ and less damping (just solve the last two equations) they are not so important. In a similar manner, one can solve for the oscillations of the perturbed distribution f_1, but this is left for the reader. The relations between ω and $|\gamma|$ as $k \to \infty$ are seen retrospectively to hold, but only in a logarithmically asymptotic manner.

The main result of this analysis is to introduce Landau's powerful technique for determining how realistic collective perturbations evolve, and to show that in a stellar system with a Maxwellian distribution, modes shorter than the Jeans wavelength decay. Other initial distributions can also be examined easily using (16.21). In addition to examining specific distributions it is also possible to draw some general conclusions regarding conditions for stability. From the previous analysis it is clear that the criterion for the system to be linearly stable is that the dispersion relation has no roots with Re $p > 0$ for real values of k. In plasma physics, Noerdlinger (1960) and Penrose (1960) have developed this criterion using Nyquist's method, and the results are discussed in many texts. Analogous results hold for gravitating systems, although a major difference results from the opposite sign of the gravitational dispersion relation. Thus, where the Penrose criterion predicts stability for plasmas, it predicts instability for gravitating systems. For example, distribution functions with a single hump, such as the Maxwellian, produce stable plasmas, but lead to Jeans' instability in gravitating systems. Stability criteria do not describe the detailed nature of the growing modes. For this, a full scale analysis such as the one applied to rotating systems by Lynden-Bell (1962) is needed.

If the linear analysis indicates a system to be stable to infinitesimal perturbations, a non-linear extension of this analysis, including the higher order expansion terms, also indicates stability. This makes sense because in order to grow into the non-linear regime an infinitesimal instability must first pass through the linear regime. However, there is another type of instability which behaves quite differently. This is one which starts out in the non-linear regime with a finite amplitude. A globular cluster passing through a galaxy is an obvious example. A large enough perturbation can change the background distribution function so much that its response becomes quite unlike the linear modes. In turn, this strongly affects the nature of the instability.

Although we have dealt with a single mass-component system, it is simple to generalize to many masses. If each mass component has an unperturbed distribution f_α, then the dispersion relation is

$$\sum_\alpha \frac{4\pi G m_\alpha}{k^2} \mathbf{k} \cdot \int \frac{\partial f_\alpha/\partial \mathbf{v}}{ip - \mathbf{k}\cdot\mathbf{v}} d^3\mathbf{v} = 1. \tag{16.24}$$

Exchanging the summation and integration, we see that this is equivalent to using a reduced distribution function

$$f(\mathbf{v}) = \frac{1}{M}\sum_\alpha m_\alpha f_\alpha(\mathbf{v}), \tag{16.25}$$

which may be normalized to the largest mass M component (or any convenient average mass). All the previous results will then apply if we substitute this reduced distribution function. They can also be generalized readily to cases when the different mass distributions are given different initial perturbations. This could occur in systems where stars or galaxies of different mass form with different distributions.

16.3. Other damping mechanisms

Linear Landau damping is not the only process which causes collective motions to decay. Here I will mention some other mechanisms briefly as a prelude to more detailed discussions in later sections. These mechanisms are generally more complicated and difficult to describe analytically.

Non-linear Landau damping would seem to be the next step toward complexity. This is the case for electron plasmas, but gravitating systems are again rather different. To examine non-linear Landau damping in plasmas, second and third order corrections are included in the perturbed distribution function and potential. At this level there are still wave solutions, but the waves are coupled and can interact with one another. In particular, two waves can beat against each other. The resultant wave moving at the beat frequency can in turn exchange energy with resonant particles near its phase velocity. This is the normal form of non-linear Landau damping. By now the reader has probably realized that the reason it does not apply directly to gravitating systems is that for these the damping timescale is so short, or $|\gamma| \gg \omega$. So waves do not propagate long enough to interact and produce beat frequencies in a relatively straightforward manner.

Nevertheless, there are some non-linear mechanisms whose qualitative effects can be understood simply. Capture by a clump of stars is the analog of non-linear trapping in plasma waves. Capture can happen in several ways. As a clump forms, it contracts. A star passing through the clump feels the overall gravitational potential well becoming deeper. The well it tries to climb out from is deeper than when it entered, and the star is caught in the cluster. This process is mean field trapping. Even if the mean field does not trap it, the star can exchange energy with members of the cluster through long-range encounters. Cumulative effects of distant random encounters may deplete the star's energy enough to trap it in the cluster. These encounters may involve scatterings from collective modes in the system, as well as from individual stars. A third type of trapping can occur through strong few-body interactions. The orbits of a few close neighbors may, by chance, conspire together to capture an interloper. Of course, the time reversed effect in which close or distant encounters eject a star from a cluster may also happen. The net result, on a given timescale, will depend on the details of orbits in a particular cluster. In general, these processes have to be treated by detailed numerical computations, although we shall see in later sections that under certain conditions they are amenable to analytic calculations.

Another class of non-linear interactions is the tidal disruption of clusters. Section 1 described how the differential force of an external mass can distort, and even disrupt, a cluster. When strong density inhomogeneities affect each other tidally, they tend to smooth one another out. An extreme example is when two clusterings penetrate each other with enough kinetic energy that most of each cluster emerges intact, but some stars are ripped out into a long tidal trail. Eventually the trail may disperse. In the most extreme case, individual galaxies are observed to collide leaving streams of luminous matter behind.

Phase mixing is still another type of non-linear interaction. Its essence is contained in a simple one-dimensional example. Suppose the trajectories of stars did not change. Then each would move in a straight line $x = x_0 + vt$ with constant velocity, and the perturbed distribution function would be $f_1(x,v,t) = f_1(x_0 + vt, v, t)$. For instance, an initial sinusoidal perturbation $f_1 \propto \sin(kx_0 + kvt - \omega t)$ would propagate ballistically along the orbits. Perturbations of different velocities would have different phases. To find the net density perturbation at a given position at a given time (perhaps as a prelude to finding the gravitational potential perturbation), we would integrate f_1 over all velocities and get a result proportional to $k^{-1}t^{-1}\cos[kx_0 + kvt - \omega t]_{v_{\min}}^{v_{\max}}$. After times long compared to the time for a typical star to cross the wavelength k^{-1}, the net density and potential perturbation nearly vanishes. Orbits of stars with different velocities have co-mingled their phases. Similar effects occur for non-linear initial perturbations, making allowance for the forces changing the initial orbits. We see also how the distribution function can oscillate rapidly, without the oscillation appearing in the potential.

Indeed, readers who took to heart the suggestion after Equation (16.23) will have noticed that, in addition to all the poles of ϕ_1, the distribution function $F_1(\mathbf{v}, p)$ has an extra pole at $p = -i\mathbf{k}\cdot\mathbf{v}$. It produces an additional term $\propto e^{-i\mathbf{k}\cdot\mathbf{v}t}$ in $f_1(t)$ due to free streaming. This term does not decay, but carries along the memory of the initial perturbation forever. However, it oscillates faster and faster in velocity space as $t \to \infty$ and contributes negligibly to the potential. In some situations it is possible that these oscillations are not entirely lost. If another perturbation is added to the system its distribution function oscillations may interfere constructively with those of the first perturbation. For a time the resonant terms would noticeably perturb the density, leading to a gravitational echo, analogous to plasma echoes.

Although the memory of initial conditions propagates forever in the collisionless Boltzmann approximation, it is damped when higher order encounters are considered. Thus, the additional fluctuations present in the Fokker–Planck or BBGKY descriptions can also cause waves to decay. A simple illustration is the free-fall collapse of a cluster of stars, starting with only radial inward motions. It is essentially a converging, spherical, non-linear wave. If the stars moved just in the average spherically symmetric mean field, they would all collide at the center. They avoid this fate because the local graininess of the field perturbs the orbits. Fluctuations add fairly random tangential and radial components to the orbits and in a few initial crossing times the system relaxes to a quasi-equilibrium state with a roughly Maxwellian velocity distribution.

This evolution also provides an example of a large class of non-linear damping mechanisms loosely called 'violent relaxation'. Despite applying mainly to in-homogeneous systems, violent relaxation is worth mentioning here since it will be very important in later sections. For violent relaxation to occur, the system's density and velocity distributions must be very far from their quasi-equilibrium forms. This generally means the presence of clumps in position and velocity space. Bound subsystems comprise only a small fraction of these clumps. Turmoil pervades the system. Stars scatter off clumps and clumps scatter off clumps. Old clumps dissolve while new ones arise. Any coherence is temporary, to be replaced by new collective modes. So strong are these interactions that all the non-linear processes described in this section compete on time scales of order $(G\rho)^{-1/2}$. After several of these relaxation times the initial clumps are redistributed and inhomogeneities smoothed out. What is left is a system in quasi-equilibrium with a roughly Maxwellian velocity distribution. No direct collisions are responsible for this end result – unlike the case of a perfect gas – only the non-linear encounters among collective modes.

17

Star–gas interactions

17.1. Gas dynamical processes

Many phenomena occur when stars plunge through clouds of gas. Among the most dramatic is the formation of shocks and ionized wakes, especially around stars which emit strongly in the ultraviolet. The ionizing radiation can be produced either by the star directly, or from the bow shock accompanying supersonic motion. Although we will not usually include gas dynamic and radiative processes, this one is an exception since it is important. So we give a brief general discussion of the phenomenon. Then we will describe collisionless accretion, the slowing down of stars by gas, and modifications of the Jeans and two-stream instabilities.

Suppose a star moves supersonically through a cloud of hydrogen. (The role of heavier atoms is mainly to increase and complicate the radiation processes.) In the direction of motion there is a bow shock which embraces the star more tightly at high Mach numbers (v_*/v_{sound}). To estimate the temperature immediately behind the shock front, equate the thermal energy $3kT/2$ to the kinetic energy of an atom, giving $T \approx 10^5 v_{\perp 100}^2$ K, where $v_{\perp 100}$ is the inflow velocity (in units of $100\,\text{km s}^{-1}$) normal to the shock surface. As this shock-heated gas flows behind the star, it expands and cools. Part of the cooling will be from free–free radiation, and part from the expansion itself. From standard formulae for free–free emission by an ionized gas, one learns that the timescale for substantial radiation is $\tau_{\text{ff}} \approx 10^6 v_{100} n^{-1}$ yr, where n is the gas number density cm^{-3}. Other radiative processes (e.g., plasma oscillations, line emission from heavy atoms in a more realistic gas composition) may reduce this timescale appreciably. However, it is much longer than the free-expansion timescale $\tau_{\text{exp}} \approx r_*/v \approx 3 \times 10^{-4}(r_*/r_\odot)(v_{100})^{-1}$ yr. So the main cooling process will be the adiabatic expansion.

The shape of the expanding wake can be estimated very roughly by a simplified picture in which the star sets off a series of cylindrical self-similar detonation waves (see Saslaw & De Young, 1972). These occur at intervals of the order of r_*/v. Since the radius of a cylindrical self-similar detonation increases as $t^{1/2}$, the wake profile will increase roughly as $a^{1/2}$ where $a \gg r_*$ is the distance from the star measured back along the wake. This result requires a homogeneous external medium, and will not apply near the star where gas is ablated into the stream.

Part of the star's atmosphere may be driven off by Kelvin–Helmoltz and other instabilities in the post-shock flow. In this way a turbulent stream forms, containing hot gas, magnetic field, high energy particles, and other debris from the star. A general energy argument shows that the mass that can be ablated is less than the mass of interstellar gas that flows through the shock in the star's atmosphere. To escape, the ablated material must gain energy $\sim Gm_*m_{abl}/r_*$. This energy comes from the post-shock flow and must be less than the kinetic energy of the interstellar gas with respect to the star, since ablation is not 100% efficient: $Gm_*m_{abl}/r_* \lesssim m_{int}v^2$, or $m_{abl}/m_{int} \lesssim E_{kin}/E_{grav} \approx v^2/v_{esc}^2$. It is reasonable to suppose that the star's kinetic energy, E_{kin}, is less than its own self-gravitational binding energy E_{grav}. This is certainly the case for stars accelerated coherently by gravitational forces. Hence $m_{abl} < m_{int}$, an inequality that is usually satisfied by a rather large amount. For example, a solar size star moving at a Mach number of 10 through an HII region ($T \approx 10^4$ K) would lose less than a few per cent of the mass sweeping by it. Cumulative effects of ablation can become important.

Very hot stars moving through neutral gas can ionize it dramatically. Even in neutral hydrogen, the HII region produced by a rapidly moving source of ionizing radiation is rather complicated, but it may be crudely visualized by the following approximation. A stationary source, which suddenly turns on, produces a spherically expanding HII shell whose radius is given by (e.g., Splitzer, 1968)

$$R_{HII}^3 \approx \frac{S_{(0)}t}{4n_H}. \tag{17.1}$$

Here $S_{(0)}$ is the total number of Lyman continuum photons emerging from the surface each second, n_H is the number density of ions plus neutral hydrogen atoms, and t is the effective time for which a given region of gas has been irradiated. This holds for times short compared with the time required to establish an equilibrium, known as a 'Stromgren sphere', when the photo-ionizations just balance the collisional recombinations.

When the star moves, the ionization front generated at a particular place initially expands faster than the star's velocity. Eventually the front slows down as residual neutral atoms diminish the photon flux. After the star reaches a position D, say, it catches up with the front generated when it was at $D - vt$. Until this time, we may assume that the front expands approximately spherically; after this time the expansion essentially stops and a new spherical region begins. Such an idealization works best for regions far from the trajectory of the star, which include most of the volume of the radiating gas. Thus, we use $R_{HII} = vt$ to eliminate t in (17.1), obtaining for the approximate maximum radius of the HII region

$$R_{HII} \approx \frac{1}{2}\left[\frac{S_{(0)}}{n_H v}\right]^{1/2}. \tag{17.2}$$

Taking the surface to be a blackbody of temperature T, the number of ionizing

Lyman continuum photons per second which emerge is

$$S_{(0)} = 2 \times 10^{12} \beta(T) r_*^2 T^3. \tag{17.3}$$

Allen (1963) provides a convenient table for $\beta(T) = N_{0-912\text{Å}}/N_{0-\infty}$, the ratio of Lyman continuum photons to the total number of photons. This ratio is a very strong function of temperature. Ordinary, late-type stars have little effect. But luminous, hot, early-type stars, and especially supermassive stars or compact massive objects surrounded by radiating disks of gas will completely change the ionization structure of clouds they pass through. One important possibility is that the ionization front produces a pressure wave which may steepen into a shock as it moves through the cloud. Since the radiation can change rapidly and the geometry is irregular, a complex system of shocks may occur which compresses part of the cloud, now subject to thermal and gravitational instability. Thus, a sufficiently massive star or other object may catalyze star formation, leaving behind a wake of bright stars that illuminate the diffuse gas.

Other primarily gas dynamic processes, just to mention in passing, are stellar winds. They exchange material between the star and its surroundings. Stellar winds vary greatly in importance, being strongest in massive young stars where they can even prevent accretion. Winds may be driven by both thermal gas and radiation pressures, as well as by centrifugal and magnetic forces. The resulting mass loss need not be isotropic. Some stars can lose gas in one direction and accrete it in another, particularly if the background medium is inhomogeneous or the star is surrounded by a disk.

All these gas dynamical processes form an enormous subject in their own right. This brief description was just to sketch the scene beyond the boundary of purely gravitational interactions, to which we now return.

17.2. Accretion and momentum decrease

Accretion of gas by stars is a very complicated affair when pressure, viscosity, and radiation transfer are taken into account. This is necessary when the mean free path of the gas for atomic collisions and photon scattering (elastic and inelastic) is small. We will deal with the opposite case when gravity dominates. It applies to many important examples such as accretion of diffuse low density gas or dust, accretion of individual, small, high density gas clouds, and the collisional capture of another star or solid body.

Consider, for example, an atom or a mote of dust moving past a star. To determine if the grain collides with the star we need to know its distance of closest approach. If the grain starts a very long distance away with velocity v, we know from energy conservation that if it grazes the surface of the star at radius r_*, its velocity v' is given by

$$v'^2 - \frac{2Gm_*}{r_*} = v^2. \tag{17.4}$$

We also know that its original and final angular momenta are equal

$$Rv = r_* v'. \tag{17.5}$$

Since R is the original impact distance for the grain's deflected orbit to graze the star, it is the radius of the effective collision cross section. Solving for R,

$$\frac{R^2}{r_*^2} = 1 + \frac{2Gm_*}{r_* v^2}. \tag{17.6}$$

Thus grains moving so fast that the star's gravity hardly changes their orbit see just the geometrical cross section $\pi R^2 \approx \pi r_*^2$. But slowly moving grains see a much larger cross section. For $v = 0$, the cross section is indeed infinite: wherever the grain is, it falls radially into the star.

From this cross section, we estimate the rate of accretion, ignoring the complications mentioned before. When a grain hits the outer 'edge' of a star its energy is presumed to be dissipated and its mass added to that of the star, giving

$$\frac{dm_*}{dt} = \pi R^2 v\rho = \pi r_*^2 \rho v \left(1 + \frac{2Gm_*}{r_* v^2}\right) \approx \frac{2\pi G\rho m_* r_*}{v} \tag{17.7}$$

for a density ρ of grains. The last expression is for low velocity accretion. If both colliding bodies are about equally extended, the effective geometric collision radius is a factor ~ 2 greater and both the cross section and accretion rate increase by a factor ~ 4. A star moving rapidly will double its mass during the time, τ_{fast}, that it sweeps through an equivalent amount of material, whereas a slowly moving star doubles on a time, $\tau_{\text{slow}} \approx \tau_{\text{fast}} E_{\text{kin}}/E_{\text{grav}}$, which is reduced by the ratio of the star's kinetic energy to its internal gravitational energy.

The previous result assumed that all the particles far from the star had just the relative velocity v and no random component of peculiar motion. There is no difficulty in allowing for a more general velocity distribution. In fact from (17.7) one can already make a good guess at the modified result in many cases. For example, a star at rest in a Maxwellian gas should accrete at a rate found from replacing v in (17.7) by the velocity dispersion of the gas, and it very nearly does. A moving star is slightly more complicated. Let $N(v, \theta, \phi)$ be the number density of particles moving with velocity v in the θ and ϕ directions at large distances from the star. Then if particles are not deflected on average by collisions, the total accretion rate is found by multiplying each particle by its accretion cross section and integrating over all particles moving toward the star

$$\frac{dm_*}{dt} = \pi r_*^2 \rho \int_0^{2\pi} \int_0^\pi \int_0^\infty v\left(1 + \frac{2Gm_*}{r_* v^2}\right) N(v, \theta, \phi) dv d\theta d\phi. \tag{17.8}$$

A star moving at velocity v_* through a Maxwellian distribution of particles produces

$$N(v, \theta, \phi) = (2\pi)^{-3/2} \rho \langle v^2 \rangle^{-3/2} \exp[-(v^2 + v_*^2 - 2vv_* \cos\theta)/2\langle v^2 \rangle] v^2 \sin\theta. \tag{17.9}$$

This can be easily expanded and integrated term by term by the reader for large and small values of $v_* \langle v^2 \rangle^{-1/2}$ to see how the exact accretion rate compares with (17.7).

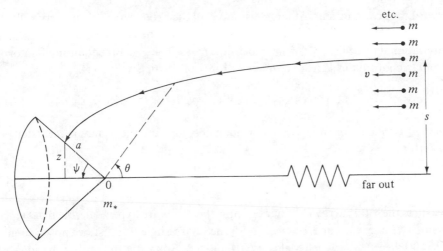

Fig. 20. Orbits of accreted stars or gas particles.

For the particles to be essentially collisionless, their mean free path should be larger than, say, ten times the effective gravitational radius, Gm_*v^{-2}, of the star

$$\lambda = (n\sigma)^{-1} \gtrsim 10Gm_*v^{-2} \approx 10^{13}(m_*/m_\odot)(v_{100})^{-2}\,\mathrm{cm}. \qquad (17.10)$$

Taking a cross section $\sigma \approx 10^{-16}$ for neutral atoms shows this is satisfied for $n \lesssim 10^3\,\mathrm{cm}^{-3}$ in the vicinity of a solar mass star moving at $100\,\mathrm{km\,s}^{-1}$. If there were no magnetic field (rather unrealistic), ionized atoms would be scattered mainly by cumulative effects of distant Coulomb interactions and their orbits would change significantly on a timescale analogous to the relaxation time τ_R in Equation (2.9) for stars (replacing the gravitational force by the electrostatic force). Then the Coulomb relaxation timescale must be much longer than the time to cross the gravitational radius of the accreting star. Dust grains would have similar types of timescales if they were not subject to significant radiation pressure. Conditions will obviously vary considerably among different applications. For stars, we have directly from (2.9) the requirement that $\tau_R \gtrsim 10\,Gm_*v^{-3}$, or for solar mass stars with velocity dispersions and relative velocities of $100\,\mathrm{km\,s}^{-1}$, $n \lesssim 5 \times 10^{-42}$ stars cm^{-3}. If this seems to be a funny unit of density, remember it is equal to about 10^{14} stars per cubic parsec. This result is very sensitive to velocities and masses, being proportional to $\langle v^2 \rangle^{3/2}v^3m^{-2}m_*^{-1}$, where the field stars have mass m and velocity dispersion $\langle v^2 \rangle^{1/2}$, and the accreting star of mass m_* moves through them with velocity v. Even so, for all but the most extreme conditions in galactic nuclei, the accreted stars are effectively collisionless.

Under these collisionless conditions, we can go further and examine the density distribution around the accreting star. Since the collisionless Boltzmann equation is, as we saw in Section 7, equivalent to the equation of the orbits, we could use either approach. The orbit equations lead more directly to the answer, so they are the ones to employ here. As an example, the accreting 'star' could be a massive black hole.

Suppose, as in Figure 20, an infinite stream of stars or collisionless atoms moves toward the accreting star. At large distances all the orbits have velocity v parallel to the x axis. (Velocity dispersions will be added later.) The accreting star at 0 exerts a gravitational force $Gm_*r^{-2} \equiv \mu r^{-2}$ per unit mass. The general equation of an orbit in polar coordinates is (e.g., Landau & Lifshitz, 1976, Section 15).

$$\frac{1}{r} = \frac{\mu}{s^2v^2}[1 + e\cos(\theta + \theta_0) = 1 + e\cos\theta_0\cos\theta - e\sin\theta_0\sin\theta]. \quad (17.11)$$

Here m_* is much more massive than what it accretes, and the squared eccentricity of the orbit

$$e^2 = 1 + \frac{2Es^2v^2}{\mu^2m} \quad (17.12)$$

measures the total energy E. In our example the orbits are hyperbolic, the mass m has a velocity v at $r = \infty$, and $E = mv^2/2$. So the phase angle θ_0 must be chosen to give $r^{-1} \to 0$ when $\theta \to 0$, which yields $e\cos\theta_0 = -1$. Squaring this, using the identity $e^2\sin^2\theta_0 + e^2\cos^2\theta_0 = e^2$, and substituting the value of E into the eccentricity shows that $e\sin\theta_0 = -sv^2\mu^{-1}$. The negative sign ensures that $\sin\theta \to s/r$ for small θ. Thus the equation of motion (17.11) becomes

$$\frac{1}{r} = \frac{\mu}{s^2v^2}(1 - \cos\theta) + \frac{\sin\theta}{s}. \quad (17.13)$$

Along these orbits, stars which intersect the cone of angle ψ at a distance from m_* between a and $a + da$ will have come originally from the cylindrical shell between s and $s + ds$. We need to relate the density of stars entering the cone to the density leaving from s. This is easily done. From Figure 20 and Equation (17.13) we see that a and s are related by

$$\frac{1}{a} = \frac{\mu}{s^2v^2}(1 + \cos\psi) + \frac{1}{s}\sin\psi \quad (17.14)$$

and therefore

$$\frac{da}{a^2} = \frac{2\mu ds}{s^3v^2}(1 + \cos\psi) + \frac{ds}{s^2}\sin\psi. \quad (17.15)$$

Let the number density of stars with velocity v at infinity be η, so their initial flux is ηv per unit area per unit time. The total number coming from an element of the distant cylindrical shell of radius s is $2\pi s\eta_s vds$ per second. They leave with angular momentum sv. Since this is also their angular momentum $av_\perp$ on arriving at the cone, they enter it with a normal velocity sv/a. The element of area around this part of the cone is $2\pi zda$. But since the vertical distance $z = a\sin\psi$, this area is $2\pi a$ $\sin\psi da$. The number entering da is therefore $\eta_a^+(sv/a)(2\pi a\sin\psi da)$. This must be equal to the number leaving ds, giving

$$\eta_a^+ = \frac{2\pi s\eta_s vds}{\dfrac{sv}{a}(2\pi a\sin\psi\,da)} = \frac{\eta_s}{\sin\psi}\frac{ds}{da}. \quad (17.16)$$

Eliminating s from the two previous equations and using the half-angle formula

$\sin\psi = 2\sin\frac{1}{2}\psi\cos\frac{1}{2}\psi$ extensively, shows that

$$\eta_a^+ = \frac{1}{2}\eta_s\left(\frac{2\mu}{av^2} + \sin^2\frac{\psi}{2}\right)^{-1/2}\left\{\frac{\mu}{av^2} + \sin^2\frac{\psi}{2}\right.$$

$$\left. + \left(\frac{2\mu}{av^2} + \sin^2\frac{\psi}{2}\right)^{1/2}\sin\frac{\psi}{2}\right\}\cosec\frac{\psi}{2} \quad (17.17)$$

stars per unit volume are entering the cone at a and ψ. Stars also leave the cone there, and a similar analysis shows departing stars to have a number density

$$\eta_a^- = \frac{1}{2}\eta_s\left(\frac{2\mu}{av^2} + \sin^2\frac{\psi}{2}\right)^{-1/2}\left\{\frac{\mu}{av^2} + \sin^2\frac{\psi}{2}\right.$$

$$\left. - \left(\frac{2\mu}{av^2} + \sin^2\frac{\psi}{2}\right)^{1/2}\sin\frac{\psi}{2}\right\}\cosec\frac{\psi}{2}. \quad (17.18)$$

Therefore the total number density of stars at (a, ψ) is

$$\eta_a = \eta_a^+ + \eta_a^- = \eta_s\left(\frac{2\mu}{av^2} + \sin^2\frac{\psi}{2}\right)^{-1/2}$$

$$\times \left(\frac{\mu}{av^2} + \sin^2\frac{\psi}{2}\right)\cosec\frac{\psi}{2}. \quad (17.19)$$

Notice the strong increase of density for small values of ψ in the wake of the star. At $\psi = 0$ the density becomes infinite and it is clear from (17.16) that the analysis breaks down here because we are dividing by zero area. This is also region where stars coming from $s = 0$ intersect stars deflected onto the axis. If the accreted material were a collisional gas it would form a shock along the axis. Moreover, from the axial symmetry of the situation it is clear that when ψ is below the axis we must use its absolute value to avoid negative densities. Since the volume of the conical wake is proportional to $\sin^2(\frac{1}{2}\psi)$, the infinite density still leads to a finite total number of stars. For very slow initial relative velocities, greater gravitational deflection decreases the density by about a factor of two below its corresponding value at high initial velocities. This is because the accreted stars move past the accreting star with higher final velocity and, on average, spend less time behind it.

Using these results we can consider the more realistic example when the incoming stream has some velocity dispersion $\langle v^2 \rangle^{1/2}$ and a distribution function $N(v, \theta, \phi)$ such as the Maxwellian (17.9). It is easiest to calculate the resulting density on the axis of symmetry at a distance a from the accreting star, as indicated by Figure 21. We expect from (17.19) that collisionless accretion will have its highest density along this axis. The stars which reach A will have some initial range of positions, speeds and directions. To find the number in a given initial range $d\theta d\phi dv$ which contributes to the density at A we replace η_s in (17.19) by $N(v, \theta, \phi)$ from (17.9). The total density at A is then the integral over all contributing initial ranges

$$\frac{\rho(a)}{\rho} = (2\pi)^{-1/2}\langle v^2 \rangle^{-3/2}\int_{v=0}^{\infty}\int_{\theta=0}^{\pi}\left(\frac{\mu}{av^2} + \sin^2\frac{\theta}{2}\right)\left(\frac{2\mu}{av^2} + \sin^2\frac{\theta}{2}\right)^{-1/2}$$

$$\cdot\exp\{-(v^2 + v_*^2 - 2vv_*\cos\theta)/2\langle v^2\rangle\}\sin\theta\cosec\frac{\theta}{2}v^2 d\theta dv. \quad (17.20)$$

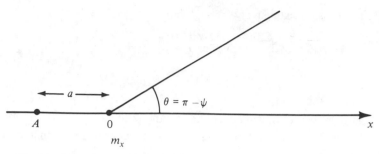

Fig. 21. Accretion on the axis of stars initially moving at an angle θ.

The extra factor of 2π comes from the azimuthal integral over $d\phi$. Thus we now have the density on the leeward half of the axis.

On the windward half of the axis, we expect the density to be lower. This is because the stars which are deflected onto it have a velocity $v + v_*$, and there are fewer of these in the Maxwellian distribution. A similar analysis shows that the windward density at a distance a along the axis is given by (17.20) with v_* replaced by $-v_*$, as expected from the symmetry. This difference of density on the two sides of the star is yet another illustration of the physical cause of dynamical friction.

To regain the results of Section 14 using this approach, we could introduce fluctuations in the density upstream of the perturbing star and see how these fluctuations propagate and change the star's orbit. Moreover, by extending this analysis to see how temporary condensations develop in the wake, we could obtain a more quantitative description of grexons. This is left for the more curious reader to attempt.

We next turn briefly to the situation when a star moves through gas or dust so rapidly that the gravitational deflection is negligible. The gas does not have enough time to respond gravitationally before the star is off and away. So the geometrical term dominates accretion in (17.7). Nevertheless, the star will still slow down significantly by transferring momentum directly to the gas it collides with. The timescale for such slowing down is approximately the time needed for the star to sweep through its own mass of interstellar gas

$$\tau \approx \frac{m_*}{\pi r_*^2 \rho v} \approx \frac{3 \times 10^{20}(m_*/m_\odot)}{(r_*/10^{11}\,\text{cm})^2(\rho/10^{-24}\,\text{g cm}^{-3})(v/10^7\,\text{cm s}^{-1})}\text{yr.} \quad (17.21)$$

This timescale is very long, except under extreme conditions which may apply in dense galactic nuclei. There we may find $r_* \approx 10^{12}, \rho \approx 10^{-14}, v \approx 10^9$, giving $\tau \approx 3 \times 10^6\,\text{yr}$. However, more dramatic slowing down processes. such as stellar collisions, may intervene on shorter timescales.

17.3. Jeans and two-stream instabilities

In Section 15.2 we saw that a large gas cloud can become unstable and fragment on length scales over which gravitation dominates gas pressure. If we increase the mass by an amount δm in a region whose characteristic size is λ, then the change in kinetic energy this produces is $\delta E_{kin} \approx \frac{1}{2}c^2(\delta m)$, where c is the sound speed. But the gravitational energy changes by an amount $\delta E_{grav} \approx (\delta m)(\rho_0 \lambda^3)G/\lambda$, where ρ_0 is the average background density. Equating these two energy changes gave the critical Jeans length for instability.

Now consider what will happen in a system whose background density is dominated not by the gas, but by stars imbedded fairly uniformly throughout the gas. An example is a large gas cloud in a galactic nucleus. If the gas is perturbed its change in kinetic energy is the same as before, but its gravitational energy changes by $\delta E_{grav} \approx (\delta m)(\rho_* \lambda^3)G/\lambda$ where ρ_* is now the stellar density. Thus the result is to replace the gas density by the stellar density in the expression (15.11) for the Jeans wavenumber.

Two new caveats apply to this situation. First, we must ensure that the stars do not heat the gas significantly as it collapses. From (17.21), converted into an energy deposition timescale, we see that motions induced by stars moving through the gas are generally unimportant as a source of heating. Applying (17.1)–(17.3) to ordinary stars, the reader can show that they are not usually important energy sources either. However, the presence of young massive stars or even a few supermassive stars producing soft X-rays, low energy cosmic rays and winds can alter the picture considerably. These are important for constructing detailed models of galactic nuclei and perhaps during the formation of galaxies as well.

The second caveat is that new instabilities will occur as the size of the shrinking cloud becomes of the same order of magnitude as the separations of the stars. While the cloud remains larger than the stellar separation, it is much more stable against internal fragmentation. This is because a large density perturbation within the cloud does not significantly change the average gravitational force, due mainly to the stars. But as the cloud shrinks between the stars its situation changes dramatically. Not only can it fragment more readily, but the tides from nearby stars may rip it apart before it has a chance to condense much further. All these problems which are necessary to understand star formation in galactic nuclei are far from being solved.

Conditions are also quite different if the gas and stars have a net bulk motion relative to one another, instead of stars just moving at random within a cloud. Now there are opportunities for microscopic instabilities, such as the two-stream instability, to fragment the gas. To analyze a system with collisional (gas) and collisionless (star) components, one uses the fluid Equations (15.2), (15.4) and (15.9) for the gas, and the collisionless Boltzmann equation (15.26) for the stars, coupling them through Poisson's equation (15.4) with ρ the total density. This set of equations is linearized and examined for growing and damping modes in the manner described in Sections 15 and 16. Sweet (1963) has carried through the analysis which is now much more intricate and complex.

For a stream of gas moving at subsonic speed u through a system of equal mass stars, there is a critical wavelength which may be written as

$$\lambda_{crit} = \frac{(c^2 - u^2)^{1/2}}{[\langle v^2 \rangle(\rho_g/\rho_*) + 2(c^2 - u^2)]^{1/2}} \lambda_{j*}. \tag{17.22}$$

This holds for non-zero values of ρ_*. Here c is the speed of sound in the gas which has the density ρ_g and streams with velocity u relative to the stars whose density is ρ_* and velocity dispersion is $\langle v^2 \rangle$. The stars have a Jeans length given by $\lambda_{j*}^2 = \pi \langle v^2 \rangle (G\rho_*)^{-1}$ analogous to (15.11). Perturbations *smaller* than λ_{crit} grow by the two-stream instability. For a small ratio of ρ_g/ρ_* the instability criterion is similar to (15.40) for two cold streams of stars, but now the velocity dispersion replaces the streaming velocity in (15.40). As the streaming velocity u in (17.22) approaches the sound speed of the gas, the critical wavelength becomes very small and the gas becomes unstable on nearly all wavelengths. When the relative motion is supersonic, it turns out that all wavelengths are also unstable. The e folding time for a density perturbation to increase is approximately $\lambda / \langle v^2 \rangle^{1/2}$, which is how long a typical star takes to pass through the perturbation.

Fragmentation of gas streams produced in this manner has many astronomical applications. A globular cluster passing through a smooth region of gas in the galactic plane should leave a wake of clouds behind. Large clouds passing through regions densely populated by stars should break up. Smooth streams of intergalactic gas running through stellar haloes of galaxies should fragment. Then these fragments can follow orbits described in the previous section. Can their concentration in the wake condense into a child galaxy?

18

Problems and extensions

Some basic aspects of many-body theory remain to be developed in future sections where they will be connected more closely to their astronomical applications. Here I give several extensions which can be worked out as problems by the reader, or can be used to enter the literature. Other suggestions for practice problems are sprinkled lightly throughout the text.

18.1. The point mass approximation

So far we have usually supposed that gravitating bodies, from dust grains to galaxies, can be treated as point masses. This is clearly an idealization. Geometric collisions make this approximation fail; so does tidal disruption. A third reason for failure is angular momentum transfer from the orbit to the spin of an object. Consider two solid ellipsoids, each of mass m and semimajor axes a,b where $a > b$. Show that, as they pass by one another at average distance r and velocity v, each acquires an angular momentum of order $G(mae)^2/r^2v$, where $e^2 = 1 - (b/a)^2$. How close do they have to pass to transfer an appreciable fraction of their orbital energy into rotational energy? Is this a plausible process for stars? For galaxies? What happens if the masses are not approximated as solid bodies? What residual internal circulation would result? Is there a net vorticity? How does the transferred energy change the size of the galaxy? (See Harrison, *MNRAS*, **154**, 167, 1971.)

18.2. Plummer's model

H.C.Plummer (e.g.,*MNRAS*,**76**, 107, 1915) found that the radial density distribution

$$\rho(r) = \frac{3M\alpha^2}{4\pi} \frac{1}{(r^2 + \alpha^2)^{5/2}}, \tag{1}$$

where the scale length

$$\alpha = \left(\frac{3M}{4\pi\rho_c} \right)^{1/3}, \tag{2}$$

with ρ_c the central density, gives a good fit to the density in globular clusters. Although Plummer used it for globular clusters (where the fit is not actually so good at the center), it is often taken as a convenient representation for galaxies and clusters of galaxies as well. Show that the mass distribution is

$$M(r) = \frac{M}{\left(1 + \dfrac{\alpha^2}{r^2}\right)^{3/2}}, \tag{3}$$

the potential is

$$\phi(r) = \frac{GM}{(\alpha^2 + r^2)^{1/2}}, \tag{4}$$

and the self-gravitational potential energy for an infinite radius is

$$|E_{grav}| = \frac{3\pi}{32}\frac{GM^2}{\alpha}. \tag{5}$$

18.3. Solutions of the master equation

To derive the Fokker–Planck equation from the master equation (sometimes called the Kolmogorov–Feller equation, from its development for probability theory), it was necessary to ignore close encounters giving large-angle scattering. For small stellar systems, where $\ln N$ is not much greater than order unity, this is a poor approximation. Show that the master equation can be solved by distribution functions of the form

$$f(v, t) = \sum_{i=1}^{\infty} c_i T_i(t) g_i(v). \tag{1}$$

Find $T_i(t)$ and the eigenfunction equation and boundary conditions satisfied by $g_i(v)$. Show that all the eigenvalues are positive. Which eigenvalue will dominate in $f(v, t)$ as $t \to \infty$? Show that this eigenvalue represents the fractional loss rate in a quasi-steady state for a spherical system. (Retterer, Astron. Journal, **84**, 370, 1979 discusses these separable solutions in detail with numerical computations.)

18.4. Self-similar solutions of the collisionless Boltzmann equation

Often it is possible to find systems whose only change at different times is an increase or decrease of scale. Otherwise its properties remain the same. Then the temporal evolution of such self-similar systems is completely contained in a set of scaling relations, which can be extracted from the equation of motion. Usually this approach is most successful under conditions of high symmetry.

Consider a large (infinite if one makes the 'Jeans swindle') spherically symmetric collisionless gravitating system. Its distribution function $f(t, r, \theta, \phi, R, \Theta, \Phi)$ depends on time and the six position and velocity coordinates. First show that in spherical polar coordinates, the collisionless Boltzmann equation has the form

$$\frac{\partial f}{\partial t} + R\frac{\partial f}{\partial r} + \frac{\Theta}{r}\frac{\partial f}{\partial \theta} + \frac{\Phi}{r\sin\theta}\frac{\partial f}{\partial \phi} + \left(\frac{\Theta^2 + \Phi^2}{r} - \frac{\partial\Psi}{\partial r}\right)\frac{\partial f}{\partial R}$$

$$+ \left(\frac{\Phi^2}{r}\frac{\cos\theta}{\sin\theta} - \frac{R\Theta}{r} - \frac{\partial\Psi}{r\partial\theta}\right)\frac{\partial f}{\partial\Theta}$$

$$- \left(\frac{R\Phi}{r} + \frac{\Theta\Phi}{r}\frac{\cos\theta}{\sin\theta} + \frac{1}{r\sin\Theta}\frac{\partial\Psi}{\partial\phi}\right)\frac{\partial f}{\partial\Phi} = 0, \tag{1}$$

where Ψ is the gravitational potential. (For a method of deriving this see the comment after Equation (60.7).) For spherical symmetry the spatial angular derivatives are zero and f depends only on t, r, R, and the square of the tangential velocity component $T = \Theta^2 + \Phi^2$. Change to polar coordinates in velocity space

$$\Theta = T^{1/2}\cos\alpha, \tag{2}$$

$$\Phi = T^{1/2}\sin\alpha, \tag{3}$$

with $-\infty \le \Theta, \Phi \le \infty$, $0 \le T \le \infty$ and $0 \le \alpha \le 2\pi$. Show that under these conditions Equation (1) reduces to

$$\frac{\partial\tilde{f}}{\partial t} + R\frac{\partial\tilde{f}}{\partial r} + \left(\frac{T}{r} - \frac{\partial\Psi}{\partial r}\right)\frac{\partial\tilde{f}}{\partial R} - \frac{2RT}{r}\frac{\partial\tilde{f}}{\partial T} = 0, \tag{4}$$

where

$$\tilde{f}(t, r, R, T) \equiv \pi f(r, t, R, \Theta, \Phi). \tag{5}$$

Show also that Poisson's equation reduces to

$$\frac{\partial\Psi}{\partial r} = \frac{G}{r^2}M(r, t), \tag{6}$$

where

$$M(r, t) = \int_{x=0}^{r} 4\pi x^2 \int_T \int_R \tilde{f}\,dR\,dT\,dx \tag{7}$$

is the mass within r.

The next step is to find a transformation which leaves Equations (4), (6), and (7) invariant. Since the variables of these equations appear as powers, a good bet is a multiplicative transformation

$$\tilde{f} = a^\alpha f^*, \quad t = a^\beta t^*, \quad R = a^\gamma R^*, \quad r = a^\delta r^*, \quad T = a^\varepsilon T^*, \quad M = a^\eta M^*, \tag{8}$$

where a is a parameter. To determine the exponents, substitute these relations into (4), (6) and (7) to find that the six exponents are constrained by the four relations

$$\gamma = \delta - \beta,$$
$$\varepsilon = 2(\delta - \beta),$$
$$\eta = 2(\delta - \beta) + \delta,$$
$$\alpha = -3\delta + \beta. \tag{9}$$

This leaves just one degree of freedom, since the relations are homogeneous.

To turn these results into the invariants of the transformation we combine all the variables into five sets of two each. Each combination must be invariant. Normalizing the time variable t to its initial value t_0 and expressing $(\alpha, \beta, \ldots, \eta)$ in terms of $k = 1 - \delta/\beta$, show that a set of five invariants is

$$\frac{\tilde{f}}{(t/t_0)^{3k-2}}, \quad \frac{r}{(t/t_0)^{1-k}}, \quad \frac{R}{(t/t_0)^{-k}}, \quad \frac{T}{(t/t_0)^{-2k}}, \quad \frac{M}{(t/t_0)^{1-3k}}. \tag{10}$$

The value of the free parameter k is determined by the need to conserve mass. What is it?

These results form the starting point for finding new solutions of the collisionless Boltzman equation which scale as functions of the invariants. So far, however, they have turned out to be useful only in rather special cases. (See Munier, *et al. Astron. Astrophys.* **78**, 65, 1979.)

18.5. Wave propagation in an inhomogeneous system

Since homogeneous systems are rather special, it is natural to ask how an inhomogeneity modifies the propagation and damping of waves. To do a linear perturbation analysis the inhomogeneity must itself be stable, at least on a timescale longer than the wave takes to change. The simplest case is when the inhomogeneity is the system itself. (A fragment within a fragment is more complicated.) An example is a spherical cluster of stars. Its kinetic pressure balances gravity. This determines the density and mean field. (Stars evaporating slowly from the system are ignored.)

Start with the linearized collisionless Boltzmann and Poisson equations for such a system. The equilibrium distribution function is $f^{(0)}(\mathbf{r}, \mathbf{v})$. Consider perturbations whose wavelengths are much less than the radius of the system. By taking Fourier–Laplace transforms, show that the 'dielectric function' (perhaps digravic function would be more suitable) of a cluster is

$$D(\mathbf{k}, z) = 1 + \frac{4\pi i G}{ksE} \int_{-\infty}^{\infty} du \int_{-s\infty}^{u} du' e^{i[\alpha(u') - \alpha(u)]} \frac{\partial \tilde{f}^{(0)}(u')}{\partial u'}. \tag{1}$$

Here $u = \mathbf{k} \cdot \mathbf{v}/k$, $\tilde{f}^{(0)}(u) = \int f^{(0)} dv_\perp$, $v_\perp$ is the component of $\mathbf{v}$ perpendicular to $\mathbf{k}$, $E = |\mathbf{F}^{(0)} \cdot \mathbf{k}|/k$, $s = \text{sign } \mathbf{F}^{(0)} \cdot \mathbf{k}$, $\mathbf{F}^{(0)}(\mathbf{r})$ is the equilibrium gravitational force, and $\alpha(u, z) = sE^{-1}(\frac{1}{2}ku^2 - uz)$ with z the Laplace transform parameter. Show that it is easier for waves to move inward toward the center of the cluster than for them to move out. (Severne & Kuszell, *Astrophys. and Space Sci.*, **32**, 447, 1975.)

18.6. Accretion of gas by stars

When a neutron star moves through a dense gas cloud its accretion is strongly influenced by pressure and radiation. Estimate the size and equilibrium density of the accretion column when pressure is important. If most of the kinetic energy of the infalling gas is radiated when it hits the surface of the star, what is the resulting luminosity? (Carlsberg, *Ap. J.*, **220**, 1041, 1978).

19

Bibliography

Of making many books there is no end; and much
study is a weariness of the flesh.

Ecclesiastes

In listing references, I have only included those most directly related to the text. The subject advances so fast that, at time t, there is not much point in listing a bibliography which is up-to-date as of a time $t - \tau$, for it will be out of date by about $t + \tau$. And these days the most informative complete bibliography is found in *Astronomy and Astrophysics Abstracts*. So with apologies to everyone, past and present, who is not mentioned, here are the references on which parts of the text are based, or to which the text refers directly. Other related references can be found in the sections on problems and extensions. They provide an entry into the literature rather than a summary of it.

There are a number of classical texts on stellar dynamics, orbit theory and celestial mechanics. Representative examples are:

Arnold, V.I., 1978. *Mathematical Methods of Classical Mechanics* (New York: Springer-Verlag).

Chandrasekhar, S., 1960. *Principles of Stellar Dynamics* (New York: Dover).

Chandrasekhar, S., 1961. *Hydrodynamic and Hydromagnetic Stability* (London: Oxford UP).

Chandrasekhar, S., 1969. *Ellipsoidal Figures of Equilibrium* (New Haven: Yale UP).

Hagihara, Y., 1970, 1972. *Celestial Mechanics* (Cambridge Mass: MIT Press).

Hamilton, W.R., 1834. On a general method in dynamics *Phil. Trans. Roy. Soc.*, Pt. II, **124**, 247.

Hayli, A. (ed.), 1975. *Dynamics of Stellar Systems* (Boston: D. Reidel).

Jeans, J. H. 1919. *Problems of Cosmogony and Stellar Dynamics* (London: Cambridge UP).

Jeans, J. H., 1928. *Astronomy and Cosmogony* (London: Cambridge UP).

Kurth, R., 1957. *Introduction to the Mechanics of Stellar Systems* (London: Pergamon).

Lecar, M. (ed.), 1972. *Gravitational N-Body Problem* (Boston: D. Reidel).

Mihalas, D. & Binney, J., 1981. *Galactic Astronomy* (San Francisco: W.H. Freeman).

Ogorodnikov, K.F., 1965. *Dynamics of Stellar Systems* (New York: Pergamon).

Peebles, P.J.E., 1980. *The Large Scale Structure of the Universe* (Princeton: Princeton UP).

Smart, W.M., 1938. *Stellar Dynamics* (London: Cambridge UP).

Szebehely, V., 1967. *Theory of Orbits* (New York: Academic Press).

These are also referred to elsewhere throughout this book.

Descriptions of the plasma analog of some gravitational many-body processes can be found in most modern texts of plasma physics. No detailed historical account of the development of our subject exists although Ogorodnikov gives a very brief summary of some early aspects. It would make an interesting story, especially regarding cross fertilizations with celestial mechanics, plasma physics, and other many-body problems.

Section 2:

Cox, D.R. & Miller, H.D., 1965. *The Theory of Stochastic Processes* (London: Chapman and Hall).

Jeans, J.H., 1913. 'On the 'kinetic theory' of star-clusters', *MNRAS*, **74**, 109.

Lecar, M. & Cruz-Gonzales, C., 1972. 'A numerical experiment on relaxation times in stellar dynamics', in '*Gravitational N-Body Problem*', M. Lecar (ed.), p. 131 (Boston: D. Reidel).

Rybicki, G.B., 1972. 'Relaxation times in strictly disk systems', in *Gravitational N-Body Problem*, M. Lecar (ed.), p. 22 (Boston: D. Reidel).

Section 3:

Einstein, A., 1955. *Investigations on the Theory of the Brownian Movement* (New York: Dover).

Fox, R.F., 1978. 'Gaussian stochastic processes in physics', *Phys. Reports*, **48**, 179.

Ito, K., 1951. 'On stochastic differential equations'. *Mem. Amer. Math. Soc.*, No. 4.

Langevin, P., 1908. 'On the theory of Brownian motion', *C.R. Acad. Sci. Paris*, **146**, 530.

Section 5:

Landau, L.D. & Lifshitz, E.M., 1976. *Mechanics* (New York: Pergamon).

Michie, R.W., 1961. 'Structure and evolution of globular clusters', *Ap. J.*, **133**, 781.

Michie, R.W., 1963. 'On the distribution of high energy stars in spherical stellar systems', *MNRAS*, **125**, 127.

Rosenbluth, M.N., MacDonald, W.H. & Judd, D.L., 1957. 'Fokker–Planck equation for an inverse-square force', *Phys. Rev.*, **107**, 1.

Section 8:

Jeans, J.H., 1915. 'On the theory of star-streaming and the structure of the universe', *MNRAS*, **76**, 70.

Jeans, J.H., 1919. *Problems of Cosmogony and Stellar Dynamics*, pp 230–6.

Jeans, J.H., 1922. 'The motions of stars in a Kapteyn-universe', *MNRAS*, **82**, 122.

Section 10:

Chandrasekhar, S. & Lee, E. P., 1968. 'A tensor virial-equation for stellar dynamics', *MNRAS*, **139**, 135.

Grad, H., 1958. 'Principles of the kinetic theory of gases', in *Handbook der Physik*, S. Flugge (ed.) Vol. XII (Berlin: Springer-Verlag, p. 205).

Section 11:

The BBGKY hierarchy is described in many plasma physics texts. The approach of this section follows:
Saslaw, W.C., 1972. 'The kinetics of gravitational clustering', *Ap. J.*, **177**, 17, which contains further references.

Section 12:

Haggerty, M. & Severne, G., 1976. In *Advances in Chemical Physics*, Vol. 35, p. 119, I. Prigogine & S. Rice (eds.) (New York: Wiley and Sons).
Prigogine, I., 1962. *Nonequilibrium Statistical Mechanics* (New York: Interscience).

Section 13, 14:

Saslaw, W.C., 1975. 'The ejection of massive objects from galactic nuclei: gravitational scattering of the object by the nucleus', *Ap. J.*, **195**, 773.
White, S.D.M., 1978. 'Simulations of merging galaxies', *MNRAS*, **184**, 185.

Section 15:

Chandrasekhar, S., 1961. *Hydrodynamic and Hydromagnetic Stability*, p. 588 (London: Oxford UP).
von Hoerner, S. & Saslaw, W.C., 1976. 'The evolution of massive collapsing gas clouds', *Ap. J.*, **206**, 917.
Jeans, J.H., 1928. *Astronomy and Cosmogony*, p. 345 (London: Cambridge UP).
Lebedev, V.I., Maksumov, M.N. & Marochnick, L.S., 1966. 'Collective processes in gravitating systems, I.', *Sov. A.J.*, **9**, 549.

Section 16:

Landau, L.D., 1946. 'On the vibrations of the electronic plasma'. *J. Phys.* (USSR), **10**, 25.
Lynden-Bell, D. 1962. 'The stability and vibrations of gas of stars', *MNRAS*, **124**, 23.
Marochnick, L.S., 1968. 'A test star in a stellar system', *Sov. A.J.*, **11**, 873.
Noerdlinger, P.D., 1960, 'Stability of uniform plasmas with respect to longitudinal oscillations', *Phys. Rev.*, **118**, 879.
Penrose, O., 1960. 'Electrostatic instabilities of a uniform non-Maxwellian plasma', *Phys. Fluids*, **3**, 258.

Section 17:

Allen C.W., 1963. *Astrophysical Quantities* (London: Athlone Press).
Danby, J.M.A. & Camm, G.L., 1956. 'Statistical dynamics and accretion', *MNRAS*, **117**, 50.
Eddington, A.S., 1926. *The Internal Constitution of the Stars* (London: Cambridge UP).

Ikeuchi, S., Nakamura, T. & Takahara, F., 1974. 'Collective instabilities of self-Gravitating systems – infinite homogeneous case', *Prog. Theoret. Phys.*, **52**, 1807.

Landau, L.D. & Lifshitz, E.M., 1976. *Mechanics* (New York: Pergamon).

Mathews, W.G., 1972. 'Collapse in galactic nuclei', *Ap. J.*, **174**, 101.

Saslaw, W.C. & De Young, D.S., 1972. 'The ejection of massive objects from galactic nuclei: interactions between the massive object and the galactic gas', *Astrophys. Lett.*, **11**, 87.

Splitzer, L., 1968. *Diffuse Matter in Space* (New York: Wiley Interscience).

Sweet, P.A. 1963. 'Cooperative phenomena in stellar dynamics', *MNRAS*, **125**, 285.

PART II

Infinite inhomogeneous systems – galaxy clustering

> And much harder it is to suppose all the particles in an infinite space should be so accurately poised one among another as to stand still in a perfect equilibrium. For I reckon this as hard as to make, not one needle only, but an infinite number of them (so many as there are particles in an infinite space) stand accurately poised upon their points.
>
> *Newton*

Having introduced the basic descriptions of gravitational many-body physics, it is time to attend to some astronomical applications. Many basic processes still remain to be explored, but they are best introduced in their astrophysical contexts.

Occasionally in Part I I have mentioned that infinite homogeneous systems – indeed all homogeneous gravitational systems – are anomalies, idealizations which do not exist in nature. However convenient they may be for mathematical analyses they are too unstable to represent anything we see, except as a first approximation. Newton recognized this and described it qualitatively in his letter to Bentley. Jeans (see Section 15) formulated it quantitatively for a static universe. Important complications, and a new richness of results, occur in the expanding Universe. This is a fundamental problem, for it begins to describe how matter is distributed around us on the largest scales.

20

How does matter fill the Universe?

20.1. General description

One of the major mysteries of the Universe is the inhomogeneous distribution of matter throughout space. Another major mystery is the homogeneous distribution of matter throughout space. This apparent contradiction–puzzle–paradox typifies our uncertainty about the origin of structure in the Universe. Both statements are true. This may make them a 'great truth' in Niels Bohr's sense of the phrase: a statement whose opposite is also true.

To measure the inhomogeneities of matter we need only look around us. Indeed, the closer we look the greater the density contrasts usually seem to be. We, ourselves, mainly made of water, are about as dense as our planet and a little denser on average than our sun. Each of these is about 10^{23} times as dense as the average smoothed out density in the solar neighborhood, which is a typical average density within large galaxies. In turn, the average density within large galaxies is about 1000 times that within large clusters of galaxies. If we average over regions containing many large clusters the density drops by another factor of 1000. These relationships are illustrated by Table 2 and Figure 22, both from excellent surveys by de Vaucouleurs (1970, 1971). Although the masses and radii astronomers obtain for large systems are often uncertain by factors of at least two, such uncertainty is hardly visible over the enormous range of these logarithmic scales. The main result is clear: average density decreases with increasing volume.

Does this density decrease ever stop? Can we ascribe an average density to the entire visible Universe? In a fundamental sense, we do not yet know the answer to this question. For as we look further into space, and earlier in time, the quality and rate of our information both decrease. We receive relatively fewer, more redshifted photons from the distant objects. Linear resolution is poorer. More photons have been scattered or absorbed by intervening gas and dust. So we cannot measure their properties as accurately as those of nearby objects in the time available on large telescopes. If we live in a hierarchial Universe with distinct clusters on every scale, its average density would tend to zero as the scale was increased. We know there are

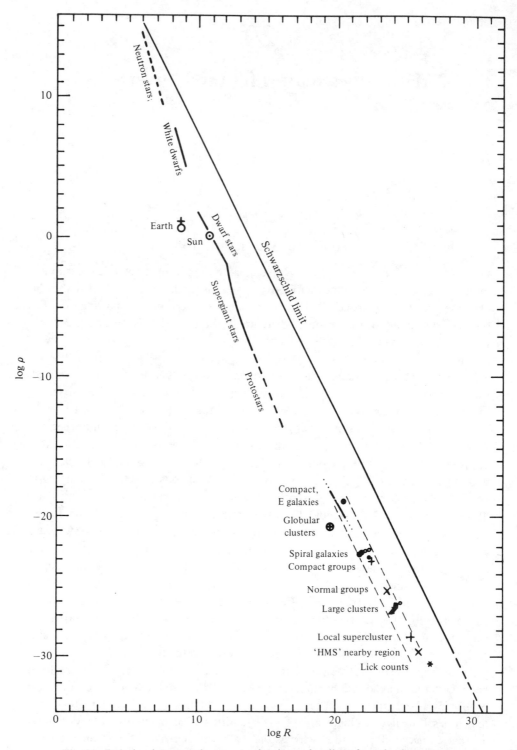

Fig. 22. Relation between the average density and radius of gravitating systems.
Thin dashes on the lower right show the range of densities obtained by applying
the virial theorem to clusters of stars and galaxies. (From de Vaucouleurs, 1970.)

Table 2. *Mass–radius–density data (from de Vaucouleurs, 1970, who also gives detailed references)*

Class of objects	Examples	log M (g)	log R (cm)	log ρ (g cm⁻³)
Neutron stars	pulsars	33.16	5.93	14.75
		32.54	7.44	9.60
White dwarfs	L930–80	33.45	8.3	7.93
	αCMaB	33.30	8.77	6.37
	vM2	32.90	9.05	4.13
Main sequence stars	dM8	32.2	9.95	1.76
	Sun	33.30	10.84	0.15
	A0	33.85	11.25	− 0.55
	05	34.9	12.1	− 2.0
Supergiant stars	F0	34.4	12.65	− 4.2
	K0	34.4	13.15	− 5.7
	M2	34.7	13.75	− 7.2
Protostars	IR	35.3?	16.2?	− 13.9?
Compact, dwarf, elliptical galaxies	M32, core	41.0	19.5?	− 18.1
	M32, effective	42.5	20.65	20.0
	N4486-B	43.4	20.5	− 18.75
Spiral galaxies	LMC	43.2	21.75	− 22.65
	M33	43.5	21.8	− 22.5
	M31	44.6	22.3	− 22.9
Giant elliptical galaxies	N3379	44.3	22.0	− 22.35
	N4486	45.5	22.4	− 22.3
Compact groups of galaxies	Stephan	45.5	22.6	− 23.1
Small groups of spirals	Sculptor	46.2	24.1	− 26.7
Dense groups of ellipticals	Virgo E, core Fornax 1	46.5	23.7	− 25.2
Small clouds of galaxies	Virgo S Ursa major	47.0	24.3	− 26.5
Small clusters of galaxies	Virgo E	47.2	24.3	− 26.3
Large clusters of ellipticals	Coma	48.3	24.6	− 26.1
Superclusters	Local	48.7	25.5	− 28.4

departures from homogeneity on scales of at least ∼ 50 Mpc and possibly much larger. But then the evidence blurs.

On the largest scales a new type of evidence enters. It concerns not the distribution of galaxies but the uniformity of the cosmic microwave background. This radiation has an essentially blackbody spectrum and is presumed to be a relic of an early, hot, dense phase of the expanding Universe. Although it once dominated the total energy density of the Universe, it now contributes little. Even so, it would still be perturbed by the gravitational potential wells of large scale inhomogeneities in the matter distributions. Fractional density perturbations of order $\delta\rho/\rho$ should lead to temperature perturbations of order $\delta T/T$ in the microwave background. These are not found at the 0.1% level, but may exist at lower levels. The angular scales over which these measurements are made involve contributions from very large linear scales. This suggests quite strongly that there is an average density over scales that are a substantial fraction of the size of the visible Universe.

What value could this average density have? The estimate from galaxy counts on the largest scale, beyond superclusters, is roughly $\rho_{gal} \approx 3 \times 10^{-31}$ g cm^{-3}. Now it becomes useful to have a standard of comparison for this density. In the standard Einstein–Friedmann cosmologies there is a critical density,

$$\rho_c = \frac{3}{8\pi G} H^2, \tag{20.1}$$

which divides closed universes which eventually collapse from open universes which expand forever. In this expression, derived in most elementary cosmology texts, $H = \dot{R}/R$ is the Hubble constant. The radius of curvature R is a measure of the scale of the Universe and is just a function of time in the models we consider here, which are homogeneous averaged over large scales. The density of the Universe determines the strength of gravity on a global scale, and thus is related to the expansion rate.

The Hubble constant (which generally depends on time) can be measured by looking at the velocity field of galaxies. Galaxies at distances $r \ll R$ from any point in space move away from that point with a radial velocity $V_r = rH$ as a result of the cosmic expansion (see Section 23). Generally, galaxies have local peculiar velocities as well as the expansion velocity. It is not very easy to separate peculiar velocities out from the total flow, nor is it easy to measure r independently of v in order to determine H. Elementary astronomy texts discuss the determination of H in detail. We will use $H = 50$ km s^{-1} Mpc^{-1} throughout this book. Very little of what we discuss depends on its exact value, which probably lies between 40 and 100 km s^{-1} Mpc^{-1}. For $H = 50$, Equation (20.1) gives $\rho_c \approx 4 \times 10^{-30}$ g cm^{-3}, which is about ten times the observed density on the largest scales.

At this stage of the argument, astronomers seem to divide into two groups. One group believes that since ρ_c is the only density especially singled out by general relativistic world models, it is likely to be the case in our actual Universe. The critical density is thus 'predicted' by the special properties of those models – properties such as Mach's principle, and the possibility that the Universe can be recycled through a series of expansions and contractions. An infinite series of cycles would eliminate the problem of an origin. Besides, universes with different properties might emerge from each singular dense phase and there would be many chances to produce life like us, contemplating its origin. Then, too, the question of whether entropy can be prevented from increasing indefinitely as the cycles continue lends itself to much amusing speculation involving changes in the direction of time. This is all such fun that one wished it were true. The result has been to stimulate searches for the 'missing mass' which are reminiscent of medieval quests for the holy grail. A difference may be that the holy grail could only be found by a knight displaying true modesty and humility.

The more astronomers look for the missing mass, the more it isn't there. Few possibilities have been left unplowed. Intergalactic gas (ionized and molecular), dwarf galaxies, intergalactic stars, rocks, black holes, massive neutrinos, undetected elementary particles, haloes around galaxies – for none is there yet compelling evidence that it accounts for the missing mass. Some day, perhaps, we will find it. Meanwhile a second group of astronomers is prepared to accept that $\Omega \equiv \rho/\rho_c < 1$ and follow its implications.

Whatever the value of Ω, we have the problem of explaining inhomogeneity on small scales (relative to $R(t)$) and homogeneity on large scales. Here, also, two general points of view have emerged. One considers the early Universe to have been completely homogeneous. During its expansion, instabilities arose to produce the structure we now see. The trouble with this view is that we do not know any physical instabilities which arise naturally and are strong enough to do the job in standard cosmological models. Section 21 will show how Jeans instability is weakened by the expansion. The alternative, 'practical' point of view is that sufficient perturbations were already present in the initial conditions, for reasons we do not understand.

Initial perturbations may be isothermal, adiabatic, or have an arbitrary combination of these properties. Figure 23 illustrates various possibilities schematically and we next describe them briefly. Detailed reviews (e.g., Weinberg, 1974, Peebles, 1980) are readily available. The material filling the early Universe, at redshifts $z \gg 10^3$, is presumed to be a continuous nearly equilibrium fluid of matter and radiation. The matter may consist of many species – baryons, electrons, neutrinos, and at very high temperatures and densities during the first few minutes of the expansion from the singular state, more exotic particles. Electromagnetic radiation has an essentially blackbody spectrum, and gravitational radiation may also be present. For the first few minutes all physical interactions – strong, weak, electromagnetic, and gravitational are important; later, gravity and electromagnetism dominate.

Isothermal perturbations have an inhomogeneous particle density embedded in a uniform radiation background. In most realistic cosmological models at redshifts $z \gg 10^5$, radiation dominates the energy density and so determines the temperature. Since this temperature is little affected by baryon inhomogeneities, such perturbations are called isothermal.

Adiabatic perturbations, on the other hand, involve both matter and radiation fluctuating together with the same phase. The entropy per baryon, $\propto T^3\rho^{-1}$, is constant so that $3\delta T/T = \delta\rho/\rho$. Incidentally, this entropy per baryon, which is also the number of photons per baryon, remains constant during equilibrium expansion of the Universe. The radiation temperature decreases as R^{-1} and the density as R^{-3}. Observations give this dimensionless constant a value of about 10^8. No-one really

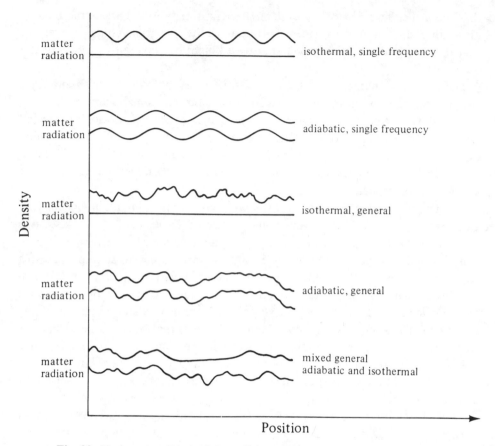

Fig. 23. Various possible initial conditions for gravitational perturbations.

knows why it should have this value – far from the other dimensionless cosmological quantities which are of order 1, 10^{40} or 10^{80}. Perhaps it supplies a critical clue to the earliest state of the Universe.

Adiabatic and isothermal perturbations evolve differently as the Universe expands. For both, the period around $z \approx 10^3$ when the Universe cools below about 3000 K is a critical juncture. Above this temperature, matter is ionized and free electron scattering tightly couples changes in the gas and radiation. Below this temperature, neutral atoms form and matter decouples from the radiation. The timescale over which decoupling occurs is about a Hubble time H^{-1} at $z \approx 10^3$.

Before decoupling (often called recombination, even though there has never been significant previous combination of these electrons and ions), isothermal perturbations are locked in place. Overdense regions cannot move relative to the general expansion because electron scattering against the uniform radiation quickly damps any peculiar velocities. After decoupling, the isothermal perturbations are free to respond to gravitational and pressure forces. Gas perturbations which exceed the Jeans length (15.11) at $z \approx 10^3$ involve a mass $M_J \approx 10^6 \, \mathrm{M}_\odot$. These begin to contract. Because the initial spectrum of these perturbations is unknown – too much dissipation occurred between $z \approx 10^3$ and the present – many models with differing assumptions have been constructed. Most of these agree on one basic result. Clumps of about $10^6 \, M_\odot$ form first. Their mutual gravity then causes them to aggregate into larger and larger systems. Eventually they form galaxies and clusters of galaxies. The internal development of each level of clumping, however, is completely model dependent. In some models supermassive stars form and explode, reheating surrounding gas. In other models fairly normal star formation occurs. So far, there do not seem to be any really robust and definitive observational tests to distinguish among these internal developments. Many have enough flexibility to be made consistent with the observations, but none is uniquely demanded by the observations. All of them, though, must gravitate. So we will examine the nature of their clustering more closely.

Adiabatic initial perturbations have a more complicated behavior. They do not grow before decoupling, but tend to oscillate if they are smaller than the causal horizon (approximately if $\lambda < cH^{-1}$). In some cosmological models when the radiation energy density ($\propto R^{-4}$) becomes less than the matter density ($\propto R^{-3}$) before decoupling, adiabatic perturbations can evolve to resemble isothermal ones. For both cases the main perturbation in the total energy density is in the matter rather than the radiation. If this does not occur, then the radiation viscosity can damp adiabatic perturbations during the decoupling era. Viscous damping operates faster on shorter length scales, so only very large perturbations survive. The survivor's size is model dependent, but typically involves masses between $10^{13} - 10^{15} \, M_\odot$. Thus clusters of galaxies would form first in this picture. They would fragment into galaxies which, in turn, fragment into stars. Many models have been made of this process, but the gas dynamical processes are more difficult to calculate definitively than the simpler gravitational clustering processes.

Naturally, if both isothermal and adiabatic perturbations are initially present, then both clustering and fragmentation can occur. The situation may be made as complicated as one wishes. At this early stage of our understanding, however, it seems more useful to ask: 'What are the minimum initial conditions we must impose on the early Universe to account for the structure we see today?' To add adiabatic perturbations which decay during decoupling is a bit like growing a flower garden by planting seeds of weeds and flowers, then adding a heavy dose of weed killer. To justify this we would have to ask whether we see any weeds today. Shapes of the largest scale structures may provide important clues, particularly if they cannot be produced by gravitational clustering. This is another reason for understanding simple gravitational clustering in an infinite system.

20.2. Quantifying the distribution

To develop the theory of gravitational clustering, and compare its results with observations, we need a quantitative description of the large scale distribution of matter. One main approach to this problem creates models of the distribution and then finds the best fit of the models' parameters to the observations. The other main approach deals with n-particle correlation functions between galaxies. Both these approaches are related and neither proves a completely useful description by itself. We shall see that although correlation functions are closely related to the underlying gravitational physics, they are awkward for depicting large scale properties of inhomogeneities. On the other hand, models suitable for large scale descriptions have little direct relation to the basic physics. As a result, I shall opt to emphasize the physical understanding of correlation functions in the next few sections and discuss models occasionally when they seem useful.

First, to clarify the difference between the spirit of modeling and of correlation analysis, consider some simple examples of models. In the late 1930s Holmberg (1940) introduced the multiplicity function. For a given sample, one counts the number of single galaxies, doubles, triples and so on to derive relative frequencies of different size groups. A moment's reflection shows that such a frequency distribution depends critically on the territory ascribed to a group and on the magnitude limit of the sample. Such are the parameters of even a simple model. Quite separate is the question of whether these groups are gravitationally bound.

A more complicated type of model starts by specifying the number of clusters containing between N and $N + dN$ galaxies. Then it parametrizes the distribution of cluster radii and the density runs within clusters, perhaps by Gaussians or power laws. For added information, it need not assume spherical clusters, and can supply them in a variety of shapes. Finally it is necessary to give the distribution of cluster centers. A three-dimensional distribution with the desired parameters is then computed, projected onto the sky, and compared with what we see. Varying the model's properties gives a feel for its accuracy and robustness. Uniqueness would be too much to hope for. Although any such model could describe an instantaneous 'snapshot' of galaxy clustering, it is not easily tested for dynamical consistency. Could it have evolved from a realistic initial state? The only way to find out is to evolve many initial states and see if their statistics resemble those of the model.

The very simplest model – a uniform Poisson distribution – is often used for comparisons. As an initial state, to be evolved dynamically, it has the virtue of presupposing the least starting information. As a comparison with what we see, it provides a standard to measure global non-uniformity. For example, divide the distribution of galaxies on the sky down to some specified limiting magnitude into m equal areas. Each area contains n galaxies, and the average is $\langle n \rangle$. For large values of $\langle n \rangle$, the Poisson distribution approaches a Gaussian for the number of fields which contain n galaxies

$$N(n) = \frac{m}{(2\pi \langle n \rangle)^{1/2}} e^{-(n-\langle n \rangle)^2/2\langle n \rangle}, \tag{20.2}$$

with the total number of galaxies being $N_T = m\langle n \rangle$. The dispersion of this distribution, $\langle n \rangle^{1/2}$, may be compared with the dispersion of the m observations:

$[\Sigma_{i=1}^{m}(n_i - \langle n \rangle)^2/m]^{1/2}$. Their ratio measures the departure from a uniform random distribution. We will develop a fundamental theory based on this approach in Section 34.

Another measure, the standard χ^2 test, uses information from the entire distribution by calculating

$$\chi^2 = \sum_{n=1} \frac{(N(n)_{\text{obs}} - N(n)_{\text{expec}})^2}{N(n)_{\text{expec}}}. \tag{20.3}$$

This measures the 'integrated' departure of the observed histogram $N(n)_{\text{obs}}$ from the expected one (e.g., for the Gaussian distribution). If the two histograms are the same, $\chi^2 = 0$. The greater their difference, the larger the value of χ^2. As the value of χ^2 increases, the probability that the observed distribution can represent a random sample from the expected Poisson or Gaussian distribution decreases rapidly, as discussed in elementary statistics texts. If the maximum value of χ^2 depends sensitively on the size and shape of the areas used for counting, it may give an idea of these characteristic parameters for clustering.

Many variants of these tests were applied to the observed galaxy distribution in the period from about 1930 to 1970. They all indicated significant departures from a uniform distribution. Details of these departures, however, were obscured by two main effects. The first, as mentioned before, was the wide range of parameters available for descriptive models. The second, equally fundamental, is the interstellar obscuration caused by dust. Photographs of nearby galaxies had long showed their irregular, clumpy distributions of gas and dust. There was no reason to suppose our galaxy to be qualitatively different. But it took astronomers a long time to come to grips with the effects of inhomogeneous interstellar extinction on the counts of external galaxies. This was chiefly due to lack of detailed information.

Although the problem is now fully realized, there is no complete solution to the disentanglement of real clumping from the apparent clumping caused by obscuration. Apart from decreasing the magnitudes of external galaxies, dust also makes them appear smaller by removing the low surface brightness parts of their images. This makes faint galaxies hard to distinguish from stars. These problems are ameliorated somewhat by independent estimates of the dust distribution from ratio frequency observations of their associated molecular clouds. These, as well as star counts, show there is usually much less dust at high galactic latitudes, well above the plane of our galaxy. Even though these clearings are incomplete, and depend on galactic longitude, galaxy counts at high galactic latitude are generally more reliable measures of their intrinsic distribution. Studies of these high latitude counts in the last 30 years also showed significant clumping.

Having found this clumping it became necessary to measure it more precisely, even if less completely. Correlation analyses to do this were introduced by Neyman, Scott & Shane (1953) for the two-dimensional distribution and by Kiang & Saslaw (1969) for the three-dimensional distribution. The basic idea is quite simple. First find the average density over a very large scale. Next, divide the area (or volume) into

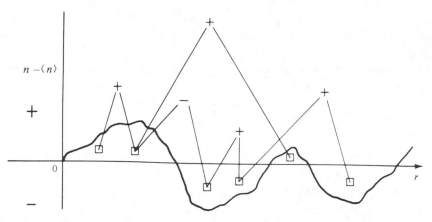

Fig. 24. Schematic illustration of the signs of density correlations.

many small cells of the same size. Then find the difference between the density in each cell and the average – the net density. Multiply the net densities of both cells in a pair of given separation and average this over all pairs of the given separation. Finally, do this for all separations which have sufficient pairs within the sample for reasonable statistics. The result is the pair correlation as a function of distance. It is the probability – over and above the uniform random probability – of finding a galaxy in a region a distance r from another region in which there is already known to be a galaxy.

Similarly, one can measure the three, four, and higher galaxy correlation functions of a sample. If there is a clump or cluster in the sample, a density excess in one part of the cluster will tend to correlate positively with the density excesses of nearby parts. Beyond the scale of the clump, correlations will tend to be negative as comparisons occur between regions of enhanced and decreased density. A uniform random distribution would give zero correlation. Density deficits within holes correlate positively. This is shown schematically in Figure 24. Thus the scale of the pair correlation function characterizes the scale of clustering.

But there is a subtlety in this approach which sometimes causes utter confusion. It is that clustering is different from inhomogeneity. A homogeneous system can be clustered! All this requires is that the clustering be the same everywhere throughout the system. Perhaps the simplest case to imagine is an infinite one-dimensional lattice with point galaxies at unit intervals along the line (or at the vertices of unit cubes in the slightly more realistic three-dimensional case). Then the average density of galaxies is zero (because in the limit where the region is a point there are an infinite number of points without galaxies) and the correlation function has a series of δ function spikes at $r = 1, 2, 3$, etc. Averaged over scales $\gg 1$, the system would be homogeneous, and yet it has correlations over all scales. Now suppose every galaxy were to split into a variable number of galaxies, and all of them remain close ($\Delta r < \frac{1}{2}$) to the site of their original galaxy. The correlation spikes would broaden and become finite. Regions of negative correlation would appear, but homogeneity on large scales would be preserved. Moreover, providing the splitting and positions of the smaller galaxies are random (within the Δr constraint), the system will also appear statistically homogeneous on smaller scales. The three-dimensional plasma continuum analogs are of Debye spheres where electron positions are positively correlated in a homogeneous plasma. Everywhere you look you have the same probability of finding the center of a Debye sphere. Similarly, if the pair correlation function of galaxies does not depend on where you look, then in this sense the Universe can be both clustered and homogeneous.

Because the pair correlation function contains very limited information about the galaxy distribution and is hard to observe over large distances, it is not really a very stringent measure of homogeneity. A few large clumps, each with many galaxies, do not contribute much to the averaged pair correlation function. The nature of high order correlation functions, say the 100-galaxy correlation function, determines large scale homogeneity. So we have come full circle: to describe inhomogeneity we need to return to models of clusters!

Despite its limitations, the pair correlation function over short distances ($\lesssim 20$ Mpc) has proven to be very useful in gaining theoretical understanding of self-gravitating systems. Subsequent sections describe these theoretical results. First we turn briefly to its observational analysis.

20.3. Measurement of the galaxy pair correlation function

In 1969, Totsuji and Kihara determined the two-galaxy correlation function for $r \lesssim 20$ Mpc. Subsequent analyses (see Peebles, 1980, for details) confirmed and extended their results. Here I sketch their approach and summarize several later results.

We start with the quantity which is measured most directly: the angular two-galaxy correlation function on the sky. This two-dimensional covariance, $W(\theta)$, is defined by

$$P_{12}(\theta)d\Omega_1 d\Omega_2 = \langle n \rangle^2 [1 + W(\theta)] d\Omega_1 d\Omega_2. \tag{20.4}$$

Here $P_{12}d\Omega_1 d\Omega_2$ is the joint probability of finding a galaxy in the solid angle $d\Omega_2$ and another in $d\Omega_1$ separated by an angle θ, and $\langle n \rangle$ is the average surface density. If the joint probabilities were independent, $W(\theta)$ would be zero. P_{12} is essentially the number of galaxy pairs an angular distance θ apart. So, for a sample which divides the sky into equal area cells, W can be estimated from

$$W(\theta) = \frac{\langle n_i n_j \rangle_\theta}{\langle n \rangle^2} - 1. \tag{20.5}$$

The angular brackets with θ subscript indicate an average over all cells i and j whose centers have separations between $\theta - \Delta\theta/2$ and $\theta + \Delta\theta/2$, with $\Delta\theta \ll \theta$ chosen for statistical convenience. Analogous estimators are sometimes used if the sample is divided into rings.

The obvious normalization $\langle n \rangle^2$ of (20.5) is not the only, or even the best, choice.

Instead of normalizing by this average surface density for the whole sample, one often uses $\langle n_i \rangle \langle n_j \rangle$ which normalizes only to the region of the sample used to find $W(\theta)$ at that separation. Another alternative local normalization would be $\langle \frac{1}{2}(n_i + n_j) \rangle^2$, or any reasonable average function of order n^2. These have the advantage of making some (weighted) correction for sample inhomogeneities due to observing interstellar dust or systematic errors in galaxy counting across a plate, for example. More sophisticated methods of doing the counting would be to divide the area of the sample up in many different ways and average the results of $W(\theta)$ over this ensemble.

Measurements of $W(\theta)$ must also contend with the boundary effects of a finite sample (see Kiang & Saslaw, 1969, Hewett, 1982). The safest way to eliminate boundary effects would be to exclude all pairs with a member closer to the boundary than the largest separation θ_{max}. Unfortunately the remaining sample is usually so small that its statistical noise becomes high: different samples of the same underlying distribution would differ substantially. A compromise approach would be to use only cells farther than θ from the boundary for each value θ. Thus the statistical noise would be less for smaller θ. Each of these corrections has the disadvantage of varying sensitivity to inhomogeneities of differing scales. A third approach renormalizes the number of galaxy pairs by the number of cell pairs within the sample boundary area. This formally decreases the statistical noise by assuming that the information contained in the boundary area is the same as that already obtained from the interior.

Clearly, none of the approaches to estimating $W(\theta)$ is entirely satisfactory. It is best to have a sample which is large and true, without extraneous inhomogeneities. Lacking idealized data, it is still possible to estimate the bias in the real data (Sharp, 1979). The idea is to cross correlate the data with a set of points having a Poisson distribution over the same area. The cross correlation function $W_{pg}(\theta)$ between the points and the galaxies is given by

$$P_{pg}(\theta)d\Omega_1\,d\Omega_2 = \langle n_p \rangle \langle n_g \rangle [1 + W_{pg}(\theta)]d\Omega_1\,d\Omega_2, \qquad (20.6)$$

analogous to (20.4). Now P_{pg} is the joint probability of finding a point in $d\Omega_1$ and a galaxy in $d\Omega_2$. If the two species have unrelated distributions, then $W_{pg} = 0 = W_{gp}$. Were we to center on each galaxy and find $W_{gp}(\theta)$ by averaging over the whole sample of galaxies, we would expect $W_{gp}(\theta) = 0$ since the points are distributed uniformly randomly around each galaxy no matter what the distribution of galaxies is. Departures of $W_{gp}(\theta)$ from zero therefore describe the bias introduced into $W(\theta)$ by the analysis technique (form of cells, treatment of boundary areas). Similarly, by centering on each point and finding $W_{pg}(\theta)$ – which generally is not expected to be zero since it is a measure of the inhomogeneity of the galaxy distribution – we can estimate the bias which introducing a cell structure produces. Thus by forming a corrected covariance,

$$W_{corr}(\theta) = W(\theta) - W_{pg}(\theta) - W_{gp}(\theta), \qquad (20.7)$$

the results of different analysis techniques can be compared and generally brought into agreement.

The spatial (three-dimensional) galaxy pair correlation function is more directly related to the physics of galaxy clustering, but somewhat harder to determine. The uncertainties I've just described are compounded with the uncertainty of galaxy distances. Larger and more complete redshift samples are becoming available, so this is not now a particularly serious problem. Without redshift information we would have to rely on luminosity distances, as did the early determinations of galaxy correlations.

Determinations of the spatial pair correlation function usually make the simplifying assumptions that it is isotropic and depends only on the relative separation r of two galaxies. Given that there is a galaxy at some point in space, the probability of finding a second galaxy a distance r to $r + dr$ away can then be written

$$P(r)dr = 4\pi \langle (n(t)) \rangle [1 + \xi(r)]r^2 dr. \qquad (20.8)$$

As the Universe expands, its average spatial density of galaxies $\langle n(t) \rangle$ changes. Equation (20.8) is also a bit simpler than its analog (20.4) because we have written it for the conditional probability rather than the joint probability. We should bear in mind that these assumptions of isotropy and homogeneity – a statement of the cosmological principle on small scales – are only idealizations for making the analysis easier. They leave out a lot of information contained in the actual distribution, or in more detailed models. However, the resulting $\xi(r)$ does characterize some essential properties of gravitating systems.

An estimator for $\xi(r)$, analogous to (20.5) can be constructed by examining the fluctuations of density from their average value, $\Delta n_1 = n(\mathbf{r}_1) - \langle n \rangle$. Averaging these fluctuations over all pairs of cells, we may write

$$\langle \Delta n_1 \Delta n_2 \rangle = \langle n \rangle \delta(\mathbf{r}_1 - \mathbf{r}_2) + \langle n \rangle^2 \xi(r), \qquad (20.9)$$

with $r = |\mathbf{r}_1 - \mathbf{r}_2|$. To see the meaning of this relation, suppose first that $\mathbf{r}_1 = \mathbf{r}_2$, $\xi = 0$, and integrate over $d\mathbf{r}_1 d\mathbf{r}_2$. The result is $\langle (\Delta N)^2 \rangle = \langle N \rangle$, where $\langle N \rangle$ is the average number of galaxies in a cell. These are the 'root N' fluctuations of Poisson statistics. For non-zero ξ, the integral $\langle n \rangle^2 \int \int \xi(r) d\mathbf{r}_1 d\mathbf{r}_2$ measures the average correlation $\langle \Delta N_1 \Delta N_2 \rangle$ between fluctuations of two volumes a distance r apart.

Many analyses of galaxy counts, starting with Totsuji & Kihara in 1969, have estimated $\xi(r)$. They find that for small separations $\xi(r)$ has the form of a power law

$$\xi(r) = ar^{-\gamma}. \qquad (20.10)$$

If r is measured in Mpc, a representative result is

$$\xi(r) \approx 50r_{\text{Mpc}}^{-1.8} \qquad (20.11)$$

between $0.2 \lesssim r \lesssim 20$ Mpc. This is for a Hubble constant of $H = 50 \, \text{km s}^{-1} \, \text{Mpc}^{-1}$. For other values of H the amplitude scales as $(50/H)^\gamma$ and the limits as $(50/H)$. The results vary from sample to sample with amplitudes ranging between 30 and 100 and exponents between 1.6 and 2.2. These differences arise from different identification and measuring techniques, sample size and completeness, and assumed luminosity functions, as well as from possible intrinsic differences. Some samples are restricted to certain types of galaxies and the presence of large clusters may affect the smaller samples. Nevertheless a value of $\gamma \approx 2$ seems to be a fundamental property of the distribution of galaxies.

At small distances $\xi \gg 1$ and the non-linear aspects of clustering dominate. A uniform random description fails completely. At large distances, ξ ceases to be a power law, but since samples have fewer pairs at these separations, the detailed behavior of ξ becomes hard to extract from the noise. Also at large separations, higher order correlation functions become relatively more important for describing the distribution. These also are hard to measure, but some progress has been made (see Peebles, 1980).

Although our largest telescopes easily find the positions of myriads of galaxies, our understanding of how to characterize their distribution is in its infancy. This is one of the most challenging problems of pattern recognition. We need to know, for example, how to characterize long chains of galaxies in an objective, quantitative – rather than impressionistic – manner. Then we could compare their structure with the structures which evolve dynamically from different sets of initial conditions and world models, providing some insight into their origin.

21

Gravitational instability of the infinite expanding gas

They [atoms] move in the void and catching each
other up jostle together, and some recoil in any
direction that may chance, and others become
entangled with one another in various degrees
according to the symmetry of their shapes and
sizes and positions and order, and they remain
together and thus the coming into being of composite things is effected.

Simplicius

What types of structure arise from the instability of homogeneous gravitating systems? This is the fundamental question we can now tackle by extending the techniques of Part I. Eventually we will compare the results with the observed features described in Section 20, and see how these simple initial conditions can account for the main properties of galaxy clustering, but not for the origin of the galaxies themselves.

Many detailed calculations, starting in the mid-1960s but based on the pioneering work of Lifshitz (1946) and Bonnor (1957), have examined the growth of local gaseous instabilities in expanding cosmological models. They treat interactions between gas and radiation, showing mainly how electron scattering damps the growth of large amplitude isothermal perturbations before the era of decoupling, and how radiation viscosity damps small adiabatic perturbations during decoupling. They treat interactions of gas with massive collisionless particles such as heavy neutrinos, and with seeds of finite amplitude gravitational instabilities such as black holes or other condensed pregalactic objects. They treat thermal instabilities of the gas, both during decoupling and after reheating by the first generation of pregalactic objects to form. They treat primordial turbulence in the gas as well as in the radiation field. They treat instabilities caused by quantum fluctuations and elementary particle phase transitions when the Universe was a mere 10^{-33} s 'old'. All this, and more, is behind the effort to understand the origin of galaxies. So far, it has not led to any generally accepted explanation.

This section contains a simple calculation whose purpose is to illustrate two points. The first is to show the fundamental difficulty in forming galaxies from a homogeneous gas. The second is to extend the results of Section 15.2, so we can compare them with the instabilities of a grainy system which describes galaxy clustering.

Section 15.2 showed that perturbations longer than the Jeans length, given by

(15.11), in a stationary homogeneous gas, have amplitudes which increase exponentially. If this were all that was needed, galaxy formation would be a snap. For then, the initial $\sqrt{N}$ fluctuations (which cause initial perturbations of amplitude $\delta\rho_0/\rho_0 \approx 10^{-34}$) in the $\sim 10^{68}$ atoms forming a galaxy, would become very non-linear after about 100 e folding times. Their growth could have begun at about 1% of the present age of the Universe. The essential thing wrong with this argument is that the Universe expands.

To see how expansion fights gravitational instability we consider conditions, after decoupling, which are otherwise most favorable for instability. Perturbations are not coupled to a radiation field, as radiation drag is negligible. Moreover, their scale is much less than the size $\sim ct$ of the Universe, their velocities are small, $v \ll c$, and their gravitational energy is less than their rest mass energy $GM/Rc^2 \ll 1$. Under these conditions Newtonian theory applies. We can take over Equations (15.2)–(15.4) and (15.9) directly. The density and pressure perturbations are now written in dimensionless form,

$$\rho = \rho_0(t)[1 + \rho_1(\mathbf{r}, t)], \tag{21.1}$$

$$P = P_0(t)[1 + P_1(\mathbf{r}, t)], \tag{21.2}$$

$$\phi = \phi_0(t) + \phi_1(\mathbf{r}, t), \tag{21.3}$$

$$\mathbf{v} = \mathbf{r}H(t) + \mathbf{v}_1(\mathbf{r}, t), \tag{21.4}$$

to normalize out their changing background values. Notice the essential difference between (21.4) and (15.8); this adds the expansion of the gas. $H(t) = \dot{R}/R$ is the Hubble parameter, with $R(t)$ the scale length for distances in the Universe. Fully relativistic universes are consistent with non-zero values for the zero order quantities and give the same results as these Newtonian calculations for the basic perturbations we examine here.

Linearizing the equations of momentum and mass conservations and of the gravitational potential (15.2)–(15.4) by substituting (21.1)–(21.4) and retaining just the first order terms gives

$$\frac{\partial \mathbf{v}_1}{\partial t} + H\left(\mathbf{v}_1 + r\frac{\partial \mathbf{v}_1}{\partial r}\right) = \nabla\phi_1 - \frac{\gamma P_0}{\rho_0}\nabla\rho_1 \tag{21.5}$$

since (15.9) is $P_1 = \gamma\rho_1$ in terms of the new dimensionless first order quantities;

$$\frac{\partial \rho_1}{\partial t} + \nabla\cdot\mathbf{v}_1 + Hr\frac{\partial \rho_1}{\partial r} = 0, \tag{21.6}$$

remembering to apply (15.2) to the zero order terms; and

$$\nabla^2\phi_1 = -4\pi G\rho_0\rho_1, \tag{21.7}$$

by inspection. Next take the divergence of (21.5) and eliminate ϕ_1 and $\mathbf{v}_1$ by substituting (21.6), (21.7) and the vector identity

$$\nabla\cdot\left(r\frac{\partial \mathbf{v}_1}{\partial r}\right) = r\frac{\partial}{\partial r}(\nabla\cdot\mathbf{v}_1) + \nabla\cdot\mathbf{v}_1. \tag{21.8}$$

The result,

$$\left(\frac{\partial}{\partial t} + 2H + Hr\frac{\partial}{\partial r}\right)\left(\frac{\partial \rho_1}{\partial t} + Hr\frac{\partial \rho_1}{\partial r}\right) = 4\pi G\rho_0\rho_1 + \frac{\gamma P_0}{\rho_0}\nabla^2\rho_1, \qquad (21.9)$$

describes the evolution of the fractional density perturbation $\rho_1(\mathbf{r}, t)$.

Part of this evolution is caused by the general expansion, and part is intrinsic to the instability. Separating these two effects becomes possible by transforming to a reference frame whose scale increases with the general expansion. Distances $\mathbf{x}$ in these co-moving coordinates are just the distances in the static frame normalized continuously by the expansion scale

$$\mathbf{x} = \frac{\mathbf{r}}{R(t)}. \qquad (21.10)$$

So if all the growth of size in the original static frame is due to the general expansion, then there is no growth in the co-moving frame. Since the co-moving frame measures only residual growth, we would expect the evolution equation (21.9) to become simpler in this frame, and it does

$$\frac{\partial^2 \rho_1(\mathbf{x}, t)}{\partial t^2} + 2\frac{\dot{R}}{R}\frac{\partial \rho_1}{\partial t} - 4\pi G\rho_0\rho_1 + \frac{\gamma P_0}{\rho_0 R^2}\nabla^2\rho_1(\mathbf{x}, t). \qquad (21.11)$$

In (21.11) ρ_1 is a function of $\mathbf{x}$ and its evolution follows from (21.9) and (21.10) bearing in mind that

$$\left(\frac{\partial}{\partial t}\right)_r = \left(\frac{\partial}{\partial t}\right)_x + \left(\frac{\partial \mathbf{x}}{\partial t}\right)_r\left(\frac{\partial}{\partial \mathbf{x}}\right)_t = \frac{\partial}{\partial t} - \mathbf{x}\frac{\dot{R}}{R}\frac{\partial}{\partial \mathbf{x}}. \qquad (21.12)$$

Comparison with (15.10) shows that the effects of expansion are to decrease the contribution of spatial gradients to the growth of ρ_1 as the Universe expands, and to decelerate growth through the $\partial\rho_1/\partial t$ term. This second effect is caused by the momentum which expanding material carries away from the perturbation.

It is clear by inspection, that, unlike (15.10), Equation (21.11) does not have exponentially growing separable solutions like (15.16). So we see immediately that expansion destroys the simple exponential growth of perturbations. To go further, we can Fourier analyze (21.11) in the co-moving frame where all wavelengths $\lambda = 2\pi R/k$ expand with the Universe. Then, writing

$$\rho_1(\mathbf{x}, t) = \rho_1(t)e^{i\mathbf{k}\cdot\mathbf{x}}, \qquad (21.13)$$

we have

$$\frac{d^2\rho_1(t)}{dt^2} + 2\frac{\dot{R}}{R}\frac{d\rho_1}{dt} = \left(4\pi G\rho_0 - \frac{4\pi^2\gamma P_0}{\rho_0\lambda^2}\right)\rho_1. \qquad (21.14)$$

Consider an initial perturbation $\rho_1(t_1) > 0$ which starts growing at t_1 so $\dot{\rho}_1(t_1) > 0$. In order for this perturbation to stop growing, it must reach a maximum value and at that time it will have $\dot{\rho}_1 = 0$ and $\ddot{\rho}_1 < 0$. But then the right hand side of (21.14) must be less than zero, and this will never occur if

$$\lambda^2 > \frac{\pi}{G\rho_0}\frac{\gamma P_0}{\rho_0}. \qquad (21.15)$$

Therefore, long wavelengths satisfying (21.15) keep increasing their amplitude. Since the second factor on the right hand side is the squared sound speed of the unperturbed gas, Equation (21.15) is just Jeans' criterion (15.11) in the co-moving frame. At any time t, the critical wavelength in the 'laboratory' frame is related to its initial value by $\lambda_J(t) = \lambda_J(t_1)R(t)/R(t_1)$. Small wavelengths oscillate and their amplitude decreases. Expansion does not change the basic instability criterion.

To isolate the expansion's effect on the growth rate, consider very long wavelength perturbations which are essentially pressure-free. These contract the most easily, according to the first three terms of (21.14). Next we need to specify the rate of expansion and how it alters the average density $\rho_0(t)$. Suppose we keep the expansion quite general, and use the simple relations

$$R \propto t^m, \rho_0 \propto t^{-q}, \tag{21.16}$$

where m and q are positive but need not be integers. Some of the general relativity models are special cases of (21.16) and all the isotropic general relativity models can be represented by it for short times during part of their history. However, we are not confined to general relativity models but can examine a generally expanding Universe. Since there is no specific evidence that the strong field equations of general relativity actually apply to our Universe (even though no other theory is better), the generality of (21.16) may be useful. Shortly we will see what constraints general relativity imposes. In cosmologies which conserve mass the values of q and m are related. If the Universe is filled with dust (the technical term for pressure-free gas, not necessarily referring to interstellar or household dust), then $\rho_0 \propto R^{-3}$ and $q = 3\,m$. In radiation dominated eras (which are not discussed here) $\rho_0 \propto R^{-4}$ and $q = 4\,m$.

At time t_1, which can be taken to be the time when matter and radiation decouple, the perturbation begins to grow. From (21.14) and our previous discussion of Jeans' instability, the natural timescale for this growth is

$$\tau_1 = [4\pi G\rho(t_1)]^{-1/2}. \tag{21,17}$$

We define an important constant which characterizes the expansion by

$$f^2 \equiv 4\pi G\rho(t_1)t_1^2 = t_1^2/\tau_1^2. \tag{21.18}$$

The value of f measures the mismatch between the age of the Universe and its instantaneous gravitational timescale. It is determined by details of the cosmological model. Old models with high density have large values of f.

With these relations the growth of perturbation amplitudes, (21.14) becomes

$$t_*^q \frac{d^2\rho_1}{dt_*^2} + 2mt_*^{q-1}\frac{d\rho_1}{dt_*} - f^2\rho_1 = 0 \tag{21.19}$$

in terms of a dimensionless time $t_* \equiv t/t_1$. When $q = m = 0$ we recover the exponential growth in a static background. Exact solutions can be obtained for any values of q, m, and f. For $q = 2$, Equation (21.19) has the Cauchy linear form and its solutions are (Saslaw, 1972)

$$\rho_1 = c_1 t_*^{r_1} + c_2 t_*^{r_2}, \qquad r_1 \neq r_2, \tag{21.20}$$

or

$$\rho_1 = t_*^r(c_1 + c_2 \ln t_*), \qquad r_1 = r_2 = r, \tag{21.21}$$

or

$$\rho_1 = t_*^a[c_1 \cos(b \ln t_*) + c_2 \sin(b \ln t_*)], \qquad r_{1,2} = a \pm ib, \tag{21.22}$$

where

$$r_{1,2} = \frac{1-2m}{2} \pm (m^2 - m + f^2 + \tfrac{1}{4})^{1/2}. \tag{21.23}$$

The particular case of $q = 2$, $m = \tfrac{2}{3}$, $f^2 = \tfrac{2}{3}$ is the one which was originally treated in most detail. It represents the Einstein-de Sitter universe as well as the initial rate of expansion from the singularity in several other isotropic Friedmann models. In this case $r_{1,2} = +\tfrac{2}{3}, -1$. Since $\rho_1 \to 0$ as $t_* \to 0$, the constant $c_2 = 0$ leaving $\rho_1 \propto t_*^{2/3} \propto R(t)$. Thus a growth factor of about 10^3 characterizes present linear perturbations which started to grow at decoupling when $z \approx R(\text{now})/R(t_1) \approx 10^3$. The fluctuations $\rho_1 \approx 10^{-34}$ mentioned at the beginning of this section would make no impression at present. Only $\approx 0.1\%$ perturbations at decoupling could now be non-linear in these models. We know what we need, but why should such large inhomogeneities be present at $z \approx 10^3$? This is the paradox of galaxy formation.

To complete the solution, for values of $q \neq 2$, we make the complex transformations

$$z = \frac{2if}{2-q} t_*^{(2-q)/2}, \tag{21.24}$$

$$\rho_1 = z^{(1-2m)/(2-q)} \chi(z) \tag{21.25}$$

and discover that in this case (21.19) is really a Bessel equation in disguise. It now has the general solution

$$\rho_1(t_*) = t_*^{(1-2m)/2} \mathrm{Re}[c_1 J_{(1-2m)/(2-q)}(z) + c_2 Y_{(1-2m)/(2-q)}(z)], \tag{21.26}$$

taking the real part of complex Bessel and Weber functions (see Abramowitz & Stegun, 1964). Initial conditions determine the complex constants c_1 and c_2. The situation is favorable for growth when the background expands slowly, as when $m < \tfrac{1}{2}, q < 2$. (The case $m > \tfrac{1}{2}, q < 2$ requires the boundary condition that ρ_1 is finite at $t_* = 0$ and decreases as $t_* \to \infty$.) If we require $\rho_1 \to 0$ as $t_* \to 0$, and put $\rho_1(t_* = 1) = \rho_1(1)$, then $c_2 = 0$. The case without expansion follows from the uniform asymptotic expansion of the Bessel function for large imaginary z, explicitly showing the exponential density growth again.

However, the value $q = 2$ has another interesting feature. Making the transformation

$$v \equiv \rho_1 t_*^m \tag{21.27}$$

and writing

$$f_{\text{now}}^2 = 4\pi G \rho_{\text{now}} t_{\text{now}}^2, \tag{21.28}$$

Equation (21.19) takes the form

$$\frac{d^2 v}{dt^2} = \left[f_{\text{now}}^2 \left(\frac{t}{t_{\text{now}}} \right)^{2-q} - m(1-m) \right] \frac{v}{t^2}. \tag{21.29}$$

For a given value of m, the perturbation's amplitude accelerates most rapidly for $q = 2$ in those cases, $q \leq 2$, favorable for long term growth. In this sense $q = 2$ gives the most rapid growth of perturbations in expanding models.

Suppose the Universe were such that $f_{now}^2 \gg |m^2 - m + \frac{1}{4}|$ and $|m - \frac{1}{2}| \ll 1$, as a simple example to see how the timescale mismatch, represented by f_{now}, affects the growth rate. Then the growing solution ((21.20), (21.23)) is

$$\rho_1(t_{*now}) = \rho_1(1)(t_{*\,now})^{f_{now}}. \tag{21.30}$$

Values of $\rho_{now} = 2 \times 10^{-29}\,\mathrm{g\,cm^{-3}}$ and $t_{now} = 5 \times 10^{17}\,\mathrm{s}$ give $f_{now} = 2.0$. So if t_{*now} is $10^4 - 10^5$ the present growth factor is about 10^9, still too small to produce galaxies from statistical fluctuations. However, if f_{now} were a factor ~ 4 greater, statistical fluctuations could produce galaxies. This could mean that ρ_{now} would be about an order of magnitude greater than the maximum value accepted at present, or the age of the Universe must be about four times its currently accepted value, or they could have smaller values such that the product $(\rho_{now} t_{now}^2)^{1/2}$ is about four times its conventional upper limit. This higher value does not seem to contradict any observations directly. The oldest stars provide only a lower limit to the age of the Universe. Interpreting the inverse Hubble constant as an 'age' of the Universe is sensitive to the particular cosmological model assumed. With regard to its density, the Universe may contain large amounts of non-luminous matter such as gravitational waves, massive neutrinos or other types of elementary particles unknown to us now.

The potential catch in this argument is that large values of f_{now} are not consistent with homogeneous general relativistic models. Although there is no direct evidence that such models describe the global properties of our universe, many astronomers believe they do. A large class of these models expand according to (see Ellis & MacCallum, 1969, or many textbooks of cosmology)

$$\left(\frac{\dot{R}}{R}\right)^2 = \frac{1}{3}\Lambda + \frac{8}{3}\pi G\rho - \frac{3c^2 k}{R^2} + \frac{a^2}{R^6}, \tag{21.31}$$

where Λ is the cosmological constant, ρ the total energy density (including radiation), k the curvature constant ($k = 1, 0, -1$ for closed, flat, and open universes, respectively), and a is a measure of anisotropy. If $a \neq 0$, R is a geometric average of the scale lengths along the principal anisotropy axes. Effects of anisotropy would diminish rapidly with expansion. If any term other than $8\pi G\rho/3$ dominates the right hand side, then q will be substantially different from 2 and $\sqrt{N}$ fluctuations would not usually grow significantly. A period of rapid growth would be possible if a combination of terms were to simulate evolution with $q \approx m \approx 0$ or with $q \approx 2$ for some time. This happens in some versions of Lemaitre models where negative Λ, corresponding to a cosmic repulsion, just balances the gravitational attraction. But this has to be very finely tuned and is usually unstable to changes in the equation of state produced by radiation from the contracting perturbations. As perturbations radiate, the Lemaitre models begin expanding faster, making it harder for the perturbations to continue contracting.

On the other hand, if the term $8\pi G\rho/3$ dominates, we could have $q \approx 2$. However, it is easy to verify from (21.28) that this case would give $f_{now} \backsim 0.6$. In particular, for the Einstein-de Sitter universe which is just closed, $\Lambda = k = a = 0$ and $\rho_0 \propto R^{-3} \propto t^{-2}$, giving $f_{now} = \frac{2}{3}$. Perturbations grow slowly. The small values of f occur because the expansion time since the singularity is approximately the inverse Hubble constant in a general relativity model whose evolution is dominated by its energy density. There is little mismatch between this age and the model's gravitational response time.

Our present understanding of galaxy formation is in a most unsatisfactory, indeed, paradoxical, state. The theories, ingenious though they are, yield little more than is put into them. They cannot be tested with much rigor and none has gained general acceptance so far. If galaxies did not exist we could easily explain their absence.

22

Gravitational graininess initiates clustering

The single atoms each to other tend,
Attract, attracted to, the next in place
Form'd and impell'd its neighbour to embrace.
Alexander Pope

Parallel to the comparison developed in Sections 15 and 16, we next turn to new phenomena which graininess introduces into the growth of perturbations. In a major application of this theory, the grains are galaxies. Despite our ignorance of their formation, we can ask how gravity causes galaxies to cluster. Does the clustering we expect explain what we see?

The simplest result can even be anticipated from our analysis in the last section. Consider a gas of galaxies. Let it be uniform except for $\sqrt{N}$ fluctuations. If we treat it, for a moment, as a fluid, then the growth rate of perturbations in a standard cosmology, $\rho_1(t) \propto R(t) \propto t^{2/3}$, tells us that observed clusters with $\sim 10^4$ galaxies should be able to form easily starting at redshifts of $10^2 - 10^3$. So galaxy clustering promises to be more understandable than galaxy formation, although it still has its mysteries.

Realizing the relative ease of galaxy clustering, it becomes natural to push the process back a step. Could galaxies themselves be the result of an earlier clustering? Suppose that isothermal perturbations of the Jeans mass at decoupling, $\sim 10^6 \, M_\odot$, started (somehow) with large amplitudes and formed bound systems. It would take about 10^4 of these to form a small galaxy. Without any larger scale residual perturbations (which would enhance clustering), it would require an expansion factor of $\sim 10^2$, to $z \approx 10$, to form galaxies. This would leave just about enough time for the galaxies to form their many small clusters. A few large ones could collect around exceptionally large statistical fluctuations. The order-of-magnitude estimates make this a plausible idea. Coupled with the fact that gravitational clustering is unavoidable, whatever else happens, we are led to examine this process in some detail.

Since the fluid (gas) description averages over scales large compared with the interparticle separation, it does not contain any information about clustering within those scales. To extract this 'microscopic' information about galaxy correlations, we have to return to the orbit equations. These are represented compactly by Liouville's equation (10.38).

The two-point correlation function is hiding in Liouville's equation. To find it we must first define it in terms of distribution functions. Recall from (11.3) that the distribution function $f^{(n)}dV^{(n)}$ is the joint probability for finding any n galaxies in the differential position – velocity volume element $dV^{(n)}$ of phase space. The single-particle distribution function $f^{(1)}$, when averaged over a volume containing many galaxies, is the fluid density. Since we do not take that average here, it applies as a probability density to any size volume. The form of Equation (20.6) suggests defining a sequence for volume distributions in which each joint probability is related to lower order joint probabilities according to

$$f^{(1)}(1) \equiv f(1),$$ (22.1)

$$f^{(2)}(1,2) \equiv f(1)f(2)[1 + g(1,2)],$$ (22.2)

$$f^{(3)}(1,2,3) \equiv f(1)f(2)f(3)[1 + g(1,2) + g(2,3) + g(3,1) + h(1,2,3)],$$ (22.3)

etc.

Equation (22.2) is formulated a bit differently from (10.29) so that $g(1,2)$ will eventually lead directly to $\xi(r)$ in the usual notation. Normalizations are the same as in (11.1) and (11.2). So far there are no restrictions on the size of the correlation functions.

Theories should first be developed to describe the simplest conditions containing the essential physics of a problem. Assume, therefore, that the galaxy distribution is initially uniform, as also suggested by the near isotropy of the microwave background. Then the single-particle distribution function does not depend on position. When clustering starts it has an equal probability of occurring anywhere. 'Macroscopically' the system remains uniform and its early clustering is described by the higher order correlation functions. Initially these higher order correlations are zero. This cosmological situation is opposite to the case of non-equilibrium plasmas, where correlations start large and decrease as the plasma relaxes to a uniform Maxwellian distribution. Cosmic correlations grow from nothing, through a linear regime, then into great density inhomogeneities. After the system becomes non-linear it is no longer plausible to expect uniformity with $f^{(1)}(1) = f^{(1)}(2)$. Our goal here is to find what happens in the linear regime, and how long the system takes to pass through it.

Evolution of distribution functions is described by the BBGKY hierarchy (11.4). To find the two-point correlation function with uniform initial conditions we need only consider the first two equations of the hierarchy. Substituting (22.1)–(22.3) into (11.4) and rearranging terms, the first two equations of the hierarchy become

$$\left[\frac{\partial}{\partial t} + \mathbf{v}_1 \cdot \frac{\partial}{\partial \mathbf{x}_1} - \frac{\partial \Phi(1)}{\partial \mathbf{x}_1} \cdot \frac{\partial}{\partial \mathbf{v}_1} \right] f(1) = \mathscr{I}(1)$$ (22.4)

and

$$f(1)f(2)\left[\frac{\partial}{\partial t} + \sum_{i=1}^{2} \left(\mathbf{v}_i \cdot \frac{\partial}{\partial \mathbf{x}_i} - \frac{\partial \Phi(i)}{\partial \mathbf{x}_i} \cdot \frac{\partial}{\partial \mathbf{v}_i} \right) \right] g(1,2)$$

$$= \sum_{i \neq j}^{2} [\mathscr{I}(i,j) + \mathscr{K}(i,j)].$$ (22.5)

Equation (22.4) has been used in deriving (22.5), and it contains the following abbreviations

$$\Phi(i) \equiv \int f(3)\phi(i,3)\mathrm{d}\mathbf{x}_3\,\mathrm{d}\mathbf{v}_3, \tag{22.6}$$

$$\mathscr{I}(i) \equiv \int f(3)\frac{\partial\phi(i,3)}{\partial\mathbf{x}_i}\cdot\frac{\partial}{\partial\mathbf{v}_i}[f(i)g(i,3)]\mathrm{d}\mathbf{x}_3\,\mathrm{d}\mathbf{v}_3, \tag{22.7}$$

$$\mathscr{I}(i,j) \equiv \left[f(j)\frac{\partial f(i)}{\partial\mathbf{v}_i} + f(i)f(j)\frac{\partial g(i,j)}{\partial\mathbf{v}_i} + g(i,j)f(j)\frac{\partial f(i)}{\partial\mathbf{v}_i}\right]\cdot\frac{\partial\phi(i,j)}{\partial\mathbf{x}_i} \tag{22.8}$$

$$\mathscr{K}(i,j) \equiv f(i)\int f(3)\left\{g(i,3)\frac{\partial f(j)}{\partial\mathbf{v}_j} + \frac{\partial}{\partial\mathbf{v}_j}[f(j)h(i,j,3)]\right.$$

$$\left. - g(i,j)\frac{\partial}{\partial\mathbf{v}_j}[f(j)g(j,3)]\right\}\cdot\frac{\partial\phi(j,3)}{\partial\mathbf{x}_j}\mathrm{d}\mathbf{x}_3\,\mathrm{d}\mathbf{v}_3. \tag{22.9}$$

There are a couple pages of straightforward algebra in this derivation which can provide a useful exercise.

The terms in these BBGKY equations have been separated according to their physical import. In the left hand side of (22.4) is the familiar convective time derivative (following the motion) and the mean field gravitational force which accelerates galaxies. This much is the collisionless Boltzmann equation of Section 7. On the right hand side is the analog of a 'collision integral' which couples changes in the single-body velocity – density distribution with changes in the two-body correlations. On the left hand side of (22.5) is the complete derivative of the two-body correlation function. It is summed over $i = 1, 2$ because the changes occur at two places in phase space. The source terms on the right hand side are of two types. Those in $\mathscr{I}$ involve the distribution functions, the correlations, and the gravitational forces at each point. They represent local inhomogeneities in the system due to the discrete nature of the mass particles and are referred to as 'graininess terms'. Terms in $\mathscr{K}$, on the other hand, are integrated over the entire system and represent global inhomogeneities in the correlations of particles. They are known as 'self-inducing' terms since they describe a back-reaction of the averaged correlation on its growth at a particular place. The second self-inducing term also couples with the three-body correlation function.

We can see why the graininess terms do not appear in the continuum approximation. (This approximation is sometimes called the 'fluid limit' but should not be confused with taking moments of the collisionless Boltzmann equation to derive the fluid equations.) In the continuum limit the mass of each particle $m \to 0$ and the total number of particles $N \to \infty$ in such a way that the total mass mN remains finite. (The analogous statement often used in a plasma is that the total volume $V \to \infty$ so that the density N/V remains finite.) Taking the continuum limit, we see that $\phi(i,j) \to 0$ and since the distribution functions retain finite normalized values the graininess terms $\mathscr{I}$ vanish. However, the self-inducing terms $\mathscr{K}$ are larger by a factor N because of the integral and they remain finite in the continuum limit.

They are the terms which would describe a generalization of Jeans' perturbation analysis of a gas to take correlations of the perturbations into account.

When there are no correlations to start with, $\mathscr{K}$ and all but the first term of $\mathscr{J}$ vanish. This remaining term is a remarkable one. It represents the graininess of the single-particle distribution function and causes spontaneous growth of two-particle correlations. No initial perturbations, not even infinitesimal ones, are necessary to produce instability. The seeds of galaxy clustering are contained in the discrete nature of the galaxies themselves.

Once g has begun to grow, its graininess will also promote the growth of correlations through the second and third terms of $\mathscr{J}$. However, by that time, terms in $\mathscr{K}$ will make an even larger contribution. This is because the integrals of $\mathscr{K}$ effectively extend over the entire system to contribute self-inducing terms of order N times the graininess terms. So the last two terms in $\mathscr{J}$ are not important.

By the time most galaxies become bound into clusters the higher order correlations will become important. However, in the early stages, we may neglect the self-inducing term involving h in comparison with the others since initially h grows more slowly than g. Mathematically, this is because the source terms of the third hierarchy equation vanish when there are no correlations, so h cannot grow until g grows. Physically, this is because with a two-particle potential the triple correlations are more likely to form through the correlation of a pair of correlated galaxies with a third uncorrelated galaxy than through the simultaneous correlation of three initially uncorrelated galaxies. The third term of $\mathscr{K}$ is also small compared with the first during the early clustering since it is quadratic in g.

Thus we are left with the first term of $\mathscr{J}$ and the first term of $\mathscr{K}$. Introducing the definition

$$\Psi(i,j) \equiv \int f(3)g(i,3)\phi(j,3)\mathrm{d}\mathbf{x}_3\,\mathrm{d}\mathbf{v}_3, \qquad (22.10)$$

we can write the first term of $\mathscr{K}(i,j)$ as

$$f(i)\frac{\partial f(j)}{\partial \mathbf{v}_j}\cdot\frac{\partial \Psi(i,j)}{\partial \mathbf{x}_j}. \qquad (22.11)$$

Comparing this with the first term of $\mathscr{J}(i,j)$ shows that a sufficient condition for the graininess term to dominate is

$$|\Psi(1,2)| < |\phi(1,2)| \qquad (22.12)$$

for all $\mathbf{x}_1$ and $\mathbf{x}_2$. Section 23 explores this regime and shows how long it lasts.

Initially the collision integral, $\mathscr{I}$, of the Boltzmann equation (22.4) is zero, and the system is strictly collisionless. Comparing $\mathscr{I}$ with the first term of $\mathscr{K}$ shows that the system remains approximately collisionless as long as condition (22.12) holds.

Therefore, the early evolution of an initially uncorrelated system is described by the simplified equations

$$\left[\frac{\partial}{\partial t} + \mathbf{v}_1\cdot\frac{\partial}{\partial \mathbf{x}_1} - \frac{\partial \Phi(1)}{\partial \mathbf{x}_1}\cdot\frac{\partial}{\partial \mathbf{v}_1}\right]f(1) = 0 \qquad (22.13)$$

and

$$f(1)f(2)\left[\frac{\partial}{\partial t} + \sum_{i=1}^{2}\left(\mathbf{v}_i \cdot \frac{\partial}{\partial \mathbf{x}_i} - \frac{\partial \Phi(i)}{\partial \mathbf{x}_i} \cdot \frac{\partial}{\partial \mathbf{v}_i}\right)\right]g(1,2)$$

$$= \sum_{i \neq j}^{2} f(i)\frac{f(j)}{\partial \mathbf{v}_i} \cdot \frac{\partial \phi(i,j)}{\partial \mathbf{x}_i}. \tag{22.14}$$

This is a basic result. Using techniques developed in Section 4 we will be able to find the exact solution for $g(1,2)$. This will answer our question: How long does it take correlations to become significant and what do they look like?

23

Growth of the two-galaxy correlation function

Two are better than one; because they have a good
reward for their labour.

Ecclesiastes

If we look out from any point in an isotropic, homogeneous Universe it must appear to expand in the same way. This expansion, mentioned in Section 21, has the form $H(t)\,\mathbf{x}$ with $H = \dot{R}/R$ when the coordinate system is chosen to coincide with a point of zero systematic motion.

To see this pretend to be a 'fundamental observer' at point O, moving with the average flow. Looking out at time t to an arbitrary point P along the position vector $\mathbf{x} = OP$, you see the velocity $\mathbf{v}(\mathbf{x}, t)$ of P relative to you. Another fundamental observer is moving with the flow at O' and his velocity relative to you is $\mathbf{v}(\mathbf{s}, t)$ where $\mathbf{s} = OO'$. (The reader may find it helpful to draw a diagram.) He measures the velocity of the same point P, located at $\mathbf{x}' = \mathbf{x} - \mathbf{s}$ in his coordinate system, and finds it to be $\mathbf{v}'(\mathbf{x}', t) = \mathbf{v}'(\mathbf{x} - \mathbf{s}, t) = \mathbf{v}(\mathbf{x}, t) - \mathbf{v}(\mathbf{s}, t)$. Now, since the Universe is homogeneous and isotropic, $\mathbf{v}'$ must be the same function of $\mathbf{x}'$ and t that $\mathbf{v}$ is of $\mathbf{x}$ and $\mathbf{t}$. Therefore, $\mathbf{v}(\mathbf{x} - \mathbf{s}, t) = \mathbf{v}(\mathbf{x}, t) - \mathbf{v}(\mathbf{s}, t)$. By inspection the solution of this functional equation is that the velocity is a linear function of position, so it has the form $\mathbf{v} = f(t)\mathbf{x} = \dot{\mathbf{x}}$. Writing $\mathbf{x} = R(t)\mathbf{x}_0$ shows that $f(t) = \dot{R}/R$. Thus the system's expansion or contraction is uniform. A reminder: this analysis applies only to regions with a light crossing time small compared to $H^{-1} = R/\dot{R}$. Such Newtonian results are perfectly adequate for scales on which galaxies are correlated.

The detailed form of $H(t)$, as in (21.31), follows from the background cosmology and the time that clustering starts. Whatever its form, momentum conservation holds so it satisfies

$$\frac{\mathrm{d}}{\mathrm{d}t}(H\mathbf{x}_i) + \frac{\partial \Phi(i)}{\partial \mathbf{x}_i} = 0, \tag{23.1}$$

with

$$\frac{\mathrm{d}}{\mathrm{d}t} \equiv \frac{\partial}{\partial t} + H\mathbf{x}_i \cdot \frac{\partial}{\partial \mathbf{x}_i} \tag{23.2}$$

being the usual convective derivative.

It will be more direct to consider the distribution and correlation functions as depending on the peculiar velocities,

$$\mathbf{u}_i = \mathbf{v}_i - H(t)\mathbf{x}_i, \tag{23.3}$$

left after subtracting the background expansion. With this change of variables, Equation (23.1) removes the mean potential Φ from (22.13) and (22.14), which become

$$\left[\frac{\partial}{\partial t} + (\mathbf{u}_1 + H\mathbf{x}_1)\cdot\frac{\partial}{\partial\mathbf{x}_1} - H\mathbf{u}_1\cdot\frac{\partial}{\partial\mathbf{u}_1}\right]f(1) = 0 \tag{23.4}$$

and

$$f(1)f(2)\left\{\frac{\partial}{\partial t} + \sum_{i=1}^{2}\left[(\mathbf{u}_i + H\mathbf{x}_i)\cdot\frac{\partial}{\partial\mathbf{x}_i} - H\mathbf{u}_i\cdot\frac{\partial}{\partial\mathbf{u}_i}\right]\right\}g(1,2)$$

$$= \sum_{i\neq j}^{2} f(i)\frac{\partial f(j)}{\partial\mathbf{u}_j}\cdot\frac{\partial\phi(i,j)}{\partial\mathbf{x}_j}. \tag{23.5}$$

We can further reduce the mathematical complications caused by the expansion by using a set of co-moving coordinates similar to (21.10) and a new time variable

$$\mathbf{x}'_i \equiv \mathbf{x}_i\frac{R_0}{R(t)}, \quad \mathbf{u}'_i \equiv \mathbf{u}_i\frac{R(t)}{R_0}, \quad \frac{\partial t'}{\partial t} \equiv \frac{R_0^2}{R^2(t)}. \tag{23.6}$$

The new correlation functions are obtained by equating them with their unprimed counterparts, in order to keep the same normalization

$$f'(\mathbf{x}'_i, \mathbf{u}'_i, t) \equiv f(\mathbf{x}_i, \mathbf{u}_i, t), \tag{23.7}$$

$$g'(\mathbf{x}'_1, \mathbf{u}'_1, \mathbf{x}'_2, \mathbf{u}'_2, t) \equiv g(\mathbf{x}_1, \mathbf{u}_1, \mathbf{x}_2, \mathbf{u}_2, t). \tag{23.8}$$

Also, t_0 is a fiducial time which we take to be the time when the system was correlation-free, and $R_0 \equiv R(t_0)$.

Consider f. Dropping the particle label 1 for simplicity of notation and changing to co-moving coordinates, Equation (23.4) becomes

$$\left(\frac{\partial}{\partial t'} + \mathbf{u}'\cdot\frac{\partial}{\partial\mathbf{x}'}\right)f' = 0. \tag{23.9}$$

Since we have assumed the system to be isotropic on the average, we can work in terms of a radial coordinate $r' \equiv |\mathbf{x}'|$, a radial velocity component u'_r, and a tangential velocity component u'_t. In terms of these coordinates, Equation (23.9) takes the form (see Equation (1) of problem 4 in Section 18)

$$\left(\frac{\partial}{\partial t'} + u'_r\frac{\partial}{\partial r'} + \frac{u'^2_t}{r'}\frac{\partial}{\partial u'_r} - \frac{u'_r u'_t}{r'}\frac{\partial}{\partial u'_t}\right)f' = 0. \tag{23.10}$$

This equation states that if we follow the motions of galaxies in a co-moving system, the total derivative of their distribution function is zero. The phase density is constant along the orbits. In this sense it is an integral of the motion. We will return to this interpretation, which is a special case of Jeans' theorem, in Section 39. Here it is useful because it tells us that if f is an integral of the motion it must (at least in normal topologies) be a function only of other integrals of the motion. Physical intuition tells us that the energy and angular momentum

$$E \equiv u_r'^2 + u_t'^2, \quad L = r'u_t$$
(23.11)

are integrals of the motion, and this can be verified by substituting them into (23.10). Are there any others? Yes,

$$K \equiv \frac{r'u_r'}{u_r'^2 + u_t'^2} - t'.$$
(23.12)

How is this found? By integrating along the characteristic curves of (23.10), as discussed in Section 4. The four characteristic equations give the three independent integrals of the motion: E, L, and K.

Another constraint on f' – that it be homogeneous – limits its dependence to the energy $E = \mathbf{u}'^2$; it is independent of t'. As the simplest illustration, we will use a Maxwellian form for the initial velocity distribution. At present the observations provide little residual evidence of the initial velocity distribution of galaxies. Too much phase mixing, clustering, ejection, and large scale perturbation have been at work on the velocities we now see. The actual distribution may, to some extent, reflect the processes by which galaxies formed, but this also is an open question. In such a state of ignorance, the Maxwell distribution is perhaps the most natural choice in that it is the only distribution for which the three Cartesian components of velocity are independent random variables with an expectation value of zero (see Section 8). The reader might try other initial distributions to see if they can produce long chains of galaxies.

For the Maxwellian distribution,

$$f'(\mathbf{u}') = n_0 \left(\frac{\beta_0}{2\pi}\right)^{3/2} \exp[-\beta_0 u'^2/2],$$
(23.13)

where n_0 is the constant co-moving density and β_0 is a constant characterizing the co-moving velocity dispersion. In more familiar non-co-moving coordinates, Equation (23.13) is

$$f(\mathbf{u}, t) = n(t) \left[\frac{\beta(t)}{2\pi}\right]^{3/2} \exp[-\beta(t)u^2/2],$$
(23.14)

where

$$\frac{n(t)}{n_0} = \frac{R_0^3}{R^3(t)}, \quad \frac{\beta(t)}{\beta_0} = \frac{R^2(t)}{R_0^2}.$$
(23.15)

Assuming a Maxwellian distribution is consistent with the requirement that the system's 'temperature' vary as $R^{-2}(t)$ during the free adiabatic expansion characterizing initially unclustered galaxies.

Returning to the equation for g, introduce (23.14) into (23.5) and use the fact that all two-particle functions depend only upon relative spatial coordinates in a homogeneous system

$$\left[\frac{\partial}{\partial t} + (\mathbf{u}_{12} + H\mathbf{x}_{12}) \cdot \frac{\partial}{\partial \mathbf{x}_{12}} - H \sum_{i=1}^{2} \mathbf{u}_i \cdot \frac{\partial}{\partial \mathbf{u}_i}\right] g(1,2) = -\beta(t)\mathbf{u}_{12} \cdot \frac{\partial \phi(1,2)}{\partial \mathbf{x}_{12}},$$
(23.16)

where $\mathbf{x}_{12} \equiv \mathbf{x}_1 - \mathbf{x}_2$ and $\mathbf{u}_{12} \equiv \mathbf{u}_1 - \mathbf{u}_2$. In co-moving coordinates this equation becomes

$$\left(\frac{\partial}{\partial t'} + \mathbf{u}'_{12}\cdot\frac{\partial}{\partial \mathbf{x}'_{12}}\right)g'(1,2) = -\beta_0\frac{R'(t')}{R'_0}\mathbf{u}'_{12}\cdot\frac{\partial \phi'(1,2)}{\partial \mathbf{x}'_{12}}, \tag{23.17}$$

where primed functions of primed variables are obtained by equating them with their unprimed counterparts, e.g., $R'(t') = R(t)$ and $R'_0 = R'(t'_0)$.

The evolution Equation (23.17) for g' can be simplified further. When the self-inducing terms are not important the growth of $g'(1,2)$ involves only $\mathbf{x}'_{12}$ and $\mathbf{u}'_{12}$ so $g'(1,2)$ will depend on velocities only through $\mathbf{u}'_{12}$. (As clustering becomes important this will cease to be true. Then the three-body character of the self-inducing terms makes $g'(1,2)$ depend on a center-of-mass velocity as well as $\mathbf{u}'_{12}$. Neglect of this more general dependence has made several attempts to extend the BBGKY hierarchy to more clustered situations and higher order correlations rather poor approximations.) Similarly, the right hand side of (23.17) is a function of $\mathbf{x}'_{12}\cdot\mathbf{u}'_{12}$ and $|\mathbf{x}'_{12}|$, so $g'(1,2)$ must be a function of r', u'_r, u'_t, and t' only, where $r' \equiv |\mathbf{x}'_{12}|$ and u'_r and u'_t are the radial and tangential projections of $\mathbf{u}'_{12}$ onto $\mathbf{x}'_{12}$. (Again, this also fails when self-inducing terms become important.) In these coordinates, (23.17) becomes

$$\left(\frac{\partial}{\partial t'} + u'_r\frac{\partial}{\partial r'} + \frac{u'^2_t}{r'}\frac{\partial}{\partial u'_r} - \frac{u'_r u'_t}{r'}\frac{\partial}{\partial u'_t}\right)g'(r', u'_r, u'_t, t')$$

$$= -Gm\beta_0\frac{R'(t')}{R'_0}\frac{u'_r}{r'^2}. \tag{23.18}$$

This is the basic equation describing the growth of correlations in an expanding initially uncorrelated system of galaxies which starts with a uniform spatial distribution and a Maxwell–Boltzmann velocity distribution. Next, we solve it exactly and find its range of validity.

The basic equation is a first order, linear, partial differential equation. Its solution is generated by the now-familiar method of characteristics (as used in Section 4 and Equations (23.11), (23.12)). For (23.18) the characteristics are the integral curves of the set of ordinary differential equations

$$dt' = \frac{dr'}{u'_r} = \frac{du'_r}{u'^2_t/r'} = \frac{du'_t}{-u'_r u'_t/r'} = \frac{dg'}{-a'(t')u'_r/r'^2}, \tag{23.19}$$

with

$$a'(t') \equiv Gm\beta_0\frac{R'(t')}{R'_0}. \tag{23.20}$$

To simplify these equations, introduce the integrals (23.11) and (23.12) of the corresponding homogeneous equation. Then (23.19) reduces to

$$dg' = -\frac{Ea'(t')(t' + K)}{[E(t' + K)^2 + L^2/E]^{3/2}}\,dt'. \tag{23.21}$$

Integration of this quadrature gives the exact solution. All we need to do is specify the expansion $a'(t')$.

Many relativistic models have expansion rates given by (21.31). For definiteness we will examine the Einstein-de Sitter case which has

$$\frac{R(t)}{R_0} = \left(\frac{t}{t_0}\right)^{2/3} \tag{23.22}$$

and

$$\frac{8}{3}\pi G\rho(t) = H^2(t) = \frac{4}{9t^2}. \tag{23.23}$$

In fact, as will become apparent, the initial growth of clustering is not strongly affected by the use of other general relativistic models. Physically this is because graininess starts acting most strongly over small scales where the expansion velocities are also smaller. Only when correlations start emerging over large scales does cosmology become very important to them.

Combining (23.22) with the definition (23.6) of co-moving time gives

$$t' - t_0' = 3t_0[1 - (t_0/t)^{1/3}], \tag{23.24}$$

so that

$$R'(t') = R_0'[1 - (t' - t_0')/3t_0]^{-2}. \tag{23.25}$$

During the early stages of clustering we can expand this in terms of

$$\tau \equiv \frac{t - t_0}{t_0} \tag{23.26}$$

and use the first order approximation

$$a'(t') = Gm\beta_0[1 + 2(t' - t_0')/3t_0]. \tag{23.27}$$

We will see that this approximation is consistent with neglecting the self-inducing terms of (22.5).

Substituting (23.27) into (23.21) and integrating gives

$$g' = Gm\beta_0[I(t_0') - I(t')] \tag{23.28}$$

with

$$I(s) = \frac{2}{3t_0\sqrt{E}}\,\text{Arcsh}\,[E(s + K)/L] - \frac{1 + 2(s - t_0')/3t_0}{[E(s + K)^2 + L^2/E]^{1/2}}, \tag{23.29}$$

since, by assumption, there are no correlations at time t_0. In terms of positions and velocities (23.28) is

$$g'(r', u_r', u_t', t') = Gm\beta_0\left[\,[1 + 2(t' - t_0')/3t_0]/r' - \{[r' - u_r'(t' - t_0')]^2\right.$$

$$+ [u_t'(t' - t_0')]^2\}^{-1/2} + \frac{2}{3t_0(u_r'^2 + u_t'^2)^{1/2}}$$

$$\times \left.\left\{\text{Arcsh}\left[\frac{u_r'}{u_t'} - (t' - t_0')\frac{u_r'^2 + u_t'^2}{r'u_t'}\right] - \text{Arcsh}\left(\frac{u_r'}{u_t'}\right)\right\}\right] \tag{23.30}$$

This is the whole solution; no recombining of Fourier components is necessary. It satisfies (23.18) and vanishes on the three-dimensional r', u'_r, u'_t hyperplane when $t' = t'_0$. It also satisfies the condition,

$$g' \to 0 \text{ as } r' \to \infty \text{ for all } u'_r, u'_t, r', \tag{23.31}$$

expected from its physical interpretation. We can see how the factor $\frac{2}{3}$ from the expansion rate has carried through, so small modifications for different expansion rates are easy to make. Equation (23.30) contains complete velocity and position information about the early growth of correlations.

With the full correlation function in hand we can extract the spatial correlation $v'(r', t')dr'$. This is the conditional probability of finding, at time t', any galaxy in a small volume at comoving distance r' from some other galaxy, irrespective of velocities. To find it multiply (23.30) by $n_0^{-1} f'(1) f'(2)$ and integrate over $\mathbf{u}'_1$ and $\mathbf{u}'_2$. The result (after some algebra) is

$$v'(r', t') = n_0 a'(t') \frac{1}{r'} \text{erfc}(\kappa' r'), \tag{23.32}$$

where

$$\kappa' \equiv \frac{\beta_0^{1/2}}{2(t' - t'_0)}, \tag{23.33}$$

and erfc denotes the usual complementary error function. Equation (23.32) is exact when $a'(t')$ is given by (23.27) which, in turn, approximates $a'(t')$ of the Einstein-de Sitter model for small τ. Thus, for small values of $\kappa' r'$, correlations grow at the same rate as $R'(t')$ similar to linear fluid perturbations.

Although the initial gravitational clustering is described most naturally in terms of co-moving coordinates, it is interesting to express it in terms of non-co-moving coordinates and the original time variable. Expanding (23.24) shows that $t' - t'_0 = t - t_0$ to first order in τ. Remembering to transform the volume element as well, (23.32) becomes

$$v(r, t) = Gm\beta(t)n(t)r^{-1} \text{erfc}[r/\lambda(t)] \tag{23.34}$$

to this order, with

$$\lambda(t) \equiv \frac{2(t - t_0)R(t)}{R_0 \sqrt{\beta_0}}. \tag{23.35}$$

The correlation function diverges as r^{-1} for small r/λ. This singularity, typical of gravitating systems, does not lead to infinite values for physically important quantities. For large r/λ the standard asymptotic series for erfc (e.g., Abramowitz & Stegun) gives

$$v(r, t) \approx \pi^{-1/2} Gm\beta(t)n(t)\lambda(t)r^{-2} e^{-r^2/\lambda^2} \left[1 - \frac{\lambda^2(t)}{2r^2} + \cdots \right]. \tag{23.36}$$

Thus, the correlation length is effectively $\lambda(t)$. Were there no universal expansion, λ would grow linearly in time at a rate proportional to the galaxies' mean thermal speed $(2/\beta_0)^{1/2}$. In other words, a given galaxy's sphere of influence at time t would extend just as far, on the average, as another initially nearby galaxy could have traveled since time t_0.

Expansion, however, causes the correlation length to increase. This occurs because in the early stages almost none of the galaxies are bound into hard binaries and so, to a greater or lesser degree, correlated galaxies, like uncorrelated galaxies, separate with the general expansion. This natural extension of the correlation length is not compensated by the decrease in thermal velocities caused by the expansion. This expansion decrease refers to velocities measured by local observers at different points along a galaxy's trajectory, whereas the relevant velocity for the propagation of correlations is with respect to the point of initial correlation. (Effects of expansion on the peculiar velocity distribution are discussed further in Section 28.)

Now we can return to the two approximations used to derive (23.34). The first, and most fundamental, was neglecting self-inducing terms – justified as long as condition (22.12) holds. Having computed the pair correlation function, we can estimate $\Psi(1, 2)$ and compare it with $\phi(1, 2)$. Using (23.34) gives

$$\Psi(1, 2) \equiv \int f(3) g(1, 3) \phi(2, 3) dx_3 dv_3$$

$$\approx - \int v(|\mathbf{x}_1 - \mathbf{x}_3|) \frac{Gm}{|\mathbf{x}_2 - \mathbf{x}_3|} dx_3$$

$$= - Gm \int \frac{v(|\mathbf{x}|)}{|\mathbf{x} - \mathbf{x}_{12}|} dx. \tag{23.37}$$

The step to the last equation involves a change of variables and the relative coordinate $\mathbf{x}_{12} \equiv \mathbf{x}_1 - \mathbf{x}_2$. Differentiating with respect to $\mathbf{x}_{12}$

$$\nabla^2 \Psi(r, t) \approx 4\pi Gm v(r, t), \tag{23.38}$$

where $r \equiv |\mathbf{x}_{12}|$. The solution of this equation is

$$\Psi(r, t) \approx Gmr^{-1} A(t) F[r/\lambda(t)], \tag{23.39}$$

with

$$A(t) = 2\pi Gmn(t) \beta(t) \lambda^2(t), \tag{23.40}$$

and

$$F(x) = \tfrac{1}{2} \operatorname{erf} x + x\pi^{-1/2} e^{-x^2} - x^2 \operatorname{erfc} x. \tag{23.41}$$

Since $|F(x)| \leq \tfrac{1}{2}$, the condition that $|\Psi(1, 2)| < |\phi(1, 2)|$ for all $\mathbf{x}_1$ and $\mathbf{x}_2$ is simply that $A(t) < 2$ or

$$\pi Gmn(t) \beta(t) \lambda^2(t) \lesssim 1. \tag{23.42}$$

Using the density–time relation (23.23) for the Einstein-de Sitter model, this inequality takes the simple form

$$\tau^2 \left(\frac{t}{t_0}\right)^{2/3} \lesssim \frac{3}{2}. \tag{23.43}$$

As long as t/t_0 is of order unity the requirement that graininess terms remain the dominant source of correlations is $\tau \lesssim 1$. This is also the second approximation used to derive (23.34). Thus the two approximations are consistent.

If clustering begins when the Hubble age is t_0 these results are valid for about one more Hubble time. A lot can happen in a Hubble time. Looking back would correspond to a redshift of order unity. During this short period we will see that clustering can grow to dominate the galaxy distribution.

24

The energy and early scope of clustering

The observed correlation function $\xi(r, t)$ is simply related to $v(r, t)$ by

$$\xi(r, t) = \frac{v(r, t)}{n(t)} = Gm\beta(t)r^{-1}\operatorname{erfc}[r/\lambda(t)]. \tag{24.1}$$

Figure 25 illustrates the shape of the early correlations with a graph of $x^{-1}\operatorname{erfc} x$ and compares this with highly evolved correlations of the form x^{-2}, where at any particular time $x \equiv r/\lambda(t)$. Normalizations of these curves change with time, but their shape has also evolved considerably judging from an approximate representation of the observations ((20.11), and following discussion). Moreover, as the Universe expands from its initially uncorrelated state, the scale determined by $\xi = \xi_0 \approx 1$, over which clustering becomes significant, also increases. For small r/λ, Equation (24.1) shows that $r(\xi_0) \propto R^2(t)$.

If at a time t_* in an Einstein-de Sitter universe we were to examine the earlier growth of clustering from the initial state at t_0, so $t_0 \leq t \leq t_*$, then (24.1) takes the form

$$\xi(r, t) \approx 0.1 \, mu_*^{-2}(t/t_*)^{4/3}r^{-1}\operatorname{erfc}(r/\lambda), \tag{24.2}$$

with

$$\lambda(t) \approx (t - t_0)\left(\frac{tt_*}{t_0^2}\right)^{2/3} u_* \text{ Mpc.} \tag{24.3}$$

In these two equations r is measured in megaparsecs, $u_* = (3/\beta_*)^{1/2}$ is the velocity dispersion of unclustered galaxies in units of 100 km s^{-1} at time t_*, m is the mass of a galaxy in units of $10^{11} \text{ M}_\odot$, and all times are measured in units of 10^{10} yr. Taking the fiducial values of these quantities, and the maximum value of $t = t_*$ to be the present time, makes it clear that galaxy clustering must have begun at $z \gg 1$ if galaxies were initially unclustered and ξ is to have its currently observed amplitude $\xi_* \gg 1$. This is just what we expected from the introduction to Section 22. We also see how a low initial velocity dispersion enhances the early clustering significantly.

Physical properties of clustering are also brought out by other measures of correlation. One such quantity is the correlation energy. Since $v(r, t)$ represents the excess density of galaxies around a given galaxy above the density of an uncorrelated

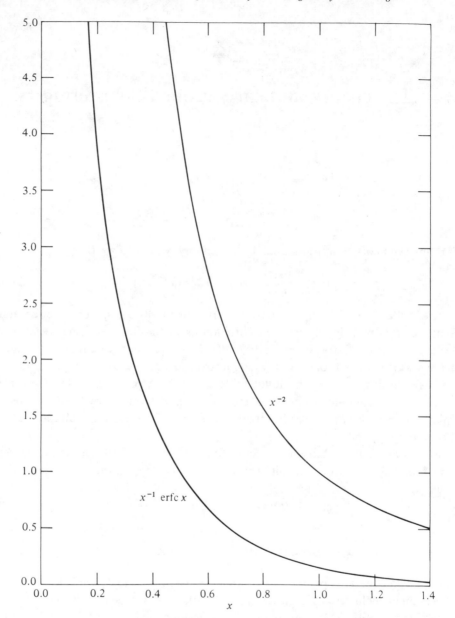

Fig. 25. Shapes of early and late correlation functions.

state, the excess potential energy per galaxy above that of the uncorrelated state is simply

$$U(t) = -\frac{1}{2} \int_0^\infty \frac{Gm^2}{r} v(r,t) 4\pi r^2 \mathrm{d}r = -2\sqrt{\pi G^2 m^3 \beta(t) n(t) \lambda(t)}. \qquad (24.4)$$

The factor $\frac{1}{2}$ accounts as usual for the sharing of energy between pairs of galaxies, and $v(r,t)$ comes from (23.34). Thus correlation energy increases linearly with time.

   To conserve energy in a gravitational system the kinetic energy of thermal motion must increase to compensate the decrease of correlation potential energy. When the magnitude of the correlation energy becomes of the same order as the mean thermal energy, the correlations dominate the system's behavior. Our neglect of the coupling integral in (22.4) prevents correlations from altering the evolution of the single-particle distribution function and therefore from affecting the mean kinetic energy. However, until correlations actually dominate, it is reasonable to approximate the mean kinetic energy per galaxy by $3m/\beta(t)$. Therefore the condition that correlations *not* dominate becomes

$$\tfrac{4}{3}\pi^{1/2}G^2m^2\beta^{3/2}n(t)(t-t_0)R^2(t)/R_0^2 \lesssim 1. \qquad (24.5)$$

Another way to understand the condition (24.5) is to rewrite it in terms of the initial impact parameter b_0 needed to produce a large gravitational deflection (see equation (2.1)). For the Einstein-de Sitter cosmology we then have

$$\frac{(b_0/u_0)}{t_0}\tau\left(\frac{t}{t_0}\right)^{4/3} \lesssim 9\sqrt{\pi}. \qquad (24.6)$$

So the shorter the orbital time b_0/u_0 during which a close encounter deflects a galaxy significantly, the less important are correlations. Consequently, correlations take longer to grow when the random initial velocities u_0 are larger.

Perhaps the simplest measure of clustering which can be derived from the pair correlation function is the excess number of galaxies expected in some neighborhood of a given galaxy compared with the number in the corresponding uncorrelated state. The number expected within a sphere of radius r about a given galaxy at time t is

$$N_c(r,t) = \int_0^r v(s,t)4\pi s^2 ds = 2\pi Gmn(t)\beta(t)\lambda^2(t)D[r/\lambda(t)], \qquad (24.7)$$

where

$$D(x) = x^2\operatorname{crfc}x - x\pi^{-1/2}\exp(-x^2) + \tfrac{1}{2}\operatorname{erf}x. \qquad (24.8)$$

In the limits of small and large spheres

$$N_c(r,t) = \pi Gmn(t)\beta(t)2r^2, \quad r \ll \lambda(t)$$
$$= \pi Gmn(t)\beta(t)\lambda^2(t), \quad r \gg \lambda(t). \qquad (24.9)$$

The excess number can be compared with $N_0(r,t) = 4\pi r^3 n(t)/3$, the number expected when no correlations exist. A natural radius to use for this comparison is the mean interparticle separation $\sim n^{-1/3}(t)$. From (24.9), for small λ/r, this ratio becomes

$$\frac{N_c}{N_0} = 3Gmn(t)(t-t_0)^2\left(\frac{R(t)}{R_0}\right)^4. \qquad (24.10)$$

When this ratio approaches or exceeds unity the system will begin to appear highly clustered. For the Einstein-de Sitter cosmology the condition that clustering not be important becomes

$$\tau^2\left(\frac{t}{t_0}\right)^{8/3} \lesssim 2\pi. \qquad (24.11)$$

This is very similar to the condition (23.43) that self-inducing terms be negligible in the kinetic equations.

Fluctuations are closely related to correlations, as shown by Equation (20.9). In order that a system have just $\sqrt{N}$ fluctuations on a length scale r requires the correlation contribution $N_c(r, t) \ll 1$. Comparing (24.7) with (23.42) and using the fact that $0 \leq D(x) < \frac{1}{2}$, we see that the condition for an expanding gravitating system to have $\sqrt{N}$ fluctuations on all scales is also equivalent to condition (23.43). So this condition is a multiple measure of when gravitational clustering is important.

Finally, we will see how the rate of clustering is simply related to graininess. At any time t and radial distance r, $N_c(r, t)/N_0(r, t)$ is proportional to m and independent of $n(t)$. Thus a system with galaxies of mass αm will cluster α times faster than a system with mass m. The rate of clustering will also be greater for the more massive galaxies within a system. This was to be expected since the graininess source term in (23.18) is also proportional to m and vanishes in the fluid limit.

All these results are strictly valid only for $\tau = (t - t_0)/t_0 \lesssim 1$. Even in this short time clustering can become significant starting from an initially unclustered distribution of galaxies. For $\tau \gtrsim 1$ correlations should grow even faster than for $\tau \lesssim 1$ since the self-inducing terms spring into action. These source terms in the kinetic equation correspond to the correlation of galaxies with groups of already clustered galaxies, a more efficient process than the early correlation of uncorrelated pairs of galaxies. At this state, $\xi(r, t)$ becomes much greater than unity and the clustering becomes non-linear.

How are we to analyze this non-linear era, which we know is demanded by present observations? At first it would seem tempting to add the self-inducing terms and step up a rung on the ladder of the BBGKY hierarchy. But the work is enormous and the rungs are very slippery.

At the next level the mathematics becomes so intractable that many major and uncertain approximations are necessary, even for numerical solutions. The main problem is that as gravitational clustering becomes non-linear its translational and rotational symmetries are broken on ever-enlarging scales: inhomogeneities dominate. When the correlations are small $g(1, 2) = g(r, u_r, u_t, t)$, but when they are large $g(1, 2) = g(\mathbf{x}_1, \mathbf{x}_2, \mathbf{u}_1, \mathbf{u}_2, t)$. The number of necessary variables increases from four to 13, and that is just for the two-point function!

Another major problem with extending the BBGKY approach is that higher order correlations become important in the non-linear regime. This can be seen from the hierarchy equation for the three-point correlations which start growing when the $g(i, j)$ grow. Soon after clustering begins, N-body experiments (to be described in Sections 26–28) tell us that clumps of increasingly many galaxies accumulate. It would therefore seem that several-point correlation functions become important very quickly. Thus the next stage, after $\xi(r, t)$ becomes of order unity, is that *all* the low level correlation functions become large and pursuit of just two or three of them will not give a convergent description of clustering. After a point there is no point to the many-point correlation functions. There must be a better way!

25

Later evolution of cosmic correlation energies

25.1. The cosmic energy equation

A more useful approach to non-linear correlations would be to derive results from the BBGKY hierarchy which do not depend either on $g(1, 2)$ being small, or on the absence of higher order correlations. In the limit as all correlations become non-linear such results would follow from Liouville's equation itself. Since Liouville's equation is equivalent to the equations of motion such results should also follow from orbit theory.

Conservation of energy applies to non-linear correlations over large regions just as it applies locally. However, in an expanding system we must allow for two modifications. First, expansion causes the kinetic energy of peculiar motions within a given region to decrease. This opposes the growth of irregularities which increase peculiar velocities. Second, correlations extend across the boundaries of any volume. What happens within the volume depends also on clustering around the region. Unlike a perfect gas, an arbitrary subvolume is not self-contained.

The cosmic energy equation describes this situation. It follows directly from the galaxies' equations of motion (Irvine, 1961; Layzer, 1963). Having set up the apparatus of the BBGKY hierarchy in the previous sections we can use it to derive this cosmic energy equation by following the approach of Fall & Severne (1976). From the two lowest equations of the hierarchy we obtain a powerful result valid to all orders for any irregularity in the distribution of galaxies, whether linear or not. The only assumption is that the single-galaxy distribution function $f(1)$ remains uniform so the Universe is homogeneous over very large scales. On these large scales translationally invariant correlation functions describe the clustering. This is just the assumption, also common to direct derivations from the equation of motion, which we have using all along.

Consider an arbitrary volume of space in the co-moving frame. Peculiar velocities and spatial correlations give average kinetic and potential energies per galaxy of

$$T' \equiv \frac{1}{2} \frac{m}{n_0} \int u_1'^2 f'(1) \mathrm{d}\mathbf{u}_1' \tag{25.1}$$

and

$$U' \equiv -\frac{1}{2n_0} \int \frac{Gm^2}{r'} \gamma'(1,2) \mathrm{d}\mathbf{u}_1' \, \mathrm{d}\mathbf{u}_2' \mathrm{d}\mathbf{r}', \tag{25.2}$$

where

$$\gamma'(1,2) \equiv f'(1)f'(2)g'(1,2) \tag{25.3}$$

and the peculiar co-moving velocity is again

$$\mathbf{u}_i' = \frac{R(t)}{R_0}(\mathbf{v}_i - H(t)\mathbf{x}_i) \tag{25.4}$$

and n_0 is the constant co-moving density. Equations (23.6) show that in the observer's non-comoving frame, these energies are

$$T(t) = \frac{R_0^2}{R^2(t)} T'(t') \tag{23.5}$$

and

$$U(t) = \frac{R_0}{R(t)} U'(t'). \tag{25.6}$$

Changes of the correlation energies in the two frames are therefore related by

$$\frac{\mathrm{d}}{\mathrm{d}t}(T+U) + H(2T+U) = \left(\frac{R_0}{R(t)}\right)^3 \left[\frac{R_0}{R(t)}\frac{\mathrm{d}T'}{\mathrm{d}t'} + \frac{\mathrm{d}U'}{\mathrm{d}t'}\right]. \tag{25.7}$$

Now the **BBGKY** hierarchy rides to the rescue of (25.7) from its co-moving derivatives. Using the transformations of Sections 22 and 23, and making the homogeneity assumption that $f'(1)$ is independent of $\mathbf{x}_1'$ and the spatial dependence of $g(1,2)$ only involves relative coordinates $\mathbf{x}_{12}' = \mathbf{x}_1' - \mathbf{x}_2'$, the reader can show that (22.4) and (22.5) have the co-moving form

$$\frac{\partial f'(1)}{\partial t'} = \frac{R'(t')}{R_0'} \int \Theta_{12}' \gamma'(1,2) \mathrm{d}\mathbf{x}_2' \, \mathrm{d}\mathbf{u}_2' \tag{25.8}$$

and

$$\left[\frac{\partial}{\partial t'} + (\mathbf{u}_1' - \mathbf{u}_2')\cdot\frac{\partial}{\partial \mathbf{x}_{12}'}\right]\gamma'(1,2) = \frac{R'(t')}{R_0'}\Theta_{12}'[f'(1)f'(2) + \gamma'(1,2)]$$

$$+ \frac{R'(t')}{R_0'}\int [\Theta_{13}' \ f'(1)\gamma'(2,3)$$
$$+ \Theta_{23}'f'(2)\gamma'(1,3)$$
$$+ (\Theta_{13}' + \Theta_{23}')h'(1,2,3)]\mathrm{d}\mathbf{x}_3' \, \mathrm{d}\mathbf{u}_3', \tag{25.9}$$

where

$$\Theta_{ij}' \equiv \frac{\partial}{\partial \mathbf{x}_i'}\left(\frac{-Gm}{|\mathbf{x}_i' - \mathbf{x}_j'|}\right)\cdot\left(\frac{\partial}{\partial \mathbf{u}_i'} - \frac{\partial}{\partial \mathbf{u}_j'}\right) \tag{25.10}$$

and $h'(1,2,3)$ is our previous three-point co-moving correlation function multiplied by $f'(1)f'(2)f'(3)$. These include all the collision, graininess, and self-inducing terms – all the source terms.

Substituting (25.8) into the co-moving time derivative of (25.1) gives

$$\frac{dT''}{dt'} = \frac{1}{2}\frac{m}{n_0}\frac{R'}{R_0'}\int u_1'^2 \Theta_{12}'\gamma'(1,2)du_1'du_2'dx_2'$$

$$= \frac{Gm^2R'}{2n_0R_0'}\int (\mathbf{u}_1' - \mathbf{u}_2')\cdot\frac{\partial}{\partial \mathbf{x}_{12}'}\frac{1}{x_{12}'}\gamma'(1,2)dx_{12}'du_1'du_2'. \qquad (25.11)$$

To get the second line, one replaces $u_1'^2$ by $(u_1'^2 + u_2'^2)/2$ and makes the integral symmetric in $\mathbf{u}_1'$ and $\mathbf{u}_2'$, then integrates by parts. Similarly, substituting (25.9) into the co-moving time derivative of (25.2) gives

$$\frac{dU'}{dt'} = -\frac{Gm^2}{2n_0}\int\frac{1}{x_{12}'}\left[-(\mathbf{u}_1' - \mathbf{u}_2')\cdot\frac{\partial}{\partial \mathbf{x}_{12}'}\right]\gamma'(1,2)dx_{12}'du_1'du_2'. \qquad (25.12)$$

All the graininess and self-inducing terms of (25.9) vanish over the boundaries of the velocity integrations. This removes any effects of higher order correlations. Integrating (25.12) by parts over $\mathbf{x}_{12}'$ shows immediately that the right hand side of (25.7) vanishes.

Thus we are left with the cosmic energy equation in the form

$$\frac{d}{dt}(T + U) + H(2T + U) = 0. \qquad (25.13)$$

It is expressed in non-co-moving coordinates where, in terms of observed quantities,

$$T = \frac{1}{2}\frac{m}{n(t)}\int(\mathbf{v}_1 - H\mathbf{r})^2 f(1)d\mathbf{v}_1, \qquad (25.14)$$

$$U = -\frac{1}{2}Gm^2n(t)\int\frac{1}{r}\xi(r)d^3\mathbf{r}, \qquad (25.15)$$

with $r = |\mathbf{x}_{12}|$ now the standard radial coordinate. Again, this result applies exactly to the full non-linear evolution of correlations, so long as the density is homogeneous when averaged over scales much larger than typical correlation lengths. Note that in regions of galaxy anti-correlation $\xi(r) < 0$ and $U > 0$.

If the correlated galaxies are bound together and are in virial equilibrium, then $2T = -U$ and their total energy $T + U$ is conserved. This is because bound structures do no work against the expansion of the Universe. For example, the expansion of the Universe does not affect binary stars. Similarly, if there is no expansion, $\dot{R} = 0$ and total energy is conserved. On the other hand, if there were no gravity, or no potential correlation energy U, the peculiar kinetic energy would decrease as $R^{-2}(t)$. Peculiar velocities in a freely expanding adiabatic gas decrease as R^{-1} (see also (23.6)), so this is just what we expect. In general, (25.13) shows how the change in total energy caused by expansion is governed by the degree to which the galaxy cluster participates in the expansion of the Universe.

25.2. Stability of $\xi(r) \propto r^{-2}$

We can use the cosmic energy equation to obtain a rather remarkable result about the stability of $\xi(r) \propto r^{-\gamma}$ for the special case $\gamma = 2$, close to what is observed. First we will need to find another relation between T and U. Writing quite generally

$$T = -\beta(t)U, \qquad (25.16)$$

we would expect $\beta(t)$ to be a complicated function of time. For long times, however, it has a simple but important asymptotic property.

To prove this property, suppose the cosmic expansion has the form

$$R(t) \propto t^p. \qquad (25.17)$$

This form is true for at least a short time in any cosmology, and for all times in the Dirac ($p = \frac{1}{2}$), Einstein–de Sitter ($p = \frac{2}{3}$) and Milne ($p = 1$) universes. In general, $p = \mathrm{d} \ln R / \mathrm{d} \ln t$ will be time dependent, but we can get a good idea of the effects of expansion by supposing p to be a constant and taking different values for it (as in Section 21).

Substituting (25.16) and (25.17) into the cosmic energy equation (25.13) gives

$$\frac{\mathrm{d}}{\mathrm{d}t}(1 - \beta)U + \frac{p}{t}(1 - 2\beta)U = 0, \qquad (25.18)$$

whose solution is

$$U = U_0(1 - \beta(t))^{-1} \exp\left[\int \frac{p(2\beta - 1)}{(1 - \beta)t} \mathrm{d}t \right]. \qquad (25.19)$$

From this it is clear that for a growing irregularity in which $|U(t)|$ increases, $\frac{1}{2} < \beta < 1$. Physically, all this says is that such a region is somewhere between virial equilibrium and free expansion. More interestingly, Equations (25.17) and (25.19) show immediately for constant or slowly changing β, that any perturbation with growth $U \sim R(t)$ must have $(2\beta - 1)/(1 - \beta) = 1$, or $\beta = \frac{2}{3}$, regardless of cosmological model.

Further insight into the development of non-linear irregularities can be got by supposing $U^{-1}\mathrm{d}U/\mathrm{d}t = q/t$. Then (25.18) can be solved for β to give, for p and q independent of time:

$$\beta(t) = \frac{p + q}{2p + q} + \frac{[(2p + q)\beta_0 - (p + q)]}{2p + q}\left(\frac{t_0}{t}\right)^{2p + q}, \qquad (25.20)$$

where $\beta(t_0) = \beta_0$. For long times, the memory of the initial value is lost, suggesting that an equilibrium or stationary state is reached. Indeed, this is readily seen to be the case in any expanding universe since $\int t^{-1}p(t)\mathrm{d}t$ increases with time. The 'dissipation' is caused by orbital relaxation and phase mixing. Therefore,

$$\lim_{t \to \infty} \beta(t) = \frac{p + q}{2p + q}. \qquad (25.21)$$

In a stationary (but not static) state $q = 0$ and $\beta \to \frac{1}{2}$ regardless of the expansion rate. For the Einstein–de Sitter case with $p = q = \frac{2}{3}$, the asymptotic value of β is $\frac{2}{3}$. These results are in general agreement with N-body simulations, described in Section 27. In practice, the 'asymptotic' state often may be reached rather rapidly in one or two dynamical crossing timescales.

Since $\beta(t)$ changes very slowly with time after its early relaxation, and for most of its history has values between 0.50 and about 0.67, we can replace $\beta(t)$ by some average value $\bar{\beta}$, say, in (25.18). Then the correlation energy evolves according to

$$\frac{dU}{dt} = \frac{(2\bar{\beta} - 1)}{(1 - \bar{\beta})} \frac{p}{t} U. \tag{25.22}$$

Next, assume the correlation function to be a general power law

$$\xi(r, t) = a(t) r^{-\gamma} \quad \text{for} \quad r_0(t) \leq r \leq r_1(t), \tag{25.23}$$

with r in dimensionless units of megaparsecs. This form is suggested by the fact that gravity and the quasi-Newtonian cosmological background are scale-free within the volume over which correlations occur. Later sections discuss this assumption further. Substituting Equations (25.15) and (25.23) into (25.22) gives the evolution of $\gamma(t)$

$$\frac{d}{dt} \ln (na) + \left(\frac{1}{2 - \gamma} - \ln r_1 \right) \frac{d\gamma}{dt} + (2 - \gamma) \frac{d \ln r_1}{dt} = \frac{(2\bar{\beta} - 1)}{(1 - \bar{\beta})} \frac{p}{t} \tag{25.24}$$

for $\gamma < 2$ and $r_0 \ll r_1$. If $\gamma > 2$, this result has the same form except that terms with r_1 change sign and r_1 is replaced by r_0. (In both cases just the dominant terms in either r_1 or r_0 are retained.)

Although $a(t)$ is not known exactly in any particular case, when t is large we can write approximately

$$\frac{d \ln (na)}{dt} = -\frac{p'}{t}. \tag{25.25}$$

Since $n \propto R^{-3}(t)$ it would be possible to have $p' < 0$ if the correlation amplitude a were to grow faster than $R^3(t)$, but such fast growth does not occur in N-body experiments (see Section 27). Moreover, it seems a reasonable (but unproven) conjecture that the growth of low order correlation functions slows down when higher order correlations, characterizing the non-linear regime, begin growing rapidly. We do not know $r_1(t)$ in detail either; fortunately, however, when $|\gamma - 2| \ll 1$ it does not matter.

Examining the stability of γ close to 2, we set $\gamma(t_0) = \gamma_0$ and solve (25.24) to find the important result

$$\gamma = 2 - (2 - \gamma_0) \left(\frac{t_0}{t} \right)^m, \tag{25.26}$$

where

$$m = \frac{(2\bar{\beta} - 1)}{(1 - \bar{\beta})} p + p'. \tag{25.27}$$

This holds for either $\gamma_0 > 2$ or $\gamma_0 < 2$. It is a result in the fully non-linear regime of clustering.

Thus the correlation exponent $\gamma = 2$ is dynamically stable when approached either from above or below (as long as $m > 0$). We already know from the linear regime that $\gamma = 0$ is highly unstable. Since the equations do not distinguish any other value of γ it is reasonable to conjecture that $\gamma = 2$ is its only stable value.

Over a wide range of conditions this stability does not depend significantly on details of either initial conditions or the cosmological expansion. These seem to be secondary influences; although very contrived conditions might alter the stability of $\gamma = 2$. In a more rapidly expanding universe (large p), the value $\gamma = 2$ is reached more rapidly for larger m. Moreover, for clustering close to virial equilibrium, γ changes more slowly, finally ceasing to evolve when $\bar{\beta} = \frac{1}{2}$ and $p' = 0$.

26

N-body simulations

What in the Universe I know can give directions
how to go?

W.H. Auden

Because the classical interaction between gravitating point masses is totally understood – the subject has had its Newton – it is possible to solve the clustering problem completely, in principle. All that is necessary is to integrate the equations of motion with the desired initial conditions and read off the answer. Non-linear regimes are dealt with as easily as linear ones. From the time fast computers became available, about 1960, to the present day, a considerable industry has grown up around these numerical solutions. Beginning with the earliest computations for a couple of dozen particles, the power of computers and techniques has improved by more than an order of magnitude each decade. Now it is possible to follow several thousand galaxies for tens of relaxation times with direct integrations.

New numerical techniques are always being developed. Their general goal is more information about increased numbers of particles for longer times. Compromises are often worked, usually sacrificing information for more particles. Since direct integrations retain the most information, and computers always get better, this approach will continue to flourish.

The direct integration method integrates the equations of motion explicitly to retain complete information about all the galaxy positions and velocities at any time. In co-moving coordinates, $\mathbf{x} = \mathbf{r}/R(t)$ (see (21.10)), these equations of motion can be derived from the Lagrangian

$$L = \sum_i \frac{1}{2} m R^2 \dot{x}_i^2 - \frac{m}{R} \phi_i(x_i), \qquad (26.1)$$

with

$$\nabla^2 \phi = 4\pi G[\rho(x, t) - \rho_0], \qquad (26.2)$$

where $\rho_0 = R^3 \bar{\rho}$. (To see how the kinetic energy term is formed, calculate $\dot{r}$ and subtract the background expansion $r\dot{R}/R$.) Then the Euler–Lagrange variations give

$$\ddot{\mathbf{x}}_i + 2\frac{\dot{R}}{R}\dot{\mathbf{x}}_i = -R^{-3}\nabla\phi_i \qquad (26.3)$$

for the *i*th galaxy. The main problems in these direct integrations are how to treat close encounters, how to mimic realistic boundary effects, how to make the program run quickly and use the minimum memory storage for maximum N, and, finally, how to retrieve the vast amount of information it generates in a useful form.

Close encounters can be treated in three ways. For very precise calculations in densely populated N-body systems the equations of motion can be regularized. This is a form of renormalization in which the singularities at $r = 0$ are removed by transforming to new independent variables (Kustaanheimo & Stiefel, 1964; Aarseth, 1971; Aarseth and Zare, 1974; Heggie, 1974). It is especially suited to cases where the galaxies start off with negligible random motion and the system goes through a period in which it generates very eccentric binary orbits. A second technique is to automatically insert an analytic Keplerian orbit into the numerical analysis during the time of close encounter and remove it after the galaxies have separated. This requires a time-consuming check to find if close encounters occur. Normally, it is simpler and much faster just to use a softened gravitational potential proportional to $(r^2 + \varepsilon^2)^{-1/2}$. The softening parameter ε represents the finite size of a galaxy and is sometimes a more realistic approximation for a particular problem than the r^{-1} potential. Typically, $0.01 \lesssim \varepsilon \lesssim 0.05$ (in units scaled to the total radius). This technique is also used for systems of stars, even though they are more like point masses than galaxies. By running models with different values of ε, one can pick out results not affected by this parameter.

Boundary conditions are more important, especially when small values of N are used to represent much larger systems. For finite clusters of stars and galaxies the free boundaries arise naturally and are no problem. For an infinite system we must prescribe what happens to a galaxy that reaches the edge of the finite simulation. One technique is to assume a periodic structure with cubic unit cells. As a galaxy leaves one side it reappears moving inward from the opposite side. Another approach reflects departing galaxies specularly from the receding boundary. On average this decreases the total energy. To avoid this decrease galaxies can be reflected from the boundary with random velocities which satisfy average energy and momentum constraints. The importance of these effects, including the growth of fluctuations initiated at the boundary, does not seem to have been studied in detail. Presumably they become less important as N increases, but for $N \lesssim 10^3$ their role is unclear.

Speed and storage problems are alleviated by using an integration scheme which treats the forces due to near neighbors and distant galaxies separately (Ahmad & Cohen, 1973; Aarseth, 1984). For each given galaxy, the forces on it from distant galaxies change more slowly than the forces from nearby galaxies. So these distant forces can be recomputed at longer intervals than the closer forces; the number of computer operations per time step decreases significantly.

Extracting relevant information from the computer experiments is fairly easy if one knows what type of information is useful beforehand. Results such as correlation functions, catalogs of groups, velocity distributions, degree to which subsystems satisfy the virial theorem, effects of mass spectra, formation of voids, etc., are readily available. By storing critical information after one-half of each dynamical timescale, or so, a record of the experiment can be preserved. As new questions arise all that is needed is to 'observe' the old experiment from a new point of view. In this way, time-lapse films can also be produced to provide an attractive picture of the evolution. When these pictures artificially rotate the system about its center the parallax in the eyes of the viewer gives a dramatic three-dimensional effect. Digital computers here on earth can simulate the great analog computer in the sky.

Direct integration techniques have their advantages and disadvantages. Their most important advantage is that few assumptions are interposed between a physical problem and its solution. Moreover, their completeness produces great flexibility regarding the questions we can ask of simulations. On the other hand, these techniques have two disadvantages. One is the relatively small value of N. This is no problem when treating star or galaxy clusters, but is does surface in the boundary effects when treating a representative volume of a much larger distribution. Second, computer round off errors make it impossible to reproduce individual orbits exactly. Different computers and programs also produce different orbits after long integration times. Despite this result, numerous studies (see Aarseth & Lecar, 1975, for a review) have concluded that macroscopic and statistical properties which are averaged over many particles or systems are determined accurately. For example, although the same galaxy might not be ejected in different computations of the same initial cluster as it evolves, the average rate of ejection of galaxies in the different computations will be very similar.

Situations sometimes arise where it is useful to have simulations with very large values of N. Galaxy clustering is one example; galaxy formation from many-star clusters or galaxy collisions are others. Various techniques are available which sacrifice exactness or detailed information to the great god N. One approach, suitable for collisionless systems, is to replace the Boltzmann equation by a set of its moments. Truncating the moment equations introduces the main physical assumptions which sacrifice information. A more sophisticated approach simulates the solution of the Fokker–Planck equation with reasonable assumptions about the transition elements and their associated diffusion coefficients (Spitzer, 1974). Then there are approaches which model the N-body problem by averaging mass motions over large volumes, often using spherical shells. These models are fast, cheap, and easy; usually they are designed to answer limited specific questions.

The alternative approach which is closest in spirit to direct integrations is the particle-mesh scheme. (See Hockney & Eastwood's (1981) book for a detailed discussion of these and related methods.) Basically the idea is to divide the force acting on each particle into two parts. The rapidly fluctuating forces of nearest neighbors (with a softened potential) are summed directly. The second part, which is long range and varies smoothly, is represented by a mesh. This is a rectangular lattice of fixed points in the cubic box containing the galaxies. Each mesh point is assigned a mass which represents the current density distribution. This defines a smooth force field which is interpolated for each particle position. After moving the particles in their total force field for a timestep, the lattice is updated and the cycle continues. This smoothing technique enables tens of thousands of galaxies to be moved about with good accuracy over long times.

All these approaches have been used for many (not quite countless) astronomical experiments. Some results which illustrate the basic physics of a variety of gravitating systems are described in subsequent sections. We start with results related to the two-point correlation function $\xi(r)$ and to higher order correlations.

Out of computers,
Endlessly flowing
Numbers come spinning
Their webs around science.

27

Evolving spatial distributions

The curtain of the Universe is rent and shattered,
the splendor-winged worlds disperse like wild
doves scattered.

Shelley

Many N-body simulations have shown how galaxies disperse and cluster as the universe expands. We shall just examine some representative examples. The 4000-body simulations of Aarseth and his collaborators (see bibliography for references) were designed to see how different initial conditions and cosmological models, within reasonable ranges, alter the clustering of galaxies. Experiments whose initial velocities had a pure Hubble flow behaved much like those which also had peculiar Maxwellian velocities smaller than the expansion velocity. Gravitational graininess built up the self-consistent velocity field after about one initial Hubble expansion in either case.

Somewhat more important is the galaxies' mass distribution. In the computation this does not change with time, although small groups which form can mimic the distant effects of larger mass galaxies at later stages of clustering. Various simulations have shown that the more massive galaxies have a greater tendency to cluster. Their two-point correlation amplitude, for example, increases faster than for less massive galaxies. Linear analysis (Equation (24.1)) leads us to expect this, at least during the initial clustering when the velocity distribution does not depend strongly on mass. Experiments also show that many results, especially regarding spatial distribution, do not depend strongly on detailed properties of the mass spectrum, so long as it continuously spans a range of at least an order of magnitude. The 4000-body experiments choose a mass spectrum of the form

$$N(m)\mathrm{d}m \propto \frac{m_*}{m}\mathrm{e}^{-m/m_*}\mathrm{d}m. \tag{27.1}$$

The total scaled mass range is $0.037 < m/m_* < 25.2$. This is roughly compatible with the luminosity function of galaxies, although their mass–luminosity relation is very uncertain. Scaling the fiducial mass m_* to a physical value using the estimated mass–luminosity function gave $m_* \approx 5 \times 10^{12}\,(\Omega/0.1)\,\mathrm{M}_\odot$, where $\Omega = \rho/\rho_c$ is the present ratio of the mass density to the critical density necessary for a parabolic ($q_0 = \frac{1}{2}$) Friedmann cosmology. However, m_* is not the most probable galaxy mass, which is smaller, since $N(m)$ is heavily skewed toward low masses with $\langle m \rangle = 1$ (in units of m_*).

The starting spatial distribution is especially important. Experiments to simulate the Universe often begin with a homogeneous distribution, although examples with large scale 'built-in' structures are possible. The homogeneous distribution may, of course, be clustered. Usually this clustering is specified by the value of n, where initial density fluctuations involving total mass M have the form

$$\frac{\delta\rho}{\rho} \propto M^{-1/2 - n/6}. \tag{27.2}$$

Poisson distributions have $n = 0$, while the value $n = -1$ is an initially more clustered distribution.

Experiments with $n = 0$ are easy to set up since they simply assign random coordinates to galaxies within an initial sphere. Starting an experiment with a specified value of $n \neq 0$ is harder. Little exact theory is available to relate an algorithm specifying galaxy positions to the value of n which results. It is particularly hard to do this without building favored scale lengths into the correlation function. In the 4000-body experiments, clustered starting conditions are obtained essentially as follows. First define a z axis pointing at random in the initial sphere. Place a galaxy at random in the sphere in the orthogonal x–y plane. Place more galaxies at specified integral separations along a rod within the sphere. Then find another z axis and repeat the process until the desired total number of galaxies is reached. Although this introduces a scale length into the correlations, it approximates a value $n = -1$ for initial fluctuations. Traces of the initial scale length remain in the two-point correlation function, but cease to dominate after the Universe has doubled or quadrupled in radius. Other experiments set up initial anti-clustering by dividing a volume into little cells and placing one galaxy at random within each cell (Efstathiou & Eastwood, 1981). This naturally inhibits the subsequent growth of correlations; results using it disagree strongly with the observed slope of $\xi(r)$.

Finally, an experiment must specify the cosmological model to determine $R(t)$. Two rather extreme cases, with present values of $\Omega = 0.1$ or 1.0, are usually used. They span the conventional range associated with observations.

Figure 26 shows a typical two-dimensional projection of cluster evolution, taken from the 4000-body computations of Aarseth, *et al.* (1979). It is for the case $n = -1$, $\Omega = 0.1$. Four 'snapshots' of the projected spatial distribution correspond to redshifts of about 31, 8, 2 and the present time. This is a useful illustration of many properties since it seems compatible with present observations. Qualitatively it resembles most other simulations, but we shall see that some quantitative features differ significantly.

In order to fit on the page the four systems are shown in co-moving coordinates. The current diameter, at $R = 31.9$ of the model, is 100 Mpc, corresponding to a present value of the Hubble constant $H = \dot{R}/R = 50 \text{ km s}^{-1} \text{ Mpc}^{-1}$. The reason for starting the clustering at a redshift $z = (R_{\text{now}}/R_0) - 1 \approx 31$ was to allow enough time

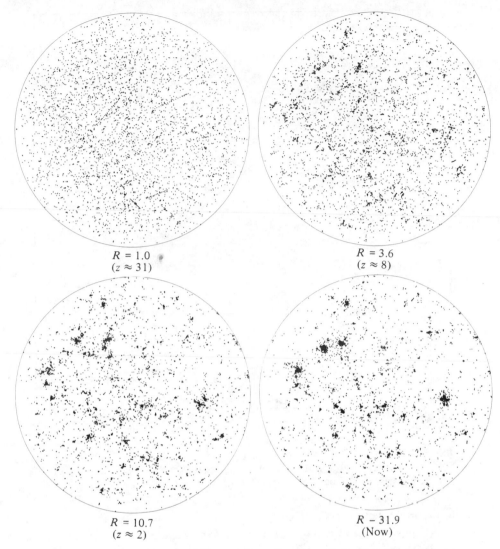

Fig. 26. Four 'snapshots' of the evolving projected spatial galaxy distribution in the $\Omega = 0.1$, $n = -1$ model. Relative radii correspond to redshifts of 31, 8, 2, and the present time.

for the amplitude of $\xi(r)$ to become comparable with its observed value (more about this later).

Initially the distribution is quite homogeneous, although some chains are evident. This 'supergraininess' helps promote early clustering and the chains soon disappear. By a redshift of about 8, small groups and clusters of groups are already well developed. Most of these groups are gravitationally bound and form the nuclei around which more galaxies will cluster. At $z \approx 2$ the clusters become very pronounced. Between then and the present the clusters shrink and become denser

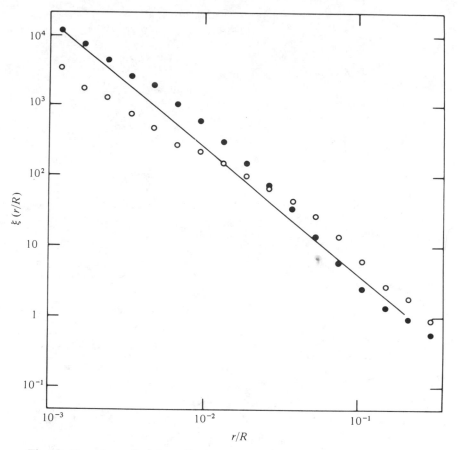

Fig. 27. Covariance functions $\xi(r)$ for experiments with $\Omega = 1$, $n = -1$ at $R = 10.7$, open circles; and with $\Omega = 0.1$, $n = -1$ at $R = 32$, filled circles. For comparison the line shows $\xi(r) = 70\, r_{\mathrm{Mpc}}^{-1.8}$. (From Gott, *et al.*, 1979.)

and the number of field galaxies decreases. In some places this depletion leaves large regions completely devoid of galaxies. The whole process is reminiscent of a phase transition. Ordinary atomic or molecular phase transitions (e.g., liquid–gas) change the system from being completely in one state to completely in another. Gravitational phase transitions are different. The system changes from a homogeneous state into a state with two components. Clusters with negative total energy break the symmetry of the galaxy distribution on a new larger scale. Between the clusters are higher energy galaxies. These have removed binding energy from clusters (a type of dissipation) to form a new component of the system. Moreover, unlike chemical phase transitions, gravitational phase transitions do not reach a final equilibrium state. They strive forever toward a more clustered arrangement, although eventually the timescale for change may become very long.

A glance at Figure 26 suggests that neither the center nor the regions near the boundary are especially distinguished. Clusters do not particularly seem to favor

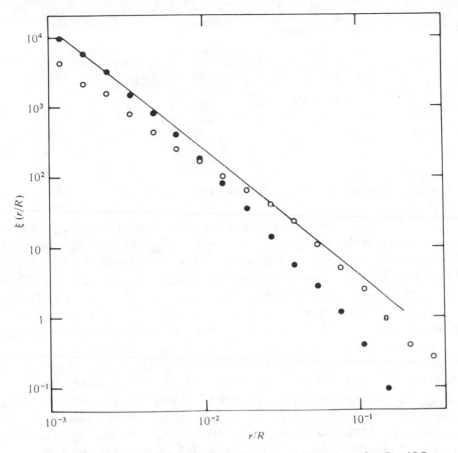

Fig. 28. Covariance functions $\xi(r)$ for experiments with $\Omega = 1$, $n = 0$ at $R = 10.7$, open circles; and with $\Omega = 0.1$, $n = 0$ at $R = 22.6$, filled circles. For comparison the line shows $\xi(r) = 70\, r_{\mathrm{Mpc}}^{-1.8}$. (From Gott, *et al.*, 1979.)

one area or the other. Quantitative measures such as the two-point correlation function confirm this impression that boundary conditions are of minor importance. Just to be sure, the correlation functions for a distance *r* are usually determined using only those galaxies at least a distance *r* from the boundary.

The two-point correlation function is one of the objective quantities which can be compared directly with observations and theory. Since experimental information is complete, the spatial correlations can be computed easily without the distance uncertainties which characterize observations. On the other hand, the range over which $\xi(r)$ can be accurately determined is limited. The average interparticle distance imposes a lower bound for distinguishing between Poisson and other initial conditions. Large fluctuations occur over smaller distances and make it very hard to specify most non-Poisson initial conditions. The radius of the sample imposes an upper bound. Actually, the number of pairs available decreases rapidly as *r* approaches *R*, causing the statistical uncertainty of ξ to grow (see Kiang & Saslaw,

Table 3. $\xi(r) = a r_{\mathrm{Mpc}}^{-\gamma}$ *for*
4000-*body experiments*
(*Gott, et al.*, 1979)

Ω	n	a	γ
0.1	-1	90	1.9
1.0	-1	80	1.4
0.1	0	30	2.3
1.0	0	40	1.7

1969 for details). In practice $r \approx 0.5R$ is about the upper limit for systems containing $\sim 10^4$ galaxies.

The main result is that $\xi(r)$ starts small and grows until it becomes a power law approximately of the form $\xi(r) \propto r^{-2}$ for large ξ. Figures 27 and 28 (from Gott, Turner & Aarseth, 1979) show this for four 4000-body experiments with different values of Ω and n. In each experiment galaxy clustering starts from a cold Hubble flow at a time which is selected to make the amplitude $a(t)$ of $\xi(r, t)$, and the Hubble constant, consistent with presently observed values ($a \approx 50$ and $H = 50$). Table 3 (also from Gott, *et al.* 1979) shows the resulting present amplitudes and exponents of $\xi(r, t)$. What initially seemed remarkable about these results is their relative indifference to Ω and n. This was disappointing to theorists who had hoped to use the slope of ξ, one of the few well-determined observed features of the Universe, to discover the value of Ω. More extensive experiment with more galaxies and different integration techniques, e.g., the particle-mesh method, have confirmed this general conclusion (see Efstathiou & Eastwood, 1981). There has even been one calculation by Aarseth (unpublished) which initially placed the galaxies randomly on the six sides of a cube in the expanding Universe. After one or two initial expansion timescales they filled the spherical volume. Their two-point correlation function was again a very good fit to r^{-2}. Yet traces of the original cube were still visible, showing that high order correlations take longer to relax, as expected. So it does not seem that the value of γ has much to say about the conditions of galaxy clustering.

Other experiments have started with strongly structured initial conditions such as pancakes, large amplitude sinusoidal perturbations, or perturbations with several power spectra and various cut-offs. Naturally, the clustering which results is closely related to the dominant initial information when these systems do not have the ability or time to relax completely. Whether such systems represent our own Universe is an open question.

Experiments also explore the growth in amplitude of $\xi(r, t) = a(t)r^{-\gamma}$ during its non-linear development. A 1000-body computation (Efstathiou, 1979) with Poisson initial conditions for $\Omega = 1$ showed that while R increased from 2.6 to 14.4, $a(t)$ increased by a factor of about 30. For the model with a current value of $\Omega = 0.1$, an increase of R from 2.6 to 31 increased $a(t)$ by about 30. Thus aR^{-3} is a decreasing function of time, as discussed after (25.25) in regard to the cosmic energy equation.

Another point involving the cosmic energy equation was the relation $\beta(t)$ between kinetic and potential correlation energies (25.16). A 400-body experiment (Miyoshi & Kihara, 1975) shows that after several relaxation timescales $\beta(t)$ settles down to fluctuate between about 0.6 and 0.7. Although the result (25.21) would suggest $\beta \approx \frac{2}{3}$ for $p \approx q$, the simple form for the time dependence of $U(t)$ probably does not apply at this stage (an unsettled question). Nevertheless, the result that β stops evolving significantly seems reasonable since in the non-linear regime most of the correlation energy is near virial equilibrium and new correlations over large scales contribute relatively less to the total.

Two-point correlation functions are clearly a very helpful and informative measure of clustering, but since they are far from a complete specification of clustering, we must press beyond them. What next? The obvious approach would be to examine the 3-, 4-, 5-,..., 20-,... (etc.) point correlations or the closely related Fourier spectrum of the density distribution. Considerable effort has gone into this approach, but neither the results nor their theoretical interpretation, so far, seem to have been particularly useful. Apart from being hard to measure and harder to interpret (see comments at the end of Section 24), the higher order correlations are closely bound up with one another in non-linear clustering. Beyond the two-point correlations, there is no stopping!

The situation calls for a bold leap. One possibility is to return to fitting models of clusters and the multiplicity function for groups. But then, just what is a group? Ambiguities in defining their boundaries (velocity as well as space) lead to a variety of models. Which would be the best characterization of the experimental density distribution? Would this model also be best for understanding the many-body dynamics? These questions have not yet been answered.

While awaiting better understanding, we can measure high order clustering from an extreme point of view, basically opposite to everything discussed so far. Instead of concentrating on clusters of galaxies, consider the voids where there are no galaxies. An empty region is relatively simple to find and measure. If its size would have contained n galaxies in a smooth distribution, then the hole is related to the $n + 1$- galaxy correlation function. This correlation function describes the n missing galaxies and the one galaxy defining the scale of the void. This scale is set by the galaxy nearest to that spatial point in the void which is farthest from the boundary. The first and most clear-cut step of this approach is to consider holes to be spherical. Later refinements could take more details of their shapes into account. Further extensions, as in Section 34, would also consider regions containing one galaxy, two galaxies, three galaxies, etc., until the description joins up with the descriptions in terms of groups and high order correlation functions.

As a first step in this direction the 4000-body experiments referred to earlier were analyzed for the evolution of voids (Aarseth & Saslaw, 1982). In the initial Poisson distribution it is easy to calculate the probability that a given size volume is empty.

This is related to the nearest neighbor distribution. Picking an arbitrary point of space we ask for the probability that its nearest galaxy is found between r and $r + dr$. This probability is

$\omega(r)dr = $ (probability of no galaxy between 0 and r) $\times$ (probability of a galaxy at r)

$$= \left[1 - \int_0^r \omega(r)dr \right] 4\pi nr^2 dr. \tag{27.3}$$

To solve this integral equation, simply rewrite it as

$$\frac{d}{dr}\left[\frac{\omega(r)}{4\pi nr^2} \right] = -\omega(r) = -\left[\frac{\omega(r)}{4\pi nr^2} \right] 4\pi nr^2, \tag{27.4}$$

whence

$$\omega(r)dr = 4\pi nr^2 \exp\left[-4\pi nr^3/3 \right] dr. \tag{27.5}$$

Now the probability that the nearest galaxy is between r and $r + dr$ is also equal to the probability P_0 that there is no galaxy in the volume V between 0 and r, times the probability ndV that there is a galaxy in the volume element V to $V + dV$

$$\omega(r)dr = nP_0 dV. \tag{27.6}$$

Comparing the last two equations shows that the probability of a void of volume V is

$$P_0(V) = e^{-nV} \tag{27.7}$$

for a Poisson distribution.

When clustering becomes significant it is possible to develop formal generalizations of (27.7) (see White, 1979). Unfortunately, such results are difficult to apply because their low order expansions do not appear to converge. Like the BBGKY hierarchy the nasty dominance of high order clusters rears up from the equations. We will see how to tackle this problem in Section 34. Meanwhile, we may seek guidance from N-body experiments.

Even the N-body experiments are not completely straightforward because the statistics of holes depend on their shapes. Non-spherical shapes are hard to analyze, but the question of determining the distribution of spherical holes is at least well-defined. They can be interpreted as the probability distribution of the minimum dimensions of irregularly shaped holes. Although the 4000-body simulations have a radius of 50 Mpc, the hole distribution is obtained from the inner 40 Mpc to minimize boundary effects. Throughout this inner volume select 12 000 spatial points uniformly at random. For each of these random points find the distance to its nearest galaxy. Let $N(r_{nn} > r)$ be the number of points for which the distance to the nearest neighbor galaxy is greater than r. Then the quantity

$$g(r) = \frac{1}{N_{tot}} N(r_{nn} > r) \tag{27.8}$$

is a good estimate of the probability $P_0(r)$ for finding an empty hole of volume $V = 4\pi r^3/3$. (A larger void contains many smaller voids.) It represents the cross correlation between galaxy positions and random spatial points.

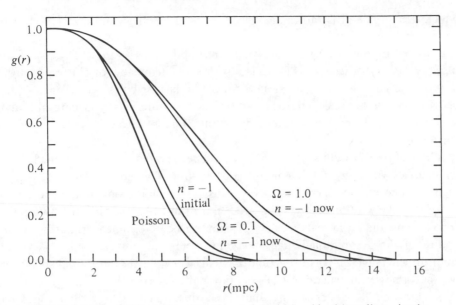

Fig. 29. The probability $g(r)$ of finding a spherical void with radius r in the distribution of galaxies with mass $m/m_* > 1$ in the 4000-body experiments. The probabilities are shown initially and at the present time for cosmologies with $n = -1$ and $\Omega = 1.0$ or 0.1. Initially both cosmologies have the same distribution. For comparison, the theoretical result for a Poisson distribution is also shown. (From Aarseth & Saslaw, 1982.)

Observations containing galaxies at known distances can also be analyzed directly using this algorithm. Clearly, the sizes of holes will depend on the cutoff in the mass function (the luminosity function in the observations). Even though the mass–luminosity relation for galaxies is very poorly known, the experiments and observations can be normalized to the same average number density.

Figure 29 (from Aarseth & Saslaw, 1982) shows the resulting probability for finding a hole of radius r in the $\Omega = 1$ and $\Omega = 0.1$ models with $n = -1$ and $m/m_* > 1$. About 650 galaxies are counted in each model with this mass cut-off, corresponding to an average density of about 2.5×10^{-3} galaxies Mpc^{-3}. Equation (27.7) with $N = 650$ is also plotted for comparison; it fits the initial Poisson conditions (not plotted) very well. Not unexpectedly, the initial distribution for $n = -1$ is somewhat more clustered than for $n = 0$, as evidenced by the higher probability of finding large holes. As the system evolves, holes grow on larger and larger scales until the distribution takes the present form in the diagram. Departures from Poisson statistics become substantial and are evident over all distances greater than one or two megaparsecs. Graphs with $n = 0$ are very similar to these.

The radius of the most probable hole is given by the peak of the differential distribution. This is the number of holes with radii between r and $r + dr$; it is essentially the derivative of the curve of Figure 29. Since the differential distributions are found to be fairly symmetric this peak occurs approximately at the value of r for which $g(r) = 0.5$ and is rather flat over two or three megaparsecs. The effect of changing the mass (luminosity) cutoff on the size of the most probable hole is mainly determined by the mean intergalaxy distance and scales as the cube root of the number density of galaxies. (There is also a slight tendency for clusters to form around the most massive galaxies.) Sizes of the largest holes on the tails of the distributions do not scale quite as well with the mean intergalaxy spacing, although it is a good first approximation, and are rather more affected by details of the clustering.

A hole of 10 Mpc radius, for example, would be expected to have contained about ten galaxies with $m/m_* > 1$ in a uniform distribution. Therefore, holes involve quite high order correlation functions. Despite this, the cosmology and initial conditions make relatively little difference to the clustering. So, as with the two-point correlations, the large scale clustering seems to be determined primarily by the nature of the gravitational many-body problem and only secondarily by the environment and origin of the fluctuations. This will make the observed distribution of holes a powerful general test of gravitational clustering theories.

Is there a measure of clustering which is sensitive to cosmology? So far we have examined only the development of clustering in position space. A complementary evolution is occurring in velocity space. To this we now turn.

28

Evolving velocity distributions

Naturally, peculiar velocities must grow consistently with increased spatial clustering. Again, N-body simulations provide valuable insight, especially into the non-linear regime. Since velocity distributions depend strongly on the amount of clustering it is important to separate galaxies according to the density contrast in their local neighborhood.

For the 4000-body experiments, for example, it is possible to use the integration scheme itself to determine the number density inside the nearest neighbor sphere surrounding any galaxy. This sphere is essentially where the fluctuating component dominates the gravitational force. The result is a measure of the local density contrast, which varies from about 10^4 or 10^5 in the centers of rich clusters to less than 0.5 in the field galaxies. A subset of field galaxies may be selected by also requiring them to be separated from their nearest neighbor by at least twice the average separation that a uniform random distribution would have. These most isolated galaxies are called extreme field galaxies. They are almost always well outside the haloes of clusters.

As an indication of the relative degree of isolation these criteria imply, at $R = 32$ (the present time) in the $\Omega = 0.1$, $n = -1$ model, about 23% of all the galaxies are in clusters (density contrast > 100), about 62% are intermediate ($100 >$ contrast > 0.5), 15% are field galaxies and 5% are extreme field galaxies. Various algorithms can be used for specifying concentration for both experiments and observations. Their important aspect is the relative degree of isolation.

Hubble streaks are the velocity analog of spatial clusters. To see them, first find the peculiar velocity v of each galaxy in the radial direction

$$v = v_r - rH \tag{28.1}$$

and in the tangential directions. A mythical observer can be imagined at the center of the sphere since the numerical simulations represent a typical region of the Universe. Plotting each galaxy's peculiar radial velocity versus its radial distance from the observer develops the Hubble streaks. Figure 30 (from Saslaw & Aarseth, 1982) shows them for the cluster galaxies at the present time in the $\Omega = 0.1$, $n = -1$ model.

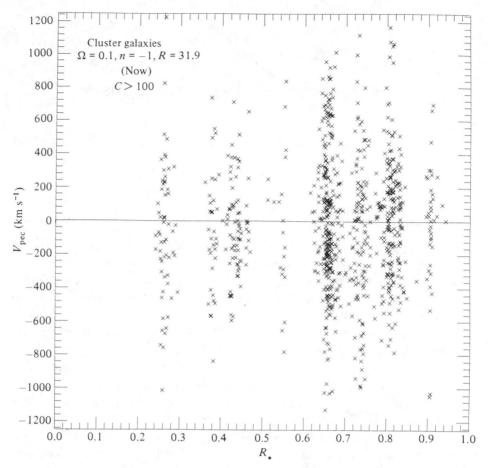

Fig. 30. The peculiar radial velocities of cluster galaxies at a given scaled radius $(R_* = r/R)$ in the $\Omega = 0.1$, $n = -1$ model at the present time. (From Saslaw & Aarseth, 1982.)

Fewer streaks are present at earlier times. Field galaxies are also much more uniformly distributed. Each Hubble streak contains the velocity dispersions of all the clusters at a given radius, creating an impression of the average strength of velocity clustering. Figure 26 with $R = 31.9$ shows the corresponding spatial clustering.

As the Universe expands there is a tendency for peculiar velocities to decay adiabatically: $v \propto R^{-1}$. In an extreme case, if the most isolated field galaxies were to behave like a gas of non-relativistic particles, they would be affected only by the smooth average background gravitational field. Then the adiabatic pressure–density relation, $P \propto \rho^{5/3} \propto R^{-5}$, combined with the relation between pressure and energy density, $P \propto Nv^2/R^3$, gives $v \propto R^{-1}$. Alternatively, one can derive this result directly from the equation of motion (26.3) for the co-moving peculiar velocity $v = \dot{r} - rH = R\dot{x}$. For a smooth background $\nabla\phi = 0$ and (26.3) becomes $\mathrm{d}(Rv)/\mathrm{d}t = 0$, whence $v \propto R^{-1}$.

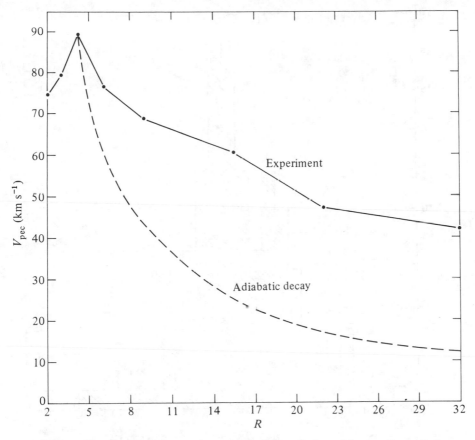

Fig. 31. Comparison between the experimental evolution of the peculiar radial velocity dispersion of extreme field galaxies and the adiabatic decay rate for the $\Omega = 0.1$, $n = -1$ numerical 4000-body simulation. (From Saslaw & Aarseth, 1982.)

Both these arguments are wrong because their premise is false. *N*-body experiments provided the insight that adiabatic expansion is not the dominant term in the motions. The non-uniform forces of distant forming clusters are more important, even for the most isolated field galaxies.

Figure 31 shows the discrepancy between the experimental results and adiabatic decay for the $\Omega = 0.1$, $n = -1$ simulation (others are similar). The numbers of extreme field galaxies at $R = 3$, 9 and 32 are 202, 197, and 176 respectively; that population does not change much. A velocity dispersion of about 90 km s^{-1} builds up rapidly from the initially cold flow. Then it decreases much more slowly than in adiabatic decay.

Once the departure from adiabatic decay for the extreme field peculiar velocities is recognized it can be understood simply. Perturbations of extreme field galaxies by large (though distant) clusters can nearly counteract the effects of adiabatic expansion. To see this in the simplest case, consider the co-moving equation of motion (26.3) for an extreme field galaxy being perturbed by just one cluster of mass

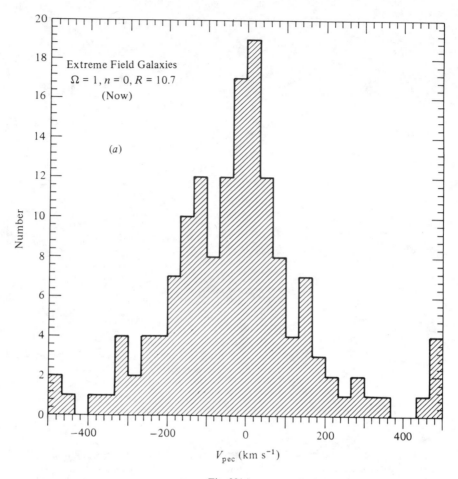

Fig. 32(*a*)

M a co-moving distance r away

$$\frac{\mathrm{d}(Rv)}{\mathrm{d}t} = \frac{GM}{Rr^2}. \tag{28.2}$$

For an approximate solution, once significant clustering has begun, regard $M(t)$ and r^{-2} as average values and integrate using a given expansion $R(t)$. For example when $\Omega = 0.1$ the Universe expands nearly as an empty model for $R \gtrsim 3$. (Recalling that $R = 3$ corresponds to a redshift $z \approx 10$, observe that the density term in the right hand side of Equation (21.31) becomes unimportant so $R \propto t$.) Therefore,

$$v \approx \frac{R_0 v_0}{R} + 5 \times 10^3 \ln\left(\frac{R}{R_0}\right) \langle M_{14} \rangle \langle r_{\mathrm{Mpc}}^{-2} \rangle t_{10} \ \mathrm{km \, s^{-1}}. \tag{28.3}$$

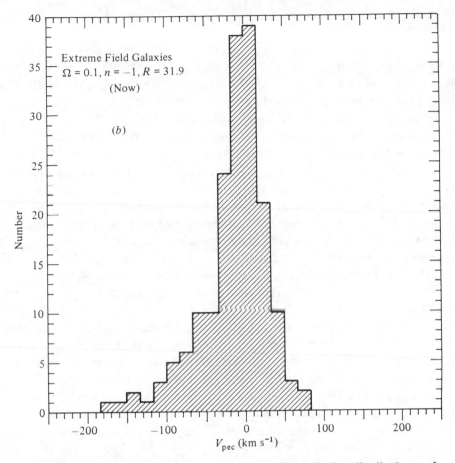

Fig. 32(*b*) The different present peculiar radial velocity distributions of extreme field galaxies in the high density $\Omega = 1$ and low density $\Omega = 0.1$ models, which both fit the observed two-galaxy spatial correlation function well. (From Saslaw & Aarseth, 1982.)

Here R_0 and v_0 are initial values, the time average cluster mass is measured in units of 10^{14} M $_\odot$, and the present age of the Universe t is in units of 10^{10} yr. The current physical distance, given by $\langle r_{\text{Mpc}}^{-2} \rangle = R^{-2}(t)\langle r^{-2} \rangle$ is measured in megaparsecs.

The first term is just the adiabatic decay. With astronomically reasonable values of the parameters it is therefore easy to build up peculiar velocities of ~ 50 km s^{-1} by the perturbation of the nearest large cluster. Of course, this gives only a guide to the non-linear development, in rough agreement with Figure 31.

Results of *N*-body experiments do not have to be accepted on faith alone. Often it is possible to think of checks for internal consistency. Galileo suggested an appropriate check in this case. We want to be sure that the velocity dispersion of extreme field galaxies is produced by perturbations of large clusters rather than by neighbor galaxies or ejection from a small group. One way is to see if the velocity dispersion of subsets of extreme field galaxies with different masses is essentially independent of their mass. It is. Over a mass range of about 20:1, the velocities differ by only about 20%.

A second check is to actually follow the orbits of extreme field galaxies and see if they passed through groups. In the $\Omega = 0.1$ model, very few have been ejected from a group. However, in the $\Omega = 1.0$ model, nearly half of the ten highest velocity galaxies passed through groups. This might lead to an interesting way to determine Ω if only we understood the consequences (distorted galaxies?, strange rotation curves?, peculiar attacks of star formation?) of belonging to a group. This result is related to the faster rate of group formation in higher Ω cosmologies, but it is not understood very well.

Other methods of determining Ω, more amenable to observations, also emerge. Since the whole history of clustering does depend significantly on Ω, and since the peculiar velocities of extreme field galaxies accumulate as a direct result of this clustering, these peculiar velocities should also depend strongly on Ω. This turns out to be the case. Field galaxies have a memory which stores the history of clustering.

Figure 32, for example, shows the present extreme field velocity histograms of the $\Omega = 0.1, n = -1$ and the $\Omega = 1.0, n = 0$ experiments. Both models have about 160 of these galaxies and both models have two-point correlation functions in agreement with the observations. So not much difference there. But the velocity distributions are completely different! The dispersion in the $\Omega = 1$ case is $185 \, \mathrm{km \, s^{-1}}$, while in the $\Omega = 0.1$ case it is $42 \, \mathrm{km \, s^{-1}}$. This difference also shows up, to a lesser degree, in the less isolated galaxies. The velocity dispersion of field galaxies is an excellent measure of very large scale spatial correlations.

As a check we can also examine the velocity dispersion among the centers of mass of different galaxy clusters. To some extent they should also act as test particles and share in this phenomenon. Results here are somewhat harder to interpret because the clusters and groups change their membership significantly between $z \approx 1$ and $z \approx 0$ (unlike the extreme field galaxies). This is an active period of cluster growth. Nevertheless, it is possible to follow particular clusters as they acquire members. In the $\Omega = 0.1, n = -1$ model, the velocity dispersion of the centers of mass of the 30 most concentrated clusters decreases from $62 \, \mathrm{km \, s^{-1}}$ at $z = 1$ to $50 \, \mathrm{km \, s^{-1}}$ at $z = 0$. In the $\Omega = 1, n = -1$ model it increases from $140 \, \mathrm{km \, s^{-1}}$ to $220 \, \mathrm{km \, s^{-1}}$ during this period, indicating a rapid build up of cluster peculiar velocities as the clusters form. So for cluster centers, too, the perturbations of neighboring clusters and the added velocity as clusters accrete members both swamp adiabatic decay. Again the velocity distribution of field objects is a powerful discriminator between different histories of spatial clustering.

These are exciting times for the observations as well. New techniques are making rapid redshift surveys possible. Large complete samples of field galaxies will help us understand contributions from large scale flows in the local supercluster of galaxies. Better secondary distance indicators, such as neutral hydrogen measures or spectral features related to absolute optical luminosity, will establish more accurate peculiar velocities. It is much too early to say what the final results will be but the preliminary trends in the data point to low densities of matter in clusters of galaxies. For the Universe to be closed it would then have to contain a nearly uniform invisible distribution of matter which did not exert significant net gravitational forces on the field galaxies.

29

Short review of basic thermodynamics

29.1. Concepts

So far, we have followed two broad avenues of insight into the instability and clustering of infinite gravitating systems: linear kinetic theory and numerical N-body simulations. Now we turn onto a third avenue: thermodynamics. Classical thermodynamics is a theory of great scope and generality. It survived the relativity and quantum mechanical revolutions of physics nearly intact. In part, this was because among all theories of physics thermodynamics has the least physical content. Its statements relate very general quantities which must be defined anew, through equations of state, for each specific application. With this view, it is natural to ask whether thermodynamics also subsumes gravitating systems.

The answer is yes, with certain caveats and qualifications. Results of gravitational thermodynamics – gravithermodynamics, or GTD for short – are often surprising and counter-intuitive compared to the thermodynamics of ordinary gases. Specific heats, for example, can be negative and equilibrium is a more distant ideal. Basically, these differences are caused by the long-range, unsaturated (unshielded) nature of gravitational forces. As a result, rigorous understanding of GTD is less certain than for ordinary thermodynamics. The present situation is a bit similar to the early thermodynamic gropings of Watt, Carnot, Kelvin and Joule.

Straightforward introduction of gravity into thermodynamics leads again to the Jeans instability from a new point of view. It links up with linear kinetic theory and provides new insight into non-linear clustering. Later, in Part III, we will see how it also describes important aspects of stability and evolution in finite star clusters.

This section reviews ideas and relations from ordinary thermodynamics which we will need as we go along. It makes no attempt to be exhaustive, wide ranging, or subtle, for many texts discuss this rich subject in great detail. (The bibliography gives some examples.) Our goal here is mainly to establish a clear description and consistent notation. Thermodynamics sometimes seems more complicated than it really is because different sets of dependent and independent variables are useful in different situations. Lack of attention to such details can cause utter confusion.

A basic premise of classical thermodynamics is that systems in equilibrium can be characterized by a finite set of macroscopic parameters. These may be averages of microscopic parameters which are not known in detail and which vary in nature and degree among different types of systems. To discover that a real system is in equilibrium would be to disturb it. Therefore, equilibrium is always an idealization and means that macroscopic disturbances occur on timescales very long compared to microscopic relaxation timescales.

Macroscopic parameters which describe the system normally include the total internal energy U, the entropy S, the volume V which the system occupies, and the number of objects N it contains. If it contains more than one type of object each species is characterized by its number N_i. Number and volume are obvious parameters. The reason for using the internal energy is that it is a conserved quantity in the system, usually the only one.

Entropy, however, is not so obvious. A second basic premise of thermodynamics is that one can imagine an ensemble of similar systems, each with different values of their macroscopic parameters. If systems in the ensemble are allowed to interact very weakly, just enough to reach equilibrium, then the most probable value of the macroscopic parameters follows from maximizing the entropy. Thus entropy will be a function of the other parameters, $S = S(U, V, N)$. This relation is called the fundamental equation of thermodynamics in the entropy representation. It can be inverted to give the fundamental equation in the energy representation: $U = U(S, V, N)$. It is always important to remember which representation is being used.

Entropy, like the conductor of a symphony orchestra, determines the roles of the other players. The idea that the maximum value of one thermodynamic function can determine the most probable value of other parameters evolved from variational principles in classical mechanics. Gibbs' bold leap from one system to an ensemble created the statistical mechanical foundation of thermodynamics.

A small change in the parameters of the system is represented by the differential of the fundamental equation. In the energy representation this is

$$dU = \frac{\partial U}{\partial S}\bigg)_{V,N} dS + \frac{\partial U}{\partial V}\bigg)_{S,N} dV + \frac{\partial U}{\partial N}\bigg)_{S,V} dN = TdS - PdV + \mu dN. \quad (29.1)$$

Quantities held constant in a partial derivative are often written explicitly for clarity, and as a reminder of the representation used. The three derivatives here are the temperature, pressure (conventionally defined with a minus sign so that energy decreases as volume increases), and the chemical potential. Similarly, in the entropy representation

$$dS = \frac{\partial S}{\partial U}\bigg)_{V,N} dU + \frac{\partial S}{\partial V}\bigg)_{U,N} dV + \frac{\partial S}{\partial N}\bigg)_{U,V} dN = \frac{1}{T}dU + \frac{P}{T}dV - \frac{\mu}{T}dN, \quad (29.2)$$

noting that

$$\left.\frac{\partial S}{\partial V}\right)_{U,N} = -\frac{(\partial U/\partial V)_{S,N}}{(\partial U/\partial S)_{V,N}}. \tag{29.3}$$

Of course, (29.2) could have been derived from (29.1) just by transposing terms, but that would have been neither general nor instructive.

Equation (29.3) is a simple instance of relations among implicit partial derivatives. Since these relations pervade thermodynamics it is worth pausing to review them. Consider an arbitrary continuous function $A(x, y, z)$ of, say, three variables. We can relate the partial derivatives of A since given values of $A(x, y, z)$ define implicit relations among its independent variables. When A changes, its total differential is

$$dA = \left.\frac{\partial A}{\partial x}\right)_{y,z} dx + \left.\frac{\partial A}{\partial y}\right)_{z,x} dy + \left.\frac{\partial A}{\partial z}\right)_{x,y} dz. \tag{29.4}$$

Since this relation holds for all values of dA, we may choose $dA = 0$ and, in addition, keep one of the variables constant, $dz = 0$ say. The result, dividing through by dx, is

$$\left.\frac{\partial y}{\partial x}\right)_{A,z} = -\frac{(\partial A/\partial x)_{y,z}}{(\partial A/\partial y)_{z,x}}, \tag{29.5}$$

which has the form of (29.3). Similar relations follow by setting $dx = 0$ and $dy = 0$, which is equivalent to cyclic permutation of x, y, and z, i.e., letting $x \to y$, $y \to z$ and $z \to x$. If we divide (29.4) through by dy to give the inverse of (29.5), we see

$$\left.\frac{\partial x}{\partial y}\right)_{A,z} = \frac{1}{(\partial y/\partial x)_{A,z}}. \tag{29.6}$$

Multiplying the three forms of (29.5) together and using (29.6) shows that

$$\left(\frac{\partial x}{\partial y}\right)_{A,z} \left(\frac{\partial y}{\partial z}\right)_{A,x} \left(\frac{\partial z}{\partial x}\right)_{A,y} = -1. \tag{29.7}$$

Another relation which is often helpful is a chain rule when x, y, and z are themselves functions of a parameter w, so that $dx = (dx/dw)dw$, etc. Substituting these three differential relations into (29.4) for $dA = dz = 0$, and comparing the result with (29.5), gives

$$\left.\frac{\partial y}{\partial x}\right)_{A,z} = \frac{(\partial y/\partial w)_{A,z}}{(\partial x/\partial w)_{A,z}}. \tag{29.8}$$

Returning to (29.1) we see that the temperature, pressure, and chemical potential (whose physical justifications are left to the standard textbooks) are all derivatives of the fundamental equation. Thus these variables play a different role than S, V and N. This shows up in the relation between internal energy, heat, and work.

Heat and work are forms of energy transfer. Unlike internal energy a system does not contain a certain quantity of work or heat. These are processes. In fact, the amount of heat and work needed to transfer a given quantity of energy depends strongly on how the transfer is made. An early triumph of experimental calorimetry was to show that a small quasi-static transfer (i.e., between states very close to equilibrium) of heat, đQ, can be associated with a quantity whose increase is independent of the transfer method. The trick was to divide the heat change by the temperature

$$\frac{\text{đ}Q}{T} = \text{d}S, \tag{29.9}$$

and the resulting quantity is the entropy. The symbol đQ indicates an 'imperfect differential', one whose integrated value depends on the path of integration in a thermodynamic S–V–N phase space. Heat is energy transfer into the relative 'random' microscopic motions of objects in the system.

The total internal energy transferred quasi-statically is the sum of the heat, mechanical work and chemical work done on the system

$$\text{d}U = \text{đ}Q + \text{đ}W_{\text{m}} + \text{đ}W_{\text{c}} \tag{29.10}$$

Substituting (29.9) and comparing with (29.1) shows that the mechanical and chemical quasi-static work are

$$\text{đ}W_{\text{m}} = -P\text{d}V \tag{29.11}$$

and

$$\text{đ}W_{\text{c}} = \mu\text{d}N. \tag{29.12}$$

Since dU and PdV can be determined experimentally, (29.10) provides a way to measure the amount of energy transferred as heat. It also makes clear how the choice of fundamental variables determines that the mechanical work is $-P$dV. Otherwise, confusion could have arisen from a simple expectation that dU might also contain a VdP term.

Another important difference between the fundamental and derivative variables is their extensive–intensive nature. Suppose we imagine two systems which do not interact. Then the values of the fundamental variables, U, S, V and N, for the combined system are the sum of their values for the individual systems. These quantities are 'extensive'. On the other hand, combining non-interacting systems does not change the values of their derivative variables, T, P, and μ. They are 'intensive'. For the special case of perfect gases a stronger definition of extensive is possible. Instead of imagining non-interacting systems we consider an actual system and subdivide it into two parts with an imaginary wall. Then the extensive and intensive properties apply to interacting regions of actual systems. This applies because the interaction is so weak – just strong enough to create equilibrium. Imperfect gases with long-range correlations, or gravitating systems with long-range forces, do not always satisfy this more restrictive definition of extensive, so one must be a bit careful in drawing consequences from it.

With ordinary thermodynamics, the equality of T, P, or μ between two actual subsystems means they are in thermal, mechanical, or chemical equilibrium respectively. This follows from the entropy maximum principle. For example, consider two subsystems which have a constant total energy $U = U_1 + U_2$ and are separated by a rigid wall which transmits heat but not matter. The total entropy $S = S_1 + S_2$. Maximizing total entropy at constant V_1, V_2, N_1 and N_2 then gives, from (29.2),

$$\mathrm{d}S = 0 = \mathrm{d}S_1 + \mathrm{d}S_2 = \frac{1}{T_1}\mathrm{d}U_1 + \frac{1}{T_2}\mathrm{d}U_2 = \left(\frac{1}{T_1} - \frac{1}{T_2}\right)\mathrm{d}U_1, \qquad (29.13)$$

whence $T_1 = T_2$ at equilibrium. Results for P and μ follow similarly. The extensive property of entropy and the conservation of energy are essential to this line of argument.

29.2. Interrelations

Our discussion so far has summarized the framework for describing thermodynamic information. It is useful to emphasize one or another part of this framework in different situations. Therefore we now summarize the interrelations between different descriptions which, however, all contain the same basic information.

From (29.1) we obtain the intensive parameters as functions of the basic extensive variables: $T = T(S, V, N)$, $P = P(S, V, N)$, $\mu = \mu(S, V, N)$. These are the 'equations of state'. In terms of the basic variables, each equation of state is a partial differential equation. Therefore, it cannot contain as much information about the system as the fundamental equation: a function of integration, at least, is missing.

The set of all equations of state can, however, be equivalent to the fundamental equation in information content. Since the fundamental equation depends on extrinsic variables, it is first order homogeneous

$$U(\lambda S, \lambda V, \lambda N) = \lambda U(S, V, N). \qquad (29.14)$$

Differentiating both sides with respect to λ and then selecting the particular value $\lambda = 1$ gives the Euler equation

$$U = TS - PV + \mu N. \qquad (29.15)$$

So knowledge of $T(S, V, N)$, $P(S, V, N)$ and $\mu(S, V, N)$ enables us to retrieve the fundamental equation $U(S, V, N)$.

An equation which complements (29.1) and involves differentials of the intensive variables can now be derived by taking the differential of (29.15) and subtracting (29.1) from it. The result is called the Gibbs–Duhem relation

$$S\mathrm{d}T - V\mathrm{d}P + N\mathrm{d}\mu = 0. \qquad (29.16)$$

Thus variations of the intensive parameters are not completely independent. By substituting $S(T, P, N)$, $V(T, P, N)$ and $\mu(T, P, N)$ into (29.16) and integrating we can find a relation between T, P, and μ; alternatively we can relate T, P and μ by eliminating S, V and N from the equations of state. Note that the relation between P, V, T and N, which for perfect gases is called *the* 'equation of state', is somewhat of a hybrid representation and does not contain complete information about the thermodynamics.

Everything so far can also be restated in the entropy representation which uses

$S(U, V, N)$ as the fundamental equation. Then the Gibbs–Duhem relation, for example, becomes

$$U\mathrm{d}\left(\frac{1}{T}\right) + V\mathrm{d}\left(\frac{P}{T}\right) - N\mathrm{d}\left(\frac{\mu}{T}\right) = 0. \tag{29.17}$$

Some thermodynamic derivatives have an obvious physical meaning and are often used for convenient expressions of relationships. Among these are the isothermal compressibility

$$\kappa_{\mathrm{T}} \equiv -\frac{1}{V}\frac{\partial V}{\partial P}\bigg)_{N,T}, \tag{29.18}$$

the coefficient of thermal expansion

$$\alpha \equiv \frac{1}{V}\frac{\partial V}{\partial T}\bigg)_{N,P}, \tag{29.19}$$

the specific heat at constant volume

$$CV \equiv \frac{T}{N}\frac{\partial S}{\partial T}\bigg)_{N,V} = \frac{1}{N}\frac{\mathrm{d}Q}{\mathrm{d}T}\bigg)_{N,V}, \tag{29.20}$$

and the specific heat at constant pressure

$$C_{\mathrm{P}} \equiv \frac{T}{N}\frac{\partial S}{\partial T}\bigg)_{N,P} = \frac{1}{N}\frac{\mathrm{d}Q}{\mathrm{d}T}\bigg)_{N,P}. \tag{29.21}$$

Each involves the fractional change of an important variable under specified conditions.

Other thermodynamic derivatives can often be conveniently reformulated by making use of the equality of mixed partial derivatives of the fundamental equation. For example, from

$$\frac{\partial^2 U}{\partial V \partial S} = \frac{\partial^2 U}{\partial S \partial V} \tag{29.22}$$

we readily get

$$\frac{\partial T}{\partial V}\bigg)_{S,N} = -\frac{\partial P}{\partial S}\bigg)_{V,N}. \tag{29.23}$$

This is an instance of the 'Maxwell relations' of which there are myriads from each basic representation and choice of fundamental variables.

The fundamental variables S, V and N may not always be best for a particular situation. For example, we may want to rewrite the fundamental equation for two systems in thermal equilibrium in terms of their common temperature T, along with V and N, particularly since T is more readily measurable than their entropy S. Naive substitution of $S(T, V, N)$ into $U(S, V, N)$ to produce $U(T, V, N)$ does not work. Although it produces a perfectly correct relation $U(T, V, N)$ is a first order partial differential equation whose integral contains an undetermined function. Thus the information contents of $U(S, V, N)$ and $U(T, V, N)$ are not equivalent.

Finding an equivalent formulation is easy with the Legendre transform. The basic idea is to replace a point-by-point description of a curve (or surface) by a specification of its tangent lines (or planes). A simple illustration with one variable exemplifies this. Suppose the internal energy was just a function of entropy $U(S)$ and we wanted to replace this with a function of temperature $F(T)$ containing equivalent information. Instead of the point-by-point specification $U(S)$ we consider its tangent whose slope at any point is

$$T(S) = \frac{dU}{dS}. \tag{29.24}$$

But to specify the tangent fully requires its intercept F, as well as its slope. Thus in the U–S plane, the tangent is (see Figure 33)

$$T = \frac{\Delta U}{\Delta S} = \frac{U - F}{S - 0}. \tag{29.25}$$

The key idea is to use the intercept to specify the new function of T, or, rearranging (29.25),

$$F = U - TS = U - \frac{dU}{dS}S. \tag{29.26}$$

Finally, using $U(S)$ to eliminate U in (29.26) and substituting $S(T)$ from inverting (29.24) gives $F(T)$. This is the Legendre transform of $U(S)$. Starting with $F(T)$ we can recover $U(S)$ by inverting the process, except at a critical point where $d^2F/dT^2 = 0$. This last condition is related to the requirement that the system be thermodynamically stable; mathematically it ensures that T depends on S.

The symbols of this example were chosen with knowledge aforethought. $F(T, V, N)$ is the free energy and is derived from $U(S, V, N)$ as in (29.26) by keeping the other variables constant. The free energy is one example of a thermodynamic potential. Other Legendre transforms lead to other potentials. If pressure is the relevant variable, then the enthalpy

$$H(S, P, N) = U + PV \tag{29.27}$$

is useful. If both temperature and pressure are specified, we can use the Gibbs function

$$G(T, P, N) = U - TS + PV \tag{29.28}$$

and so on. The equivalence of partial second derivatives of the thermodynamic potentials breeds whole new sets of Maxwell relations. Legendre transformations may also be applied in the entropy representation where the results are called generalized Massieu functions rather than thermodynamic potentials. An example is $S(P/T) = S - (P/T)V$ which is a function of U, P/T and N.

A major role of thermodynamic potentials is to describe systems in equilibrium at constant temperatures, pressure, or chemical potential. For example, if a system can exchange energy with a reservoir (another system of such great heat capacity that its temperature remains unchanged), then the equilibrium values of unconstrained

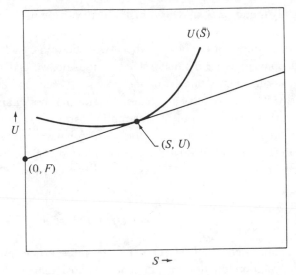

Fig. 33. Specification of the curve $U(S)$ by means of the slope and intercept of its tangent.

internal thermodynamic parameters in the system minimize the free energy at the constant temperature of the reservoir. To see this, recall from standard thermodynamics texts that in stable equilibrium the total system and reservoir energy is a minimum

$$d(U + U_r) = 0 \qquad (29.29)$$

and

$$d^2(U + U_r) = d^2 U > 0 \qquad (29.30)$$

when the entropy is constant so that

$$d(S + S_r) = 0 \qquad (29.31)$$

The reason $d^2 U_r - 0$ is that all the differentials of intensive parameters vanish, by definition, for the mighty reservoir. Thus

$$d(U + U_r) = dU + T_r dS_r = 0 \qquad (29.32)$$

and, since $dS_r = - dS$ and $T_r = T$, we get

$$d(U - TS) = dF = 0. \qquad (29.33)$$

Finally (29.30) shows that $d^2 F > 0$. Similarly, systems in mechanical contact with a pressure reservoir have constant enthalpy.

29.3. Connections with kinetic theory and statistical mechanics

Although thermodynamics relates various quantities it does not tell us what those quantities *are* for any particular physical system. We need to find the actual equations of state, or the fundamental equation. Links with kinetic theory and statistical mechanics make this step possible.

We have already found one link with kinetic theory through the relation (25.15) between gravitational potential energy and the two-point correlation function $\xi(r)$. This provides an example of confusion when two notations, each standard in a different subject, come together. To help remedy confusion, we shall continue to use U for the total internal energy in Part II, and now let W become the gravitational correlation energy. We will not have much more use for thermodynamic 'work' here, which only occurs in an incomplete differential form anyway.) Through the complete single-particle distribution function we also have a link with the kinetic energy, Equation (25.14), and thus with the total internal energy. For an ordinary homogeneous gas, whose random velocities are represented by a temperature T, the internal energy is

$$\frac{U}{NkT} = \frac{3}{2} + \frac{n}{2kT} \int_0^\infty \phi(r)\xi(r, n, T)4\pi r^2 \mathrm{d}r, \qquad (29.34)$$

where k is Boltzmann's constant. Here $\phi(r)$ is the interparticle potential energy. Generally, the correlation function may depend on density $n = N/V$, and temperature as well as separation.

This result is not a fundamental equation of the system since it is not of the form $U(S, V, N)$. Nor is it an equation of state in the energy representation since it is not of the form $T(S, V, N)$. But it is an equation of state, $T^{-1}(U, V, N)$, in the entropy representation since $T^{-1}(U, V, N) = \partial S(U, V, N)/\partial U$.

To find the fundamental equation and develop the thermodynamics we need two more equations of state. From physical intuition we would expect the pressure to equal that of a perfect gas minus a term proportional to the interaction energy density. Detailed discussions by Fowler (1936, Chapters 8 and 9) show this is indeed the case

$$\frac{P}{kT} = n - \frac{n^2}{6kT} \int_0^\infty r \frac{\mathrm{d}\phi(r)}{\mathrm{d}r} \xi(r, n, T)4\pi r^2 \mathrm{d}r. \qquad (29.35)$$

(Section 30.2 discusses one derivation of this result further.) Now we can integrate the Gibbs–Duhem equation (29.17) to find the third equation of state

$$\frac{\mu}{T} = \frac{\mu}{T}(U, V, N) = \frac{\mu}{T}(u, v) \qquad (29.36)$$

where the equations are all expressed in terms of $u = U/N$ and $v = V/N$. The three equations of state can then be substituted into the entropy representation of the Euler equation,

$$S(U, V, N) = \frac{1}{T}U + \frac{P}{T}V - \frac{\mu}{T}N \qquad (29.37)$$

to give the fundamental equation. This procedure forges the link between thermodynamics and kinetic theory. As an exercise the reader can apply it when $\xi = 0$ to obtain the entropy of a perfect monotomic gas

$$S = \frac{N}{N_0}S_0 + NR \ln\left[\left(\frac{U}{U_0}\right)^{3/2}\left(\frac{V}{V_0}\right)\left(\frac{N}{N_0}\right)^{-5/2}\right], \qquad (29.38)$$

where the fiducial entropy is

$$S_0 = \frac{5}{2} N_0 R - N_0 \left(\frac{\mu}{T} \right)_0. \tag{29.39}$$

There are equivalent, alternative paths to the fundamental equation. Two other paths are especially useful. One is to find the specific entropy $s = S/N$ directly by integrating

$$ds = \frac{1}{T} du + \frac{P}{T} dv. \tag{29.40}$$

The second is to calculate the Legendre transform of the entropy. Since all our equations of state are expressed in terms of temperature, the appropriate Legendre transform is the Massieu function

$$S(1/T, V, N) \equiv S(U, V, N) - \frac{\partial S}{\partial U} U = S - \frac{1}{T} U = -\frac{F}{T}. \tag{29.41}$$

This Massieu function is trivially related to the free energy. We call it the free entropy. Since S is a function of U, N and V, Equation (29.41) gives the relation

$$\frac{\partial (F/T)}{\partial (1/T)} \bigg)_{N,V} = U. \tag{29.42}$$

Therefore, an integration provides another version of the fundamental equation for thermodynamics of kinetic systems.

Next we turn to links between thermodynamics and statistical mechanics. These provide other routes toward equations of state, routes which are sometimes more convenient or easier to follow. The basic principles are available in a multitude of texts, so again we just recall the most important ones here.

To each reservoir, and its associated thermodynamic potential, there corresponds an ensemble of statistical mechanics. These ensembles are imaginary collections whose members each replicate the macroscopic properties of a system. Microscopic properties of the system – e.g., which star is where – generally differ from member to member. An ensemble whose members are completely isolated and have well-defined energies, volumes and numbers of objects, is called the 'microcanonical ensemble'. Its fundamental equation depends just on these variables: $S(U, V, N)$. If the members of the ensemble are in thermal equilibrium with a heat reservoir they define a 'canonical ensemble' whose fundamental equation is the free energy $F(T, V, N)$. If the members of the ensemble are in equilibrium with both thermal and particle reservoirs they define the 'grand canonical ensemble'. Its fundamental equation is the grand canonical potential $U(T, V, \mu) = U(S, V, N) - TS - \mu N$.

Boltzmann and Planck provided the fundamental statistical plank underlying thermodynamics. The entropy of a macrostate of the microcanonical ensemble is related to the number Ω of microscopic configurations the ensemble has for that macrostate by

$$S = k \ln \Omega. \tag{29.43}$$

In this basic equation k is Boltzmann's constant. That something like this relation holds is plausible because entropy is additive while the probabilities of microscopic states are multiplicative. The amount of mystery (another variable of statistical mechanics?) in this remark is reduced by considering two independent systems. Their combined entropy is $S_{12} = S_1 + S_2$. Their combined microscopic probability is $\Omega_{12} = \Omega_1\Omega_2$. If S is to be an extensive function $f(\Omega)$ just of Ω, then $f(\Omega_1\Omega_2) = f(\Omega_1) + f(\Omega_2)$. The general solution of this functional equation has the form of (29.43).

To apply Boltzmann's equation (29.43) we must learn how to count microstates. Each microstate is assumed to have an equal *a priori* probability of occurring. Some macrostates of the system, however, are consistent with more arrangements of microstates. These macrostates have greater entropy and a higher probability of occurring in the ensemble. When the counting is done for a classical gas, standard texts on statistical mechanics show that

$$S = k \ln Z + \frac{U}{T}, \tag{29.44}$$

where Z is the partition function

$$Z = \sum_i e^{-\varepsilon_i/kT}. \tag{29.45}$$

The sum is taken over all the different energy states ε_i of the system. If the number of objects in each energy state is n_i, then $\Sigma_i n_i \varepsilon_i = U$ and $\Sigma_i n_i = N$. It is now possible, but usually complicated, to express T in terms of U, V, N to get $S(U, V, N)$. Instead it is more natural using this approach to deal with the fundamental equation for the free energy

$$F = U - TS = -kT \ln Z \tag{29.46}$$

from the two previous equations. Generally, the energy states depend on the volume of the system. Knowledge of the partition function, therefore, tells us all the thermodynamic parameters. For example

$$U = F + TS = -T^2 \frac{\partial}{\partial T}\left(\frac{F}{T}\right)_V = \frac{kT^2}{Z}\frac{\partial Z}{\partial T}, \tag{29.47}$$

using (29.46) for the last step. Combining this result with (29.44) expresses the entropy directly in terms of Z. The rest of thermodynamics follows by the methods described earlier. Of course, the partition function must be worked out for each specific system.

When working out these details it is often useful to take the 'thermodynamic limit' of an infinite system. The reason is that the energy states, which make up the partition function, generally depend on the size and shape of the system. Hidden in (29.45) is an implicit dependence on volume. Thus, the resulting intensive parameters could depend on the shape of the system. Moreover, they may also depend on which ensemble is used for their calculation. The thermodynamic limit avoids these problems, ordinarily, by calculating Z in the limit $N \to \infty$, $V \to \infty$ with $N/V \to$ constant. It is the same general idea we have met before in going from the kinetic description to a fluid description. This limit has been proved rigorously to exist in many, but not all, conditions. Originally the main restrictions were that the interparticle potential decrease faster than r^{-3} as $r \to \infty$ and increase faster than r^{-3} as $r \to 0$. The repulsive core prevents collapse, while the long-range decay prevents global properties from influencing the thermodynamics. Later the proofs were extended to inverse square forces under certain conditions, to be described in Section 30.

29.4. Fluctuations and phase transitions

Equilibrium thermodynamics deals with average macroscopic quantities. In reality, objects are always buzzing around inside the system and, in any subregion, the macroscopic quantities will fluctuate around their average values. A remarkable aspect of equilibrium thermodynamics is its ability to describe these fluctuations. Although fluctuations are normally small, they can become enormous in gravitational systems. Giant fluctuations are forerunners of instability.

A straightforward way to calculate fluctuations is to imagine many closed systems of the same fixed N, V, and T in equilibrium with a heat reservoir. The ebb and flow of heat between systems and the reservoir makes the energy of each system fluctuate. The average internal energy is, using (29.45),

$$ U = \bar{\varepsilon}_i = \frac{\sum\limits_i \varepsilon_i e^{-\varepsilon_i/kT}}{\sum\limits_i e^{-\varepsilon_i kT}}. \tag{29.48}$$

Differentiating this with respect to $1/kT$ without doing any external work on the system ($\varepsilon_i = \text{constant}$) gives the mean square fluctuation of U

$$ \frac{\partial U}{\partial(1/kT)}\bigg)_{N,V} = -kT^2 \frac{\partial U}{\partial T}\bigg)_{N,V} = -\frac{\sum \varepsilon_i^2 e^{-\varepsilon_i/kT}}{\sum e^{-\varepsilon_i/kT}} + \left(\frac{\sum \varepsilon_i e^{-\varepsilon_i/kT}}{\sum e^{-\varepsilon_i/kT}}\right)^2 \tag{29.49}$$

or

$$ kT^2 \frac{\partial U}{\partial T}\bigg)_{N,V} = \overline{\varepsilon_i^2} - \bar{\varepsilon}_i^2 = (\Delta U)^2. \tag{29.50}$$

Equation (29.20) shows that the fluctuations are closely related to the specific heat at constant volume. When a large change of internal energy for a system of given N and V can be accomplished with a small temperature change, large fluctuations can grow. By using other ensembles, fluctuations of N and V can be calculated similarly. More powerful techniques are available for mixed and higher order moments (Callen, 1966).

Many systems are partially described by an equation of state $P(T, V, N)$, so a particularly useful measure of their fluctuations is the mean square volume change

$$ (\Delta V)^2 = -\frac{\partial V}{\partial(P/kT)}\bigg)_{1/T,N} = -kT \frac{\partial V}{\partial P}\bigg)_{T,N} = kTV\kappa_T. \tag{29.51}$$

Similarly, holding the volume fixed and allowing the number of objects within it to fluctuate gives

$$(\Delta N)^2 = -\frac{kTN^2}{V^2}\frac{\partial V}{\partial P}\bigg)_{T,N} = \frac{kTN}{V}\frac{\partial N}{\partial P}\bigg)_{T,V}. \tag{29.52}$$

For a perfect gas, $PV = NkT$, we recover $\sqrt{N}$ fluctuations in the fractional volume. Both these cases are for systems in mechanical equilibrium with a pressure reservoir.

Infinite reservoirs of the intensive parameters are, of course, idealizations. In practice, it is usually possible to put an imaginary boundary around a subregion within a system, and consider the rest of the system to be the reservoir. Then these fluctuations apply to subregions within the system of interest. They can apply either to one subregion at different times, or to variations among many subregions at the same time.

For certain regimes of the thermodynamic parameters a system may lose its 'elasticity'. If the kinetic energy of the bodies is close to balancing their interaction energy, for example, a substantial change in the volume of a subregion may not affect the pressure very much. Then large density fluctuations are possible. These subregions find that an unusually wide range of microstates become equally probable. Some of these microstates may have lower free energy – or higher entropy – than the earlier microstates at slightly different temperature, pressure, or density. Once the subregion falls into such a microstate the earlier microstates become much less probable. The subregion becomes trapped in its new state. A trapped subregion, in turn, modifies the thermodynamic functions of the surrounding system. Many such subregions may form nuclei and their new microstate may propagate throughout the entire system. A phase transition occurs.

A multitude of fluctuations are constantly probing the possibility of a phase transition. Only when conditions are just right do they succeed. Then the consequences are dramatic and legion. Ice turns to water, water to steam. Crystals change structure, magnetic domains collapse.

Thermodynamics cannot portray systems evolving through phase transitions. The situation is too far from equilibrium; fluctuations are as large as their average quantities. But thermodynamic descriptions contain enough information to predict their own end. After the transition is over, a new thermodynamic description may apply, perhaps with a modified equation of state. Often it is possible, moreover, to find equations of state which bridge the transition, in a phenomenological, approximate way.

30

Gravity and thermodynamics

30.1. Statistical mechanical approach

The difficulties of statistical mechanics, even in 'ordinary' laboratory systems, are sufficiently great that results tend to fall into one of two types. One type is derived rigorously from fundamental principles, albeit for restricted and artificial systems. These results are useful in providing precisely understood examples of special phenomena. The other type is derived less rigorously and more intuitively for more realistic systems. These provide a tentative understanding of more interesting phenomena. Experiments or numerical calculations can check this understanding. Occasionally an overlap brings forth a rigorous result for a realistic system and there is general rejoicing.

Gravitational thermodynamics is no different. Both types of results lead to useful understanding. First we will illustrate what can be learned exactly for special systems. A standard trick for finding simplified systems is to work in less than three dimensions. The gravitational partition function in two dimensions can be solved exactly and leads to an important general feature of the equation of state.

Two-dimensional systems must not be confused with objects confined to a plane and interacting with the r^{-1} gravitational potential. That would just be a two-coordinate system in three dimensions. Rather, we solve the two-dimensional Poisson equation for a point mass at the origin. By working this out analogously to the standard three-dimensional calculation for the Coulomb potential, the reader can show that

$$\phi_{ij} = Gm^2 \ln r_{ij}, \tag{30.1}$$

where $r_{ij} = |\mathbf{r}_i - \mathbf{r}_j|$. (Try the one-dimensional case too.) This potential is the same as for an infinite uniform density rod in three dimensions (another exercise for the reader). Thus the motion of mass points in two dimensions can be visualized as the motion of a collection of infinite, parallel, mass lines observed in a plane which cuts perpendicularly across them. The two-dimensional potential, and its related force $\propto r^{-1}$, are long range and do not saturate. These basic properties, also true in three dimensions, alter the statistical mechanics profoundly.

The two-dimensional pressure equation of state for a time independent distribution follows readily from the partition function (29.45). First, rewrite it for a continuum of energy states by changing the sum to an area integral over the two-dimensional coordinates and momenta

$$Z = \int e^{-U/kT} d\mathbf{x}_1 d\mathbf{p}_1 \dots d\mathbf{x}_N d\mathbf{p}_N. \tag{30.2}$$

The total internal energy U is the same as the Hamiltonian of the complete system

$$U = \sum_{i=1}^{N} \frac{p_i^2}{2m} + \frac{1}{2} \sum_{i=1}^{N} \sum_{j \neq i}^{N} \phi(r_{ij}). \tag{30.3}$$

All the masses are the same, for simplicity.

Now the pressure is related to the free energy $F(T, N, V)$ by $P = -(\partial F / \partial V)_{T,N}$ and thus to the partition function through (29.46)

$$P = kT \frac{\partial \ln Z}{\partial V} \bigg)_{T,N}. \tag{30.4}$$

In two dimensions, the area A just replaces the volume; think of area as two-dimensional volume. Therefore, we want to separate out the area dependence in (30.2). To do this, change to dimensionless spatial coordinates x_1' by writing

$$x_i = A^{1/2} X_i'. \tag{30.5}$$

Note also, from the form of U, that the spatial and momentum parts of Z separate. For (30.4) we will only need to consider the spatial part (also denoted by U in the next three equations) involving the interaction energies. Thus

$$Z \propto A^N \int e^{-U/kT} d\mathbf{x}_1' \dots d\mathbf{x}_N'. \tag{30.6}$$

Taking the derivative,

$$\frac{\partial Z}{\partial A} \bigg)_{T,N} = NA^{-1} Z - \frac{A^N}{kT} \int e^{-U/kT} \frac{\partial U}{\partial A} d\mathbf{x}_1' \dots d\mathbf{x}_N'. \tag{30.7}$$

Remembering that $r_{ij} = A^{1/2} r_{ij}'$ and using (30.3) gives

$$\frac{\partial Z}{\partial A} \bigg)_{T,N} = \frac{N}{A} Z - \frac{A^N}{kT} \int e^{-U/kT} \frac{1}{2} \sum_{i \neq j} \frac{r_{ij}}{2A} \frac{\partial \phi(r_{ij})}{\partial r_{ij}} d\mathbf{x}_1' \dots d\mathbf{x}_N'. \tag{30.8}$$

Inserting the potential (30.1) enables us to perform the summation trivially, for it now contains just $N(N-1)$ identical terms which do not depend on r. The remaining integral is just $\propto ZA^{-N}$. So it all becomes quite simple when substituted into (30.4)

$$PA = NkT \left[1 - \frac{(N-1)Gm^2}{4kT} \right]. \tag{30.9}$$

This is the exact equation of state for a two-dimensional gravitating system (Salzberg, 1965).

Imagine that the gravitational coupling of the particles can be varied at will. Then as $G \to 0$, or equivalently as $m^2/T \to 0$, we recover the equation of state for a perfect gas. On the other hand, as the number of particles is increased, it reaches a critical value $\approx 4kT/Gm^2$ at which the mutual gravitational energy becomes so large that the pressure vanishes. If the particles are now pushed close together so that ϕ_{ij} and, consequently, U, take on large negative values, then the integral for the partition function (30.6) no longer converges. At this point the entire thermodynamic description breaks down. A phase transition occurs and the system collapses. However, details of the collapse, with possible fragmentation, are beyond the power of equilibrium thermodynamics.

This equation of state (30.9) also illustrates another important general aspect of gravitational thermodynamics. When interactions are important the thermodynamic parameters may lose their simple intensive and extensive properties for subregions of a given system. This modification is also true for the ordinary thermodynamics of imperfect gases. As we consider larger and larger subregions of a given system of constant density the pressure decreases due to its non-linear dependence upon N. Although this behavior may, at first sight, seem unfortunate, it does not turn out to be so important. There are two related reasons for this. First, we can always go back to a Gibbs ensemble of weakly interacting systems. That will preserve the intensive–extensive distinction, as discussed in Section 29.1. Second, we will see in the next subsection that even subregions of a given system may interact only weakly if they are essentially uncorrelated. Then the subregions behave like a Gibbs ensemble and the thermodynamic parameters remain extensive (or intensive) among uncorrelated subregions.

Moving into three dimensions we can see why matters become more complicated. For the r^{-1} potential, the derivative of the partition function depends on the exact positions of all the objects. In the next subsection, we will see that this can still be written in a very useful form. But first we mention some exact three-dimensional results applying to a related problem.

In order for the thermodynamic limit to exist rigorously a system must have an equilibrium ground state. For such a state to have a minimum free energy it should be 'saturated'. This means that an increase of N would leave the free energy per particle unaltered. Saturated free energy is therefore proportional to N. If the free energy increases with a higher power of N, then, in the thermodynamic limit as $N \to \infty$, the binding energy per particle also becomes infinite. Thus the ground state would become more and more condensed as N increases. From a slightly different point of view, N would increase and V decrease indefinitely, making it impossible to achieve a thermodynamic limit with $\rho \to$ constant.

Gravitational systems, as often mentioned earlier, do not saturate and so do not have an ultimate equilibrium state. A three-dimensional purely gravitational system in virial equilibrium (not a state of ultimate equilibrium) has a binding energy $\propto N^2$. This result is modified if we consider a combination of gravitational and quantum mechanical forces to determine the ground state (Levy-Leblond, 1969; Hertel, Narnhofer & Thirring, 1971). When both forces are important the total energy is

$$U(N) \approx N(p^2/2m) - N^2(Gm^2/2r). \tag{30.10}$$

If the minimum distance between two particles is $r \gtrsim h/p$ from the uncertainty relation, then minimizing U with respect to momentum p and substituting this value of p into U gives the lowest energy state $U_0(N) \approx -N^3(G^2m^5/8h^2)$. However, if the particles are fermions satisfying an exclusion principle, then they cannot be within each other's de Broglie wavelength h/p so the N particles would fill a volume $\sim N(h/p)^3$ with average interparticle separation $r \gtrsim N^{1/3}(h/p)$. This leads similarly to a slightly more saturated ground state with $N^{7/3}$ replacing N^3 in the previous result. These heuristic results, applied to relativistic fermions, were found in the early studies by Landau and Chandrasekhar of the stability of white dwarfs and neutron stars. They have been made more exact and derived rigorously by examining the thermodynamic limit of quantities with the form $N^{-7/3}F(T, V, N)$ in the ground state. Non-existence of the ordinary thermodynamic limit for gravitating fermions just tells us again that gravitating systems are essentially unstable. We will see in the next subsection how this still permits a modified thermodynamic description to be useful.

As with the two-dimensional case, three-dimensional gravitational thermodynamics is a useful description if the system is not in a phase of rapid instability. Thermodynamics can be applied to an ensemble of slowly evolving gravitational systems, much like the concepts of ordinary equilibrium thermodynamics can be applied to ordinary slowly evolving systems (see de Groot & P. Mazur, 1962). The main difficulty is that in three dimensions no one knows how to sum the partition function exactly in a closed form. So we shall have to make do with a more open form which connects the partition function with kinetic theory.

30.2. Kinetic theory approach

We have already met the connection with kinetic theory in the relation (29.35) between pressure and the two-point correlation function. Although this relation can be derived purely from kinetic theory we can now see that it has a deeper connection with statistical mechanics.

This close connection is readily understood by deriving the pressure equation of state in three dimensions in a manner parallel to the derivation of (30.9) in two dimensions. Thus we can make the earlier derivation serve a double duty. The volume derivative of the partition function is similar to (30.8) except that V replaces A and the factor of $\frac{1}{2}$ from transforming to dimensionless area (volume) coordinates is replaced by $\frac{1}{3}$. Now, although the derivative of $\phi(r_{ij})$ gives summation terms which depend on r_{ij}, these terms are again all of the same form. Again there are $N(N-1)$ of them, representing all pairs of objects. The configuration integral can, therefore, be taken just over two representative pair coordinates, $dx_1 \, dx_2$, transforming back to standard coordinates. With the relation between local density and the two-point correlation function, Equation (29.35) results.

The reason for the great importance of $\xi(r)$ now begins to emerge. It determines the separations of pairs of objects whose mutual potential energy tells how gravity modifies the thermodynamic functions. The fact that gravity is a binary interaction also shows, through (29.35), why only the two-particle correlations are needed for thermodynamics. Higher order correlations would add no new information about the interactions. Of course, $\xi(r)$ itself cannot be derived from equilibrium statistical mechanics. So we must adjoin the kinetic theory of previous sections to provide a closed description. Actually the kinetic theory, in turn, is connected with the statistical mechanics through Liouville's theorem. At that level, tantamount to knowing all positions and velocities and, therefore, being able to sum the partition function exactly, all descriptions of gravitating systems merge together.

Equation (29.35) now also illustrates why the requirements for gravitational thermodynamics are not as restrictive as might seem at first sight. Although the integral over the correlation function formally has an infinite limit, the correlation function will generally be non-zero only between finite limits. These regions are often small compared with other astronomical scales. In an infinite system such regions will not interact strongly, and over large scales thermodynamic parameters can retain their extensive and intensive properties. Averaged over regions greater than those where $\zeta \neq 0$ a thermodynamic description exists. This form of statistical shielding is somewhat analogous to Debye spheres in a plasma. It is a bit more subtle, however, because the gravitational long-range correlations are not reduced by shielding of opposite charges as in the plasma. In cosmological applications, such as galaxy clustering, the initial correlations may have had a short range, and may not have had enough time to grow to larger scales. At early times, for example, the value of $\xi(r, t)$ from (24.1) can be substituted into (29.35) to obtain the modified pressure. We will follow this kinetic development in more detail in Section 32.

As correlations, clumps and clusters evolve and grow in intensity and scale, the thermodynamic quantities must be averaged over larger and larger scales. On these scales the fluctuations from average volume to average volume are small. Nevertheless, the thermodynamics derived from large scales is reflected in the properties of the system on smaller scales, since all scales commingle.

In addition to averaging out long-range correlations, gravitational thermodynamics ignores the very short-range correlations where two objects occupy the same point of configuration space. In practice such collisions with zero angular momentum happen rarely and do not populate phase space significantly. It is more likely that two objects become bound as a result of close encounters involving three or more bodies. Then the bound objects can be treated as a single object of greater mass. This is a form of mass renormalization. It is a small scale version of the previous long-range averaging where highly correlated regions are essentially renormalized into single objects.

Like ordinary thermodynamics, gravitational thermodynamics assumes that thermodynamic changes in a system can be made on timescales during which the underlying distribution (correlation) function does not change significantly. The physical reasons permitting this are different in each case. In ordinary thermodynamics short-range interactions (e.g., collisions in gases) keep the distribution function close to its equilibrium value. These short-range interactions act on a short timescale and thermodynamic changes occur over relatively longer times, so the underlying distribution is always relaxed. In gravitating systems this intrinsic stability is absent. However, many astronomical systems change their underlying distribution so slowly that we can imagine thermodynamic transformations occurring on much shorter timescales during which the underlying distribution changes very little. Thus $\xi(r, t)$, for example, is effectively 'frozen' at a given value of t for the thermodynamic transformations. In Section 31, we will determine conditions under which this approximation breaks down.

30.3. Model approach

Models are attempts to find solvable mathematical descriptions which contain the essential physics of a complicated problem. In early descriptions of an imperfect gas the van der Waals equation of state provided a simple model which could be related to phase transitions. This model represents a large class of physically similar, but more mathematically complicated, equations of state for uniform systems. So it is natural to ask how it is related to gravitating systems.

Most thermodynamics texts discuss the van der Waals equation of state which is frequently written

$$\left(p + a\frac{N^2}{V^2}\right)\left(1 - b\frac{N}{V}\right) = \frac{NkT}{V}. \tag{30.11}$$

Often this form provides a good fit to laboratory gases. Values for the constants a and b are measured experimentally. These values differ for different substances but can be related to the underlying physics. The value of b refers to the short-range, hard-core part of the particle interaction potential which gives rise to the 'excluded volume' of the particles. The value of a refers to the long-range part of the potential, producing a 'reduced pressure'.

For gravitating systems of point masses, $b = 0$ since there is no hard core. Comparison with (29.35) then suggests that the van der Waals form might apply with

$$a = \frac{1}{6}\int_0^\infty r\frac{d\phi(r)}{dr}\xi(r, n, T)4\pi r^2\,dr. \tag{30.12}$$

But we see this is a reasonable approximation only when the two-point correlations do not depend strongly on density and temperature. Equation (24.1) for $\xi(r,t)$ shows that when correlations are linear at any given time, ξ does not depend on density, but it has a temperature dependence which must be taken into account. Thus even slightly clustered gravitating systems are a bit more complicated than a simple van der Waals gas. For a better approximation we can replace a in (30.11) by a/T.

More complicated model approaches may be developed to include, for example, the effects of density gradients. A class of such models developed by Landau assumes that the local free-energy density of a system can be expanded as an analytic perturbation series around its homogeneous equilibrium state. Then the free energy explicitly includes terms involving $(n - \langle n \rangle)^2$ and $[\nabla(n - \langle n \rangle)]^2$, multiplied by constant coefficients. However, it is harder to relate these more complicated phenomenological models directly to the underlying 'microscopic' description of gravitating systems.

In Part III we will describe models for properties of finite systems. Along with different boundary conditions they must be treated differently from infinite systems because they can evolve in a much wider variety of ways. The possibility of highly asymmetric finite configurations invokes a whole new range of energy states for the partition function. These states are suppressed in infinite systems which are constrained to be homogeneous over large scales.

31

Gravithermodynamic instability

There were great oscillations
Of temperature.... You knew there had once been
warmth;
But the Cold is the highest mathematical Idea...
the Cold is Zero...
The Nothing from which arose
All Being and all variation

Edith Sitwell

31.1. The vanishing of sound speed

Earlier, in Section 15, we saw how Jeans' fluid analysis and subsequent kinetic analyses predict gravitational instability. When the gravitational energy of a perturbation becomes roughly equal to its kinetic energy the perturbation becomes unstable and contracts. In those analyses the propagation of self-gravitating waves is compared with the propagation they would have in a non-gravitating medium. As gravity becomes more important, waves propagate slower and slower until they contract rather than travel.

A thermodynamic point of view can describe this behavior, and more. By putting gravity into the equation of state for the system we can see directly how the sound speed vanishes. In general the adiabatic sound speed is given by

$$c^2 = \frac{\partial P}{\partial \rho}\bigg)_S,\tag{31.1}$$

the derivative being taken at constant entropy. Equations of state such as (29.35) do not involve the entropy directly, so it is best to rewrite (31.1) in terms of more convenient derivatives such as $(\partial P/\partial V)_T$.

To rewrite the derivative we utilize the techniques of Section 29. Applying (29.5) and (29.8) to (31.1) gives

$$\frac{\partial P}{\partial \rho}\bigg)_S = -\frac{(\partial S/\partial \rho)_P}{(\partial S/\partial P)_\rho} = -\frac{\dfrac{\partial S}{\partial T}\bigg)_P \dfrac{\partial T}{\partial \rho}\bigg)_P}{\dfrac{\partial S}{\partial T}\bigg)_\rho \dfrac{\partial T}{\partial P}\bigg)_\rho} = \frac{\dfrac{\partial S}{\partial T}\bigg)_P}{\dfrac{\partial S}{\partial T}\bigg)_\rho}\frac{\partial P}{\partial \rho}\bigg)_T = -\frac{V^2}{mN}\frac{\partial S/\partial T)_P}{\partial S/\partial T)_V}\frac{\partial P}{\partial V}\bigg)_T.\tag{31.2}$$

Here we have used $\rho = mN/V$ and suppressed the N subscript since it is kept constant throughout. The last equality contains the ratio of specific heats at constant pressure and volume, usually denoted by γ in the kinetic theory of ordinary gases. However, we will have to examine this ratio in more detail to make it suit or purposes better.

A keen eye will notice that the thermodynamic derivatives in the last equality of (31.2) use different sets of independent variables. In $\partial S/\partial T)_P$, the independent variables are (T, P), while the other derivatives use (T, V). For consistency we want everything in terms of one set, say (T, V). The simplest and most powerful way to make these transformations is with the Jacobian formalism, already familiar from Section 7 (see Equation (7.8) and (7.9)) in a dynamical context.

Thermodynamic Jacobians are defined in terms of determinants, as in (7.9), except now we do not restrict ourselves to absolute values. To transform $(u, v, \ldots, w)$ to $(x, y, \ldots, z)$ involves

$$\frac{\partial(u, v, \ldots, w)}{\partial(x, y, \ldots, z)} = \begin{vmatrix} \dfrac{\partial u}{\partial x} & \dfrac{\partial u}{\partial y} & \cdots & \dfrac{\partial u}{\partial z} \\ \vdots & & & \\ \dfrac{\partial w}{\partial x} & \dfrac{\partial w}{\partial y} & & \dfrac{\partial w}{\partial z} \end{vmatrix} \tag{31.3}$$

It is easy to check that Jacobians have the property

$$\left.\frac{\partial u}{\partial x}\right)_{y, \ldots, z} = \frac{\partial(u, y, \ldots, z)}{\partial(x, y, \ldots, z)}, \tag{31.4}$$

which turns out to be very useful for re-expressing thermodynamic derivatives. Moreover, from the properties of determinants,

$$\frac{\partial(u, v, \ldots, w)}{\partial(x, y, \ldots, z)} = -\frac{\partial(v, u, \ldots, w)}{\partial(x, y, \ldots, z)} \tag{31.5}$$

and

$$\frac{\partial(u, v, \ldots, w)}{\partial(x, y, \ldots, z)} = \frac{\partial(u, v, \ldots, w)}{\partial(r, s, \ldots, t)} \frac{\partial(r, s, \ldots, t)}{\partial(x, y, \ldots, z)} \tag{31.6}$$

and

$$\frac{\partial(u, v, \ldots, w)}{\partial(x, y, \ldots, z)} = 1 \Big/ \frac{\partial(x, y, \ldots, z)}{\partial(u, v, \ldots, w)}. \tag{31.7}$$

These last two properties are easy to remember since they are formally the same as ordinary derivatives. We will meet these simple transformations again in Section 33 in yet a different context.

Now we proceed along these lines to make the transformation $(T, P) \to (T, V)$ in (31.2)

$$\left.\frac{\partial S}{\partial T}\right)_P = \frac{\partial(S, P)}{\partial(T, P)} = \frac{\partial(S, P)}{\partial(T, V)} \frac{\partial(T, V)}{\partial(T, P)} = \frac{\partial V}{\partial P}\Big)_T \left[\frac{\partial S}{\partial T}\Big)_V \frac{\partial P}{\partial V}\Big)_T - \frac{\partial S}{\partial V}\Big)_T \frac{\partial P}{\partial T}\Big)_V \right]$$

$$= \frac{\partial S}{\partial T}\Big)_V \left[1 - \frac{\partial V}{\partial P}\Big)_T \frac{\partial P}{\partial T}\Big)_V \frac{\partial T}{\partial S}\Big)_V \frac{\partial S}{\partial V}\Big)_T \right]. \tag{31.8}$$

The third equality follows by expanding the determinant. The last term in (31.8) involves derivatives with respect to four quantities, and since we only need two independent variables to specify the thermodynamics for $N = $ constant, we should be able to simplify this term further. Applying (29.15) to the last entropy derivative would not eliminate S as the fourth variable. To transform this derivative we need something stronger – a Maxwell relation. This relation should involve the second derivative of something whose first derivative at constant volume is the entropy and which also has a volume derivative at constant temperature. Thus this something must be a function of temperature and volume at constant N. The free energy $F(T, V, N)$ comes to mind. Indeed, $F = U - TS$ so $dF = dU - TdS - SdT$ and substituting (29.1) for dU at constant N shows $dF = -SdT - PdV$. Thus

$$S = -\left(\frac{\partial F}{\partial T}\right)_V \quad \text{and} \quad P = -\left(\frac{\partial F}{\partial V}\right)_T \tag{31.9}$$

and so

$$\left(\frac{\partial S}{\partial V}\right)_T = -\frac{\partial^2 F}{\partial V \partial T} = -\frac{\partial^2 F}{\partial T \partial V} = \left(\frac{\partial P}{\partial T}\right)_V, \tag{31.10}$$

which is our desired Maxwell relation. Inserting this into (31.8) and combining with (31.1) and (31.2) gives finally

$$c^2 = -\frac{V^2}{mN}\left(\frac{\partial P}{\partial V}\right)_T\left[1 - \left(\frac{\partial V}{\partial P}\right)_T\left(\frac{\partial T}{\partial S}\right)_V\left(\frac{\partial P}{\partial T}\right)_V^2\right]. \tag{31.11}$$

Here the derivatives are taken with respect to two independent variables, T and V (remember the entropy derivative can be inverted) so this is as simple as the sound speed can get in the general case for constant N.

The sound speed has the entertaining possibility of vanishing if the equations of state satisfy

$$\left(\frac{\partial P}{\partial V}\right)_T = \left(\frac{\partial T}{\partial S}\right)_V\left(\frac{\partial P}{\partial T}\right)_V^2. \tag{31.12}$$

This would never happen for a perfect gas whose isothermal compressibility is always negative and whose specific heat at constant volume is always positive.

Gravitating systems however have the two equations of state (29.34) and (29.35)

$$U(T, V, N) = \tfrac{3}{2}NkT - \frac{3N^2}{V}a(T, V, N) \tag{31.13}$$

and

$$P(T, V, N) = \frac{NkT}{V} - \frac{N^2}{V^2}a(T, V, N), \tag{31.14}$$

where

$$a(T, V, N) = \frac{2\pi}{3}Gm^2\int_0^\infty r\xi(r, T, V, N)dr \tag{31.15}$$

is essentially the correlation energy divided by the number density.

It is understood here that in a medium which is homogeneous on large scales, ξ contains V and N only in the combination $n = N/V$. Note also that the standard relation, $P = 2U/3V$, for perfect gases no longer holds and the ratio of pressure and energy density more generally depends on the form of the potential and the integral over the covariance function. This integral converges for gravitating systems if $\xi(r)$ decreases faster than r^{-2} as $r \to \infty$. Then $a(T, V, N) \to a(T, n)$. If the integral does not converge, then the thermodynamic functions of finite systems have an additional direct dependence on the volume (and also the shape) of the system. The non-convergent case can still provide a useful description of 'global' thermodynamic and stability properties. However, then one must use some of the thermodynamic interrelations, such as the Euler equation which depends on homogeneity of the fundamental equation, with great care. Unless otherwise stated we will assume the integral converges. This will generally be the case in infinite systems if correlations have not had time to grow over all scales, or if there is sufficient shielding from regions where $\xi(r)$ is negative. In finite systems close to equilibrium, $\xi(r)$ may also have a range much less than the size of the system.

When $a(T, V, N)$ is important it may be possible to satisfy the condition (31.12) for $c^2 = 0$. To check this, we need $\partial S/\partial T)_{V,N}$ which follows either from dividing the relation $\Delta U - T\Delta S$ by ΔT and taking the differential limit, or, more formally, from

$$\left.\frac{\partial S(U, V, N)}{\partial T}\right)_{V,N} = \frac{\partial S}{\partial U}\frac{\partial U}{\partial T} = \frac{1}{T}\left.\frac{\partial U}{\partial T}\right)_{V,N}. \tag{31.16}$$

Since $U(T, V, N)$ is already in the required form, substituting (31.13)–(31.16) into (31.12) gives

$$-\left(1 - \frac{2N}{kV}\frac{\partial a}{\partial T}\right)\left(1 - \frac{2Na}{kTV} + \frac{N}{kT}\frac{\partial a}{\partial V}\right) = \frac{2}{3}\left(1 - \frac{N}{kV}\frac{\partial a}{\partial T}\right)^2. \tag{31.17}$$

This is the condition the correlation function must satisfy in order that sounds cease to propagate and perturbations become permanent. Although the definition of the sound speed itself follows from linear equations of motion, changes of the correlation function in (31.17) can be large and non-linear. In this sense there is more information in (31.17) than in a linear dispersion relation. Scale information is implicit in the shape of $\xi(r)$, which does not have to be Fourier decomposed into different wavelengths.

For an illustration, when $a(T, V, N) = a_0$ is a constant, Equation (31.17) requires $Na_0/kTV = \frac{5}{6}$ when $c = 0$. Then replacing the integral over $\xi(r)$ by a squared scale length, equating kT with $mv^2/3$ and comparing with (15.11), shows that the essential physics of this simple case is the same as for the Jeans criterion.

A more realistic case which yields a new result occurs if $\xi(r, T)$ does not depend on n. This describes linear clustering (Equation (24.1)) and may also apply to non-linear clustering whose form is self-similar at different densities. Then (31.17) becomes a quadratic equation in the derivative $\partial a/\partial T$. It is possible to obtain general solutions by using Legendre transforms, but the essential physics emerges without too much work. First we note that the physical requirement that $\partial a/\partial T$ be a real quantity leads to either $Na/kTV \geq \frac{5}{6}$ or $Na/kTV \leq \frac{1}{2}$. Next, the consistency of the sign of the first factor of (31.17) determines an inequality for $\partial a/\partial T$ in each case. Combining these inequalities in each case yields the constraint that either

$$\frac{\partial \ln a}{\partial \ln T} < \frac{3}{5} \quad \text{or} \quad \frac{\partial \ln a}{\partial \ln T} > 1, \tag{31.18}$$

when the sound speed vanishes. Moreover, we see directly from (31.17) that as $Na/kTV \to \frac{5}{6}$, the derivative $\partial a/\partial T \to 0$ and a small change in velocity dispersion has no effect on the gravitational clustering. The system is in a neutral state, ready to become unstable.

31.2. Spatial fluctuations

A most important characteristic of these neutral states is that fluctuations become very large. These fluctuations are constantly probing phase space for the existence of higher entropy states into which they can settle. Substituting the equation of state (31.14) into (29.52) shows that the second moment of the density fluctuations has the form

$$\left(\Delta N\right)^2 = N\left(1 - \frac{2Na}{kTV}\right)^{-1}, \tag{31.19}$$

when $a(T, V, N) = a(T, n)$ has no explicit dependence on V. As expected, the condition for zero sound speed is essentially the same as for large fluctuations.

Formally the fluctuations become infinite. Actually we know that cannot be the case for several reasons. Apart from the lack of infinite resources in a finite system, even to gain very large numbers of objects from an infinite reservoir would take a long time – at least the linear travel time of the objects at temperature T. This process would put the system far from equilibrium and it would no longer be described by the equilibrium thermodynamics we are using. In fact the whole description breaks down when $(\Delta N)^2$ becomes of order N^2, for then the notion of an average quantity becomes thermodynamically meaningless.

Lest it be thought that the singularity is just a peculiarity of the second moment, I should mention that the higher order moments diverge even more strongly under the same conditions, the $m + 1$ moment diverging at least as fast as the $-m$ power of the expression in parentheses in (31.19); moments of many other thermodynamic parameters also diverge strongly; the disease is general; thermodynamics collapses.

Subregions of the system find their states of greater gravitational entropy and become bound into clusters. Density inhomogeneities form over larger and larger regions and the thermodynamic description, which presumes constant T, N, and V, becomes void. Attempts have been made to extend thermodynamics further into these conditions by adding the effects of gradients to the set of uniform parameters. Thus the free energy becomes an integral of a free-energy density which can be expanded away from its equilibrium value to incorporate gradients (an approach of Landau's). It appears, however, that conditions are usually too singular for this approach to add much useful insight. Its chief virtue is to determine the onset of the regime of great oscillations.

Away from this regime, but where gravity is still important, fluctuation theory makes it possible to calculate, at least in principle, the probability of any given state. This probability follows from Einstein's inversion of the Boltzmann–Planck postulate (29.43). The probability that an instantaneous microstate of the system lies in the range $d\hat{x}_0 \ldots d\hat{x}_n$ of extensive parameters is

$$\Omega d\hat{x}_0 \ldots d\hat{x}_n = \Omega_0 e^{-\Delta S/k} d\hat{x}_0 \ldots d\hat{x}_n. \tag{31.20}$$

Here Ω_0 is a normalization constant and ΔS is the difference between the entropy of the particular microstate and the entropy of the most probable (equilibrium) microstate. Carat superscripts denote instantaneous values.

Back in Section 3 we saw how the Laplace and Fourier transforms produce 'characteristic' functions which generate the moments of a distribution, such as $\Omega(\hat{x}_0, \ldots, \hat{x}_n)$. Now expanding the positive Fourier transform of Ω in the case of one fluctuating variable, say $\hat{x}_0$, gives

$$\omega(y) = \int_{-\infty}^{\infty} e^{iy\hat{x}_0} \Omega(\hat{x}_0) d\hat{x}_0$$

$$= e^{iyx_0} \int e^{iy(\hat{x}_0 - x_0)} \Omega(\hat{x}_0) d\hat{x}_0$$

$$= e^{iyx_0} \sum_{n=0}^{\infty} \frac{(iy)^n}{n!} \int (\hat{x}_0 - x_0)^n \Omega(\hat{x}_0) d\hat{x}_0$$

$$= e^{iyx_0} \sum_{n=0}^{\infty} \frac{(iy)^n}{n!} \langle (\hat{x}_0 - x_0)^n \rangle, \tag{31.21}$$

where x_0 is the average value of $\hat{x}_0$. The last equality is just a sum over the fluctuation moments. Therefore, from all the moments of a fluctuating thermodynamic variable, we can compute its distribution by the inverse Fourier transform

$$\Omega(\hat{x}_0) = \frac{1}{2\pi} \int_{-\infty}^{\infty} e^{-iy\hat{x}_0} \omega(y) dy$$

$$= \sum_{n=0}^{\infty} \frac{\langle (\hat{x}_0 - x_0)^n \rangle}{2\pi n!} \int e^{-iy(\hat{x}_0 - x_0)} (iy)^n dy$$

$$= \sum_{n=0}^{\infty} \frac{(-1)^n \langle (\hat{x}_0 - x_0)^n \rangle}{n!} \delta^{(n)}(\hat{x}_0 - x_0). \tag{31.22}$$

The last expression follows from n differentiations of the Fourier integral representation of the delta function: $\delta(x) = (2\pi)^{-1} \int e^{-ixy} dy$. Successive integrations by parts show that the nth derivative of the δ function has the property $\int \delta^{(n)}(x) f(x) dx = (-1)^n f^{(n)}(0)$, where these integrals all have the standard range from $-\infty$ to $+\infty$.

With this result, any thermodynamic quantity involving an integral over the distribution can be expressed as a sum containing derivatives of the fluctuation moments evaluated at equilibrium. If several variables fluctuate the multidimensional Fourier expansion can be used. Since the fluctuation moments depend on the equations of state, and these in turn are determined by $\xi(r)$, the two-body correlation function can describe larger scale correlations. We do not gain any free information since $\xi(r)$ is itself determined by large scale correlations in the coupled hierarchy. This closes the loop.

In special circumstances one can measure $\xi(r)$ directly, say from galaxy counts, or postulate its form for certain physical conditions. Then (31.22) enables one to describe the higher order correlations consistent with a given $\xi(r)$. This works when the series converges, when the fluctuations are not too large.

31.3. Temporal fluctuations

As well as asking how properties of a gravitating system fluctuate from place to place, we can focus on a given region and ask how its properties fluctuate from time to time. The region is open to the rest of the system with respect to these fluctuating properties. As an example, consider the total number of objects $N(t)$ at any time in a fixed volume. Then the volume belongs to a grand canonical ensemble.

The history of a thermodynamic parameter such as $N(t)$ can be considered as a path in parameter–time space. Figure 34 shows a schematic path. Hydrodynamic evolution equations give the average path $N_0(t)$. Jittering around this average path are fluctuations $\eta(t)$ such that

$$N(t) = N_0(t) + \eta(t). \tag{31.23}$$

No information about $\eta(t)$ is given by the standard hydrodynamic equations. Some formulations (e.g., Simon, 1970) add fluctuating forces to supplement these equations. When the fluctuations are small, it is plausible that these forces are basically random, as in the Langevin or Fokker–Planck formulations. However, as the temporal fluctuations become stronger and more correlated, the full apparatus of higher order kinetic theory is needed to describe them. The fluctuations themselves must incorporate the effects of truncating the description – by removing the higher order or longer time kinetic correlations – in a self-consistent way.

An alternative approach is to start with a macroscopic, thermodynamic description and see what it leads to. If $\eta \ll N_0$ it is supposed that the evolutions of $N_0(t)$ and $\eta(t)$ are decoupled. For static systems of perfect gas it is usually postulated that $\eta(t)$ satisfies the linear regression equation

$$\dot{\eta} = -\eta/\kappa, \tag{31.24}$$

where κ is a positive phenomenological constant related to the decay time of a stable fluctuation. This is really a stochastic description in that it gives the most probable path (under the stated conditions) for a fluctuation to decay. Its time-reversed solution is the most probable path for a fluctuation to grow. Of course, not all fluctuations follow the most probable path, and if $N_0(t)$ is not constant the most probable path will change with time.

To generalize (31.24), we imagine that between t_a and t_b in Figure 34 there are all possible paths around $N_0(t)$ which the system could conceivably follow. It is

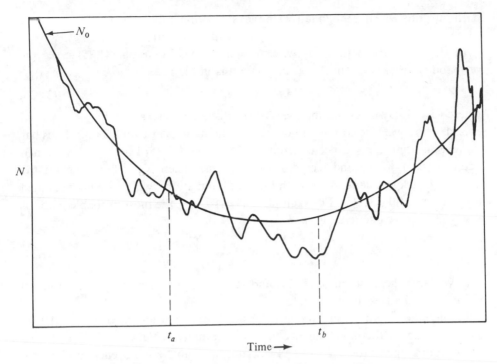

Fig. 34. A schematic path of number versus time.

plausible that away from phase transitions the probability of actually following each path decreases as the integrated departures of that path from $N_0(t)$ become larger. To quantify this we need a measure for the probability of an arbitrary path.

The clue comes from Equation (31.20). It suggests that the probability of any path (i.e., a 'complete historical state of the system') is proportional to $\exp - \int [\Delta S(t)/k\tau]\,dt$, where the entropy departure is integrated over the path and τ is a timescale. This cannot be quite right, however, since we know that for a perfect gas near equilibrium (which must be valid as a special case), a positive fluctuation is as likely as a negative one. Thus we are led to an exponential involving the integral of $(\Delta S)^2$ over the path. A system in contact with a thermal reservoir can be described by its free entropy (29.41), which is now an explicit function of time $S(1/T, V, N, t)$. Such considerations lead to the basic postulate that the probability amplitude (whose square is proportional to the probability) for any path is

$$G = \exp\left\{ -\frac{N_0(t)}{2k^2\tau} \int_{t_a}^{t_b} \left[\frac{kT\tau}{\mu N_0(t_0)} \frac{d(S - S_0)}{dt} + \frac{\partial(S - S_0)}{\partial \eta} \right]^2 dt \right\}. \quad (31.25)$$

Here $S_0 = S(N_0(t))$ and μ is the chemical potential.

Equation (31.25) gives the probability amplitude for a single path. To find the total probability amplitude for reaching the state N_b from the state N_a, we must integrate over all possible paths. The solutions require elementary techniques of path integrals which, though fairly simple, are peripheral to this book. Therefore, some of the results (Saslaw, 1970) are just mentioned here.

In the case of a perfect gas with $N_0 = $ constant, Equation (31.25) gives the probability amplitude $\psi(\eta, t)$ that the system is in the macroscopic state $\langle N, t |$

$$\psi(\eta, t) \propto e^{-t/a} e^{-\eta^2/b},\qquad(31.26)$$

where a is a decay timescale and $b \approx N_0$. The dispersion in η is $\sim N^{1/2}$, as we would expect. Moreover, the correlation of density fluctuations at two different times has the exponential decay expected from (31.24). Thus (31.25) leads to a combined description of spatial and temporal fluctuations. When N_0 is not constant the changing background modifies both types of fluctuations, which are now coupled together but still solvable. For example, if $N_0(t) \propto t^{-m}$, the most probable path for decaying density fluctuations is

$$\eta(t) = \eta(t_0) \left(\frac{t_0}{t} \right)^{m/2} \exp\left\{ -\frac{t}{\tau(m+1)} \left[\left(\frac{t}{t_0} \right)^m - \frac{t_0}{t} \right] \right\}.\qquad(31.27)$$

This replaces the solution of (31.24) and reduces to it when $m = 0$. When $m > 0$ expansion enhances the decay rate.

In the case of gravitating systems the free entropy is modified using (31.13)–(31.15). Again the thermodynamics changes profoundly as $Na/kTV \to 1$ (for which a is, of course, given by (31.15)). As gravitational and kinetic energies become comparable the probability amplitude becomes highly singular, and its singularity may be non-analytic with the form $\exp(1/x)$, where $x \to 0$. The most probable path $\eta(t)$ also diverges (although its divergence is analytic). Moreover, as the system approaches this regime, its fluctuations remain positively correlated over longer and longer intervals of time. Equivalently, for a given time interval the correlations become greater.

Paths leading to giant fluctuations and permanently collapsed subsystems become the most probable ones. In a gravitating system the contribution of gravity to the total free entropy acts oppositely to the effect of the kinetic contribution. When $Na/kTV \approx 1$ the two contributions nearly cancel, although each may be individually large. When this happens, large density increases do not cause large entropy deviations (unlike the case of a perfect gas). Therefore, they become highly probable. The system strives toward extreme values of its entropy, and, when gravity dominates, these extreme values are found in a non-homogeneous phase.

31.4. Gravitational phase transitions

> A house divided against itself cannot stand.... It
> will become all one thing, or all the other.
>
> *Abraham Lincoln*

Gravitational phase transitions qualitatively resemble other types of phase transitions in material systems, but they also differ in significant ways. All phase

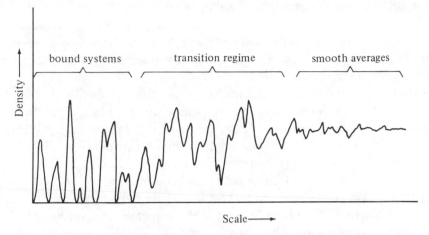

Fig. 35. Schematic illustration of density fluctuations on different length scales.

transitions refer to a change, usually abrupt, from one type of dominant symmetry to another. Often this symmetry change is connected to a change in long-range correlations, or order, within the system. Common examples are the growth of long-range order as a gas condenses to a liquid, and the growth of crystalline symmetries as a liquid freezes to a solid. Within the solid there can be phase transitions to different crystalline forms as the pressure and temperature change. Phase transitions occur also in ferromagnetic substances, superconductors, and liquid helium. During these processes, a sample's macroscopic appearance may or may not change. Laboratory transitions normally occur between two stable phases and, except along a critical line in the $P-T$ plane, the substance is all one phase or all the other.

In infinite gravitating systems, one phase is a uniform distribution. The other phase contains both clusters of many objects and a floating population of higher total energy objects between the clusters. Figure 35 shows an example. This floating population must exist in order that total energy be conserved as clusters form. Some of the vagrants have been ejected by clusters; others are just perturbed slightly by clusters they have never joined. After long periods of time, some clusters will capture vagrants of compatible energy. Clusters will also eject some of their original members. How closely these two processes balance is unknown.

Since neither gravitational phase is secularly stable the system is always in a transitional phase. The most important feature of this transition is that it occurs at different rates on different scales. Once bound clusters form, their internal orbits follow the collisionless Boltzmann equation fairly closely and evolve significantly only on timescales $\gg N/(G\rho)^{1/2}$. Considering these clusters themselves as particles of a gravitating system defines a new Jeans length. The transition to second order clustering occurs most rapidly at this new Jeans length. Although longer length scales are also unstable, they take longer to grow and averages over such scales can again be considered uniform for long periods of time. The situation is illustrated schematically in Figure 35. Over the short length scales within clusters, and the long, nearly uniform ones, we would expect thermodynamic concepts to provide a good approximate description. On the intermediate scale thermodynamic averages break down, and this is the scale on which the phase transition occurs. This description applies uniformly throughout the Universe.

As the Universe evolves, the first clusters combine into second order clusters, then the second into third order clusters and so on as a continuous hierarchy of clustering grows over larger and larger scales. In this sense also, the Universe may be said to be in a state of continuous phase transition. In a Universe of finite age, which had a nearly uniform distribution of galaxies at some initial time, the degree of clustering is directly related to the rate and amount of expansion. Some descriptions of clustering are more sensitive to expansion than others, as we have seen in Sections 25–8. In all cases, the underlying result is an increase in the gravitational contribution to the system's entropy as it evolves.

The physics of gravitational phase transitions is poorly understood compared with its laboratory counterparts. For example, the meaning of the critical point is modified by the long range nature of gravitational forces. Moreover, in laboratory phase transitions, significant insight has come from relations between singular thermodynamic quantities close to the critical point. Exponents characterizing their divergences are predicted to occur in certain combinations and satisfy certain inequalities. Much remains to be learned about gravitational phase transitions in these terms. Specially designed numerical experiments, in lieu of laboratory measurements, would be very helpful tests.

32

Thermodynamics and galaxy clustering;
$$\xi(r) \propto r^{-2}$$

If you came this way,
Taking any route, starting from anywhere,
At any time or at any season,
It would always be the same....

T.S. Eliot

Gravithermodynamics can be applied to a variety of astronomical problems. In this section we shall see how it provides insight into the asymptotic state $\xi \propto r^{-2}$ of galaxy clustering. Subsequently, in Part III, it will describe some properties of finite systems.

Numerical simulations of galaxy clustering, discussed in Section 27, show that the two-point correlation function eventually tends toward $\xi(r) \propto r^{-\gamma}$ with $\gamma \approx 2$. This asymptotic result is nearly the same, independent of whenever the clustering starts and whatever route it takes, within broad limits. Section 25 showed that the value $\gamma = 2$ is dynamically stable. Now we will find that it is asymptotically stable in a deeper thermodynamic sense.

To proceed, we need to assume a form for the two-point correlation function. Power laws are scale-free. That is, $f(\lambda r)/f(r) = \lambda^k$, which is independent of r. This may be seen, following Euler, by differentiating with respect to λ, setting $\lambda = 1$ (a valid particular case), and solving for $f(r)$. The Newtonian gravitational potential is a power law and the cosmological background is scale-free (i.e., curvature is negligible) in all models over the distances on which correlations are observed to be important. These facts suggest that the correlation function will be a power law also. On the other hand, if correlations are not present initially, they will only have had time to propagate throughout some distance $r_1(t)$ since the galaxies formed. Moreover, on some small scale r_0, galaxies will not have interacted as point masses during their evolution. These facts suggest an upper and lower break in the power law correlation. No simple argument describes the details of these breaks, although the linear analysis leading to (24.1) indicates that $\xi(r)$ decreases rapidly with large r. It appears that the essential physics is described by using a simple cut-off

$$\xi(r,t) = a(t)r^{-\gamma} \text{ for } r_0(t) \leq r \leq r_1(t)$$
$$= 0 \quad \text{otherwise.} \tag{32.1}$$

The amplitude $a(t)$ will also depend on T, N and V.

Any value of γ is consistent with our assumptions so far; the particular value $\gamma = 2$

will be selected by the thermodynamics. Substituting (32.1) into (31.12)–(32.14) gives the internal energy and pressure

$$U = \tfrac{3}{2}NkT - \frac{2\pi Gm^2 N^2 r_1^{2-\gamma} a(t)}{V(2-\gamma)}(1-h^{2-\gamma}) \quad \text{for } \gamma \neq 2$$

$$= \tfrac{3}{2}NkT - \frac{2\pi Gm^2 N^2 a(t)}{V}\ln h^{-1} \quad \text{for } \gamma = 2, \tag{32.2}$$

$$P = \frac{NkT}{V} - \frac{2\pi Gm^2 N^2 r_1^{2-\gamma} a(t)}{3V^2(2-\gamma)}(1-h^{2-\gamma}) \quad \text{for } \gamma \neq 2$$

$$= \frac{NkT}{V} - \frac{2\pi Gm^2 N^2 a(t)}{3V^2}\ln h^{-1} \quad \text{for } \gamma = 2, \tag{32.3}$$

where $h \equiv r_0/r_1 \leq 1$. Note also that $a(t)$ incorporates a scale length which can be taken to be r_1^{γ}, so that the gravitational contribution always varies as r_1^2.

To complete the thermodynamic description we could find the remaining equation of state for μ, the chemical potential. An equivalent and easier alternative, however, is to calculate the free energy using (29.42). Since $\xi(r,t)$ is scale-free in r, the entire temperature dependence must be in the amplitude $a(t) = a(t, r_1, T, V, N)$. Let

$$z \equiv 1/T \tag{32.4}$$

and expand

$$a(z) = \sum a_m z^m \tag{32.5}$$

The coefficients a_m may depend on r_1, t, V and N. (Within the summation, m is a counting index having nothing to do with mass.) Gravitational correlation effects must vanish in the limit of very large random kinetic energy, i.e.,

$$\lim_{z \to 0} a(z) = 0 \tag{32.6}$$

so only positive powers of m appear in this expansion.

Substituting (32.2) and (32.5) into (29.42) and integrating from $1/T_0$ to $1/T$ gives

$$F(T,V,N) = \frac{T}{T_0}F_0(T,V,N) - \frac{2\pi Gm^2 N^2 r_1^{2-\gamma}(1-h^{2-\gamma})}{V(2-\gamma)}$$

$$\times \sum_{m=1} \frac{a_m}{(m+1)z}(z^{m+1} - z_0^{m+1}) \quad \text{for } \gamma \neq 2$$

$$= \frac{T}{T_0}F_0(T,V,N) - \frac{2\pi Gm^2 N^2 \ln h^{-1}}{V}\sum_{m=1}\frac{a_m}{(m+1)z}(z^{m+1} - z_0^{m+1})$$

$$\text{for } \gamma = 2. \tag{32.7}$$

Note that $z \geq z_0$, by definition of the fiducial temperature. The integration of (29.42) is undefined up to a function of N and V which is included in $F_0(T,V,N)$. Since all the temperature dependence of the interaction is explicit in the summation, the F_0 term must be the free energy of a perfect gas of randomly moving galaxies without any correlations, i.e., represented by $h = 1$. Therefore, from ordinary thermodynamics (see (29.38)),

$$F_0(T,V,N) = -\frac{S_0}{N_0}NT_0 - \tfrac{3}{2}NkT_0 \ln\frac{T}{T_0} - NkT_0 \ln\left(\frac{V}{V_0}\frac{N_0}{N}\right) = F_{\text{non-interacting}}$$

(32.8)

Equations (32.2), (32.3), and (32.7) with (32.8) completely specify the thermodynamic state of a system having a power law two-point correlation function. From these, any other thermodynamic properties can be derived using the techniques discussed in Section 29. In particular, the total entropy relative to the fiducial state S_0 is, from (29.41)

$$S = -\frac{\partial F}{\partial T}\bigg)_{V,N} = \frac{S_0}{N_0}N + \tfrac{3}{2}Nk\ln\frac{T}{T_0} + Nk\ln\left(\frac{VN_0}{V_0N}\right)$$

$$- \frac{2\pi Gm^2 N^2 r_1^{2-\gamma}(1-h^{2-\gamma})}{V(2-\gamma)} \sum_{m=0} \left[\frac{m}{m+1}(z^{m+1} - z_0^{m+1}) + z_0^{m+1}\right]_{a_m}$$

(32.9)

for $\gamma \neq 2$. For $\gamma = 2$, we again just replace the factor $r_1^{2-\gamma}(1 - h^{2-\gamma})/(2-\gamma)$ by $\ln h^{-1}$. This is not the fundamental thermodynamic equation $S(U,V,N)$ which would require knowing the temperature dependence of $a(t)$ in (32.2) in order to eliminate $T(U,V,N)$. But (32.9) for $S(T,V,N)$ is the most useful representation of the entropy for our purposes. Although it could be generalized slightly by using the formal temperature integral for $a(t)$ instead of its power law expansion, the expansion makes two important results obvious immediately.

First, when $T = T_0$ (or $z = z_0$), the only additional entropy is that of the fiducial state. Departures from the fiducial state add to the gravitational entropy since $z \geq z_0$. Only in the limit $T_0 \to \infty$ (i.e., $z_0 \to 0$) is there no gravitational contribution to the fiducial entropy since all correlations vanish. Second, the term with $m = 0$ does not contribute to the entropy. Any term in the energy U, such as the effect of cosmic expansion, $\propto (r_1 H)^2 N$, which is independent of temperature, does not affect the entropy. This is because it is an ordered motion.

Gravitational contributions to the entropy and other thermodynamic quantities often have the opposite sign from the 'perfect gas' or non-interacting contributions. Now we may ask: 'What value of γ maximizes the gravitational contribution for any value of h?' Suspecting $\gamma = 2$, because it is the only value singled out, leads us to compare the two functions

$$\Gamma(\gamma \neq 2) \equiv \frac{1 - h^{2-\gamma}}{2-\gamma}$$

(32.10)

and

$$\Gamma(2) \equiv \ln h^{-1} = -\ln h.$$

(32.11)

(Note that the factor $r_1^{-\gamma}$ is cancelled by the r_1^{γ} factor in all the a_m coefficients.) At $h = 1$ (no correlations) the two functions are equal: $\Gamma(\gamma \neq 2) = \Gamma(2) = 0$. For any other value of h the slopes of the two functions are

$$\frac{\mathrm{d}\Gamma(\gamma \neq 2)}{\mathrm{d}h} = -h^{1-\gamma} = -\frac{h^{2-\gamma}}{h} \text{ and } \frac{\mathrm{d}\Gamma(2)}{\mathrm{d}h} = -\frac{1}{h}. \qquad (32.12)$$

Since $0 \leq h \leq 1$, if $\gamma < 2$ the slope of $\Gamma(2)$ is more negative than the slope of $\Gamma(\gamma \neq 2)$ so $\Gamma(2) \geq \Gamma(\gamma \neq 2)$ with equality holding only for $h = 1$. Thus within the range $0 \leq \gamma \leq 2$ the value $\gamma = 2$ always minimizes the free energy for $0 \leq h < 1$. On the other hand, for $\gamma > 2$, the free energy becomes infinitely negative in the limit $h \to 0$ for point masses or long correlation scales. Therefore, the value $\gamma = 2$ does indeed form a barrier which is singled out by the thermodynamics. Gravitating systems can minimize their free energy by evolving toward $\gamma = 2$, and apparently this is what galaxy clustering does.

In terms of entropy the value $\gamma = 2$ maximizes the gravitational contribution to the total entropy. Since this contribution has the opposite sign from the non-interacting contribution, its net result is to reduce the value of the total entropy. When gravity dominates, the total entropy may become negative. This is quite unlike the perfect gas whose entropy is a maximum at equilibrium. It is another reflection of the lack of equilibrium in gravitating systems. As $h \to 0$ the value $\gamma = 2$ produces an extremum of the entropy. So in this respect the general principle that systems tend to extremize their entropy (within given constraints) applies to both gravitating and perfect gas systems.

We can also look at this result from a somewhat different point of view. Ordinary non-equilibrium thermodynamics of perfect gases assumes that near equilibrium the thermodynamic functions are the same as in equilibrium except for an explicit time dependence. Thus the pressure and temperature in an expanding gas cloud will change with time, but the equation of state remains the same as when the gas is static. Physically this seems a reasonable assumption if the collisional relaxation timescale is less than the timescale for macroscopic changes.

For gravitating systems, as discussed in Section 30, we must imagine thermodynamic changes on timescales during which the correlation function $\xi(r)$ is 'frozen'. Thus we suppose that T, V, and N depend on time, but γ and a_m do not. Taking the time derivative of (32.9) to find $\mathrm{d}S/\mathrm{d}t$ under these conditions shows that the rate of gravitational entropy production is maximized when $\gamma = 2$.

This result contrasts strongly with its perfect gas analog. In perfect gases there is a stationary state characterized by a minimum rate of entropy production (see de Groot & Mazur, 1962, referred to in Section 30). Roughly speaking, this occurs because entropy production is associated with dissipation. States which dissipate strongly do not last long. So we would expect the equilibrium state to have minimum dissipation and, therefore, minimum entropy production. Gravitational clustering, however, does not possess a stationary equilibrium state. Once clustering begins in an infinite system it proceeds to the state of maximum gravitational entropy production. This state characterizes the system's non-linear development.

As time goes on, larger and larger scales of the system reach this asymptotic state of maximum entropy production with $\xi(r) \propto r^{-2}$. We have seen that the detailed processes causing this transition cannot be described directly by thermodynamics; they require the underlying kinetic theory. Merging these two descriptions would provide a promising path to understanding non-equilibrium thermodynamics of gravitating systems.

Fluctuations are also maximized as $\gamma \to 2$. Substituting (32.3) into (29.52), easily calculated by recalling (29.6), shows that $\Delta N/N \to 1$ as gravity becomes dominant. These large fluctuations are identified with the formation of well-defined, and sometimes bound, clusters of galaxies. Many galaxies must escape from a clustering region to carry away the binding energy. Eventually other clusters may capture these field galaxies. Thus we are led to inquire into the efficiency of the clustering process.

33

Efficiency of gravitational clustering

Some non-linear descriptions, such as the cosmic virial theorem, can be derived directly from first principles, but so far this has not been done for clustering efficiency. Here the technique is to construct a model, simple enough to solve analytically, which retains the essential physics of the problem.

Let us suppose there is a hierarchy of clustering on different well-defined levels. Consider the interaction of two adjacent levels of this hierarchy. For example, these levels could represent the clustering of globular clusters (or subgalaxies if these are formed from many globular clusters) to form galaxies. They could also represent the clustering of galaxies to form clusters of galaxies. Objects at the lower level are presumed to be tightly bound; we shall call them particles and represent them by point masses. Objects at the higher level are more loosely bound; they are the clusters.

Our analysis starts in a region which has already undergone its phase transition. The density of a cluster is large compared to the average background density. Not all the particles are in clusters. We shall consider the fate of an arbitrary vagrant particle moving through a field of clusters, and ask: 'What is the probability that vagrant particles will be captured by clusters?' This probability is a measure of the clustering efficiency.

The vagrant particle may be one which has been ejected from a forming cluster, or it may never have been bound. Therefore, its velocity is not related to the other particles and clusters in any simple equilibrium manner, but depends on fine details of its previous history. So we shall take its velocity at the start of the analysis to be just an initial condition.

At any particular time during the evolution, the situation can be described by clusters and vagrant intercluster particles with various positions, velocities, masses, and cluster radii. The binding between a particle and a cluster is determined by their relative velocity; similarly the binding between a particle and a group of clusters is determined by the velocity of the particle relative to the group's center of mass. These relative velocities will be influenced by the Hubble expansion. However, the N-body experiments and the analyses described in the previous sections show that although the expansion rate influences the time at which a particular state of clustering is reached, its importance to the eventual state after many initial Hubble times is only secondary. Once bound clusters form, the subsequent evolution of clustering at that level of the hierarchy slows down. Therefore, it appears that the essence of the situation regarding clustering efficiency can be approximated by a particle moving through a non-expanding field of clusters.

The condition that a particle be bound to some cluster or field of clusters is that its total energy (per unit mass) with respect to the clusters be negative. If we take the position $\mathbf{r}_i$ and mass M_i of the ith cluster as given, then the binding criterion for an arbitrary particle of velocity v and position $\mathbf{r}$ is

$$\frac{1}{2}v^2 - G\sum_i \frac{M_i}{|\mathbf{r} - \mathbf{r}_i|} - 4\pi G\sum_j \left(\frac{1}{|\mathbf{r} - \mathbf{r}_j|} \int_0^{|\mathbf{r}-\mathbf{r}_j|} \rho_j R^2 dR + \int_{|\mathbf{r}-\mathbf{r}_j|}^{R_j} \rho_j R dR \right) < 0.$$

(33.1)

The sum over i includes all the clusters in the field which do not contain the particle; the sum over j includes all those which do. The clusters are taken to be spherical with an internal density ρ_j and radius R_j. Several clusters may overlap. To make (33.1) dimensionless, define

$$\frac{v^2}{2GM_0/r_0} \equiv v_*^2,$$

(33.2)

$$M_i/M_0 \equiv X_i,$$

(33.3)

$$r_1/r_0 \equiv y_i,$$

(33.4)

$$\eta_i \equiv (4\pi r_0^3/M_0)\rho_i,$$

(33.5)

with M_0 and r_0 as fiducial mass and length scales. Then (33.1) becomes

$$E = v_*^2 - \sum_i \frac{X_i}{|y - y_i|} - \sum_j \left(\frac{1}{|y - y_i|} \int_0^{|y - y_i|} \eta_i y'^2 dy' + \int_{|y-y_i|}^{R_j/r_0} \eta_j y' dy' \right) < 0.$$

(33.6)

For a given v_*, the particle will be bound to one or more clusters in the field if $E < 0$. The particle is bound to a given region if $E < 0$ in that region. Thus for any values of v_*, y_i, X_i, R_j and η_i it is an easy matter to determine whether the particle is bound. Although this is not a dynamical model, we expect that a bound particle will move away from the cluster centre more slowly than in pure Hubble expansion, and eventually will be observed as a cluster member. The time required for this depends, naturally on the degree of binding and the Hubble expansion rate.

We now make an essential change in our point of view regarding (33.6). Instead of considering one field of clusters we consider an ensemble of fields. This means that we regard the values of X_i, $|y - y_i|$, R_j and η_j as being drawn from some probability distribution. Given these probability distributions we then ask for the probability that $E < 0$. This is defined to be the clustering efficiency. It measures the probability that a random cluster holds a particle in a given distribution.

Next we must specify the relevant distributions and see how they map into the distribution of E. This can be done self-consistently where complete information is available, such as in N-body simulations. Here we will just describe a simple illustration, which tends to underestimate the clustering efficiency of more realistic cases.

The simplest version of (33.6) contains just two terms

$$E = v_*^2 - X\sigma, \tag{33.7}$$

where we have written

$$\sigma(y) = \frac{1}{|\mathbf{y} - \mathbf{y}_i|}. \tag{33.8}$$

This formula can represent three different physical situations. First, it can describe a particle inside a cluster. In this case X is the cluster mass within the radius of the particle and $X\sigma$, with $\mathbf{y}_i = 0$, is just the particle's potential energy except for the contribution from the mass of the cluster between $|\mathbf{r} - \mathbf{r}_j|$ and R_j. If this contribution is large it is nearly independent of $|\mathbf{y} - \mathbf{y}_i|$ and can be included formally in v_*. If it is small it can be neglected to get a good lower limit on the clustering efficiency. The contributions from the other clusters, which would increase the clustering efficiency, are assumed to be comparatively small.

The second case described by (33.7) is for a particle outside all clusters, but whose gravitational binding is dominated by its nearest cluster. This will usually be the most interesting case. The third case applies when the entire summation over i and j can be described by a given statistical distribution. For example, if the values of all $X_i/|\mathbf{y} - \mathbf{y}_i|$ and η_i are independent of each other, and if the summations include enough terms that the central limit theorem applies, then the distribution of the sums approaches a normal Gaussian. In this case, the contributions of all the clusters to the binding would be taken into account. They are represented by the distribution of X, in which σ can be normalized to unity.

In (33.7), we may, if we wish, ascribe probability distributions to v_*, X, and σ. Again, to make matters simple, let us suppose that v_* refers to a particular particle and is given. Similarly, $\sigma(y)$ refers to a particular position and is given (we shall also see how a distribution of σ modifies the results). The mass distribution (or the distribution of sums in (33.6)) will have a probability density $f(X)$. From this we must get the probability density $f(E)$.

The transformation of probability densities is a standard matter. Suppose $X = (X_1, X_2, \ldots, X_n)$ is an n-dimensional random vector of n random variables X_i having joint probability density $f_x(X_1, X_2, \ldots, X_n)$. Let there be a deterministic, continuous, one-to-one transformation E with continuous derivatives which maps X into another n-dimensional random vector $U = (U_1, \ldots, U_n)$ having joint density $f_u(U_1, \ldots, U_n)$. The transformation E is defined by the n equations, $U_i = E_i(X_1, \ldots, X_n)$, which have the inverses $X_i = H_i(U_i, \ldots, U_n)$. To find the relation between f_x and f_u, we require that if an event X has the probability $P\{X \in A\}$ of being found in a region A, then the transformed event U has the same probability of being found in EA which is the image of A under the transformation E. (The probability of an event must be independent of the variables used to describe it.) Thus $P\{X \in A\} = P(U \in EA)$, or in terms of probability densities,

$$\int_A f_x(X_1, \ldots, X_n) dX_1 \ldots dX_n = \int_{EA} f_u(U_1, \ldots, U_n) dU_1 \ldots dU_n. \qquad (33.9)$$

Therefore, the probability densities transform in the same manner as the change of variables in a definite integral

$$f_u(U_1, \ldots, U_n) = f_x(X_1, \ldots, X_n) \left| \frac{\partial(X_1, \ldots, X_n)}{\partial(U_1, \ldots, U_n)} \right|, \qquad (33.10)$$

where the $X_i = H_i(U_1, \ldots, U_n)$, and the last form is the absolute value of the familiar Jacobian

$$\frac{\partial(X_1, \ldots, X_n)}{\partial(U_1, \ldots, U_n)} = \begin{vmatrix} \dfrac{\partial H_1}{\partial U_1} & \cdots & \dfrac{\partial H_1}{\partial U_n} \\ \vdots & & \\ \dfrac{\partial H_n}{\partial U_1} & & \dfrac{\partial H_n}{\partial U_n} \end{vmatrix}. \qquad (33.11)$$

It is simple to apply this result to (33.7), for there $n = 1$ and $U_1 = E$, giving

$$f(E) = \frac{1}{\sigma(y)} f(X). \qquad (33.12)$$

We now take the mass distribution of clusters (or the sum of the potential energy terms) to have the Gaussian density,

$$f(X) = \left(\frac{2}{\pi}\right)^{1/2} \mu^{-1} [1 + \mathrm{erf}\,(1/\mu\sqrt{2})]^{-1} \exp\left[-\frac{(X-1)^2}{2\mu^2} \right], \qquad (33.13)$$

normalized over the range $0 \le X \le \infty$. So M_0 from (33.3) is the most probable mass (or value of the potential energy sum), and μ is a measure of the mass dispersion. From Equations (33.12) and (33.13)

$$f(E) = \left(\frac{2}{\pi}\right)^{1/2} \mu^{-1} \sigma^{-1} [1 + \mathrm{erf}\,(1/\mu\sqrt{2})]^{-1} \exp\left[\frac{(v_*^2 - E - \sigma)^2}{2\mu^2\sigma^2} \right]. \qquad (33.14)$$

This is the probability that a particle with velocity v_* moving through a field of clusters which have Gaussian mass distributions has total energy E with respect to its nearest neighbor. It can also be the probability that the particle has energy E with respect to the entire field of clusters, under the circumstances discussed before. It depends on the distance of the particle from its nearest cluster (or the positions of all the other clusters). The distribution of E is also Gaussian, but has the most probable value $v_*^2 - \sigma$ and a dispersion given by $\mu\sigma$. As expected, the closer the particle is to some cluster, the larger σ is, and the smaller or more negative its most probable energy becomes.

Using (33.14), we can now calculate the clustering efficiency, or probability, that the particle is bound, since, for given values of v_*, σ, and μ

$$P\{E<0\} = \int_{-\infty}^{0} f(E)\mathrm{d}E = \left\{1 + \mathrm{erf}\left[\frac{(1 - v_*^2/\sigma)}{\mu\sqrt{2}}\right]\right\} \bigg/ [1 + \mathrm{erf}(1/\mu\sqrt{2})].$$

(33.15)

It is useful to recall that the error function has the properties

$$\mathrm{erf}\, z = 2\pi^{-1/2}z + O(z^3)\, \text{as}\, z \to 0,$$

and

$$\mathrm{erf}\, z \to 1 - \pi^{-1/2}z^{-1}\mathrm{e}^{-z^2}\left[1 - \frac{1}{2z^2} + O(z^{-4})\right] \text{as}\, z \to \infty.$$

(33.16)

The clustering efficiency has several interesting physical properties. When the nearest cluster dominates it is determined by the dispersion of cluster masses and the ratio of the particle's kinetic to gravitational energy. If the mass distribution is a very broad Gaussian, $\mu \to \infty$ and $P\{E<0\} \to 1$ for any finite v_*^2/σ. This is because there is a high probability of being close to a very massive cluster. For a very narrow mass distribution, $\mu \to 0$ and $P \to 0$ if $v_*^2/\sigma > 1$ but $P \to 1$ if $v_*^2/\sigma < 1$. For $v_*^2/\sigma = 1$, $P \geq \frac{1}{2}$ for any μ. For $v_*^2/\sigma \to \infty$, $P \to 0$ for any finite μ since $\mathrm{erf}(-z) = -\mathrm{erf}(z)$. As $v_*^2/\sigma \to 0$, $P \to 1$. All these trends accord with one's intuition. What is perhaps surprising, however, is the wide range of conditions in which P is large, say $P > \frac{1}{2}$, and capture of a vagrant particle becomes highly efficient. Analogous results hold when the distribution represents the gravitational energy of the entire field of clusters. It is also worth emphasizing that these results are independent of the detailed dynamical trajectory of the particle.

We can easily generalize these results when σ is also a random function in the ensemble of fields. . Now both masses of clusters and their distribution are given by some joint probability density $f(x, \sigma)$. Introducing the variables $X_1 = X, u_1 = E = v_*^2 - X\sigma$ and $u_2 = \sigma = X_2$, we obtain

$$f(E) = \int \sigma^{-1} f\left(\frac{v_*^2 - E}{\sigma}, \sigma\right)\mathrm{d}\sigma$$

(33.17)

from (33.10). The generalized clustering efficiency for (33.17) is then

$$P\{E<0\} = \int \sigma^{-1} \int_{-\infty}^{0} f\left(\frac{v_*^2 - E}{\sigma}, \sigma\right)\mathrm{d}E\mathrm{d}\sigma.$$

(33.18)

When there is no correlation between the spatial distribution of clusters and their masses, then

$$f\left(\frac{v_*^2 - E}{\sigma}, \sigma\right) = f_1\left(\frac{v_*^2 - E}{\sigma}\right)f_2(\sigma),$$

(33.19)

and the clustering efficiency simplifies. For the particular case $f_2(\sigma) = \delta(\sigma - \sigma_0)$, this result reduces to the previous form.

This approach can be applied with more difficulty to more complicated models (e.g., Saslaw, 1979). It is very difficult to escape from the basic result that clustering is highly efficient. The N-body computations described in Section 28 also show the high efficiency of clustering; relatively few field galaxies remain.

Now we are in a position to use the model to ask how correlations of clusters affect the clustering efficiency. After the analysis of Section 32 it will be no surprise to learn that $\xi(r) \propto r^{-2}$ maximizes the clustering efficiency. To see this it is only necessary to generalize the analysis which led to (27.7). Assume for simplicity that the particles and clusters have the same spatial distribution. The probability of finding a cluster between r and $r + dr$ distant from the particle is

$$\omega(r)dr = 4\pi nr^2(1 + ar^{-\gamma})dr, \tag{33.20}$$

as usual.

To find σ, we need to convert $\omega(r)dr$ into the distribution of nearest clusters. This can be done simply if we neglect two effects which are not essential to the result. The first is the influence of any *a priori* information that there are no clusters between 0 and r on the probability that objects exist at 0 and r. The second is the influence of three-point and higher order correlations on the position of the nearest cluster. With these simplifications, the probability $\omega_1(r)dr$ that a particle's nearest cluster is found between r and $r + dr$ follows analogously to (27.5)

$$\omega_1(r) = 4\pi nr^2(1 + ar^{-\gamma})\exp\left[-\left(\frac{4\pi nr^3}{3} + \frac{4\pi nar^{3-\gamma}}{3-\gamma}\right)\right]. \tag{33.21}$$

When $a = 0$, this reduces to (27.5) for a uniform random distribution.

To see the effect of correlations on the expectation value of the distance r_1 to the nearest cluster, consider the case $a < 1$

$$\langle r_1 \rangle = \int_0^\infty r\omega_1(r)dr = \frac{0.55}{n^{1/3}} - \frac{1}{3-\gamma}\left(\frac{4\pi n}{3}\right)^{(\gamma-1)/3}\Gamma\left(\frac{4-\gamma}{3}\right)a + O(a^2). \tag{33.22}$$

For $0 < \gamma < 3$, the expected distance to the nearest cluster decreases as a increases.

The gravitational potential energy in (33.6) shows a more interesting behavior. Its expectation value is the sum of the expectations of the energy of each of the clusters in the field. Since these terms are all negative, using only the first term for the nearest cluster gives a minimum strength for the gravitational binding. Therefore, using (33.7) with $\langle \sigma(r_1) \rangle = \langle r_1^{-1} \rangle$ provides the smallest value expected for the clustering efficiency. Since the probability in (33.20) applies to any particle in the field of clusters it does not matter where the particle is in the field (assuming for simplicity that it is not inside a cluster). With (33.21) it is easiest to treat the cases of small and large correlation separately, giving, when $\gamma < 3$,

$$\langle \sigma \rangle = \left\langle \frac{1}{r_1} \right\rangle = \int_0^\infty r^{-1}\omega_1(r)dr$$

$$= \left(\frac{4\pi n}{3}\right)^{1/3}\Gamma\left(\frac{2}{3}\right) + \frac{a}{3-\gamma}\left(\frac{4\pi n}{3}\right)^{(\gamma+1)/3}\Gamma\left(\frac{2-\gamma}{3}\right) \text{for } \xi \ll 1$$

$$= \Gamma\left(\frac{2-\gamma}{3-\gamma}\right)\left(\frac{4\pi na}{3-\gamma}\right)^{1/(3-\gamma)} \quad \text{for } \xi \ll 1. \tag{33.23}$$

As an initially uncorrelated system evolves and correlations grow we expect γ to increase from zero. When $\gamma \to 2$, both $\Gamma[(2 - \gamma)/3]$ and $\Gamma[(2 - \gamma)/(3 - \gamma)]$ diverge as $(2 - \gamma)^{-1}$, i.e., they have simple poles at $\gamma = 2$. This is our old friend, the logarithmic divergence, hiding in the integral for $\langle 1/r_1 \rangle$ when $a \neq 0$ (use (33.21) for small r_1). Thus, even the minimum value of $\langle \sigma \rangle$ due to the closest cluster becomes very large as $\gamma \to 2$. We are dealing here with a system having a definite value of σ, namely the average value $\langle \sigma \rangle$, rather than with an ensemble having a distribution of values for σ. Therefore, we just substitute $\langle \sigma \rangle$ into (33.15) to get the clustering efficiency.

Independent of most details, the model shows that the value $\gamma = 2$ maximizes the efficiency of gravitational clustering. This result is another aspect of the system's tendency to maximize the gravitational contribution to its total entropy. The distribution of galaxies observed in the sky appears to be dominated by this tendency. The small departures of γ from $\gamma = 2$ which are seen may be caused by correlations of obscuring dust in our own galaxy. Or they may be caused by residual effects of initial conditions and the cosmological expansion.

34

Non-linear theory of high order correlations

Like bubbles on the sea of Matter born,
They rise, they break, and to that sea return

Alexander Pope

34.1. Equation of state

Imagine life as it may have been a million years ago. You are in the jungle, being stalked by a tiger. Your ability to survive depends on pattern recognition. If you can only see the stripes on the tiger (small scale correlations), but not the overall effect of the tiger itself, you will be at some disadvantage. Perhaps this is how the ability of our eyes to recognize high order correlation functions developed. Similarly, restricting our understanding of galaxy clustering to just the two- or three-particle correlation functions means we miss a lot of the action. We need a simple measure of high order clustering which can also be related to basic gravitational physics.

In Section 27 we saw that gravitational clustering can be characterized by the distribution of voids. These, in turn, are related to the high order correlations which describe the galaxies which should have been in the region of the void but are not. We may generalize the idea of a void by working in terms of distribution functions $f(N)$ which give the probability of finding any number of galaxies in a volume V of arbitrary size and shape. For $N = 0$, the distribution of voids is $f(0)$, which is calculated in (27.7) for a Poisson distribution. As the clustering evolves, we must find some way to calculate the departure of $f(N)$ from its Poisson form.

The BBGKY approach is of no help here because the situation rapidly becomes very non-linear. We might hope that a thermodynamic approach will work in this regime, and indeed it does! As Section 30.2 showed, the two-particle correlation function contains much information about larger scale clustering, for it evolves self-consistently with all the higher order correlation functions. Our problem is to extract this information about higher order correlations from the two-particle function without having to solve the BBGKY hierarchy at all.

The previous several sections described how the two-particle correlation function determines the thermodynamic properties of an infinite system. Among these properties are equations of state which include gravitation. These equations of state, in turn, determine the equilibrium fluctuations. It is from these fluctuations – the temporary clusters that rise, break, and often dissipate – that we can extract the distribution functions $f(N)$. Then we will see how well the thoretical results fit N-body simulations.

Consider again our infinite system of galaxies, all having the same mass m. The expanding background implicitly removes the mean gravitational field but does not enter explicitly into the thermodynamics. Write Equations (29.34) and (29.35) for the internal energy and pressure in the form

$$U = \frac{3}{2}NT(1 - 2b) \tag{34.1}$$

and

$$P = \frac{NT}{V}(1 - b). \tag{34.2}$$

The temperature (in energy units with Boltzmann's constant equal to one) is defined by the random kinetic energy,

$$K = \frac{3}{2}NT = \frac{1}{2}\sum_{i=1}^{N} mv_1^2, \tag{34.3}$$

which is summed over the peculiar velocities of the N galaxies in a volume V. The value of b, from (29.34), measures the influence of the gravitational correlation energy, W

$$b \equiv -\frac{W}{2K} = \frac{2\pi Gm^2 n}{3T}\int_0^\infty \xi(n, T, r)r\mathrm{d}r. \tag{34.4}$$

The value of b depends on the form of the two-particle correlation function, ξ, and Section 32 showed that $\xi \propto r^{-2}$ over a finite range maximizes the gravitational entropy of clustering. Here, however, we are concerned primarily with b as a function of n and T. Therefore, we suppose the spatial integral to be done and the result to give the quasi-equilibrium $b(n, T)$ at any particular time.

The resulting $b(n, T)$ contains information about clustering on all scales through the dependence of ξ on all the higher N-particle correlations in the full **BBGKY** hierarchy. Thus, specifying $b(n, T)$ is a substitute for solving the whole coupled hierarchy with given initial conditions. Of course, this would be of no use if we were to specify an incorrect or inconsistent $b(n, T)$. However, the form of $b(n, T)$ is strongly constrained by simple physical arguments.

First, $b(n, T)$ must depend only on a special combination of n and T. There are several ways to see this. The simplest is to notice that b is a function of the dimensionless ratio of potential to kinetic energy. In an infinite, thermodynamically homogeneous system, the only characteristic length scale which enters the potential energy is the average galaxy separation $\bar{r} \approx n^{-1/3}$. So b should depend just on the ratio of the potential and kinetic energies of two typical galaxies at this separation: $Gm^2/Tn^{-1/3}$. To eliminate fractional powers, we can cube this ratio and write

$$b = b(G^3m^6nT^{-3}) = b(nT^{-3}). \tag{34.5}$$

This result also follows, more formally, from scale invariance. The value of b, and the consequent thermodynamics, is determined by averaging homogeneous systems. Again $\bar{r} = n^{-1/3}$ is the only length scale which can be formed from the thermodynamic variables which describe the system. All other lengths, including the correlation length, must be multiples of $\bar{r}$. Suppose there is a (possibly continuous) hierarchy of clustering. Homogeneity requires that b is invariant under the scale transformation $\bar{r} \to \lambda \bar{r}$. Physically this means that the value of b is the same for each level of the clustering hierarchy on scales larger than the two-particle correlations at that level. Thus $n^{-1/3} \to \lambda n^{-1/3}$ or $n \to \lambda^{-3} n$ and $b \to \lambda^0 b = b$, this last equation being the expression of scale invariance. Since b is the ratio $-W/2K \propto Gm^2 N^2 / rNT$, scale invariance requires that $T \to \lambda^{-1} T$ because N refers to a very large constant volume and does not depend on the level of the clustering hierarchy. Physically, this scaling of temperature means that clusters at a higher level of the hierarchy have lower random velocities relative to each other than smaller clusters have relative to each other. (This should not be confused with the rescaling which would occur during a physical expansion of the entire system.) The scale invariance of b therefore requires that

$$b(\lambda^{-3} n, \lambda^{-1} T) = b(n, T). \tag{34.6}$$

This is a functional equation whose solution is readily seen to be (34.5).

There is another, somewhat more subtle, constraint which again leads to the requirement $b = b(nT^{-3})$. This is the constraint that the entropy of a system in equilibrium (here assumed to apply to our quasi-equilibrium state) does not depend on the path by which the system reaches that state. In other words, the entropy must be a total differential. Substituting (34.1) and (34.2) into the relation (29.40) shows that the entropy is given by

$$d\left(\frac{S}{N}\right) = d \ln\left(n^{-1} T^{3/2}\right) - 3db + b \, d \ln\left(nT^{-3}\right). \tag{34.7}$$

In order that the entropy be a total differential, independent of how the system reached its state, we see again that (34.5) must be satisfied. Therefore we are led to the result $b = b(nT^{-3})$ from three basic related points of view.

For the distribution of voids it does not matter how b depends on nT^{-3}; a finite value of b turns out to be all that is required. However, for the probability distributions $f(N)$ with $N > 0$, we will need to know the functional form of $b(nT^{-3})$. Physical constraints which we consider next limit its nature severely.

From its definition (34.4) we see that the possible values of b lie between zero (no gravity or correlation) and one (virial equilibrium). Since b depends on the full quantity $G^3 m^6 T^{-3} n$, the value $b = 0$ applies both for no gravity and as $n \to 0$. In general, thermodynamic variables are analytic functions of each other. Therefore a power series expansion of $b(G^3 m^6 nT^{-3})$ must start off linearly with its argument. This is consistent with the requirement of little correlation at high temperature. Incidentally, this reasoning gives the Debye–Hückel theory when applied to dilute plasmas and electrolyte solutions. For gravity, we are more interested in the non-linear terms, and here the virial theorem comes to our aid. At low temperatures and high densities, more strongly bound clusters form. Such clusters tend to contain more and more galaxies and to be close to virial equilibrium, suggesting $b \to 1$ as $nT^{-3} \to \infty$. A limiting case is the hierarchy of tightly bound binary systems. The simplest form of b which has these properties is

$$b = \frac{b_0 n T^{-3}}{1 + b_0 n T^{-3}},\tag{34.8}$$

where $(Gm^2)^{-3} b_0$ is a positive dimensionless constant.

This form of b may also be viewed as the first Padé approximant to any general function $b(nT^{-3})$. A continued fraction would satisfy the physical requirements on b, but (34.8) leads to the simplest form of the N-galaxy volume distribution statistic $f(N)$ which agrees well with numerical experiments. So it seems a reasonable approximation to adopt, although its rigorous relation to the BBGKY hierarchy remains to be explored. Sometimes it is necessary to progress by cutting the Gordian knot.

This form of b has interesting thermodynamic implications. Consider, for example, the entropy it implies. Substituting (34.8) into (34.7) and integrating gives

$$\frac{S}{N} = \ln n^{-1} T^{3/2} - 3b - \ln(1 - b) + s_0.\tag{34.9}$$

As expected, S is extensive. For small b and a given average density, the mildly correlated structure decreases S relative to its value for the random Maxwellian distribution with $b = 0$ (perfect gas case). Entropy decreases here because the clustering is homogeneous and the theory, in this simple form, does not allow for entropy transfer between clustering galaxies and field galaxies, and for the changing velocity distribution. As $b \to 1$, subclusters form in virial equilibrium. Within each subcluster the distribution becomes a random Maxwellian again. With most galaxies in virialized clusters the contribution of the explicitly gravitational terms, $S_{grav} = -3b - \ln(1 - b)$, to the entropy becomes positive, and as $b \to 1$, $S_{grav} \to \infty$ with a logarithmic divergence.

Substituting (34.8) for temperature into (34.9) shows that the total entropy of a system of constant average overall density also becomes infinite as $b \to 1$. Therefore the entropy extremum in an infinite gravitating system does not occur when every particle has fallen into a point singularity. It occurs when there is a hierarchy of clustering which has reached virial equilibrium on all scales. This is quite different from the situation within a single finite cluster (see Section 43). For a finite cluster, the density is never uniform on any significant scale, and the correct counting of states differs from an infinite system capable of being averaged over large scales. The counting of states in finite systems is also much more strongly influenced by the formation of energetic binaries (Section 48). In infinite systems such binary formation is inhibited by the general expansion, and any binaries which may form have relatively local effects.

34.2. The distribution functions $f(N)$

We may now derive an analytic expression for the probability $f(N)$ of finding N galaxies in a randomly positioned volume V. Consider a large system which we suppose is described by equilibrium gravithermodynamics, at least as a first approximation. Collect an ensemble of subregions, all of the same volume, which are much smaller than the total volume. Both the number of galaxies and their total energy will vary among the members of the ensemble. This describes a reasonable way to sample the Universe. Thermodynamically each subregion can be regarded as being in contact with a large reservoir. Across the boundary both particles (galaxies) and energy can flow. This is the grand canonical ensemble characterized by a given chemical potential μ and temperature. It describes systems with variable N and U (see Section 29.3).

From statistical thermodynamics we have the result that the probability of finding N_j particles in the energy state $U_j(N, V)$ is

$$P_{N_j}(V, T, \mu) = \frac{e^{-U_j/T + N\mu T}}{Z_G(T, V, \mu)}, \tag{34.10}$$

where

$$Z_G(T, V, \mu) = \sum_{j, N} e^{-U_j/T + N\mu/T} \tag{34.11}$$

is the grand partition function (again in units with $k = 1$). This is just the generalization of (29.45) to the grand canonical ensemble. Here we see explicitly why states containing infinitesimally close binaries must be excluded since they would make an infinite positive contribution to Z_G.

This exclusion of binaries in the counting of states is reasonable because very few tightly bound binaries actually form. The rate at which they would form in a finite non-expanding spherical cluster is longer than the two-body relaxation time τ_R by a factor $\sim N$, and longer than the dynamical crossing time τ_c by a factor $\sim N^2$ (see Section 48). This is essentially because τ_c is a 'one-body' interaction, τ_R is a 'two-body' interaction and point mass binaries form by a three-body interaction in which the third body removes the excess energy and momentum. For an expanding system, the rate of binary formation is even slower. As the galaxies move apart, their interaction becomes weaker. N-body simulations confirm these results.

Associated with the grand canonical ensemble is a potential Ψ whose derivatives give the thermodynamic properties of the ensemble (see (29.46))

$$\Psi\left(\frac{1}{T}, V, \frac{\mu}{T}\right) = \ln Z_G \tag{34.12}$$

and

$$S = T\left(\frac{\partial \Psi}{\partial T}\right)_{V, \mu} + \Psi, \tag{34.13}$$

$$N = T\left(\frac{\partial \Psi}{\partial \mu}\right)_{V, T}, \tag{34.14}$$

for example. To find Ψ explicitly, note that the definition of entropy is

$$S = - \sum_{j,N} P_{N_j} \ln P_{N_j}. \qquad (34.15)$$

Substituting (34.10) into (34.15) readily gives

$$S = \frac{U}{T} - \frac{N}{T}\mu + \Psi. \qquad (34.16)$$

Combining this with the thermodynamic Euler relation (29.15) we see that

$$\Psi = \frac{PV}{T} = \bar{N}(1 - b), \qquad (34.17)$$

where the last equation follows from (34.2) and $\bar{N}$ explicitly denotes the average number in a volume V. All this is independent of the functional form of b.

The probability of finding a particular number N of particles in a volume V of arbitrary shape (which also applies to a projected area) follows from summing (34.10) over all energy states

$$f(N) = \sum_j P_{N_j}. \qquad (34.18)$$

Noting that e^Ψ gives the normalization of the sum-over-states (34.11), suggests that we can calculate the $f(N)$ without explicitly performing the summation (34.18). To do this we define a generating function (compare the characteristic functions of Sections 3 and 31.2) by averaging the quantity e^{qN} where q is an arbitrary variable

$$\langle e^{qN} \rangle \equiv \sum_N e^{qN} f(N)$$

$$= Z_G^{-1} \sum_{N,j} e^{-U_j/T + (q + \mu/T)N}$$

$$= \exp\left\{ \Psi\left(\frac{\mu}{T} + q\right) - \Psi\left(\frac{\mu}{T}\right)\right\}. \qquad (34.19)$$

We have used (34.11) and (34.12), and may now regard the arguments T^{-1} and V of Ψ as being understood since they do not enter explicitly in the following analysis.

At this stage it simplifies the algebra to regard $\Psi(x)$ as a function of e^x rather than of x and to write

$$z \equiv e^q. \qquad (34.20)$$

Then (34.19) becomes

$$\sum_N f(N)z^N = \exp\left\{ \Psi(ze^{\mu/T}) - \Psi(e^{\mu/T})\right\} = e^{-\Psi} e^{\Psi(ze^{\mu/T})}. \qquad (34.21)$$

Now expand the last exponential in (34.21) in a Taylor series around $z = 0$ and equate the coefficients of powers of z on both sides of the equation to find

$$f(N) = e^{-\Psi} \frac{e^{\mu N/T}}{N!}(e^\Psi)_0^{(N)}, \qquad (34.22)$$

with

$$(e^{\Psi})_0^{(N)} = \left[\left(\frac{d}{d(ze^{\mu/T})} \right)^N \exp \Psi(ze^{\mu/T}) \right]\Bigg|_{ze^{\mu/T}=0} \qquad (34.23)$$

This is the basic result.

We can apply it straightaway to the distribution of voids. For $N = 0$ and, consequently, $U = 0$, we see from (34.11)–(34.12) that $\Psi(0) = 0$ and therefore

$$f(0) = e^{-\Psi(\mu/T)} = e^{-\bar{N}(1-b)}$$
$$= e^{-4\pi r^3 n(1-b)/3}, \qquad (34.24)$$

using (34.17). The last expression, in particular, is the probability of finding an empty spherical volume of radius r. This reduces to the standard result, (27.7), for a Poisson distribution as $b \to 0$ and gravity becomes unimportant. Notice that although the derivation of (34.24) depends on the value of b, it does not depend on the functional form of $b(n, T)$. As $b \to 1$, the galaxies become more and more clustered, so the probability of finding an empty region of any size approaches unity in the idealized limit.

To find the full set of distributions $f(N)$ we first rewrite (34.23) in a more specific form. Replacing the dummy variable $ze^{\mu/T}$ by $e^{\mu/T}$ by $e^{\mu/T}$, we note that

$$\frac{d}{de^{\mu/T}} = e^{-\mu/T} \frac{db}{d(\mu/T)} \frac{d}{db}. \qquad (34.25)$$

To obtain $db/d(\mu/T)$, we use (34.17)

$$\frac{d\Psi}{d(\mu/T)} = \frac{d[Vn(1-b)]/db}{d(\mu/T)/db} = nV, \qquad (34.26)$$

where the last equality follows from (34.14) or (34.16) at constant V. Substituting (34.25) and (34.26) into (34.23) gives

$$(e^{\Psi})_0^{(N)} = \left[\left(ne^{-\mu/T} \frac{db}{d[n(1-b)]} \frac{d}{db} \right)^N e^{Vn(1-b)} \right]\Bigg|_{n=0}. \qquad (34.27)$$

Here $n = n(b, T)$, and the complete operator in parentheses is applied N times. To find the chemical potential we combine (34.1), (34.7), (34.16) and (34.17) to give

$$\frac{\mu(n, b)}{T} = \ln(nT^{-3/2}) - b - \int n^{-1}b \, dn. \qquad (34.28)$$

Two features of (34.27) are especially noteworthy. First, the properties of a distribution for a gravitationally interacting system are determined by its infinitely diffuse limit, $n = 0$. This also occurs in the virial description of an imperfect gas. Second, there is a form of $b(n)$ which simplifies the operators in (34.27) enormously, namely $n(1 - b) \propto b$. Remarkably, this is just the form (34.8) derived on physical grounds.

With (34.8) for $n(b)$ it is straightforward to calculate $(e^{\Psi})_0^{(N)}$ for the first few values of N to determine its general form, which may then be proved by induction (see Saslaw & Hamilton, 1984). Then substituting (34.17) for ψ and combining with

(34.22) finally gives the general formula for the distribution functions

$$f(N) = e^{-\bar{N}(1-b)-Nb}\frac{\bar{N}(1-b)}{N!}[\bar{N}(1-b)+Nb]^{N-1}.$$ (34.29)

This basic result depends only on $\bar{N}=nV$ and b, both of which can be determined independently of (34.8) for any collection of gravitating bodies. So there are no free parameters in the basic theory.

The moments of fluctuations may be calculated either directly from $f(N)$ or with the methods described in Sections 29 and 35. For example,

$$\langle\Delta N^2\rangle = \frac{N}{(1-b)^2},$$ (34.30)

$$\langle\Delta N^3\rangle = \frac{N}{(1-b)^4}(1+2b),$$ (34.31)

and

$$\langle\Delta N^4\rangle = \frac{N}{(1-b)^6}(1+8b+6b^2)+3\langle\Delta N^2\rangle^2.$$ (34.32)

In each case we see that the fluctuations become very large, formally infinite, as $b \to 1$. This indicates the formation of bound, virialized subclusters. The higher the moment, the more strongly it diverges, a result noted earlier in Section 31.2.

Since this type of theory does not follow rigorously from the particles' equations of motion it has to be tested. Clearly, its range of validity will be in a non-linear asymptotic state when fluctuations, groups, and giant clusters become 'frozen' into the expansion of the universe. Direct N-body simulations can be compared with the $f(N)$, as discussed in Section 27 for the special case $f(0)$. Figures 36 to 38 show examples of this comparison for a model expanding universe with $\Omega = 1.0$. It starts with a Poisson distribution of equi-mass galaxies moving just with the Hubble expansion. It is examined here after an expansion factor ~ 8 when the two-particle correlation function has reached approximately the presently observed form and amplitude. (At much earlier times the 'frozen equilibrium' has not been reached and, as expected, the results depart from (34.29).)

Figure 36 shows the distribution of spherical voids, $f(0)$, for both the initial Poisson distribution and the final evolved distribution. The solid lines give the result of the N-body simulation, and the dotted lines show the theoretical expression (34.29). Figure 37 shows the results for $f(4)$. To get a concise picture of the much higher order distributions we invert the procedure. Instead of asking for the probability of finding any radius sphere containing a given number of galaxies, we ask for the probability of finding any number of galaxies within a sphere of given radius. Figure 38 shows the results for a sphere of radius 0.3. (The co-moving radius of the model universe is always normalized to unity.)

Agreement between theory and experiment may be judged in two ways. First the differences of the dotted and solid curves in the evolved case may be compared with their differences in the initial case (where the difference represents noise in the

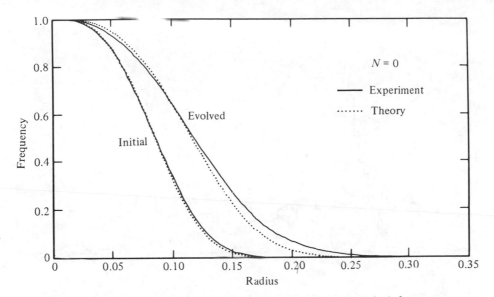

Fig. 36. Comparison between the experimental and theoretical frequency distributions $f(0)$ for spherical voids of radius r. (From Saslaw & Hamilton, 1984.)

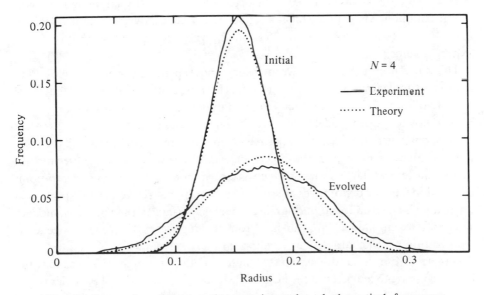

Fig. 37. Comparison between the experimental and theoretical frequency distributions $f(4)$ for spherical regions of radius r containing four galaxies. (From Saslaw & Hamilton, 1984.)

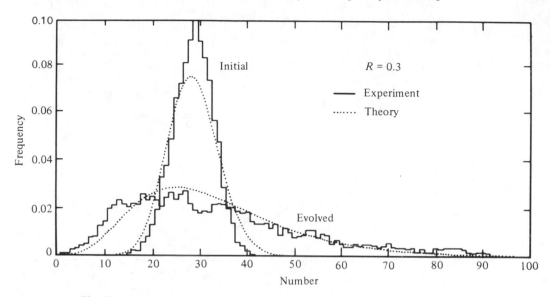

Fig. 38. Comparison between experimental and theoretical frequency distributions $f(N)$ for spherical regions of a given radius $R = 0.3$. (From Saslaw &
Hamilton, 1984.)

Poisson sample). The small systematic enhancement of the simulation for small N in
Figure 38 may be caused by more small permanently bound groups forming than
the equilibrium theory predicts. Second, the theoretical curves in all the figures are fit
by the same value of $b = 0.65$. This value could be varied by ± 0.05 and still give
good fits. But $b = -W/2K$ can also be determined directly by sampling regions of
the distribution. The value which is found by averaging over different regions is
0.74 ± 0.05. (Recall from Sections 25.2 and 27 where $\beta = \frac{2}{3} = 1/(2b)$ that this is its
expected value.)

The difference of $\lesssim 20\%$ between theory and experiment seems to show that
relatively simple equilibrium theory accounts for most, but not all, of the behavior of
$f(N)$. Evolution and inhomogeneity are certainly important but do not dominate.
This is shown by open models with $\Omega = 0.1$, where a single fitted value of b gives just
as good agreement between the experimental and theoretical curves for all the $f(N)$,
but this value of b differs from the directly determined value by $\sim 30\%$. Open models
do not cluster over large scales as much as closed models (with the same two-point
correlation function). Moreover, we saw in Section 28 that the velocity dispersion in
open models is less uniform, among clusters and field galaxies, than in closed models.
Therefore open models are further from having reached the homogeneous
temperature which characterizes equilibrium thermodynamics.

It is clear from all our discussion of infinite gravitating systems that we are really
only beginning to understand the large scale behavior of myriads of particles which
interact with a force so simple as an inverse square.

35

Problems and extensions

35.1. Gravitational instability in multiple component systems

Section 21 and subsequent calculations dealt with systems having essentially one component. Even when masses of members of the component differed they interacted in qualitatively the same way. The situation changes when two or more components interact differently but still remain coupled. Interesting new instabilities often arise.

A relatively simple example is a universe filled with baryons and a low mass collisionless fluid, perhaps neutrinos. Different instabilities occur depending on which component dominates the total mass density and which is initially perturbed. Perturbations of massive collisionless neutrinos will tend to damp on scales smaller than the neutrino Jeans mass. Show that for neutrino density perturbations in a baryon dominated universe the Jeans mass is

$$M_J \approx \frac{4\pi}{3} \langle v^{-1} \rangle^{-3} m_v \bar{n}_v \left[\frac{3}{8\pi G \rho_b} \right]^{3/2} \tag{1}$$

at any given time. What is the corresponding Jeans mass in a neutrino dominated universe? How do these values compare with the masses of stars, galaxies and clusters of galaxies for reasonable values of m_v and n_v? What difference does it make if the neutrinos are relativistic? Show that neutrino perturbations will generally grow by gravitational instability if their gravitational timescale is shorter than the time needed for an average velocity neutrino to cross the perturbation (I. Wasserman, 1981, *Ap. J.*, **248**, 1; A.D. Melott, 1983, *MNRAS*, **202**, 595).

35.2. Pancakes

Examining a spherically symmetric perturbation in the density of an expanding universe provides insight into growth processes, but is not very realistic. The same may be said for a single Fourier mode. It is likely that many perturbations with a range of wavelengths, directions, and amplitudes interacted to form galaxies. These could result in very anisotropic structures, perhaps producing strings of galaxies on the largest, least relaxed scales.

To follow such non-linear development rigorously is a very difficult problem. However, an approximation which extrapolates the Lagrangian coordinate formulation from the linear into the non-linear regime provides some insight. In these coordinates the position $\mathbf{r}$ of a particle at time t is determined by its initial position $\mathbf{q}$ and the time t, i.e., $\mathbf{r} = \mathbf{r}(t, \mathbf{q})$. If the particles are pressure-free ('dust'), then growing perturbations caused by Newtonian gravity have the form

$$\mathbf{r} = a(t)\mathbf{q} + b(t)\mathbf{p}(q). \tag{1}$$

The Hubble expansion gives rise to the first term and the perturbation to the second term. In the linear regime $a(t)$ and $b(t)$ are determined by methods like those described in Section 21. Determining $\mathbf{p}(q)$ from the initial conditions gives $\mathbf{r}$ for every particle and this is equivalent to complete knowldege of the motion.

Next, assume that this formula for $\mathbf{r}$ also gives a useful description in the non-linear regime. To examine the evolution of a group of particles starting near some value of $\mathbf{q}$, show that its deformation tensor

$$D_{ik} = \frac{\partial r_i}{\partial q_k} \tag{2}$$

can be put in the form

$$D = \begin{vmatrix} a(t) - \alpha b(t) & 0 & 0 \\ 0 & a(t) - \beta b(t) & 0 \\ 0 & 0 & a(t) - \gamma b(t) \end{vmatrix} \tag{3}$$

What is the equation governing the valucs of α, β and γ? From the equation of mass conservation, show that the density near a particle of given $\mathbf{q}$ will become infinite when $a(t) = \alpha b(t)$ for $\alpha(\mathbf{q}) > 0$, where α is taken to be the largest of α, β, γ. The physical interpretation is that compression along the most rapidly contracting axis eventually creates a dense flattened system, or 'pancake'. Its detailed non-linear evolution can only be followed, so far, by numerical simulations in simple cases. (Ya. B. Zeldovich, 1970. *Astron. and Astrophys.*, **5**, 84, A.G. Doroshkevich, *et al.*, 1980, *MNRAS*, **192**, 321; A.G. Doroskevich, *et al.*, 1983, *MNRAS*, **202**, 537).

35.3. Fluctuations and semi-invariants

Results which follow from Sections 29–31 lead to an exact expression for the fluctuations in thermodynamic systems (Callen, 1966, see the reference in Section 29). Since probabilities of fluctuations are closely related to the entropy of the fluctuation it is sensible to work in the entropy representation of thermodynamics: $S(U, V, N)$. To any extensive variable, denoted generally by X, corresponds the intensive variable F in the Massieu representation. A system which is canonical with respect to X will be in equilibrium with a reservoir of F. For example, if X is the energy U, then F is reciprocal temperature. Similarly, number fluctuations involve a reservoir with constant $-\mu/T$.

In the canonical ensemble the probability that X has a value less than some X_0 is

$$P(X < X_0) = e^{-\Psi(F)} \int_{-\infty}^{X_0} e^{-FX} dG(X), \tag{1}$$

where $\Psi(F)$ is the Legendre transform of the entropy and $dG(X)$ is the number of states the system has between X and $X + dX$. The factor e^{-FX} is proportional to the probability that the system actually is in one of its possible dG states. The factor in front of the integral is the normalization for $X_0 = \infty$ and $P(X < \infty) = 1$.

Fortunately, we do not have to know $dG(X)$ in order to calculate the fluctuations. It is just necessary to find $\Psi(F)$ from the equilibrium thermodynamics in the manner of Sections 29–31. The trick is to calculate the generating function

$$\langle e^{aX} \rangle = e^{-\Psi(F)} \int_{-\infty}^{\infty} e^{aX} e^{-FX} dG(X). \tag{2}$$

Show that this is given by

$$\ln \langle e^{aX} \rangle = \sum_{1}^{\infty} \frac{a^n}{n!} (-1)^{n-1} \frac{d^{n-1} X}{dF^{n-1}}$$

$$= \sum_{1}^{\infty} \frac{a^n}{n!} M_n(X). \tag{3}$$

The second expression defines the quantities $M_n(X)$ which are known in probability and statistics as 'semi-invariants' or 'cumulants'. Show that they satisfy the relations

$$M_n(bX) = b^n M_n(X), \tag{4}$$

$$M_n(X + b) = M_n(X) + b\delta_{n,1} \tag{5}$$

using

$$M_n(X) = \lim_{a \to 0} D^n \ln \langle e^{aX} \rangle, \tag{6}$$

where $D \equiv d/da$.

The next step is to relate these semi-invariants to the fluctuation moments. The clue comes by examining the quantity in the last equation before taking the limit $a \to 0$. Let

$$M_n(a; X) \equiv D^n \ln \langle e^{aX} \rangle = D M_{n-1}(a; X) \tag{7}$$

and therefore $M_n(X) = M_n(0; X)$. Fluctuation moments are powers of $\delta X = X - \langle X \rangle$ averaged over the ensemble, i.e., $\langle (\delta X)^n \rangle$. They are also the terms of the Taylor series expansion of

$$\langle e^{a\delta X} \rangle = \sum_{0}^{\infty} \frac{a^n}{n!} \langle (\delta X)^n \rangle. \tag{8}$$

Readily we see that $M_1(a; \delta X) = \langle \delta X e^{a\delta X} \rangle / \langle e^{a\delta X} \rangle$, and this suggests defining the more general quantity

$$\langle (\delta X)^n \rangle_a \equiv \langle (\delta X)^n e^{a\delta X} \rangle / \langle e^{a\delta X} \rangle, \tag{9}$$

with $\langle (\delta X)^n \rangle = \langle (\delta X)^n \rangle_0$. Thus show that

$$M_n(a; \delta X) = D^{n-1} \langle \delta X \rangle_a \tag{10}$$

and

$$\langle X \rangle = M_1(X) \tag{11}$$

$$\langle (\delta X)^2 \rangle = M_2(X), \tag{12}$$

$$\langle (\delta X)^3 \rangle = M_3(X), \tag{13}$$

$$\langle (\delta X)^4 \rangle = M_4(X) + 3M_2^2(X), \tag{14}$$

etc.

As a specific example, show that the average third order energy fluctuation is

$$\langle (\delta U)^3 \rangle = k^2 T^3 \left(2C_v + T \frac{dC_v}{dT} \right), \tag{15}$$

where C_v is the specific heat at constant volume.

36
Bibliography

Section 20:

Hewett, P.C., 1982. 'The estimation of galaxy angular correlation functions', *MNRAS*, **201**, 867.

Holmberg, E., 1940. 'On the clustering tendencies among the nebulae', *Ap. J.*, **92**, 200.

Klang, T. & Saslaw, W.C., 1969, 'The distribution in space of clusters of galaxies', *MNRAS*, **143**, 129.

Neyman, J., Scott, E.L. & Shane, C.D., 1953. 'On the spatial distribution of galaxies, a specific model', *Ap. J.*, **117**, 92.

Peebles, P.J.E., 1980. *The Large Scale Structure of the Universe* (Princeton: Princeton UP).

Sharp, N.A., 1979. 'Practical estimation of the angular covariance function', *Astr. & Astrophys.*, **74**, 308.

Totsuji, H. & Kihara, T., 1969. 'The correlation function for the distribution of galaxies', *PASJ*, **21**, 221.

de Vaucouleurs, G., 1970. 'The case for a hierarchial cosmology', *Science*, **167**, 1203.

de Vaucouleurs, G., 1971. 'The large scale distribution of galaxies and clusters of galaxies', *PASP*, **83**, 113.

Weinberg, S., 1974. *Gravitation and Relativity* (New York: Wiley).

Section 21:

Abramowitz, M. & Stegun, I.A., 1964. *Handbook of Mathematical Functions* (Washington: US Govt. Printing Office).

Bonnor, W.B., 1957. 'Jeans' formula for gravitational instability', *MNRAS*, **117**, 104.

Ellis, G.F.R. & MacCallum, M.A.H., 1969. 'A class of homogeneous cosmological models', *Comm. Math. Phys.*, **12**, 108.

Lifshitz, E.M., 1946. 'On the gravitational stability of the expanding universe', *Jour. Phys. USSR*, X, 116.

Saslaw, W.C., 1972. 'Conditions for the rapid growth of perturbations in an expanding universe', *Ap. J.*, **173**, 1.

Section 23:

Fall, S.M. & Saslaw, W.C., 1976. 'The growth of correlations in an expanding universe and the clustering of galaxies', *Ap. J.*, **204**, 631.

Section 25:

Fall, S.M., & Severne, G., 1976. 'Correlation dynamics in an expanding universe', *MNRAS*, **174**, 241.

Irvine, W.M., 1961. 'Local inhomogeneities in a universe satisfying the cosmological principle'. Harvard University thesis.

Layzer, D., 1963. 'A preface to cosmology, I. The energy equation and the virial theorem for cosmic distributions', *Ap. J.*, **138**, 174.

Saslaw, W.C., 1980. 'Galaxy clustering and thermodynamics', *Ap. J.*, **235**, 299.

Section 26:

Aarseth, S.J., 1971. 'Direct integration methods of the N-body problem', *Astrophys. Space Sci.*, **14**, 118.

Aarseth, S.J., 1984, 'Direct methods for N-body simulations' in *Multiple Time Scales*, J.U.B. Brackbill & B.I. Cohen (eds.) (New York: Academic Press).

Aarseth, S.J. & Zare, K., 1974. 'A regularization of the three-body problem', *Celestial Mech.*, **10**, 185.

Aarseth, S.J. & Lecar, M., 1975. 'Computer simulations of stellar systems', in *Annual Reviews of Astronomy*, **13**, 1.

Ahmad, A. & Cohen, L., 1973. 'A numerical integration scheme for the N-body gravitational problem', *J. Comp. Phys.*, **12**, 389.

Heggie, D.C., 1974. 'A global regularization of the gravitational N-body problem', *Celestial Mech.*, **10**, 217.

Hockney, R.W. & Eastwood, J.W., 1981. *Computer Simulations Using Particles* (New York: McGraw Hill).

Kustaanheimo, P. & Stiefel, E.J., 1965. 'Perturbation theory of Kepler motion based on spinor regularization', *J. Reine Angew. Math.*, **218**, 204.

Spitzer, L., 1974. 'Dynamical theory of spherical stellar systems with large N', in *IAU Symp. No.* 69, A. Hayli (ed.) (Boston: D. Reidel).

Section 27:

Aarseth, S.J., Gott, J.R. & Turner, E.L., 1979. 'N-body simulations of galaxy clustering. I. Initial conditions and galaxy collapse times', *Ap. J.*, **228**, 664.

Aarseth, S.J. & Saslaw, W.C., 1982. 'Formation of voids in the galaxy distribution', *Ap. J. Lett.*, **258**. I.7.

Efstathiou, G., 1979. 'The clustering of galaxies and its dependence upon Ω', *MNRAS*, **187**, 117.

Efstathiou, G. & Eastwood, J.W., 1981. 'On the clustering of particles in an expanding universe', *MNRAS*, **194**, 503.

Gott, J.R., Turner, E.L. & Aarseth, S.J., 1979. 'N-body simulations of galaxy clustering. III. The covariance function', *Ap. J.*, **234**, 13.

Kiang, T. & Saslaw, W.C., 1969. 'The distribution in space of clusters of galaxies', *MNRAS*, **143**, 129.

Miyoshi, K. & Kihara, T., 1975. 'Development of the correlation of galaxies in an expanding universe', *PASJ*, **27**, 333.

Turner, E.L., Aarseth, S.J., Gott, J.R., Blanchard, N.T. & Mathieu, R.D., 1979. 'N-body simulations of galaxy clustering. II. Group of galaxies', *Ap. J.*, **228**, 648.

White, S.D.M., 1979. 'The hierarchy of correlation functions and its relation to other methods of galaxy clustering', *MNRAS*, **186**, 145.

Section 28:

Saslaw, W.C. & Aarseth, S.J., 1982. 'The velocity evolution of galaxy clustering', *Ap. J.*, **253**, 470.

Section 29:

Buchdal, H., 1966. *Concepts of Classical Thermodynamics* (Cambridge UP).

Callen, H.B., 1960. *Thermodynamics* (New York: Wiley).

Callen, H.B., 1966. 'Thermodynamic fluctuations', in *Non-Equilibrium Thermodynamics, Variational Techniques, and Stability*, R.J. Donnelly, *et al.* (eds.) (Chicago: Chicago UP), p. 227.

Fowler, R.H., 1936. *Statistical Mechanics* (Cambridge UP).

Landsberg, P.T., 1961. *Thermodynamics with Quantum Statistical Illustrations* (New York: Wiley Interscience).

Lifshitz, E.M. & Pitaevskii, L.P., 1980. *Statistical Physics* (Oxford: Pergamon).

Tolman, R.C. 1936. *The Principles of Statistical Mechanics* (Oxford UP).

Section 30:

de Groot, S.R. & Mazur, P., 1962. *Non-Equilibrium Thermodynamics* (Amsterdam: North-Holland).

Hertel, P., Narnhofer, H. & Thirring, W., 1972. 'Thermodynamic functions for fermions with gravostatic and electrostatic interactions', *Commun. Math. Phys.*, **28**, 159.

Levy-Leblond, J-M., 1969. 'Nonsaturation of gravitational forces', *J. Math. Phys.*, **10**, 806.

Salzberg, A.M., 1965. 'Exact statistical thermodynamics of gravitational interactions in one and two dimensions', *J. Math. Phys.*, **6**, 158.

Section 31:

Saslaw, W.C., 1970. 'Gravithermodynamics. III. Phenomenological non-equilibrium theory and finite-time fluctuations', *MNRAS*, **147**, 253.

Simon, R., 1970. 'Fluctuations and galaxy formation in the Einstein universe', *Astron. and Astrophys.*, **6**, 151.

Section 32:

Saslaw, W.C., 1980. 'Galaxy clustering and thermodynamics', *Ap. J.*, **235**, 299.

Section 33:

Saslaw, W.C., 1979. 'The efficiency of galaxy formation and gravitational clustering', *Ap. J.*, **229**, 461.

Section 34:

Saslaw, W.C. & Hamilton, A.J.S., 1984. 'Thermodynamics and galaxy clustering: non-linear theory of high order correlations', *Ap. J.*, **276**, 13.

Finite spherical systems – clusters of galaxies, galactic nuclei, globular clusters

And if each system in gradation roll
Alike essential to th' amazing whole
The least confusion but in one, not all
That system only, but the whole must fall.

Alexander Pope

As groups of gravitating objects condense together they become more and more detached from their surroundings. Growing inhomogeneity breaks the local translational and rotational symmetry more and more strongly. Eventually systems become isolated – no longer influenced significantly by the surrounding distribution of galaxies or stars.

In Part III we consider how systems become isolated, and their resulting properties. These are spherical systems so they retain as much local symmetry as possible. Their study makes it possible to understand the basic physical processes of finite clusters. Complexities of rotation and overall asymmetry will be introduced in Part IV.

The astronomical systems which can be represented approximately as spherical clusters range from globular clusters and galactic nuclei, on a scale of parsecs, to clusters of galaxies, on a scale of megaparsecs. After setting up a general theoretical framework to describe these systems, we briefly review some of their observed dynamical properties and then go on to discuss their internal interactions and evolution in more detail.

37

Breakaway

And the world had worlds, ai, this-a-way.

Wallace Stevens

In Section 21 we were able to get some analytical understanding of linear perturbation growth. Subsequent sections used a combination of techniques, involving graininess, energy principles, numerical simulations and thermodynamics, to determine non-linear properties of clustering. In this section, we extend the general approach of Section 21 into the non-linear regime. Our aim is to see how long it takes a perturbation to begin contracting, to break away from the expansion of the surrounding universe.

If we were to extend the detailed linear Fourier perturbation technique of Section 21 into the non-linear regime, in a brute force fashion, it would rapidly become too complicated to provide much insight. Instead of that Eulerian technique it becomes simpler to take a Lagrangian point of view, as in the discussion of 'pancakes' in Section 35.

The Lagrangian technique follows the motion of a particular object, or set of objects, in the perturbation. (By contrast the Eulerian technique describes what happens as a function of spatial position). A simple inhomogeneity, whose growth typifies many more complicated inhomogeneities to order of magnitude, is just a spherical region of constant density in an otherwise uniformly expanding universe. If the inhomogeneity is to contract, and not merely expand more slowly than the rest of the universe, its density must be greater than the critical density ρ_c, just necessary to close the universe. This critical density is given by (21.31) with $\Lambda = k = a = 0$, so at any time,

$$\rho_c = \frac{3H^2}{8\pi G}.$$

(37.1)

Therefore we denote the total density of the inhomogeneity by

$$\rho_c + \Delta.$$

(37.2)

(In principle, Δ could be taken to be negative to give some indication of how a hole expands. However, the N-body simulations of Section 27 show that holes tend to be rather more irregular than clusters. Essentially this is because holes involve the boundaries of several clusters and therefore lack the coherence of a single cluster.

Thus a spherical approximation is less suitable for studying the dynamics of a hole than of a cluster.) At the initial time, denoted by subscript 'i', the average density inside a radius r_i is taken to be

$$\bar{\rho}_i = \rho_{ci} + \Delta \qquad\qquad\text{for } r_i \leq R_i$$
$$= \rho_{ei} + (\rho_{ci} + \Delta - \rho_{ei})R_i^3/r_i^3 \quad\text{for } r_i > R_i, \tag{37.3}$$

where R_i is the initial radius of the inhomogeneity and ρ_{ei} is the density of the universe external to the initial inhomogeneity. This distribution matches smoothly onto the background. The radius R of the inhomogeneity is always much smaller than the radius of curvature of the universe, so Newtonian dynamics describes the growth of the inhomogeneity.

Pressure-free matter in the inhomogeneity provides a good approximation for galaxy clustering (if not for galaxy formation) and is simple, so we will assume it. The background universe can also be taken to be simple: a standard isotropic Friedmann model with $\Lambda = 0$ filled with pressure-free 'dust'. For such models the initial value of the Hubble ratio is related to its value at the present time (no subscript) by

$$H_i^2 = H^2[2q(1 + z_i)^3 + (1 - 2q)(1 + z_i)^2] \tag{37.4}$$

and the initial external density is related to the critical density by

$$\rho_{ei} = \rho_{ci}\frac{2q(1 + z_i)}{(1 - 2q) + 2q(1 + z_i)}. \tag{37.5}$$

(See most texts on cosmology, or Sandage, 1963, for these relations.) As usual, the parameter $\dot{q} = 4\pi G\rho/3H^2$ measures the ratio of the actual density to the critical density, $q = \frac{1}{2}\Omega$. It also measures the deceleration of the universe, $q = -\ddot{a}a/\dot{a}^2 = -\ddot{a}/aH^2$, as well as its curvature: $kc^2/a^2 = H^2(2q - 1)$, with c the velocity of light and a the radius of curvature of the universe (see (21.31)). Thus q has several interpretations.

If the inhomogeneity starts at sufficiently high redshift that

$$\frac{1 - 2q}{2q(1 + z_i)} \ll 1, \tag{37.6}$$

then we can expand (37.4) and (37.5) to first order in this ratio to get

$$\frac{\rho_{ci} - \rho_{ei}}{\rho_{ci}} = \frac{1 - 2q}{2q(1 + z_i)} \tag{37.7}$$

and

$$H_i^2 = 2qH^2(1 + z_i)^3. \tag{37.8}$$

These approximations are reasonably good for galaxy clustering, for example.

With these preliminaries, we can follow the motion of spherical shells of galaxies within and outside the inhomogeneity. Initially the radius of an arbitrary shell is r_i and its evolution may be written in the form

$$r(r_i, t) = r_i a(r_i, t). \tag{37.9}$$

If there were no inhomogeneity, $a(r_i, t)$ would be uniform everywhere, and would just be the radius of curvature of the universe, $a(t)$. Then (37.9) would describe co-moving expansion. Indeed in the limit $r_i \gg R$, we expect $a(r_i, t) \to a(t)$ since the inhomogeneity makes a rapidly decreasing contribution to the total mass in the sphere. So at small r_i, the factor $a(r_i, t)$ measures the departure from a uniform Hubble expansion $\mathbf{v} = H\mathbf{r}$.

The force acting on a spherical shell initially at r_i gives it an acceleration

$$\frac{d^2 r}{dt^2} = -\frac{4\pi G \bar{\rho}_i r_i^3}{3r^2}. \tag{37.10}$$

As long as the average density $\bar{\rho}_i$ does not increase with r_i, shells initially at different r_i will not cross as the inhomogeneity contracts. Then it is straightforward to obtain the first integral of (37.10) by substituting (37.9) and multiplying through with $\dot{a}$. The constant of integration is settled by requiring (37.1) to be satisfied for the case $\bar{\rho}_i = \rho_{ci}$ initially with $a_i = 1$. Thus

$$\left(\frac{da}{dt}\right)^2 - \frac{8\pi G}{3a}\bar{\rho}_i + \frac{8\pi G}{3}(\rho_{ci} - \bar{\rho}_i). \tag{37.11}$$

Note that this is a special case of the cosmological equation (21.3), recalling that for dust $\rho \propto a^{-3}$ and not confusing the two meanings of a. Thus we expect the inhomogeneity to behave similarly to the universe as a whole, but on a different timescale due to its different average density. This similarity is also behind the close association of Newtonian and general relativistic cosmological models, and is treated in some texts on cosmology.

Each shell expands to a maximum radius which is determined by setting $\dot{a} = 0$ in (37.11)

$$u_{\max} = \frac{\bar{\rho}_i}{\bar{\rho}_i - \rho_{ci}} = \frac{1}{E}. \tag{37.12}$$

This last expression defines the initial fractional density excess of the inhomogeneity above the critical density. Equations (37.3), (37.9) and (37.12) show that all shells within the inhomogeneity reach a maximum radius proportional to their initial radius, all with the same constant of proportionality $\approx \bar{\rho}_i/\Delta$. Shells beyond the initial inhomogeneity will also reach a maximum radius and fall back onto the inhomogeneity if their initial radius is less than a critical value given by $\bar{\rho}_i = \rho_{ci}$. Thus the inhomogeneity will accrete additional mass as it contracts. This result can be shown from an Eulerian point of view (Roxburgh & Saffman, 1965) as well as from this Lagrangian approach (Gunn & Gott, 1972).

Playing around with (37.11) a bit enables us to solve it exactly. First define a dynamically scaled time by

$$\tau = \left(\frac{8\pi G \bar{\rho}_i}{3}\right)^{1/2} t, \tag{37.13}$$

and substitute this, together with E from (37.12), into (37.11) which becomes

$$a\left(\frac{da}{dr}\right)^2 = 1 - aE. \tag{37.14}$$

Next, notice that the substitution

$$a = \frac{1}{E}\sin^2 u \tag{37.15}$$

changes (37.14) into

$$\frac{da}{d\tau} = E^{1/2}\cot u. \tag{37.16}$$

We can now regard u as a parameter which relates a to τ. Hence

$$\tau = \int \frac{da/du}{da/d\tau}du = 2E^{-3/2}\int \sin^2 u\, du$$

$$= E^{-3/2}(u - \sin u\cos u) - E^{-3/2}[\sin^{-1}E^{1/2} - E^{1/2}(1 - E)^{1/2}]. \tag{37.17}$$

The constant of integration has been chosen so that when $\tau_i = 0$, $a_i = 1$, in other words $u_i = \sin^{-1}E^{1/2}$. Initially either the inhomogeneity may have a large finite amplitude, or it may be a small perturbation for which $u_i \approx E^{1/2}$. Thus this description is intrinsically non-linear.

Maximum expansion occurs when $u = \pi/2$ and $a_{max} = E^{-1}$. This happens at a dynamical time

$$\tau_{max} \approx \frac{\pi}{2}E^{-3/2} \tag{37.18}$$

for initially small perturbations when the constant term of order unity can be neglected in (37.17).

After maximum expansion, the inhomogeneity begins to contract and break away from its surroundings. Collapse is complete in this idealized illustration, when $u = \pi$ and $a = 0$. Again neglecting the constant term in (37.17), collapse is complete at

$$\tau_c \approx \pi E^{-3/2} = 2\tau_{max}. \tag{37.19}$$

Thus, if the initial perturbation is small, the expansion phase lasts just about as long as the collapse phase, as expected from the reversible nature of the motion. Figure 39 shows these results.

The collapse time may also be written in terms of more observationally oriented quantities. Using (37.3), (37.8), (37.12), (37.13), writing $\delta\rho \equiv \bar{\rho}_i - \rho_{ci}$, and recalling $\Omega = 2q$, gives the collapse time

$$t_c = \frac{\pi}{\Omega^{1/2}(1 + z_i)^{3/2}}\left(\frac{\bar{\rho}_i}{\delta\rho}\right)^{3/2}H^{-1} \tag{37.20}$$

in terms of the present Hubble time H^{-1}. Fractional perturbations must have been formed by a redshift of approximately $1 + z_i = (\pi^2/\Omega)^{1/3}(\rho/\delta\rho)$ in order to have collapsed by the present time.

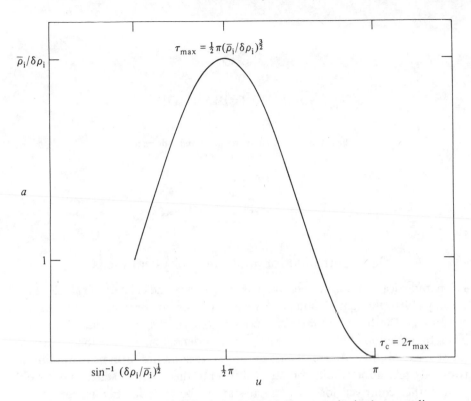

Fig. 39. The growth and collapse of a density inhomogeneity in the expanding
Universe.

Realistic inhomogeneities do not have such smooth sailing, even if they start out
spherical. Gravitational graininess destroys the phases of galaxy orbits and causes
irregularities to grow within the cluster. External tidal forces will also distort the
cluster. Even more to the point, most inhomogeneities will start with an irregularly
shaped distribution. This is apparent from the N-body simulations even when the
initial overall flow is uniform. The collapse of an irregular distribution may be
dominated by a new phenomenon, to which we now turn.

38

Violent relaxation

he...flung himself upon his horse and rode madly
off in all directions.

Stephan Leacock

38.1. Introduction and basic physical ideas

Is it paradoxical that relaxed nearly spherical galaxies and clusters of galaxies exist? Zwicky (1960) thought so and proclaimed this a major problem of extragalactic astronomy. The difficulty comes from noticing that the two-body relaxation time τ_R, given by Equation (2.11), for these systems is orders of magnitude longer than the Hubble age of the Universe. If clusters or galaxies did not form near equilibrium, two-body relaxation would not have led them to their present state. And it would seem rather contrived for their initial state to be their final state, like Athena springing from the head of Zeus.

The solution of this difficulty comes from noticing that other relaxation mechanisms exist. Indeed, the further the initial state lies from the relaxed system, the more powerful its relaxation is. King (1962) suggested qualitatively that such a relaxation process must exist and Hénon (1964) confirmed it for a restricted spherical case using numerical simulations. Lynden-Bell's (1967) work pioneered our quantitative understanding of the subject and Shu (1978) later clarified some aspects.

The basic idea behind violent relaxation is that the initial state is so far from equilibrium that large scale collective modes govern its early evolution. Since the initial distributions of masses and velocities are very irregular, individual objects (galaxies or stars) are scattered mainly by *groups* of objects, rather than by other individual objects. The groups themselves are constantly disrupting and reforming. No particular group survives violent relaxation for very long, but the statistical distribution of groups lasts longer. An object may belong first to one group, then to another, and finally to none at all, or to the whole relaxed system.

This situation is a far cry from the relatively simple case discussed in Part I where the total gravitational field acting on an object could be divided into a mean field varying slowly in space and a rapidly fluctuating field due to close neighbors. Now everything fluctuates rapidly. The amplitudes are large. Each object moves in a mean field which changes quickly in time, as well as in space. The energy of each object along its orbit is not conserved. Only the total energy of the entire system remains constant, along with other global invariants. These allow us to bypass structural details when seeking the outcome of violent relaxation.

It is easy to estimate the timescale for this collective relaxation. Consider the case where a system consists of several large irregular clusters falling together under their mutual gravitational attraction. The clusters collide on a timescale $\approx (G\rho_0)^{-1/2}$, where ρ_0 is the original average density of the whole system. Having collided, the system oscillates for several dynamical periods $(G\rho_1)^{-1/2}$, with $\rho_1 \gtrsim \rho_0$, and then settles down as Landau damping and phase mixing destroy the collective modes. The whole process may thus take place over several initial $(G\rho_0)^{-1/2}$ dynamical timescales. This is shorter than the two-body relaxation time τ_R by a factor of order $(\log N)/N$, where N is the number of objects. In a cluster of galaxies with $N \approx 10^3$, or a galaxy with $N \approx 10^{10}$ stars, this factor is very small. It is in comparison with τ_R that collective relaxation appears 'violent'.

The main result of violent relaxation can also be understood quite simply. An object achieves a final velocity which is independent of its mass. There is an equipartition of energy per unit mass, but not of energy itself. This occurs because the time-changing energy of the mean field acting on a star, $-m\partial\Psi/\partial t$, is equal to the change in the star's total energy, $md(\frac{1}{2}v^2 - \Psi)/dt$, where $\Psi(\mathbf{r}, t)$ is the gravitational potential of the stellar system. Alternatively, the acceleration is just equal to the gradient of the potential; since the scattering groups have much greater mass than the scattered object the mass of the individual object cancels out. As a corollary, this type of relaxation does not lead to mass segregation (i.e., the greater masses do not tend to clump in the central region).

What will be the distribution of these velocities? We now know it does not depend on mass, but can we say more? Suppose there are no constraints (apart from conservation of total mass, energy and momentum) and each rapidly changing collective mode produces a force which is completely uncorrelated with its predecessors, its successors, and its compatriots. Then any violently relaxing object undergoes a series of random velocity changes. We have seen in Sections 3 and 4 that such erratic forces produce a Maxwell–Boltzmann velocity distribution. Indeed the result follows directly, in this idealized case, from the central limit theorem of probability theory. A useful simple version of this theorem states that if $x_1, x_2, \ldots, x_n$ are mutually independent random variables with a common distribution whose expectation is zero and variance is σ^2, then as $n \to \infty$ the distribution of the normalized sums $S_n = (x_1 + \cdots + x_n)/\sqrt{n}$ tends to the normal distribution with density $f(x) = (2\pi)^{-1/2} \exp(-x^2/\sigma^2)$. Thus, if x_i is a one-dimensional velocity increment, the probability of finding an object whose final velocity is $x = x_1 + \cdots + x_n$ is given by the Maxwell–Boltzmann distribution. The theorem holds under more general conditions for multidimensional random variables with different distributions. The chief additional constraint is that the variance of each distribution be much less than the sum of all the variances.

Violent relaxation and more gentle two-body relaxation both produce Maxwell–

Boltzmann distributions, essentially because both rely on large numbers of independent random changes. In each case the description is rather idealized. In Part I we showed how departures from idealized two-body relaxation could be described by perturbation theory. Violent relaxation is so intrinsically non-linear that perturbation theory is not useful. An analytic understanding of departures from the idealized case has not been developed yet, although some numerical experiments may help guide the way. After examining the idealized case in more detail we will see how it may be modified in practice.

38.2. The collisionless distribution function

Now we derive in more detail the distribution function towards which violent relaxation tends. Real systems may not relax violently enough for long enough to actually reach this distribution, but it still stands as a goal. With only total momentum and energy conservation as constraints, we would expect the mass independent Maxwell–Boltzmann distribution for the reasons just described. However, additional constraints may be needed to describe the system fully, and these will modify the distribution function. The most important of these additional constraints is the system's 'collisionality'.

Collisionality is the degree to which interactions with nearby neighbors (in contrast to the mean field) alter an object's orbit. Naturally the simplest system is collisionless; only its mean field is important. Collisions become important to the extent that a system (not in equilibrium) contains many objects moving closely together in space with about the same velocity. Generally this companionship does not survive long, as each object strongly deflects the other. Sometimes, through the interaction with a third body, or with a fluctuation in the mean field, two companions can become bound to each other. Subsystems with more objects will also form, again usually temporarily. Insofar as these subsystems are important, the system has greater collisionality.

In this subsection we will constrain the violently relaxing system to be collisionless, returning to the more general case in Subsection 38.5. Mathematically this collisionless condition is expressed by saying that the six-dimensional distribution function $f(\mathbf{r}, \mathbf{v}, t)$ satisfies the collisionless Boltzmann equation (7.2). No surprise here. Since the dynamics of a violently relaxing object do not depend on its mass, any more than they depend on its color, it makes sense to consider a distribution function of velocities rather than of momenta. Writing the gravitational mean field in terms of f, the collisionless Boltzmann equation becomes

$$\frac{\partial f}{\partial t} + \mathbf{v} \cdot \frac{\partial f}{\partial \mathbf{r}} + \frac{\partial}{\partial \mathbf{r}} \left\{ G \int \frac{\int f' \mathrm{d}^3 v'}{|\mathbf{r} - \mathbf{r}'|} \mathrm{d}^3 r' \right\} \cdot \frac{\partial f}{\partial \mathbf{v}} = 0. \tag{38.1}$$

We see explicitly that the evolution of f does not depend on mass.

Since the total time derivative $\mathrm{d} f / \mathrm{d} t = 0$ following the motion, an initial region of phase cannot change its average density as it evolves. This is just the behavior in six-dimensional phase space analogous to Liouville's theorem in $6N$-dimensional phase space discussed in Section 10. In a collisionless system, objects which do not initially have the same position and velocity coordinates can never acquire them. This places an additional constraint on the distribution function: elements of the cloud of points in phase space which do not overlap initially can never overlap.

If relaxation is so violent that there is an equal *a priori* probability of finding a point anywhere in phase space, then we can use the techniques of statistical mechanics to calculate the most probable distribution of points in phase space for a given set of constraints. This is equivalent to calculating the distribution function. Usually a statistical mechanical analysis is applied directly to the physical particles, but it can also be applied to representative points in phase space. Both approaches give equivalent results (when interpreted properly) and the use of one or the other is mainly a matter of convenience for the problem in hand.

The statistical approach starts by dividing the six-dimensional phase space up into many microcells, each with a very small fixed volume $\Delta^3 v \, \Delta^3 r$. These are the smallest regions of phase space which our statistical mechanics can resolve. The validity of a statistical description for collisionless systems depends upon the existence of microcells of the right size. If they are too large they can contain so many phase points that the stars they represent have a high probability of undergoing strong two-body deflections. Then phase points will jump out of their microcells discontinuously. If they are very small, then each microcell will contain either zero or just one phase point. This is fine, but then the microcells may be so close together that the stars in two neighboring occupied microcells will interact strongly, again causing a collisional jump.

There are two ways around this dilemma. First, there may be an intermediate range in which each microcell still contains, at most, one point, but the microcells are large enough that the probability of phase points in adjacent microcells being close enough for a strong encounter is very small. Second, the microcells may again be very small, but so few are occupied that the chance of finding two neighboring occupied microcells which are close enough to represent a strong encounter is negligible. Both these situations describe collisionless systems, so we assume that at least one applies. The statistical approach allows for both possibilities. Then if each microcell starts with either zero or one phase point, it will always contain that initial number. Microcells correspond to the fine grained level of statistical mechanics.

The fine grained description is so close to the description based on collisionless, time-reversible equations of motion that it does not show any evolution. A microscope observing a single microcell in phase space, following along its path to keep it always in view, would not detect any change. Decreasing the power of the microscope brings many microcells into view. Their paths are different so the microscope cannot follow all of them. Some microcells enter the field of view, others move out. The number of occupied microcells visible at any time varies. On this scale the distribution appears to evolve. If the resolution of the microscope is now decreased so it can only detect the average properties of this enlarged field of view, then it yields a coarse grained description of the system. When this coarse grained description does not evolve, the system may be said to reach an equilibrium. It then occupies the most probable configuration in phase space; this is a fundamental assumption of statistical mechanics.

To quantify the coarse grained description, microcells are grouped into macrocells. Although each macrocell contains a very large number of microcells, it still occupies a very small fraction of the phase space representing the entire physical system. We now ask which configuration of cells in phase space is the most probable one.

The most probable macroscopic state is the one which corresponds to the greatest number of microstates. So we want to find the number of microstates for any configuration in phase space. Suppose the ith macrocell contains a total of w_i microcells and each microcell is gerrymandered to hold r phase points, where $r = 0$ or 1. Denote the number of microcells which have r phase points in the ith macrocell by $n_i^{(r)}$. Consider one macrocell. Since the phase points can be distinguished from each other (more about this in Subsection 38.5), the number of ways a microcell can be assigned to the first phase point is w_i, to the second $w_i - 1$, etc. There are a total of

$$n_i = \sum_r r n_i(r) \tag{38.2}$$

phase points in this macrocell. Therefore the number of ways of assigning microcells to all n_i phase points is

$$\frac{w_i!}{(w_i - n_i)!}. \tag{38.3}$$

A given configuration includes all the macrocells. For a given distribution of points among the macrocells, the number of possible configurations is the number of possibilities for the first macrocell times the number of possibilities for the second, etc., i.e.,

$$\prod_i \frac{w_i!}{(w_i - n_i)!}. \tag{38.4}$$

The entire system contains

$$N = \sum_i n_i \tag{38.5}$$

phase points. These N points can be distributed among the n_i macrocells in

$$\frac{N!}{\prod_i n_i!} \tag{38.6}$$

different ways. Thus the total number of microstates which are possible for a configuration denoted by the set $\{n_i\}$ is

$$W = \frac{N!}{\prod_i (n_i)!} \times \prod_i \frac{w_i!}{(w_i - n_i)!}. \tag{38.7}$$

This can be simplified slightly by taking each macrocell to have the same number of microcells, so $w_i \rightarrow w$; nothing essential is lost.

Maximizing W with respect to n_i, subject to given constraints, yields the most probable coarse grained configuration. Equivalently, we can maximize $\ln W$ since it is a monotonic function of W. These standard manipulations are easier because we can use Stirling's approximation $\ln(n!) \approx n \,(\ln n - 1)$ for $n \gg 1$ to get

$$\ln W = N(\ln N - 1) - \sum_i \{n_i(\ln n_i - 1) + (w - n_i)[\ln(w - n_i) - 1] - w(\ln w - 1)\},$$
(38.8)

which is proportional to the entropy.

An isolated system is constrained to have constant total mass and energy. If all objects have the same mass, these requirements are

$$\sum_i m n_i = N m$$
(38.9)

and

$$\sum_i m n_i(\tfrac{1}{2} v_i^2 + \Phi_i) = E,$$
(38.10)

where

$$\Phi_i \equiv \Phi(\mathbf{x}_i, t) \equiv -\frac{1}{2} \sum_{i \neq j} \frac{G m n_i}{|\mathbf{x}_i - \mathbf{x}_j|},$$
(38.11)

and $(\mathbf{x}_i, \mathbf{v}_i)$ is the center of the ith macrocell. Adding a constraint on total linear momentum would just change the inertial frame and not affect the internal dynamics of the system. If the system were to rotate, which it does not in this case, conservation of total angular momentum would be an added constraint. Collisionless systems also contain other conserved integrals of the motion (see Section 39), but these do not affect the statistical mechanics.

Incorporating the Lagrange multipliers α and β for the mass and energy constraints, and varying the entropy (38.8) with respect to the n_i at its maximum value gives

$$\delta \ln W = 0 = \sum_i [-\ln n_i + \ln(w - n_i) - \alpha m] \delta n_i$$

$$- \beta m \sum_i [(\tfrac{1}{2} v_i^2 + \Phi_i) \delta n_i + n_i \delta \Phi_i].$$
(38.12)

Equation (38.11) gives the variation in gravitational energy, and readily shows that it satisfies the identity

$$\sum_i m n_i \delta \Phi_i = \sum_i m \Phi_i \delta n_i.$$
(38.13)

Combining the last two equations and noting that the δn_i are independent, we have

$$\frac{n_i}{w - n_i} = e^{-\alpha m - \beta m \in_i}$$
(38.14)

where

$$\varepsilon_i = \tfrac{1}{2}v_i^2 + 2\Phi_i \tag{38.15}$$

is the energy per unit mass of an object in the ith macrocell. Writing the chemical potential per unit mass $\mu \equiv -\alpha/\beta$ and solving for n_i:

$$n_i = \frac{w}{1 + e^{\beta m(\varepsilon_i - \mu)}}. \tag{38.16}$$

The number of phase points in the ith macrocell is related to the coarse grained distribution function simply by averaging n_i over the macrocell's volume,

$$\bar{f}_i = \bar{f}(\mathbf{r}_i, \mathbf{v}_i) = \frac{mn_i}{w}, \tag{38.17}$$

thus completing the derivation.

This distribution has several important properties. When only a small fraction of the microcells in a macrocell are occupied, $n_i \ll w$ and equation (38.14) shows directly that the distribution has a Maxwell–Boltzmann form. Collisionless systems therefore relax to this non-degenerate limit. Alternatively, the Maxwell–Boltzmann limit is also reached when $\beta m(\varepsilon_i - \mu) \gg 1$. These refer to high energy objects whose phase points can be surrounded by large microcells, as discussed earlier.

Some degree of degeneracy may be exhibited by systems in which $\beta m(\varepsilon_i - \mu)$ is of order unity for many objects. Then $\bar{f}_i$ resembles the degenerate Fermi–Dirac distribution, except for a different normalization arising from the distinguishability of phase points. Degeneracy seems to occur at the limit of collisionless relaxation, just before two-body encounters become important. Although there is some evidence for its occurrence in numerical experiments (see references in the review by Aarseth & Lecar, 1975), its nature is not yet well understood. Experiments on one- or two-dimensional systems, for example, suppress many of the possible modes which could occur in more realistic three-dimensional cases.

During the derivation of the distribution (38.16)–(38.17), the alert reader will have noticed that the object's mass is really irrelevant. Constraints (38.9)–(38.10) could just as well be expressed in terms of constant number of objects and constant energy per unit mass, respectively. Then m would never have appeared in $f(\mathbf{r}_i, \mathbf{v}_i)$, and the Lagrange multiplier β would just be an inverse velocity dispersion. Writing $\beta' \equiv m\beta$ in (38.16) accomplishes this result. Thus the velocity dispersion is independent of mass. Associating this dispersion with a temperature through the relation $m\langle v \rangle^2 = 3kT$ works if the temperature is itself proportional to mass. In a *finite* system, such an association could not be thermodynamically rigorous because the mean gravitational field varies throughout space. Temperature, i.e., velocity dispersion, is not constant throughout the system. However, these concepts still have important heuristic value, especially if the temperature varies slowly. For example, with a slowly varying temperature proportional to mass, we would expect objects in violently relaxed systems not to show much mass segregation if their masses are different. Generally this is the case. (Section 38.4 gives an illustration.) It follows, of course, from the mean field nature of the relaxation. Extending the previous derivation to systems with different masses shows this (Shu, 1978).

38.3. Criteria for violent relaxation

How do we know if the initial conditions of a system are sufficiently far from equilibrium to make violent relaxation important? It turns out to be possible to answer this question in principle, but difficult to apply a practical test. Of course, one could do a numerical simulation to see what results, but how do we *understand* the results?

Our starting point is the by-now familiar Liouville equation (10.14)

$$\frac{\partial f}{\partial t} + \{f, H\} = 0 \tag{38.18}$$

for the $6N =$ dimensional distribution function $f = f^{(N)}$. We now write the Poisson bracket (10.15) as

$$\{f, H\} = \sum_{i=1}^{N} \left(\frac{\partial f}{\partial \mathbf{r}_i} \cdot \frac{\partial H}{\partial \mathbf{p}_i} - \frac{\partial H}{\partial \mathbf{r}_i} \cdot \frac{\partial f}{\partial \mathbf{p}_i} \right). \tag{38.19}$$

If the system is collisionless, each object of mass m interacts only with the mean gravitational potential $U(\mathbf{r}, t)$, so the total Hamiltonian is

$$H = \sum_{i=1}^{N} \left[\frac{\mathbf{p}_i \cdot \mathbf{p}_i}{2m} - mU(\mathbf{r}, t) \right]. \tag{38.20}$$

To close these equations the potential is calculated self-consistently from Poisson's equation as a functional of f,

$$U[f] = U(\mathbf{r}, t) = \frac{Gm}{V^{6N-3}} \int \frac{\int f(\mathbf{r}'_1, \dots, \mathbf{p}'_N, t) \mathrm{d}^N \mathbf{p}' \mathrm{d}^{N-1} \mathbf{r}'}{|\mathbf{r} - \mathbf{r}_{\alpha'}|} \mathrm{d}\mathbf{r}_{\alpha'}, \tag{38.21}$$

where $\mathbf{r}_{\alpha'}$ is the one coordinate not involved in the integral over f, and V is the phase space normalization for f.

Equations (38.18)–(38.21) explicitly show that the full Liouville equation has a non-linear integro-differential form. The usual procedure is to subject f to small displacements from an equilibrium state and investigate the resulting linear equations. However, this will not work for violent relaxation where the fluctuations dominate; a different approach is needed.

The procedure is to linearize Liouville's equation in a stochastic sense. The fine grained distribution function will fluctuate rapidly in time and space. Suppose the relaxation is sufficiently violent that f can be considered a stochastic function with a high degree (not necessarily infinite) of randomness. The inner integral of (38.21) is then a stochastic integral over f. This stochastic integral will coincide with the integral of the sample function of f (i.e., the standard Riemann–Stieltjes integral) so long as f is mean-square-continuous, which will be the case for any f of physical interest.

For any typical object in the system the main contribution to the mean field comes from the large number of more distant objects. Most collective modes of these objects occupy overdense regions which are small compared to the entire system. Therefore, the distribution function representing them fluctuates on a timescale short compared to the global relaxation time $\sim (G\rho)^{-1/2}$. A typical relaxing object then responds to a time average of the mean field

$$\langle U \rangle \equiv \frac{1}{\tau} \int_t^{t+\tau} U(\mathbf{r}, t) \mathrm{d}t' = \frac{Gm}{V^{6N-3}} \int \frac{\int \langle f \rangle \mathrm{d}^N \mathbf{p}' \mathrm{d}^{N-1} \mathbf{r}'}{|\mathbf{r} - \mathbf{r}_{\alpha'}|} \mathrm{d}\mathbf{r}_{\alpha'}. \qquad (38.22)$$

Note that integrations over phase space and over time commute. In general, $\langle U \rangle$ will depend upon $\mathbf{r}, t$, and the coarseness of the time grain τ. We assume, as is usual in statistical mechanics, that it is possible to find a time grain τ such that $\langle U \rangle$ exhibits stationary Markov behavior over intervals of order τ. Once this value of τ is chosen, $\langle U \rangle$ will depend only on $\mathbf{r}$. Section 10 discussed the spatial analog of this coarse graining.

If we replace U by $\langle U \rangle$ in the Hamiltonian, the Liouville equation becomes

$$\frac{\partial f}{\partial t} + \{f, \langle H \rangle\} = 0, \qquad (38.23)$$

which is now linear in f. Of course, we do not know the actual value of $\langle U \rangle$ or $\langle H \rangle$ since f has not been calculated. But it will suffice merely to know that $\langle H \rangle$ has *some* value independent of time (over intervals of order τ) and can be treated as an unknown linear operator.

Coarse graining (38.23) in time gives

$$\frac{\partial \langle f \rangle}{\partial t} + \{\langle f \rangle, \langle H \rangle\} = 0. \qquad (38.24)$$

Here $\langle f \rangle$ will depend upon time, since the value of τ which gives the functional $\langle U[f] \rangle$ a stationary Markov behavior will not, in general, do the same for $\langle f \rangle$. While nothing prevents us from performing the coarse graining (38.24) with a new τ, such that $\langle f \rangle$ and $\langle \dot{f} \rangle$ are zero, to do so would wipe out most of the information in the system. To find the most useful grain may be a delicate matter, but statistical mechanics requires only that it can be done in principle. Because (38.23) is linear, both the fine grained and the coarse grained distribution function satisfy the same equation. In general, however, this is not the case.

An alternative method for deriving the stochastically linearized Liouville Equation (38.24) is to coarse grain (38.18) directly. Then in order to obtain a linear equation of the form (38.24) we readily find that

$$\left\langle \frac{\partial U[f]}{\partial \mathbf{r}} \cdot \frac{\partial f}{\partial \mathbf{p}} \right\rangle = \frac{\partial \langle U \rangle}{\partial \mathbf{r}} \cdot \frac{\partial \langle f \rangle}{\partial \mathbf{p}} + \text{negligible correlation terms}. \qquad (38.25)$$

Physically this means that, for a given object, the largest contribution to $U[f]$ comes from distant mass whose mean field fluctuates with time in a way uncorrelated with the nearby variations in f. Moreover, orbits in the system are highly correlated for short periods of time, but are essentially uncorrelated on timescales longer than that used to coarse grain the distribution function. The collective modes against which an object scatters change rapidly with time, so that an object is not part of a particular mode for very long. As a simple illustrative example, consider a mass distribution of the form

$$\rho(r,t) \propto \frac{S_1(t)}{r+\varepsilon} + rS_2(t), \tag{38.26}$$

where ε is small, S_1 and S_2 are uncorrelated random functions of time and the maximum value of S_1 is much less than the maximum value of S_2. For small values of r the local density fluctuations are dominated by $S_1(t)$, while the mean potential and its gradients are dominated by distant objects which are governed by $S_2(t)$.

If the system is close to equilibrium, Equation (38.25) will not be satisfied since any remaining fluctuations become correlated over the entire system. These correlations grow with time and represent a damping of the violent relaxation. The degree to which (38.25) is satisfied will vary throughout the system, and change as it evolves.

The condition that Liouville's equation can be stochastically linearized, by replacing U with $\langle U \rangle$ in the Hamiltonian, turns out to be a sufficient criterion for violent relaxation (Saslaw, 1970). By generalizing the techniques described in Sections 12 and 14 it is possible to solve (38.24) for the single-particle distribution function. The result is (38.16), where the distribution is coarse grained in time. If the relaxation is sufficiently violent the ergodic hypothesis should hold, making the time-averaged distribution equivalent to the space averaged one, but this has not yet been proven. Nor do we know whether (38.25) is a necessary, as well as a sufficient, criterion for violent relaxation.

Numerical experiments can be used to check (38.25). Starting with an ensemble of systems, each in apparent turmoil, determine by a trial and error routine, whether (at a given time) there exists a time interval such that the averages $\langle U[f] \rangle$ and $\langle f \rangle$ over this interval satisfy (38.25) at a point $(\mathbf{r}, \mathbf{p})$ in phase space. If so, then a particle orbit at that point will relax violently. Examination of a representative number of points gives an idea of how much of the system is relaxing at a given time. This procedure can be continued throughout the evolution to see how long the violent relaxation lasts. Then the form of the final distribution function can be compared with the amount of violent relaxation which has occurred. This could be especially useful for understanding systems whose relaxation is produced by tidal distortion. For collisionless systems, the one-particle distribution function may replace the N-particle distribution function in (38.25). In this case it is not necessary to examine an ensemble of systems.

An analogous sufficient criterion should hold when the coarse graining is done in space rather than in time. Kadomtsev & Pogutse (1970) examined collisionless relaxation for the case of small perturbations of the distribution function. They used the methods described in Sections 15 and 16, applied to the six-dimensional collisionless Boltzmann equation. Non-wave fluctuations $(k > \omega_g/v_{thermal})$ are assumed to form macroparticles, spatial regions of size A with strong orbital correlations. Then the unperturbed spatially coarse grained six-dimensional distribution function evolves according to

$$\frac{\partial \langle f_0 \rangle}{\partial t} \propto AF\left[\langle f_0 \rangle\right]. \tag{38.27}$$

Here $F[\langle f_0 \rangle]$ is an integral functional of $\langle f_0 \rangle$ which vanishes when $\langle f_0 \rangle$ has the form of (38.16) if the collisionless exclusion principle applies. Thus there is no more violent relaxation when this equilibrium distribution is reached. Nor would the system relax violently if $A = 0$, i.e., if spatially correlated regions were absent. In actual systems, A and F will decrease with time (not yet understood theoretically), and the relaxation will damp. In many non-equilibrium astronomical situations, Equations (38.25) and (38.27) are not well satisfied and violent relaxation may be much less effective than superficial turmoil suggests.

38.4. Damping and computer simulations

Violent relaxation tends to wear itself out very quickly. There are three major mechanisms which tend to destroy the conditions conducive to its occurrence, and these mechanisms work on about the same timescale as the time needed to form the Lynden–Bell-type distribution in the collisionless system. First, there is Landau damping (Section 16) which exchanges energy between individual objects and the colllective modes, destroying the large scale structure. Second, there is phase mixing. This process does not depend directly on the gravitational interaction. It results simply from the fact that almost all orbits in the system are incommensurable, and the increasing randomization of their phases tends to destroy collective modes. Third, there is the problem that high energy objects (i.e., those whose total energy is just below zero) are ejected into elongated orbits far from the center and they do not participate fully in the relaxation.

   Thus while the statistical mechanical approach gives a very useful picture of the relaxation process, it applies in detail only to highly idealized systems. But the general principles of violent relaxation will play an important, sometimes crucial, role in systems with rapidly vibrating, inhomogeneous, gravitational fields.

   Computer simulations, which first began studying violent relaxation in the late 1960s, dealt with one-dimensional systems (Aarseth & Lecar, 1975). Computers were too small for more realistic cases. These early experiments gave mixed results. Although the theory provided a good qualitative description of degeneracy at low temperature, the values of temperature and chemical potential derived from a simple statistical mechanical treatment did not fit the experimental distribution very well. Substantial groups of high velocity stars were often flung out of the system before they had a chance to relax to the predicted statistics. Moreover, it was not very clear to what degree these experiments really satisfied the criteria for violent relaxation.

Subsequently, Aarseth & Saslaw (1972) examined the evolution of a 500-body system to check for features of violent relaxation. The objects had a mass distribution $N(m) \propto m^{-2}$ over a mass ratio 32:1. Initially they were placed uniformly at random within a sphere with a Maxwellian velocity distribution whose total energy was half the equilibrium value. The early rapid collapse was damped after two or three crossing times. If violent relaxation dominates, the velocities should be independent of mass, and this was essentially the case after two or three crossing times. However, that was not the whole story because the heavier masses had a strong tendency to gravitate towards the center of the system, consistent with two-body relaxation. Apparently both types of relaxation play a role in more realistic systems, and even their collisionless nature is in doubt.

Following these and other early attempts to simulate violent relaxation, relatively few N-body experiments were done to understand this process in detail. Today, with the latest generation of powerful computers, the time seems ripe for a new round of exploration.

38.5. Distributions for collisional systems

When a system, or part of a system, is not completely collisionless, points can accumulate in microcells of phase space. These represent groups of objects, not necessarily bound to each other, which are traveling along together with small relative velocity. Such associations may not last long and, since close gravitational encounters can deflect orbits strongly, phase points can jump in and out of their cells. If the motion is sufficiently chaotic there may be a statistical equilibrium in which points jumping into cells are balanced by those jumping out. Coarse grained distribution functions are then stationary in time. Again, such a picture is an idealization, but it is interesting to see where it leads, and how it is related to the collisionless case. It also provides some information in lieu of working out the vastly more complicated underlying kinetic theory.

Generally the number of phase points per microcell varies and the maximum number will depend on the importance of close collisions relative to more distant encounters. Furthermore, the extent to which microcells are distinguishable can vary in the most general case.

Distinguishability is a somewhat subtle problem. It depends on whether the state of one object influences *a priori* the state of another. This is related to whether the objects are localizable. There are two usual criteria by which objects can be distinguished from one another. If they belong to or represent a physical system and are interchanged (or permuted if more than two) and if the total Hamiltonian of the system is unchanged, then the objects are indistinguishable. In the second method, if it is possible to localize and label uniquely a single object and follow the continuous trajectory of its motion, then the object is distinguishable. As simple examples, classical physical particles are indistinguishable by the first criterion, but distinguishable by the second; quantum particles are indistinguishable by both criteria.

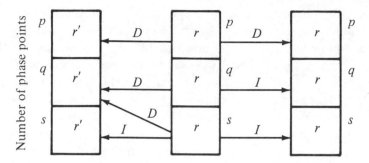

Fig. 40. Schematic illustration of the distinguishability of microcells in phase space. D denotes distinguishable and I indicates indistinguishable pairs of microcells.

For a classical gravitational system, phase points are distinguishable by the second criterion. This is evidently true for a collisionless system, but perhaps not so obvious in collisional system when the trajectories of phase points jump from one cell to another. However, even when collisional, two particles (for example) are still described by two one-particle state functions, each having six dimensions, rather than (as for quantum particles) by one two-particle state function having 12 dimensions. In the former case, one particle does not, *a priori*, influence the state of the other; in the latter case it does through symmetry or other restrictions on the combined state function.

To derive the statistical mechanics, we now take a fairly general view in which distinguishability and exclusion are treated as additional parameters. This approach has several advantages. It exhibits easily the interrelations among all the usual statistics and the special cases which apply to gravitating systems. Also, it shows that the fluctuations and correlations in gravitating systems are nearly independent of the distinguishability of microcells. Furthermore, this generalization can be used to describe statistical systems where combined state functions are necessary. Finally, I mention that this generalization is obtained at little extra cost compared to a specialized derivation, by means of a trick.

To derive the basic general distribution function, suppose that not more than p phase points can accumulate in any one microcell. This defines the collisionality of the system; if $p = 1$ it is collisionless and the collisionless Boltzmann equation holds. As $p \to \infty$, maximum collisionality is reached. The number of phase points actually in a microcell is represented by r. We suppose that the extent to which microcells are distinguishable falls, for each microcell, into one of three categories. These are illustrated schematically in Figure 40.

If $0 \leq r \leq s$, a microcell of r points is indistinguishable both from other microcells with r points and from microcells with r' points, where $0 \leq r' \leq s$, $r' \neq r$. But it is distinguishable from microcells with $r' > s$ points. These microcells with $0 \leq r \leq s$ are all the same.

Next, for $s < r \leq q$, let a microcell of r points be distinguishable from a microcell of

r' points when $s < r' \leq q$ and $r' \neq r$, but let it be indistinguishable from other microcells with r points. These microcells we call semidistinguishable. They are the usual representations of states in quantum mechanics.

Finally, for $q < r \leq p$, microcells with r points are distinguishable from other microcells of number r, as well as from those with $r' \neq r$. This is complete distinguishability and we simply call these microcells distinguishable. Further possible generalizations are to cases where microcells with r and r' points are indistinguishable from each other when $r \leq s$ and $r' > s$.

Actual values of p, q, and s depend on the observer's resolution of phase space and the system considered. There are some interesting special cases

$$
\begin{array}{llll}
\text{Lynden–Bell:} & s = 0, & q = 0, & p = 1, \\
\text{Fermi–Dirac:} & s = 0, & q = 1, & p = 1, \\
\text{Bose–Einstein:} & s = 0, & q = \infty, & p = \infty, \\
\text{Maxwell–Boltzmann:} & \text{low density limit.} & &
\end{array}
\tag{38.28}
$$

Bose–Einstein statistics are also effectively approached when p and q both equal the total number of elements in the system.

As before, we find the most probable distribution by maximizing the number of microscopic states for a given macrostate. Computing the combinational problem for these statistics is non-trivial due to the mixed type of restrictions imposed on exclusion and distinguishability. It would be quite complicated to adopt the usual statistical mechanical procedure of examining the distribution of N points among a given number of cells. Instead we use the trick of examining the distribution of the number of microcells having a given number, r, of points inside them.

Again, the microcells are grouped into macrocells. The ith macrocell contains w_i microcells and each microcell holds r phase points. The number of microcells having r points in the ith macrocell is denoted by $n_i(r)$. Hence the number of distinguishable arrangements of these w_i microcells is

$$
\frac{w_i!}{\left(\displaystyle\sum_{r=0}^{s} n_i^{(r)} \right)! \displaystyle\prod_{r=s+1}^{q} (n_i^{(r)})!}.
\tag{38.29}
$$

To find the total number of microstates corresponding to this macrocell, we multiply (38.29) by the number of ways of distributing the N_{dist} distinguishable microcells among the macrocells,

$$
\frac{N_{\text{dist}}!}{\displaystyle\prod_{r=q+1}^{p} n_i^{(r)}!}.
\tag{38.30}
$$

and take the sum of the products of all such terms. Using the largest term approximation for this sum gives for the number of microstates

$$
\Omega = \prod_i \frac{w_i! N_{\text{dist}}!}{\left(\displaystyle\sum_{r=0}^{s} n_i^{(r)} \right)! \displaystyle\prod_{r=s+1}^{p} n_i^{(r)}!}.
\tag{38.31}
$$

Notice that q does not appear in (38.31). Also the form of this equation is similar to the usual statistical mechanical form at this stage of the derivation but the interpretation of the quantities is considerably different; it is based on the number of microstates having a given number of particles, rather than the number of particles in a given microstate.

Constraints and conditions on the macroscopic state have a somewhat different form than usual. The total number of points is assumed known and constant

$$N = \sum_i \sum_{r=0} rn_i^{(r)}, \tag{38.32}$$

as is the energy

$$E = \sum_i \sum_{r=0} \varepsilon_i rn_i^{(r)}, \tag{38.33}$$

where ε_i is the energy of the ith macrocell. Although a macrocell contains many microcells it still occupies such a small volume of phase space that its energy is effectively constant throughout it. Furthermore we have the condition that the number of microcells in the ith macrocell is

$$w_i = \sum_{r=0}^{p} n_i^{(r)}, \tag{38.34}$$

and the total number of distinguishable microcells is

$$N_{\text{dist}} = \sum_i \sum_{r=q+1}^{p} n_i^{(r)}. \tag{38.35}$$

The first step to obtaining the distribution function is to maximize $\ln \Omega$, and thus Ω, by introducing Lagrange multipliers to include constraints (38.32)–(38.35)

$$\ln \Omega = \sum_i \left\{ \ln(w_i!) + \ln(N_{\text{dist}}!) - \ln\left[\left(\sum_{r=0}^{s} n_i^{(r)}\right)!\right] - \sum_{r=s+1}^{p} \ln(n_i^{(r)}!) \right.$$

$$\left. - (\alpha + \beta \varepsilon_i) \sum_{r=0}^{p} rn_i^{(r)} + \gamma_i \sum_{r=0}^{p} n_i^{(r)} + \delta \sum_{r=q+1}^{p} n_i^{(r)} \right\}. \tag{38.36}$$

To maximize (38.36) we vary the $n_i^{(r)}$ independently for $r > s$. For $r \leq s$ it is only possible to vary the total number of similar microcells, denoted by

$$n_i^s = \sum_{r=0}^{s} n_i^{(r)} \tag{38.37}$$

and containing

$$N_i^s = \sum_{r=0}^{s} rn_i^{(r)} \tag{38.38}$$

phase points. In the case where the total energy of a macrocell is linearly proportional to its mass, and the mass is constant, an equivalent procedure would be to maximize the total mass of the $n_i^{(r)}$ microcells in the ith macrocell. This explicitly shows that the distribution function depends on energy per unit mass, rather than on energy. Although in the present approach it is easier to maximize the total number of microcells in a macrocell, this must lead to the same conclusion when Lagrange multipliers α, β, γ_i and δ are computed from the full self-consistent solution. Using Stirling's approximation and performing the variation gives

$$n_i^s = e^{\gamma_i - A\zeta_i} \qquad r \le s, \tag{38.39}$$

$$n_i^{(r)} = e^{\gamma_i - \zeta_i r} \qquad s < r \le q, \tag{38.40}$$

$$n_i^{(r)} = e^{\delta + \gamma_i - \zeta_i r} \qquad q < r \le p, \tag{38.41}$$

where

$$\zeta_i = \alpha + \beta \varepsilon_i, \tag{38.42}$$

$$A = \frac{\delta N_i^s}{\delta n_i^s} \tag{38.43}$$

and

$$0 \le A \le s. \tag{38.44}$$

For $r > s$, the total number of phase points in all those microcells which contain r phase points and are located in the ith macrocell is

$$N_i^{(r)} = -\frac{\partial n_i^{(r)}}{\partial \zeta_i}. \tag{38.45}$$

The number of microcells in the ith macrocell is

$$w_i = e^{\gamma_i} \left[e^{-A\zeta_i} + \sum_{r=s+1}^{q} e^{-\zeta_i r} + \sum_{r=q+1}^{p} e^{\delta - \zeta_i r} \right] \tag{38.46}$$

and the number of phase points in those distinguishable or semidistinguishable microcells contained in the ith macrocell is

$$N_i - N_i^s = e^{\gamma_i} \left[\sum_{r=s+1}^{q} r e^{-\zeta_i r} + e^{\delta} \sum_{r=q+1}^{p} r e^{-\zeta_i r} \right]. \tag{38.47}$$

From these results and some algebraic manipulation, the coarse grained distribution function becomes

$$\bar{f} - \bar{f}^s = \frac{N - N^s}{w} = \frac{e^{-\zeta}(D_2 - D_1) - D_2}{(e^{-\zeta} - 1)[D_1 + e^{-A\zeta}(e^{-\zeta} - 1)]}, \tag{38.48}$$

where the i subscript has been dropped,

$$D_1 \equiv [(e^{-(q+1)\zeta} - e^{-(s+1)\zeta}) + e^{\delta}(e^{-(p+1)\zeta} - e^{-(q+1)\zeta})] \tag{38.49}$$

and

$$D_{n+1} = (-1)^n \frac{\partial^n D_1}{\partial \zeta^n}. \tag{38.50}$$

The coarse grained phase density $\bar{f}$ is the total number of phase points in a macrocell, divided by the phase volume of that macrocell. A microcell has an average volume normalized to unity, but other normalizations may also be used. Values of α, β, γ and δ are obtained by substituting (38.39)–(38.41) into (38.32)–(38.35) or by replacing the summations with integrals and using the continuous distribution function (38.48).

Equation (38.49) for D_1 reveals an interesting symmetry between the Fermi–Dirac statistics and those of a collisionless gravitating system. In the F–D case, $s = 0$, $q = p = 1$; therefore $A = 0$, and only terms in the left hand parentheses of (38.49) survive. In a collisionless system, $s = 0$, $q = 0$, $p = 1$; therefore $\delta = 0$, $A = 0$, and only terms in the right hand parentheses survive. Similar results hold for D_2 and the higher Ds. However, although different terms survive in each case, the form of D_1 and D_2 is invariant. Both statistics therefore give the same distribution function, as we saw previously.

In the two special cases $s = q = \delta = 0$, or $s = 0$ and $q = p$ (the first case is relevant to gravitating systems), the distribution function reduces to the much simpler form

$$\bar{f} = \frac{1}{e^\zeta - 1} - \frac{p+1}{e^{(p+1)\zeta} - 1}. \tag{38.51}$$

From this it is straightforward to see that $\bar{f}$ for the Fermi–Dirac, Bose–Einstein, and Maxwell–Boltzmann statistics emerge as special cases. The reader can also examine other interesting limits of (38.48).

With the distribution function in hand, one can readily calculate the entropy by substituting the most probable values of $n_i^{(r)}$ into $s = k \ln \Omega$ from (38.31). From the entropy, the entire thermodynamics follows, including the fluctuations and correlations of these idealized violently relaxed collisional systems. When collisions and correlations are important the entropy has a much more complicated form than the simple $S = -k \int f \ln f \, d^3\mathbf{r} \, d^3\mathbf{v}$ relation for a Maxwell–Boltzmann distribution. Unlike the collisionless case the kinetic theory underlying this more general relaxation process has not yet been worked out. Statistical mechanics is an attempt to short circuit our lack of detailed kinetic knowledge. The conditions under which these distributions apply to realistic gravitating systems remain to be understood.

39

Symmetry and Jeans' theorem

Liouville's equation in the form (10.16) shows that the total time derivative of the $6N$-dimensional distribution function $df^{(N)}/dt = 0$ following the motion of points in phase space. Therefore, $f^{(N)}$ itself is a constant of the motion. So is any function of $f^{(N)}$. This implies that there is an infinite number of constants of motion. It is natural to ask why all these constants do not lead to additional constraints on the statistical mechanics of the system.

First of all, since $f^{(N)}(\mathbf{x}^{(1)},...,\mathbf{x}^{(N)},\mathbf{v}^{(1)},...,\mathbf{v}^{(N)},t)$ is constant following the motion, it must depend only on combinations of the $\mathbf{x}$s and $\mathbf{v}$s which are themselves constant following the motion. Next, $f^{(N)}$ must satisfy the physical requirement that its logarithm be additive: the distribution function for two independent systems is the product of the two separate distributions, $f^{(N)}_{12} = f^{(N)}_1 f^{(N)}_2$. Therefore the logarithm of the distribution function is an additive quantity

$$\ln f^{(N)}_{12} = \ln f^{(N)}_1 + \ln f^{(N)}_2 \tag{39.1}$$

and can depend only on additive integrals of the motion. These additive integrals, from elementary mechanics, are the energy, linear momentum and angular momentum. Thus the logarithm of the distribution function of an independent system must be a linear combination of these conserved additive quantities

$$\ln f^{(N)} = A + BE(\mathbf{x},\mathbf{v}) + \mathbf{C}\cdot\mathbf{P}(\mathbf{x},\mathbf{v}) + \mathbf{D}\cdot\mathbf{L}(\mathbf{x},\mathbf{v}), \tag{39.2}$$

where A is a normalization constant and $B, \mathbf{C}$ and $\mathbf{D}$ are constant scalar and vector coefficients.

So we see that not all constants of motion are equivalent: some are additive, most are not. Another important distinction is that some integrals are 'isolating' and others are non-isolating. Isolating integrals are single valued. Non-isolating integrals are multi valued, usually because they depend upon a phase factor. In order for $f^{(N)}$ to be uniquely defined, a necessary physical requirement, it must depend only on isolating integrals.

Investigating these questions explicitly for collisionless systems leads us to Jeans' theorem. We saw in Section 7 that the general solution of the collisionless Boltzmann equation is an arbitrary function of the independent integrals of its characteristic curves (7.15). Integration of (7.15) gives six independent constants of the motion. Since $df/dt = 0$ following the motion, where f is now the six-dimensional distribution function, f is a function of these six constants of the motion. Alternatively, the argument can be inverted by expressing the positions and velocities of each star as a function of its initial coordinates, determining the six constants as functions of these initial values, and then re-expressing the later positions and velocities in terms of the six constants. This is, of course, independent of coordinate system.

By either route of argument the result is Jeans' theorem: in a collisionless system the phase density is an integral of the motion which depends only on other constant integrals of the motion. To go a step further, we note that since the phase density must be single valued it can only depend on isolating integrals. These are the additive integrals mentioned earlier. This is not to say, however, that the phase density must depend on all the available additive or isolating integrals.

Symmetry constraints further affect the number of integrals of motion on which f depends. If the system is in a stationary state, long after violent relaxation has done its work, then $\partial f/\partial t = 0$ and the dt term does not appear in (7.15). Only five independent integrals are left. Consider three examples of stationary systems in order of increasing spatial symmetry.

First, suppose the stationary mass distribution is arbitrary with no constraints imposed on the gravitational potential ϕ. Substituting the gradient of the potential for the acceleration in the characteristic equations (7.15) gives an equation of the form

$$v_x dv_x = \frac{\partial \phi}{\partial x} dx \tag{39.3}$$

for each of the three coordinates. Adding the three equations and integrating gives the energy integral

$$v_x^2 + v_y^2 + v_z^2 - 2\phi = v^2 - 2\phi = 2E = \text{constant.} \tag{39.4}$$

For an arbitrary distribution, no other isolating integral can be found, so $f(\mathbf{x}, \mathbf{v}) = f(E)$. Notice that this minimum symmetry in configuration space is associated with the maximum symmetry in velocity space where f has full spherical symmetry.

Second, suppose the mass distribution is rotationally symmetric. Let the z axis be the symmetry axis, so in cylindrical coordinates $\phi = \phi(r, z)$, where $r = (x^2 + y^2)^{1/2}$. Now we have the additional constraint

$$x \frac{\partial \phi}{\partial y} = y \frac{\partial \phi}{\partial x} \tag{39.5}$$

and the characteristic equations give

$$y dv_x = x dv_y. \tag{39.6}$$

Since $dx = v_x dt$ and $dy = v_y dt$, this becomes

$$d(y v_x - x v_y) = 0 \tag{39.7}$$

and, integrating

$$yv_x - xv_y = L_z = \text{constant}. \tag{39.8}$$

Thus, with rotational symmetry, $f = f(E, L_z)$. Expressing everything in cylindrical coordinates, $f = f(v_r^2 + v_\theta^2 + v_z^2 - 2\phi, rv_\theta)$. In velocity space there is symmetry only between v_r and v_z, a reduction of velocity symmetry relative to the first case. In velocity space the distribution is symmetric around the v_θ axis. In configuration space it depends only on r and z, being symmetric in θ.

Thirdly, consider the case of maximum spatial symmetry, when the potential is spherically symmetric, $\phi = \phi(r)$ with $r = (x^2 + y^2 + z^2)^{1/2}$. As an exercise, the reader can now show that $f = f(E, \mathbf{L})$. Since there is no greater spatial symmetry the fifth integral generally must be non-isolating. (There are exceptions for very simple degenerate systems such as Keplerian motion or the simple harmonic oscillator in three dimensions.) Note that f does not depend on the linear momenta since they just represent the translation of the whole system, leaving the internal distribution unaffected.

To express the velocity dependence of $f(E, \mathbf{L})$ in spherical coordinates, we note that the angular momentum dependence can only involve r, since $\phi(r)$ is spherically symmetric. Therefore, we need to find a function of L_x, L_y and L_z which depends spatially just on r. If we rotate the coordinate axes on which $\mathbf{L}$ depends, the only single valued quantity which is invariant is $\mathbf{L}^2 = (\mathbf{r} \times \mathbf{v})^2 = L_x^2 + L_y^2 + L_z^2$. (Its square root is double valued.) Using the vector identity $|\mathbf{a} \times \mathbf{b}|^2 = a^2 b^2 - (\mathbf{a} \cdot \mathbf{b})^2$, we have

$$L^2 = r^2 v^2 - (\mathbf{r} \cdot \mathbf{v})^2 = r^2(v_r^2 + v_\theta^2 + v_\phi^2) - r^2 v_r^2 = r^2(v_\theta^2 + v_\phi^2). \tag{39.9}$$

Thus the phase density is of the form

$$f(E, L^2) = f(v_r^2 + v_\theta^2 + v_\phi^2 - 2\phi(r), \ r^2(v_\theta^2 + v_\phi^2)). \tag{39.10}$$

It is symmetric in v_θ and v_ϕ and has rotational symmetry around an axis parallel to the radial vector for a given phase space point.

From Jeans' theorem we can deduce that a system with an arbitrary mass distribution cannot have any net internal motion. For then,

$$\bar{v}_x = \int\int\int_{-\infty}^{\infty} f(v_x^2 + v_y^2 + v_z^2 - 2\phi, t)v_x dv_x dv_y dv_z = 0 \tag{39.11}$$

since f is an even function of v_x, and similarly for v_y and v_z. Thus a spherical velocity distribution stays spherical, and so does a spherical spatial density distribution. These distributions can, however, expand or contract since $\overline{v_x^2}$ and $\overline{x^2}$ are not generally zero.

The collisionless Boltzmann equation may be applied in two ways. We may be given the potential ϕ, perhaps from knowledge of the spatial density, and want to find f. Sometimes this is called the Jeans problem. Alternatively, given f we may want to find ϕ – the inverse Jeans problem. No general solution exists, but special cases have been explored, especially with regard to galactic structure. In the next section we shall take up simple spherical models based on the collisionless and the Fokker–Planck approximations.

40

Quasi-equilibrium models

40.1. Polytropes and isothermal spheres

After the initial violent relaxation is over (and if it is complete the distribution function is isotropic), gravitating systems with no net rotation settle down into a spherical configuration. Subsequent dynamical evolution proceeds much more slowly and an isolated system remains for long periods in a quasi-equilibrium state. Starting with Section 43 we will examine the stability and evolution of this state, but first we describe some useful stationary models for it.

Idealized stationary spherical models are time independent and have isotropic pressure. The total change in their average stellar velocity following the motion is $d\mathbf{v}/dt = 0$. Equivalently, at every point in the system the velocity distribution is spherically symmetric. For, if it were not some stars would have an excess velocity in a preferred direction, and relaxation would cause the distribution to change in time, or the structure to evolve. Therefore, the structure of these systems is described by the last two terms of (8.20),

$$\frac{1}{\rho}\frac{dP}{dr} = \frac{d\phi}{dr}.\tag{40.1}$$

which is just the equation of hydrostatic equilibrium. More formally this follows from the condition that the distribution function depends only on angular momentum per unit mass and total energy per unit mass. Poisson's equation, here in integral form, gives the gravitational force

$$\frac{d\phi}{dr} = -\frac{GM(r)}{r^2} = -\frac{4\pi G}{r^2}\int_0^r \rho(r')r'^2 dr'.\tag{40.2}$$

Combining these two equations and differentiating eliminates the integral

$$\frac{d}{dr}\left(\frac{r^2}{\rho}\frac{dP}{dr}\right) = -4\pi G\rho r^2.\tag{40.3}$$

More information, in the form of a relation between P and ρ, is needed to solve (40.3). We know that a function $P(\rho)$ exists, since both P and ρ can depend only on r in a single-valued way under spherical symmetry. The analogous problem for the structure of a gaseous star was originally solved by Emden more than 80 years ago. He proposed using the power law

$$P = K\rho^{1+1/n}, \tag{40.4}$$

where K and n are constants. This relation led to stars of many forms, called polytropes. It is an 'effective equation of state', and n is called the polytrope index. If it is combined with the perfect gas equation, then this polytropic equation of state constrains the temperature's radial dependence. The case $n = \infty$, representing a constant temperature, is known as the isothermal sphere.

Polytropes were much studied in the field of stellar structure, especially before the 1960s when computers first made fully self-consistent numerical solutions possible. No one pretends that polytropes are real stars, or stellar systems (although sometimes more complicated models with variable n are used as better approximations). But they have several virtues. They are simple to understand and perturb. Moreover, their structures have a wide enough range to provide considerable insight into the properties of real stars. Indeed, there are astronomers who believe that major properties of real stars are not really understood until they can be reproduced with polytropes!

Eliminating pressure from (40.3) and transforming to the dimensionless quantities,

$$\rho \equiv \lambda \theta^n \tag{40.5}$$

and

$$r \equiv \alpha \zeta \equiv \left[\frac{(n+1)}{4\pi G} K \lambda^{(1-n)/n} \right]^{1/2} \xi, \tag{40.6}$$

gives Emden's equation (also called the Lane–Emden equation)

$$\frac{1}{\xi^2} \frac{d}{d\xi} \left(\xi^2 \frac{d\theta}{d\xi} \right) + \theta^n = 0. \tag{40.7}$$

The solution requires two boundary conditions. We can identify λ as the central density so that at

$$\xi = 0, \quad \theta = 1. \tag{40.8}$$

Moreover, the force at the center must go to zero, i.e., $d\phi/dr = 0$ when $\theta = 1$; from (40.1) and (40.4), we see that at

$$\xi = 0, \quad \frac{d\theta}{d\xi} = 0 \tag{40.9}$$

for any finite central density. These are the two boundary conditions. One reason for imposing boundary conditions at the center, rather than at the surface, is that the assumption of isotropic velocities is less realistic at the surface.

Solutions of Emden's equation are known in closed form for $n = 0, 1$ and 5. When $n = 0$, (40.7) can be integrated straightaway. Applying the boundary conditions gives

$$\theta_{n=0} \equiv \theta_0 = 1 - \tfrac{1}{6}\xi^2. \tag{40.10}$$

The outer boundary where $\theta_0 = 0$ occurs at $\xi = \sqrt{6}$. This is a sphere of constant density and no compressibility.

When $n = 1$ the substitution $\theta = u/\xi$ converts (40.7) into the harmonic oscillator equation for $u(\xi)$. The result is Ritter's solution

$$\theta_1 = \frac{\sin \xi}{\xi}. \tag{40.11}$$

Now the outer boundary occurs at the first zero of θ_1, namely $\xi = \pi$, and $\rho \propto \theta_1$.

When $n = 5$ the substitutions $\xi = e^{-u(z)}$ and $\theta_5 = ze^{u/2}/\sqrt{2}$, along with some fairly straightforward manipulations (an exercise for the reader), give Shuster's solution

$$\rho_5 = \theta_5^5 = (1 + \tfrac{1}{3}\xi^2)^{-5/2}. \tag{40.12}$$

At large radii, $\rho_5 \propto r^{-5}$. Although the radius is infinite the total mass converges. This is Plummer's model (see Section 18.2).

To find the mass within a sphere of radius ξ we use Emden's equation (40.7) directly

$$M(\alpha\xi) = 4\pi\alpha^3 \lambda \int_0^\xi \xi^2 \theta^n d\xi = -4\pi\alpha^3 \lambda \int_0^\xi \frac{d}{d\xi'}\left(\xi'^2 \frac{d\theta}{d\xi'}\right)d\xi' = -4\pi\alpha^3 \lambda \xi^2 \frac{d\theta}{d\xi}. \tag{40.13}$$

The mass of the whole polytrope is

$$M_n = -4\pi \left[\frac{(n+1)K}{4\pi G}\right]^{3/2} \lambda^{(3-n)/2n}\left[\xi^2 \frac{d\theta}{d\xi}\right]_{\theta_n = 0} \tag{40.14}$$

using (40.6) and the value of ξ at the smallest root of $\theta_n = 0$. When $n = 5$ there is no finite boundary, but since

$$\lim_{\xi \to \infty} \xi^2 \frac{d\theta_5}{d\xi} = -\sqrt{3} \tag{40.15}$$

the total mass is still finite. For $n > 5$ the total mass is infinite.

The potential energy of a polytrope is given as a simple exercise in Section 54.1.

Isothermal spheres, with $n = \infty$, are especially useful representations of gravitating systems. They have often been used as first approximations for spherical galaxies and clusters of galaxies. Violently relaxed systems, as well as systems resulting from two-body relaxation, would both have $P = K\rho$. If a perfect gas equation of state were used the constant K would differ (just in whether velocity dispersion depended on mass) for these two cases, but the net result would be similar. Repeating the previous procedure now gives for the normalization

$$r \equiv \alpha\xi = \left[\frac{K}{4\pi G\lambda}\right]^{1/2} \xi. \tag{40.16}$$

For the density we substitute

$$\rho = \lambda e^{-\psi}. \tag{40.17}$$

Now Emden's equation has the form

$$\frac{1}{\xi^2}\frac{d}{d\xi}\left(\xi^2\frac{d\psi}{d\xi}\right) = e^{-\psi}. \tag{40.18}$$

Again the boundary conditions are

$$\Psi = \frac{d\psi}{d\xi} = 0 \text{ at } \xi = 0. \tag{40.19}$$

No non-singular solutions of the isothermal equation are known in closed form so it must be integrated numerically, although many of its general properties have been studied (e.g., Chandrasekhar, 1939). For $\xi < 1$ we get a good idea of the behavior by expanding ψ in a power series in ξ, substituting into (40.18) and determining the coefficients as usual by equating like powers of ξ. The result for $\xi < 1$ is

$$\psi = \tfrac{1}{6}\xi^2 - \tfrac{1}{120}\xi^4 + \tfrac{1}{1890}\xi^6 + \cdots. \tag{40.20}$$

For $\xi \gg 1$ the situation is more complicated. Inspection shows that $\psi = 2\ln\xi - \ln 2$ or

$$\rho = \lambda\frac{2}{\xi^2} \tag{40.21}$$

is a solution of (40.18), but it does not satisfy our boundary conditions. Detailed analysis (e.g., Chandrasekhar, 1939) shows that this singular solution is in fact the asymptotic solution of the isothermal sphere equation for $\xi \gg 1$.

The mass distribution of an isothermal sphere is found in the previous manner to be

$$M(\xi) = 4\pi\alpha^3\lambda\xi^2\frac{d\psi}{d\xi}$$

$$= 4\pi\rho_c\left(\frac{v^2}{12\pi G\rho_c}\right)^{3/2}\xi^2\frac{d\psi}{d\xi}. \tag{40.22}$$

In the last expression we have substituted (40.16) for α, using the relation $K = v^2/3$ for an isotropic velocity dispersion v^2. Also, λ is rewritten as the central density ρ_c. Since $d\psi/d\xi = 2/\xi$ at large ξ, the total mass is infinite, increasing linearly with radius. In practice, the size could be limited by tidal interference and by departures from equilibrium.

Clearly the properties of an isothermal sphere are determined completely by its central density and its gravitational scale length α. This means that all isothermal spheres are homologous. Their scale length is closely related to the Jeans length, as it must be from dimensional considerations. But α is somewhat smaller than λ_J because relaxed systems must be inhomogeneous. Moreover, α characterizes all the wavelengths necessary to describe the density distribution, whereas λ_J is just one wavelength.

Comparisons of cluster densities with observation involve the projected density

$\eta(x)$ along the line of sight. This is easily related to the spatial density ρ by

$$\eta(x) = \int_{-\infty}^{\infty} \rho(S)\,dS, \qquad (40.23)$$

where x is the given distance from the center of the cluster and S is distance along the line of sight. Putting the origin of S at the point nearest to the center of the cluster we have $S^2 + x^2 = r^2$ so that (40.23) becomes

$$\eta(x) = 2 \int_{x}^{\infty} \frac{r}{(r^2 - x^2)^{1/2}} \rho(r)\,dr. \qquad (40.24)$$

Substituting $\rho(r)$ for a polytrope or isothermal sphere (or any other distribution, for that matter) gives the surface density. Often an integration by parts is helpful if $d\rho/dr$ is more easily determined than ρ

$$\eta(x) = -2 \int_{x}^{\infty} (r^2 - x^2)^{1/2} \frac{d\rho}{dr}\,dr. \qquad (40.25)$$

Fits of isothermal spheres to the observed projected galaxy number distributions of rich symmetrical clusters of galaxies, such as Coma, are reasonably good. Fits to the observed distribution of light in spherical galaxies or globular clusters are not so good. There may be several reasons for this, such as a spatially varying mass–luminosity ratio, a non-spherical galaxy appearing spherical in projection, or tidal interactions. Moreover, in galaxies a more fundamental difference may be that the system is not completely relaxed.

40.2. Loaded polytropes

In addition to representing smoothly distributed systems, polytropes may be used as a first approximation for systems containing a singular or very compact core. The core may be a massive object or gas cloud, or a dense stellar cluster which can be considered in relative isolation from the rest of the system. For example, a cluster of galaxies may have a very massive galaxy at the center; a galactic nucleus may contain a massive black hole. As an idealization this core is treated as a singularity in the gravitational field. It changes the surrounding distribution of objects. The result is a loaded polytrope or loaded isothermal sphere.

With a central object of mass M_0 the equation of hydrostatic equilibrium is

$$\frac{1}{\rho}\frac{dP}{dr} = -\frac{GM_0}{r^2} - \frac{GM_s(r)}{r^2}. \qquad (40.26)$$

M_s is the mass of stars (or other objects) within radius r. Now there is an added complication. The central object interacts with nearby stars. Gravitational interaction may make the central orbits more Keplerian and less random, so an isotropic pressure becomes a poorer approximation very close to the center. Moreover, a black hole or supermassive star may consume other stars that come too close; a dense gas cloud may slow them down. Clearly there will be a central region to which the simple scalar pressure of a polytrope does not apply. We can account for this effect by formally including this region as part of the central object, rather than as part of the stellar distribution. Therefore the mass of stars becomes.

$$M_s(r) = \int_{r_{min}}^r 4\pi r^2 \rho(r) \, dr, \tag{40.27}$$

where r_{min} is the inner radius of the isotropic stellar distribution. In principle the value of r_{min} could be found from a self-consistent solution of the Boltzmann equation with an anisotropic velocity distribution. But this is difficult and true progress has been slow. Meanwhile, we will have to regard r_{min} as a parameter of the model and estimate its value.

Combining these last two equations with (40.4)–(40.6), as before, gives the basic equation for loaded polytropes

$$\xi^2 \frac{d\theta}{d\xi} = -\beta - \int_{\xi_0}^\xi \theta^n \xi^2 \, d\xi, \tag{40.28}$$

where

$$\beta = \frac{M_0}{4\pi\lambda\alpha^3} \tag{40.29}$$

and

$$\xi_0 \equiv r_{min} \alpha^{-1}. \tag{40.30}$$

When $\beta = 0$ and $\xi_0 = 0$ Equation (40.28) reduces to the result for normal polytropes; its derivative becomes the Lane–Emden equation (40.7). Here, however, it is more convenient to use this integro-differential form which avoids δ functions for the central singularity.

We set the boundary conditions at ξ_0 so that $\rho(\xi_0) = \lambda$; that is,

$$\theta(\xi = \xi_0) = 1 \tag{40.31}$$

and evidently

$$\left. \frac{d\theta}{d\xi} \right|_{\xi_0} = -\frac{\beta}{\xi_0^2} \tag{40.32}$$

Already this shows an important difference from normal polytropes with $(d\theta/d\xi)_{\xi=0} = 0$: The massive object produces a cusp in the density distribution near the center.

In the isothermal case the basic equation is the same as (40.28) except that θ is replaced by ψ, θ^n is replaced by $-e^{-\psi}$, β is replaced by $-\beta$ and α now takes the form of (40.16). This is just what one expects by analogy with normal polytropes. The density is given by (40.17) and the boundary conditions become

$$\psi(\xi_0) = 0, \tag{40.33}$$

$$\left. \frac{d\psi}{d\xi} \right|_{\xi_0} = \frac{\beta}{\xi_0^2}. \tag{40.34}$$

First we find the density distributions. In the central region, where ξ is not much greater than ξ_0, the first term on the right hand side of (40.28) dominates the second term. A straightforward integration then gives the central density

$$\rho_{\text{cusp}} = \lambda \left[1 - \frac{\beta(\xi - \xi_0)}{\xi \xi_0} \right]^n. \tag{40.35}$$

Thus there is a steep cusp at the center. Beyond the cusp the density decreases much more slowly. Since (40.35) does not include the stellar mass contribution, it just describes the cusp region. In a similar manner the cusp of the isothermal sphere has the form

$$\rho_{\text{cusp}} = \lambda \exp \left[-\frac{\beta(\xi - \xi_0)}{\xi \xi_0} \right] \tag{40.36}$$

with the isothermal form of α in β from (40.29). The cusp region ends about where the stellar mass beyond ξ_0 exceeds M_0.

As with normal polytropes, there are analytic solutions for loaded polytropes with $n = 0, 1$ or 5. For $n = 5$ the solutions are in terms of elliptic integrals which are rather complicated. However, for $n = 0$ and $n = 1$, the forms are relatively simple. For $n = 0$

$$\theta_0 = 1 - \tfrac{1}{6}(\xi^2 - \xi_0^2) + \left(\frac{\xi_0^2}{3} - \frac{\beta}{\xi_0} \right) \left(\frac{\xi - \xi_0}{\xi} \right). \tag{40.37}$$

The radius is given by the first zero which, for $\xi(\theta_0 = 0) \gg \xi_0$, occurs at

$$\xi(\theta_0 = 0) = 6^{1/2} \left[1 + \frac{\xi_0^2}{3} - \frac{\beta}{\xi_0} \right]^{1/2}. \tag{40.38}$$

This differs from normal polytropes with $n = 0$ by the factor in brackets. Again the system is incompressible, so this just makes for a nice comparison rather than for a realistic model. Similarly, with $n = 1$

$$\theta_1 = \left(\xi_0 \sin \xi_0 - \frac{\beta \cos \xi_0}{\xi_0} + \cos \xi_0 \right) \frac{\sin \xi}{\xi}$$
$$+ \left(\xi_0 \cos \xi_0 + \frac{\beta \sin \xi_0}{\xi_0} - \sin \xi_0 \right) \frac{\cos \xi}{\xi}, \tag{40.39}$$

and

$$\tan \xi(\theta_1 = 0) = \frac{\sin \xi_0 - \xi_0 \cos \xi_0 - (\beta \sin \xi_0)/\xi_0}{\cos \xi_0 + \xi_0 \sin \xi_0 - (\beta \cos \xi_0)/\xi_0} \xrightarrow[\xi_0 \to 0]{} \frac{\beta \xi_0}{\beta - \xi_0}. \tag{40.40}$$

As expected, loading the $n = 1$ polytrope decreases its radius. If $\beta > \xi_0$ the radius $\xi_1(\theta = 0)$ is less than $\pi/2$; while if $\beta < \xi_0$ it is between $\pi/2$ and π.

Unlike ordinary polytropes, which just have one scale length, α, loaded polytropes have three. First, there is the region inside ξ_0 containing the point source along with those stars with highly anisotropic velocities. Second, there is the cusp within

$$r_{\text{cusp}} \approx \left(\frac{3 M_0}{4 \pi \rho_{\text{stars}}} \right)^{1/3}, \tag{40.41}$$

where the mass M_0 of the central source and its anisotropically moving companions dominates the gravitational potential. Third, there is the global scale length α which determines the extent of the broad density plateau.

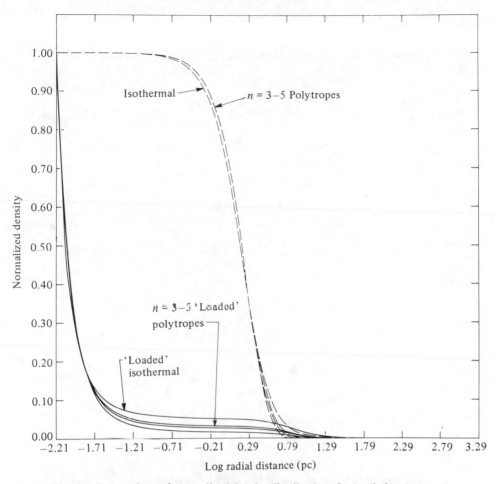

Fig. 41. Comparison of normalized density distributions for loaded and normal polytropes described in the text. The lowest polytrope curve is for $n = 3$. (From Huntley & Saslaw, 1975.)

Figure 41 illustrates the density distribution for both loaded and ordinary polytropes and isothermal spheres. The loaded examples represent galactic nuclei and have a central mass of 10^8 M$_\odot$, a central density of 10^9 M$_\odot$pc^{-3}, a central region $r_{min} \approx 6.3 \times 10^{-3}$pc and a total mass of about 10^{11} M$_\odot$. Among each set of loaded and normal systems the dependence on polytropic index is not very strong. Both types of system here are normalized to unit central density to show the cusp effect clearly. Alternatively if the curves were normalized to have the same average plateau density they would appear similar except for a sharp high spike in the loaded density at the center. Figure 42 shows the projected density for these examples, calculated from (40.24). The relative size of the projected central peak is greater by a factor of about three or four (depending on how it is defined) in the loaded systems. This is a reflection of the broader density plateau in the loaded cases.

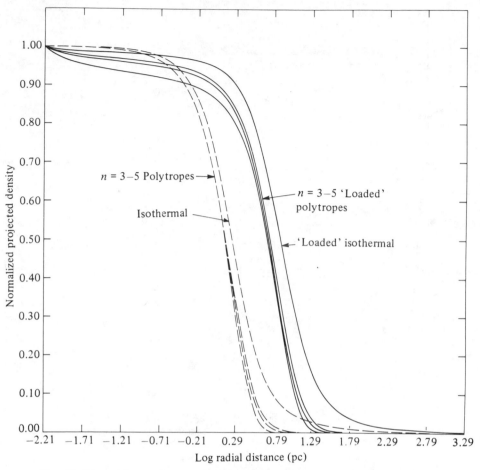

Fig. 42. Comparison of normalized projected densities for the systems in Figure 41. (From Huntley & Saslaw, 1975.)

Following a procedure similar to the one for normal polytropes gives the total mass of stars in a loaded polytrope

$$M_s(R) = -4\pi\lambda\alpha^3 \left[\xi^2(\theta=0)\frac{d\theta}{d\xi}\Big|_{\xi(\theta=0)} - \xi_0^2\frac{d\theta}{d\xi}\Big|_{\xi_0} \right]. \tag{40.42}$$

Table 4 (from Huntley & Saslaw, 1975) gives the values of these outer boundary terms for a range of loaded and normal polytropes in which the total stellar mass greatly exceeds M_0. The average of these values is about 2.54 for the loaded polytropes and 1.90 for the normal ones.

Neither loaded nor normal polytropes take into account the slow evaporation of objects from finite systems. This leads to anisotropic velocities which alter the density profile, particularly in the outer parts. For this effect, we must turn to more detailed models.

Table 4. *Comparison of boundary values for loaded and normal polytropes*

n	loaded $-\xi^2 d\theta/d\xi \big|_{\xi(\theta=0)}$	normal $-\xi^2 d\theta/d\xi \big|_{\xi(\theta=0)}$
2.5	1.331	2.187
3.0	2.010	2.018
3.5	2.481	1.891
4.0	2.836	1.797
4.5	3.135	1.738
5.0	3.468	1.732

40.3. Fokker–Planck models

The main trouble with isothermal spheres is that they are infinite. Otherwise, for a simple model they would be pretty good representations of real systems. Their main virtue is their uniform Maxwellian velocity distribution since this characterizes relaxed systems. What is the simplest way we can modify an isothermal sphere to make it finite?

We could try to require its density to go to zero at some finite radius. But this would clash with the isothermal requirement because the pressure gradient of stars near a stipulated boundary would just push the boundary farther out. This is why an isothermal sphere has no limit. Clearly, the velocity distribution must be modified at the boundary. We could envisage a model with highly anisotropic boundary velocities. In the extreme case we could require that all radial velocities at the boundary be zero or negative (i.e., inward). Or we could place a reflecting wall around the system – an artificial constraint useful for some interesting thought experiments (in Section 43) but too contrived for real systems. Less extreme anisotropies would still lead to solutions of the Boltzmann equation which are too complicated at this stage.

On the other hand, we could keep the isotropic velocity distribution by exploiting the finite escape velocity which characterizes finite systems. The escape velocity at the boundary of an isolated system just depends on its total mass and radius. For systems with companions, tidal forces also contribute to the escape velocity. Within the system the local escape velocity, i.e., the velocity needed to escape from the cluster starting from some interior point r, will depend on the local value of the gravitational potential. Stars whose velocity exceeds this amount will seldom remain in the cluster for more than a crossing time. Therefore, the velocity distribution $f(\mathbf{r}, \mathbf{v}, t)$ will be grossly depleted for $v > v_{esc}$. This is the major effect; anisotropy is generally less important.

Under these conditions we can make a heuristic guess for the velocity distribution, then verify it more formally. A Maxwellian with a cut-off has the form (5.8)

$$f(v) = A(e^{-\beta^2 v^2} - B),$$ (40.43)

where A and B are constants. Physically $f(v) = 0$ for $e^{-\beta^2 v^2} \le B$ and for this to happen at the escape velocity requires $B = e^{-\beta^2 v_{esc}^2}$. Normalizing to $f(v) = 1$ at the most probable speed $v = 0$ provides A. The result is

$$f(v) = \frac{e^{-\beta^2 v^2} - e^{-\beta^2 v_{esc}^2}}{1 - e^{-\beta^2 v_{esc}^2}}.$$ (40.44)

To derive this result more formally (King, 1965), we use the Fokker–Planck equation (5.1). In a spherical gravitating system it takes the form (Spitzer & Harm, 1958)

$$\frac{\partial f}{\partial \theta} = \frac{1}{x^2} \frac{\partial}{\partial x} \left[\alpha(x) \left(\frac{\partial f}{\partial x} + \frac{2m}{m_0} xf \right) \right].$$ (40.45)

Here

$$x \equiv \frac{v}{2^{1/2} v_0}$$ (40.46)

with v_0^2 the one-dimensional mean square velocity for stars of mass m_0, the average mass of a cluster star. Stars of mass m are interacting with the background m_0 mass stars which have a Maxwellian distribution. The dimensionless time variable is

$$\theta = \frac{3}{2^{11/2}} \frac{t}{\tau_R},$$ (40.47)

where τ_R is given by (2.9), with $m = m_0$ and $n = n_0$. Also,

$$\alpha(x) = \frac{4}{\pi^{1/2} x} \int_0^x e^{-y^2} y^2 \, dy$$ (40.48)

describes the dynamical friction.

The Maxwellian distribution is a steady state solution of the collisionless Boltzmann equation. This suggests we look for (40.44), which does not depend on time, as a steady state solution of the Fokker–Planck equation. Assume that stars escape from the cluster in such a manner as to leave the form of the velocity distribution unchanged, i.e., the depletion occurs at the same rate for all velocities. One should keep in mind that this is an idealization to simplify the problem. Thus we can write

$$f(x, \theta) = e^{-\lambda \theta} g(x),$$ (40.49)

which turns (40.45) into an eigenvalue problem

$$\frac{d}{dx} \left[\alpha(x) \left(\frac{dg}{dx} + 2 \frac{m}{m_0} xg \right) \right] + \lambda x^2 g = 0.$$ (40.50)

The eigenvalue

$$\lambda = -\frac{1}{n} \frac{dn}{d\theta}$$ (40.51)

is just the fractional loss rate during one 'relaxation time', and has a continuous range of values. Generally λ is small, so we can expand the solution in powers of λ

$$g(x) = g_0(x) + \lambda g_1(x) + \lambda^2 g_2(x) + \cdots. \tag{40.52}$$

The physical boundary conditions are that $g = 1$ and $dg/dx = 0$ at $x = 0$, and $g = 0$ at $x = x_{esc}$.

Substituting the expansion (40.52) into (40.50), and equating the coefficients of like powers of λ, gives successive approximations to the solution $g(x)$. To simplify matters we can replace $\alpha(x)$ by its limiting form proportional to x^{-1} as $x \to \infty$ for $m \neq m_0$, since the dominant contribution for $m \approx m_0$ does not depend much on this approximation and for m far from m_0 the dominant contribution comes from large x. This substitution occurs after solving the differential equations. The reader can go through this analysis, as an exercise, to show that to first order

$$g(x) = \frac{e^{-mx^2/m_0} - e^{-mx_{esc}^2/m_0}}{1 - e^{-mx_{esc}^2/m_0}} \tag{40.53}$$

and

$$\lambda = 8\pi^{-1/2} \left(\frac{m}{m_0} \right)^{5/2} \frac{e^{-mx_{esc}^2/m_0}}{1 - e^{-mx_{esc}^2/m_0}} \tag{40.54}$$

This result for $g(x)$ has the expected form (40.44). As a bonus we get the rate of evaporation, which will be discussed further in Section 46.

With the velocity distribution (40.53) we can calculate the density profile of a cluster (King, 1966). Its shape does not change with time, even though stars continually escape in this approximation. First, however, we apply Jeans' theorem (Section 39) to get a result which is very useful.

Stars escape if their velocity exceeds v_{esc}, which is a function of location in the cluster. This radial dependence enters the distribution function $f(r,v)$. The energy of an individual star is

$$E = \tfrac{1}{2}v^2 + \phi(r). \tag{40.55}$$

Now it becomes more convenient to define the zero value of $\phi(r)$ at the surface of the cluster, rather than at infinity. A star with zero total energy can then just reach the surface, or the tidal radius. So the escape velocity is

$$v_{esc}^2 = -2\phi(r). \tag{40.56}$$

Jeans' theorem says that the distribution function must depend on E (or, more generally, on E and L) in the same way throughout the entire system. At the center of the system, where the orbits are most relaxed, we can assume that (40.53) holds for $g(r = 0, v)$. Then substituting (40.46) and (40.55) gives $g(0, x)$ as a function of $E(r = 0)$. Finally, replacing $E(r = 0)$ by $E(r)$ from (40.55) gives the distribution function at *any* radius

$$f(r,v) = e^{-[\phi(r) - \phi(0)]/v_0^2} f(0,v), \tag{40.57}$$

where v_{esc} in $f(0,v)$ depends on r through (40.56). Thus the velocity distribution (40.53) has the same *form* everywhere in a steady state cluster.

Integrating $f(r,v)$ over the velocity gives the density

$$\rho = \int_0^{v_{esc}} f(r,v)4\pi v^2 \, dv \tag{40.58}$$

as a function of $\phi(r)$. To find $\rho(r)$ self-consistently, we use Poisson's equation

$$\frac{d^2\phi}{dr^2} + \frac{2}{r}\frac{d\phi}{dr} = 4\pi G\rho(\phi). \tag{40.59}$$

King's (1966) paper gives a detailed prescription for producing numerical models for spatial and surface densities. With $\phi(r)$, one can then go back to (40.57) to see how the velocity distribution changes throughout the cluster.

The results resemble isothermal spheres closely in the core of the system, but farther out the density drops rapidly to zero at the cluster (tidal) radius. The larger the ratio of the total radius to the core radius, the closer the model resembles an isothermal sphere. Comparisons with observations show that these models give reasonably good fits to the surface density of globular clusters, spherical galaxies, and the coma cluster of galaxies (e.g., King, 1966; Bailey & MacDonald, 1981; Kent & Gunn, 1982). Other, more complicated, models can be constructed to fit the surface density well, but they do not always give a good representation of the velocity field. The velocity field can often be a very powerful discriminator among models.

41

Applying the virial theorem

A theorist without practice is a tree without fruit.

Sa'di

Mass is one of the most basic properties of stellar and galactic systems, so it is natural that astronomers have developed several methods to determine it from observations. Unfortunately, results of different methods often disagree. The most notorious examples are the estimates of masses of clusters of galaxies. A voluminous, if not massive, literature has grown to surround this controversy. On the one hand, the mass can be estimated by independently finding the mass–luminosity ratio of nearby galaxies of each morphological type – spiral, elliptical, or irregular. Counting galaxies of each type in a distant cluster and measuring their luminosities then gives the total mass of the cluster galaxies. On the other hand, the mass of a cluster can be estimated by measuring the galaxy separations and velocities and applying the virial theorem. This dynamical mass usually turns out to be greater than the luminosity mass, often by a factor ~ 10 but occasionally by a factor ~ 100.

Several possible causes may contribute to this discrepancy. The mass–luminosity ratio of nearby galaxies may be intrinsically different from those in more distant groups. There, accretion, tidal interactions, or gaseous ablation, for example, may have altered galaxies even if they were similar to begin with. The mass–luminosity ratio of nearby galaxies may have been underestimated, particularly if they contain extended haloes of dark matter – gas, dust, low luminosity stars, black holes, massive neutrinos or other 'inos, pebbles, etc. The correlation between the M–L ratio and morphological type found from nearby galaxies may be unsuitable for galaxies in groups, where other morphological classifications may be more appropriate. Clusters and groups of galaxies may contain integalactic dark matter – gas, dust, etc. Clusters and groups may not satisfy the dynamical conditions needed for the virial theorem to hold.

Detailed assessments of these possibilities change with time and are beyond the scope of this book. Here we just describe how the idealized virial theorem must be modified in practice and how these modifications affect the final mass estimate.

Three major problems intervene when applying the scalar virial theorem (9.30) to astronomical clusters. We shall consider groups of galaxies as the prime example, although globular and galactic star clusters or individual galaxies themselves would do just as well. First, it is impossible in our brief astronomical lives to observe a time average of the kinetic and potential energies of a system. We are stuck with their instantaneous value. To find how representative an instantaneous value is we must inquire into the statistics of fluctuations around the average. Second, we only observe projected positions on the sky and the radial component of velocity. So we lose information about one spatial dimension and two velocity dimensions. How does this loss affect our estimate of the virial mass? Third, since telescope time is limited, only a subset of galaxies in each cluster is observed. Unlike the first two restrictions, which result from random partial sampling, this selection introduces bias. Frequently only the brighter galaxies in a cluster are measured. These tend to be the more massive ones. although sometimes they may really be a chance superposition which is then counted as one galaxy. The criteria for cluster membership introduce further bias. If a galaxy's radial velocity differs considerably from its neighbors' average, is it a foreground or background object? Or is it a cluster member which has had an unusual dynamical interaction. Where does one draw the line? Any answer already involves assumptions about relaxation and stability of clusters, the very results one hopes to determine.

The most straightforward way to examine these problems is by direct N-body simulations. Any assumptions are minimal and explicit. Moreover, for groups and clusters of galaxies, the number of objects involved can readily be handled. To see clearly where the problems arise we rewrite the virial theorem (9.30) using instantaneous values as

$$M = \frac{2RV^2}{G} \tag{41.1}$$

by putting

$$R = \frac{\left(\sum_i m_i\right)^2}{\sum_i \sum_j \dfrac{m_i m_j}{|\mathbf{r}_i - \mathbf{r}_j|}} \tag{41.2}$$

and

$$V^2 = \frac{\sum_i m_i v_i^2}{\sum_i m_i}. \tag{41.3}$$

Individual coordinates $\mathbf{r}_i$ and velocities $\mathbf{v}_i$ are taken with respect to the center of mass. The summations are taken over all objects and pairs (the factor two to account for counting each pair twice is retained in M rather than in R). The numerator of (41.2) expresses the approximation that the number of pairs $N(N-1) \approx N^2$, and should be modified by a term proportional to Σm_i^2 for small N.

In this way the virial mass, M, is expressed in terms of a mass weighted mean square velocity and a mass weighted harmonic radius. The mass weighting clearly shows how a few large close galaxies can dominate M, and how their absence in an observed sample will bias results. While it might seem more likely that large galaxies are included in the observed sample, this leads to another problem: their instantaneous departure from time-averaged values is also heavily weighted. Thus including the massive galaxies is not all to the advantage in determining M.

Projection affects all galaxies equally. Suppose the line of sight velocities are projected along the z axis to give v_z^2 and the separations transverse to the line of sight give R_{xy}. The virial mass determined from these projected coordinates is written in a form similar to (41.1) as

$$M_z' = \alpha R_{xy} V_z^2 / G. \tag{41.4}$$

The conversion factor α depends on the detailed configuration of each cluster. Since we do not know these configurations, we can only regard α as a normalization constant. *A priori*, it would be simplest to normalize to the average results for a spherical configuration with isotropic velocities. Then the masses are not correlated with positions or velocities and we would project the system equally from any angle. Let θ_i be the angle between the z axis along the line of sight and the space velocity v_i, so

$$\left(\sum_i m_i\right) V_z^2 = \sum_i m_i v_{iz}^2 = \sum_i m_i v_i^2 \cos^2 \theta_i. \tag{41.5}$$

Averaging over a spherical distribution of $\cos \theta_i$, uncorrelated with m_i or v_i

$$\langle \cos^2 \theta_i \rangle = \frac{\displaystyle\int_0^{2\pi}\int_0^{\pi/2} \cos^2 \theta_i \sin \theta_i \, d\theta_i \, d\phi}{\displaystyle\int_0^{2\pi}\int_0^{\pi/2} \sin \theta_i \, d\theta_i \, d\phi} = \frac{1}{3}. \tag{41.6}$$

Thus the spatial average of (41.5) combined with (41.3) shows that

$$\langle V_z^2 \rangle = \tfrac{1}{3} \langle V^2 \rangle. \tag{41.7}$$

This is just the equipartition of kinetic energy among its three spatial components in an isotropic system.

In the potential energy factor we write the spatial separation as $|\mathbf{r}_i - \mathbf{r}_j| = r_{ij}$. The projected separation is $r_{ij}' = r_{ij} \sin \theta_{ij}$, where θ_{ij} is the angle between $\mathbf{r}_{ij}$ and the observer's line of sight to the ith galaxy. Averaging over a spherical distribution of θ_{ij}, uncorrelated with the m_i, as in (41.6)

$$\left\langle \frac{1}{\sin \theta_{ij}} \right\rangle = \frac{\pi}{2}. \tag{41.8}$$

(Notice that the variance of R^{-1}, which involves terms with $\langle 1/\sin^2 \theta_{ij} \rangle$, is formally infinite. In practice this is of little consequence since two galaxies at nearly the same position would count as one galaxy. Moreover, the distribution in actual clusters is not usually isotropic.) Thus from (41.2)

$$\left(\sum_i m_i\right)^{-2} \langle R_{xy} \rangle \equiv \frac{1}{\left\langle \displaystyle\sum_i \sum_j \frac{m_i m_j}{r_{ij} \sin \theta_{ij}} \right\rangle} \equiv \frac{2}{\pi} \left(\sum_i m_1\right)^{-2} \langle R \rangle. \tag{41.9}$$

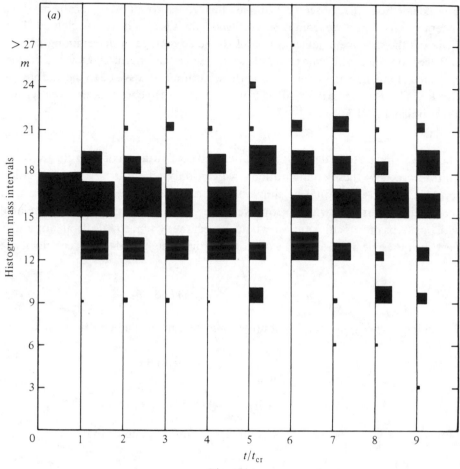

Fig. 43(*a*)

Inserting (41.7) and (41.9) into (41.4),

$$\langle M'_z \rangle = \frac{2\alpha}{3\pi G} \langle R \rangle \langle V^2 \rangle. \tag{41.10}$$

Comparing with (41.1) we see that the average conversion factor to get the true mass from observations of projected isotropic systems is

$$\alpha = 3\pi. \tag{41.11}$$

Thus the correction for projection amounts to nearly an order of magnitude.

Incomplete samples also require a correction because the velocity centroid of the sample is not the same as for the complete system. Introducing this correction $\dot{z}_c(n)$, which depends on the sample $n \leq N$, into (41.4) we have the best estimate for M in terms of observed radial velocities, $\dot{z}$, and positions on the sky

$$M = M'_z = \left(\sum_{i=1}^{n} m_i \right) \frac{3\pi \sum_{i=1}^{n} m_i [\dot{z}_i^2 - \dot{z}_c^2(n)]}{G \sum_{i=1}^{n} \sum_{j=1}^{n} \frac{m_i m_j}{[(x_i - x_j)^2 + (y_i - y_j)^2]^{1/2}}} \quad \text{with} \quad i \neq j. \tag{41.12}$$

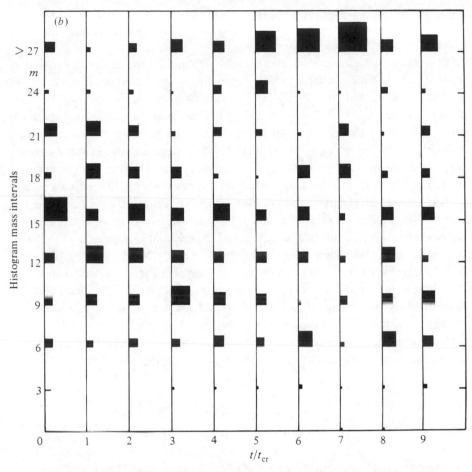

Fig. 43. Evolutionary double histograms of computed virial theorem mass for complete samples with $M \equiv N = 16$. (a) Virial mass M from complete three-dimensional information. (b) Projected virial mass M'_z. The number distribution in each mass interval is proportional to the width of the squares except for figure 43(a) at $t = 0$ when all 40 clusters give the same value. Every other sampling interval is shown. (From Aarseth & Saslaw, 1972.)

The fraction has an expectation value of unity for isotropic equilibrium configurations in the limit $n = N$ since $\dot{z}_c(N) = 0$. As n is decreased this fraction should increase above unity in order to compensate for the smaller sample mass if M'_z is to remain an accurate estimate.

At first sight there appear to be so many inherent difficulties in applying the virial theorem to observations that it seems hopeless to get a good value of M. Indeed it would seem necessary to already know all the galaxy masses in (41.12) to properly weight the observed quantities, and also to know the tangential velocities to completely correct the velocity centroid. Fortunately, however, the N-body simulations show that all this ignorance only leads to a factor ~ 2 uncertainty in the final result for M. The reason is that the uncertainties in the numerator and denominator of (41.12) strongly tend to cancel out. More fundamentally, this nice property arises because the virial theorem involves moments which average over the system.

We next examine some illustrations of this fortunate cancellation, as well as of departures from time-averaged values. The examples are based on three ensembles of clusters calculated in N-body integrations (Aarseth & Saslaw, 1972). Each cluster has a smooth distribution of galaxy masses from the continuous spectrum $n(m) \propto m^{-2}$, with the heaviest galaxy ten times the mass of the lightest. Two ensembles, each containing 40 different clusters are integrated for ten initial crossing times. One ensemble has groups with $N = 8$ galaxies; the other has $N = 16$ members in each group. A third ensemble containing 20 clusters with 32 members each is studied for one-half the time. The groups start with a statistically uniform spherical distribution and quickly relax after one or two initial crossing times. Every group is sampled at preselected intervals of $0.5\,t_{\mathrm{cr}}$, and the relevant quantities are calculated at three levels of completeness: $n = 8, 4, 2$ for the groups $N = 8$; and $n = 16, 8, 4$ for the two larger systems. Only the most massive members are included when $n < N$. This simple procedure introduces a reasonable selection effect if we assume a monotonically increasing $M(L)$ relation for all members. However, there is no attempt to distinguish between elliptical and spiral galaxies.

Departures of instantaneous values of M from the time averaged value, i.e., their true value, have average relative dispersions (for complete samples)

$$\begin{array}{cccc} & N = 8 & N = 16 & N = 32 \\ \frac{\sigma(M)}{M} = & \left\{\begin{array}{l} 0.24 \\ 0.27 \end{array}\right. & \begin{array}{l} 0.17 \\ 0.24 \end{array} & \begin{array}{l} 0.12, \\ 0.18. \end{array} \end{array} \qquad (41.13)$$

Here the upper set refers to the first half of the evolution and the lower set to the second half, for the three ensembles. Increasing N causes the deviations to decrease, very roughly by a factor of order $\sqrt{N}$. There is a slow increase of $\sigma(M)$ with time as the clusters become more inhomogeneous and eccentric binaries begin to form.

Both the instantaneous departures from equilibrium and the effects of projection for complete samples are shown in the double histograms of Figure 43. These refer to the ensemble with $M \equiv N = 16$. These histograms can be read vertically to give the distribution of total and projected virial mass estimates at any given time. They can also be read horizontally to give the evolution of departures from the true value $M = 16$. The systematic tendency, just mentioned, for $\sigma(M)$ to grow with time, is apparent even when complete information is available for M. This tendency is noticeably stronger for determinations of the projected mass, M'_z. Our expectation would be that using less information increases the dispersion, and this is borne out by the results. Even so, the combined effects of projection and instantaneous measurements usually cause the virial estimate to deviate from the true mass by a factor $\lesssim 2$, which is not too bad.

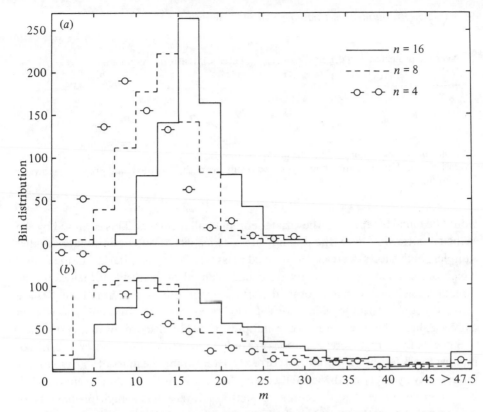

Fig. 44. Histograms of computed virial theorem mass for $N = 16$ and $n = 16, 8$ and 4. (*a*) M From three-dimensional information. (*b*) M'_z. From projected information. The last bin contains the number of all cases exceeding the limiting value. (From Aarseth & Saslaw, 1972.)

Effects of incomplete sampling are illustrated in Figure 44, again for $M \equiv N = 16$. These histograms for the distribution of individual projected and three-dimensional virial mass estimates are based on 20 successive samplings giving 800 quasi-independent estimates. These samples, taken every half crossing time, are quasi-independent in the sense that there are large variations between successive configurations, which are similar to the variations between measurements taken one crossing time apart. Projections along different axes were essentially similar within statistical fluctuations. The broadening of M due to instantaneous deviations, in the complete sample $n = 16$, is quite symmetric. Exceptionally low values of projected masses, at all levels of completeness, may be caused by nearly coincident pairs in the line of sight which dominate the potential energy. In addition, close binaries whose angular momentum axis is approximately in the direction of the observer will lead to an underestimate of the kinetic energy and a small projected length scale.

Incomplete samples show a systematic tendency to underestimate the mass which

Table 5. *Probability of finding virial mass within* 0.5N–2N

| Average projected virial mass | | | | Three-dimensional virial mass | | |
N	n_1	n_2	n_3	n_1	n_2	n_3
8...	65	48	29	98	84	61
16...	78	63	43	99	92	70
32...	81	59	43	99	92	70

Note : $n_1 = 8$, $n_2 = 4$, $n_3 = 2$ for the groups $N = 8$. But $n_1 = 16$, $n_2 = 8$, $n_3 = 4$ for $N = 16$ and $N = 32$. Units are in percentages, and results from three orthogonal projections are combined.

would be found by applying the virial theorem to all members. This is caused by the velocity correction and the bound subsystems which often dominate smaller samples; both effects decrease the derived mass (see Equation (41.12)). Large values which occur much less frequently are usually caused by excessive velocities in one direction combined with temporary departures from equilibrium. Small samples are especially subject to large fluctuations of velocity components. In small groups, such as $N = 8$, heavy binaries often occur and prevent wildly discordant mass estimates.

Since many extragalactic quantities are only known to within a factor ~ 2, it is of interest to ask for the probability of correctly determining the mass of a group to this uncertainty by applying the virial theorem. Table 5 shows the results from these N-body computations. The three-dimensional virial mass has a high probability of being correct to within a factor of two, even for small samples. But the projected mass estimate only has a roughly 50% probability of lying within this range. Thus we must generally expect an uncertainty of at least a factor two in observed estimates of virial masses for small groups.

Similar effects occur in large clusters with hundreds or thousands of members. Experiments on a cluster of $M \equiv N = 500$ galaxies gave $\langle R_{xy}/R \rangle = 0.66 \pm .08$, in very good agreement with the value $2/\pi$ of (41.9). Values of the projected virial mass were $\langle M'_z \rangle = 434$, 266 and 116 respectively for samples with $n = 250$, 50 and 10. These samples contained corresponding average actual masses of 405, 204, and 65. Again, smaller samples tend to underestimate the true mass ($M = 500$) of the total cluster. This happens even when the samples are not selected by mass, but by projected position in concentric rings around the cluster's center.

There are several techniques of data selection (see Aarseth & Saslaw, 1972) which help to minimize the bias in virial mass determinations. Generally these are difficult to apply in practice, requiring more information than is easily available. If these techniques are not applied, the net result is that the total cluster mass derived from (41.12) is systematically too small when heavy (luminous) members are included preferentially. The actual underestimate depends on density contrast, as well as the mass segregation and sample population, but can easily amount to a factor ~ 2. The alternative procedure of selecting samples from rings around the inner cluster region gives qualitatively similar results. Reliable mass derivations require many halo members to be included in both the position and velocity samples. Alternatively, a good rule of thumb would be to double the observed estimate if these corrections are not included.

At this stage it is natural to ask whether other methods might estimate the mass of clusters better than the virial theorem. Several methods have been proposed, especially for cases when there is a dark mass background (see Section 51 for further discussion), in addition to the observable point masses. These methods have in common a dependence on some model of the presumed density and velocity distributions of the cluster. Properties of these models may be postulated to take on their maximum likelihood values. This requirement relates the total mass, for example, to the galaxies' velocities and positions, as well as to structural indices of the cluster which are set by the model (see Smith, 1982). This added information often (but not always) improves the fit with N-body simulations. The method is rather strongly model dependent, however, and, in contrast with the virial theorem, has less relation to the basic physics of the cluster. It works best when you already know pretty much what you are looking for. In any event, these methods still only involve factors ~ 2.

The observed discrepancies, often of one or two orders of magnitude, between the dynamical mass and the luminosity mass in galaxy clusters are not likely to be reconciled by equilibrium dynamical models. Either there is dominant dark matter in or between galaxies, or the clusters are unbound and the virial theorem does not apply. Models of non-equilibrium clusters (Aarseth & Saslaw, 1972) lead to several methods of distinguishing between these two hypotheses:

1. After several crossing times, binary pairs can form in bound groups. Very few binaries form in unbound, expanding groups (although they could be present initially or occasionally form temporarily while two galaxies are merging).

2. Bound clusters which start with uniform density tend to become in-homogeneous and form a core–halo type of structure after several crossing times. Expanding clusters tend to remain approximately uniform.

3. Close encounters which promote equipartition of energy are much less important in expanding systems than in bound systems. Initial conditions being equal (a large 'if'!), the expanding clusters should therefore show less positional segregation of galaxies according to their mass (luminosity).

4. If significant dark matter exists in clusters its dynamical interaction will influence the position and velocity distributions of galaxies. Detailed models can show the departures from point interactions. Great scope exists for investigating various possibilities in detail.

42

Observed dynamical properties of clusters

I like reality. It tastes of bread.
Jean Anouilh

What is actual is actual only for one time and only
for one place.
T.S. Eliot

The purpose of this section is to give a very brief guide to the major dynamical properties of some real astronomical systems. It is not meant to be a review of the latest observations, for these change almost daily and often a long time must pass before they can be put into proper perspective. Specific astronomical observations, for understandable reasons, are seldom repeated. Often the pressures of time on large telescopes, or astronomical careers, are too great. The result is that, unlike laboratory physics experiments, it is difficult to gauge the real uncertainties (in contrast to the formal error limits) which surround particular observations. Another feature of astronomical observations is that a good deal of theory is usually needed to make the observation itself. This is because the systems we observe are complicated and seldom yield a basic physical quantity in a straightforward manner. The measurements we want must usually be strained through a network of interpretation.

Steady advances in observing technology mean that, at any given time, the most exciting frontier observations will often be near the limits of available instruments. Frequently these observations will just be able to rise above the 'noise level', and astronomers then speak of 'a two- or three-sigma effect', sigma being the standard error added by noise. The pressure of curiosity, and the sociology of modern science, discourage waiting until new technology is overdeveloped before using it. This is why new results, having been announced and overinterpreted, sometimes disappear. It would be an interesting study to examine how the fraction of published results, later found to be incorrect, is related to other aspects of the development of science.

The results of this Section seem to be well established, except when specifically mentioned as being uncertain. We describe three major types of roughly spherical clusters: clusters of galaxies, galactic nuclei, and globular clusters. The references are not meant to be comprehensive, just representative. They may be used to trace the literature either backward or forward (using the Science Citation Index) in time.

42.1. Clusters of galaxies

Clusters of galaxies are the largest structures known to us, apart from the Universe itself. Sometimes several clusters appear to interact together, forming a larger, more irregular, assembly. These are called superclusters, but this is mainly a semantic matter: they are just clusters on a larger scale (with subclusters!). The simplest useful way to classify clusters is by the number of galaxies in them. A quantitative form of this classification, suggested by Abell and often used, is their 'richness'. This is the number of galaxies whose magnitude lies in the interval between the magnitude, m_3, of the third brightest galaxy (in case the first brightest is peculiar) and $m_3 + 3$ (see Bahcall, 1977 & 1980, for general reviews). The number of galaxies in clusters can range from a few, in which case they are often called groups, to a few thousand.

Radii of clusters range from several hundred kiloparsecs to 2–3 megaparsecs. Typically, ~ 1 Mpc is representative of a rich cluster. The extent of the largest clusters, such as those in the Coma and Corona Borealis constellations, is rather uncertain since we do not yet know how the cluster blends into the background field of galaxies.

The average peculiar radial velocity within a cluster may range from 200 or 300 km s^{-1} to about 1500 km s^{-1}. In Coma, for example, it is $\sim 10^3$ km s^{-1}. If the velocities are isotropic the average spatial velocity would be a factor $\sqrt{3}$ greater. Small groups typically have peculiar radial velocities of a couple of hundred kilometers per second.

Optical luminosities for the richest clusters are very difficult to measure, but typical estimates are in the range $L \approx 1 - 5 \times 10^{12} L_\odot$. Consequently, estimates of their total mass, using the virial theorem, give ratios M_{VT}/L between about several hundred and a thousand. These are considerably greater than the M/L ratios of individual galaxies – thus the 'missing mass' problem for clusters (see the excellent review by Faber & Gallagher, 1979).

Several more specific types of observations are especially interesting from a dynamical point of view. The first is the detailed density distribution within clusters. The richest clusters give fairly good fits to truncated isothermal and Fokker–Planck models, as mentioned in Section 42. The Coma cluster is a favorite example (e.g., Quintana, 1979), although the southern cluster CA 0340–538, which has about half the mass of Coma, has also been studied extensively (Quintana & Havlen, 1979). In these two cases there is evidence for luminosity segregation, which probably implies mass segregation (see Section 46). In CA 0340–538, for example, the observed core radius is a function of the magnitude limit of the galaxies used to determine it. As the magnitude limit dims from $m \approx 17$ to $m \approx 19.4$, the core radius increases from 0.2 to 0.45 Mpc. The Coma cluster shows a similar segregation for moderately bright galaxies. However, the bright tail of the luminosity distribution does not share this behavior (Capelato, *et al.*, 1980).

The density distribution of rich clusters has an intriguing property which is not, however, well established yet. There is evidence, at about the 2σ level, for a secondary peak in the projected number density in several clusters, including Coma (Dekel & Shaham, 1980). It seems to occur generally at about half the gravitational radius, $R_g = 2GM/3v_r^2$ with v_r the radial velocity dispersion. If this is a real effect, it may be a fossil remnant of conditions during the formation of rich clusters.

Rich clusters are not always spherical, and those of lower mass tend to be more irregular (e.g., the Pavo cluster in the frontispiece). Some, such as A 2029 are flattened but not rotating (Dressler, 1981). This cluster, with $v_r \approx 1500 \, \mathrm{km \, s^{-1}}$, and $M_{VT}/L \approx 500$, is dominated by a central cD galaxy whose major axis is aligned with that of the cluster. The amount of flattening would require a rotational velocity about equal to its radial velocity dispersion of $1500 \, \mathrm{km \, s^{-1}}$, but this is not observed. Thus the flattening may be caused by anisotropic velocities and positions which remain from the cluster's formation, whether at early times or in a more recent merger.

Some rich clusters are so irregular that they are better represented as having two centers. These centers often show up in X-ray, as well as in optical photographs; we may really be seeing two clusters in the process of merging. Abell 98 is an example (Beers, *et al.*, 1982). It has two clumps, each about 0.7 Mpc in diameter and each with a radial velocity dispersion $\approx 1500 \, \mathrm{km \, s^{-1}}$. The ratio M_{VT}/L is about 600 in one clump and 430 in the other (comparable with the value of about 650 for Coma). Experiments with N-body computations are a useful guide to understanding these situations, even though they do not give a unique result.

Within rich clusters, there are often small subgroups of galaxies. Sometimes, as in Coma, camp followers swarm around one or two massive galaxies near the center. Other clusters seem to contain a remarkably high proportion of binary galaxies (Struble & Rood, 1981). These pairs are connected by a bridge or common luminous envelope. Typically such clusters contain about a dozen pairs, which comprise about 10–20% of all the member galaxies.

Total masses of clusters, we have seen, can be estimated from the virial theorem or from the luminosities of their individual galaxies. There is also a third method which can be applied to some rich clusters. Smaller groups move around these rich clusters. The fact that they have not been tidally destroyed puts an upper limit on the mass of the rich cluster, depending on assumptions about the orbit. Small groups around the Coma cluster, for example (Hartwick, 1981), limit its mass to $M_{Coma} \lesssim 6 \times 10^{16} M_\odot$ if the tidal limit is applied to the present position of the groups. However, if the groups are moving away from Coma and the tidal limit is applied to a closest approach at the edge of Coma, the mass is reduced to $M_{Coma} \lesssim 5 \times 10^{15} M_\odot$. This compares with estimates of 1.8 and 2.4 and 5.4×10^{15} for M_{Coma} by other astronomers (Hartwick, 1981). This range of independent estimates probably gives a better idea of the uncertainty than the formal errors attached to each estimate.

Velocity profiles within rich clusters are also very important, although they are harder to observe than density profiles and much less is known about them. Fitting the density profile to a theoretical model is far from unique, but velocity distributions are fairly good discriminators among these models. In many clusters there is a tendency for the observed velocity dispersion to decrease toward the center (Struble, 1979). This probably implies an anisotropic velocity distribution in the center, perhaps influenced by one or two massive galaxies, or by the initial conditions if the system relaxed gently as it formed.

The luminosity function of rich clusters also has dynamical implications apart from the missing mass problem. The absolute magnitudes of the brightest member of each rich cluster show a remarkably small dispersion, $\sigma \approx 0^m.4$. We do not know if this is a sign that the luminosity functions of bright clusters are similar, or if it says that the physical processes which produced just the brightest galaxies were similar (Godwin & Peach, 1979). Is it reasonable that the formation, accretion, cannabalism, etc., of galaxies led to the same luminosity for the brightest galaxies everywhere? Is it reasonable that there should be a universal luminosity function for clusters preselected to be rich?

Graphs of the cumulative logarithmic luminosity function, $\log N(m \leq V)$ versus limiting magnitude V, tend to show curves with a knee or bend (Bucknell, *et al.*, 1979). The simplest representation is two straight lines, joining at some magnitude V_0. The relation of V_0 to other properties of the clusters is an important, unsettled, question.

Clusters vary in their relative abundance of different types of galaxies, some are rich in spirals, others contain mainly ellipticals. A specially interesting group are the cD clusters – they are dense and dominated by one or two giant elliptical galaxies. The galaxy type in a cluster can also vary with luminosity. In Coma, for example (Thompson & Gregory, 1980), the brightest galaxies are ellipticals and those of intermediate brightness are usually SO's. Moreover, in the core of Coma there are few faint spirals and irregulars compared to the local galaxy luminosity function. One possible reason is that ablation and stripping have removed gas and decreased star formation in central spirals. Another possibility is that the central spirals have merged. Perhaps detailed study of their structure and stellar populations will provide a definitive answer.

Small groups of galaxies show some of the overall properties of rich clusters. Many groups also show a large discrepancy between their virial mass and their luminosity mass. This discrepancy seems to be correlated with the crossing time, $T_c \equiv R/V$, of the group (Field & Saslaw, 1971; Klinkhamer, 1980). The crossing times range from about 10^9 to 5×10^{10} yr, and groups with $T_c \lesssim 2 \times 10^9$ have the smallest mass discrepancy. Also there is a strong tendency for short T_c groups to contain mostly elliptical galaxies.

Very small, compact groups with only several galaxies may have very low M/L ratios. Two groups with four galaxies each have $M/L \approx 24$ and 4 (Rose & Graham, 1979). These values are rather uncertain since they depend on corrections for galaxy inclination, internal absorption, dynamical fluctuations, velocity errors (which depend on the position of the spectrograph slit on the galaxy), etc. Marvelously tiny groups of blue compact galaxies also exist. One example (Vorontsov-Vel'yaminov *et al.*, 1980) has six blue interacting galaxies. The total mass is about $10^{10} M_\odot$ and $M/L \approx 0.3$, implying that hot young stars dominate. The radial velocity dispersion of this group is about $44 \, \mathrm{km \, s^{-1}}$. Its total diameter is $15 \, \mathrm{kpc}$, only about one-fourth the size of our own Milky Way.

42.2. Galactic nuclei

We dance around in a ring and suppose,
But the secret sits in the middle and knows.

Robert Frost

Galactic nuclei are one of the most fascinating phenomena of the astronomical zoo. From these condensed regions, barely a few parsecs in diameter, issue vast quantities of radiation covering the entire spectrum. Brighter than a trillion suns, they shine forth as beacons across the Universe. Along with radiation they eject enormous clouds of gas, often at relativistic speeds.

Like Proteus, active galactic nuclei take many forms. A host of physical processes, prominent to varying degrees, produce the quasars, BL Lac objects, radio and emission line galaxies. All these, and others, are likely to be manifestations of galactic nuclei in different stages of activity.

Most of our knowledge of galactic nuclei comes from the radiation of hot gas inside them. The total luminosity produced by the most luminous galaxies is $\sim 10^{46} \, \mathrm{erg \, s^{-1}}$, mostly coming from its nucleus. This is the equivalent rest mass power of a couple of tenths of a solar mass per year. Quasars can have total luminosities 100 times greater. This power is roughly equally divided among the infra-red, optical, X-ray and γ-ray parts of the spectrum; the relative proportion depending on the particular objects. Diffuse optical luminosity, of low surface brightness, surrounds some quasars, especially those which are relatively nearby. Thus quasars are probably embedded in galaxies. Any galaxy around a quasar is bound to be peculiarly interesting.

The energy emerging from galactic nuclei has large scale structure which may reflect structure within the nucleus. The radio emission shows this most prominently. In many active galaxies, particularly giant ellipticals, it comes from two lobes rather symmetrically placed on either side but many galactic diameters away from the nucleus. These lobes, typically tens of kiloparsecs in diameter, have emission peaks, called 'hot spots', at their edges. Often these hot spots, which can be two or three megaparsecs apart, are aligned with the central nucleus to within several degrees. Somehow the nucleus manages to pump energy into them across these great distances. High resolution measurements, made with radio interferometers linked across continents, sometimes show structure in the nucleus on the scale of parsecs or less which is aligned with the outer hot spots (Jones, *et al.*, 1981). Bridges of radio emission are sometimes found to connect the nucleus with the distant structure, although these bridges are often seen on just one side of the nucleus. This may indicate an alternate feeding of the lobes, but the basic reason for this asymmetry within an otherwise fairly symmetric structure is mysterious.

Optical emission is also structured on scales of tens or hundreds of parsecs. Often Seyfert and other galaxies with unusually great ultraviolet emission have double or multiple regions of intense optical emission (e.g., Petrosyan, *et al.* 1979). These may be like the relatively common regions of ionized gas associated with hot young stars. Whether this star formation itself was initiated by processes within the nucleus is mysterious. On smaller scales the effects of atmospheric seeing usually smooth out structure in ground based observations (e.g., Faber, 1980). Some information is still available for bright nuclei through speckle measurements. These are made over timescales short compared with changes in seeing, and then combined. High resolution optical observations are usually best made with space telescopes.

Optical spectra, even at relatively low spatial resolution, provide important information about the gas in galactic nuclei. Abundances of most elements are not wildly different from those in the solar neighborhood. Ionization and heating processes, however, vary enormously. Generally the spectra are more consistent with a power law ionization source flux, $\propto v^{-\alpha}$, than with a thermal ionization source. Broad Balmer and other lines, with widths of several thousand kilometers per second, dominate some emission line galaxies such as type 1 Seyferts. These have narrow forbidden optical lines and are also strong X-ray emitters. Their luminosity may vary rapidly, sometimes by as much as $\sim 30\%$ from night to night (Geller *et al.*, 1979). This implies active regions on scales of ~ 1 light day $\approx 10^{16}$ cm $\approx 3 \times 10^{-3}$ pc. Other active galaxies, such as type 2 Seyferts, have narrower (~ 500–1000 km s^{-1}) Balmer lines and broader forbidden lines of about the same width as the allowed lines. The Balmer lines come from gas clouds close to the nucleus with densities $\sim 10^7$–10^{11} cm^{-3}, while the forbidden lines arise farther away where gas densities are more typically $\sim 10^3$. So in type 1 Seyferts the innermost nuclear region is more prominent whereas in type 2 Seyferts the outer part of the nucleus emits relatively more. The detailed structure of clouds and winds in galactic nuclei is much more mysterious than the weather on earth.

Unfortunately, even less is certain about the stellar dynamical components of galactic nuclei. Their starlight is usually swamped by the non-thermal emission of the gas. Absorption lines due to stars are typically broadened by velocity dispersions of ~ 200 km s^{-1} in elliptical and S0 galaxies (Tonry & Davis, 1981), but because seeing smears the images this is an average over the bulk of the galaxy. It leads to a mass–luminosity ratio in the optical of about 25 (for $H = 100$ km s^{-1} Mpc^{-1}), which varies over a factor ~ 2 among these galaxies. This ratio, however, is very model dependent, as shown by studies of the nucleus of M31, the nearest large galaxy (Tremaine & Ostriker, 1982). Here the average velocity dispersion of stars along the line-of-sight through the nucleus is about 180 km s^{-1}. But the M/L ratio is uncertain by a factor ~ 50, depending on the degree of anisotropy in the central velocities.

Averaging along the line-of-sight naturally gives heavy weight to the large number of low velocity stars relatively far from the center. If the stellar density increases sharply toward the center, we would expect from the virial theorem and

simple polytropic models to find stellar velocities of many hundreds, even
thousands, of kilometers per second. Some galactic nuclei have such steep central
luminosity spikes that we can peer more deeply into their nucleus. The radio galaxy
M87 is one of the better-studied more extreme examples. This giant elliptical has a
linear chain of optical hot spots proceeding like a jet from its nucleus. The nucleus
itself (Dressler, 1980) seems to be a cluster of stars emitting relatively little ($\lesssim 20\%$)
non-thermal light. Only an upper limit to its radius is known, but speckle
interferometry suggests it may be $\lesssim 2$ pc. The central velocity dispersion of stars is $\lesssim$
600 km s^{-1}, and probably closer to 350 km s^{-1}. It may steepen as $r \to 0$, unlike the
constant central velocity dispersion in isothermal or simple Fokker–Planck models
(Section 41). If such a steepening becomes well established, it could indicate the
presence of a massive object loading the nucleus, or of highly anisotropic central
velocities. At present the observations are not sufficient to determine a unique
model; this is a question of overriding importance.

Clearly, the central question about galactic nuclei is their basic energy source.
Several suggestions are dense stellar systems, supermassive stars, and accretion onto
supermassive black holes. These processes are not exclusive, and indeed there is a
natural progression along this sequence. At each stage, parts of the system become
more compact, and the energy which is released generates the phenomena seen in
different galactic nuclei, or perhaps in one nucleus as it evolves. With more detailed
observations, especially from space telescopes, it should become possible to choose
among the large number of models (see Saslaw, 1974) proposed to explain these
mysteries.

42.3. Globular clusters

The smallest gravitating systems which can be considered, to a first approximation,
as isolated and spherical are the globular clusters. These groups of $\sim 10^5$–10^6 stars
move together within galaxies. Those in our own galaxy are sufficiently close that
many of their detailed properties can be observed. An excellent review (Hanes &
Madore, 1980) discusses our knowledge of globular clusters at the end of the 1970s.
Here I will merely summarize some of their important dynamical properties.

We see about 150 globular clusters in our own galaxy. Some other clusters are
probably obscured, either by dust or by dense star fields. The size of a globular
cluster depends rather strongly on just how it is defined. For most clusters the
surface density fits simple isotropic pressure models fairly well, with small
discrepancies possibly caused by anisotropic velocities and mass (luminosity)
segregation. These models have three basic parameters, a core radius r_c, a tidal
limiting radius r_t and a scaling constant c; they give a surface density distribution of
the form

$$\sigma(r) = c \left\{ \left[1 + \left(\frac{r}{r_c} \right)^2 \right]^{-1/2} - \left[1 + \left(\frac{r_t}{r_c} \right)^2 \right]^{-1/2} \right\}^2 . \qquad (42.1)$$

Core radii typically range between about 2–15 pc, while tidal radii may be about a factor of ten larger. In practice, the core radius is rather model dependent, while the tidal radius is difficult to define precisely. It is affected by the uncertain background level of stars, the irregular galactic gravitational field, and the difficult determination of which stars are actually leaving the cluster. Thus using ~ 10 pc for the characteristic 'size' of a globular cluster can be no more than a rough estimate, but it sets a reasonable scale for the phenomenon.

Observational estimates of the average mass-to-light ratios range from about 1–3 for different clusters. These lead to total cluster masses in the range $10^5–10^6 \, M_\odot$. Typical velocity dispersions of stars in the clusters are between 7 and 20 km s^{-1}. The velocity needed to escape from the center of these clusters again depends on the density distribution of the model, but values between about 10 and 30 km s^{-1} seem reasonable. The main point is that these systems are very weakly bound compared with galactic nuclei or clusters of galaxies.

Relaxation times within different clusters, due to the cumulative effects of two-body interactions (Equation (2.9)), vary from $\sim 5 \times 10^6 – 10^{10}$yr at the center of the cluster. Ages for the clusters, judged from stellar evolutionary models, are usually about 10^{10} yr. So it is not surprising to find the clusters relaxed, especially since early violent relaxation or subsequent relaxation caused by irregularities in the galactic force could also have contributed on shorter timescales.

The origin of globular clusters is one of the major mysteries surrounding them. Did they form by shocks and fragmentation during the early collapse of gaseous protogalaxies? Or did they form shortly after decoupling in the early Universe at a redshift $z \approx 10^3$ and then cluster into more massive structures, eventually forming the galaxies? An observational clue may be the tendency for brighter galaxies to contain more globular clusters, and for ellipticals to contain more than spiral bulges (Hanes, 1977). Other important clues may be found in observations of their chemical composition and evolution.

Some very strange globular clusters are seen at the fringes of galaxies, and possibly between galaxies. The globular cluster Palomar 5 (Sandage & Hartwick, 1977) has a mass of about $10^4 M_\odot$ and is so loosely bound that it probably never came closer than about 15 kpc to the galactic center, at least in its present form. It has a peculiar chemical composition which may provide clues to its formation and the evolution of the outer part of our galaxy. Further afield, globular clusters with masses of $\sim 10^4 M_\odot$ and diameters of ~ 80 pc are found at distances of ~ 100 kpc (Burbidge & Sandage, 1958). Another roughly spherical example (Madore & Arp, 1979) may be about 300 kpc distant, in which case it would have a diameter of about 150 pc.

All these distant globular clusters are characterized by relatively small masses and large diameters. Their integrated surface brightness is low and shows little central concentration, unlike the nearby clusters. Are these distant objects barely bound remnants which were ripped untimely from their parent galaxy during early merging or tidal interactions? Or, are they the isolated remains of an earlier generation of gravitational clustering which were never quite absorbed into galaxies? This is the basic question.

An important clue to answering this question may lie in another class of objects which bridge the properties of distant globular clusters and normal galaxies. These are the dwarf elliptical galaxies (Hodge, 1966). Many are found in our local group of galaxies, within a megaparsec. But they are hard to find and their number is uncertain. Their low surface brightness makes it difficult to distinguish them from an irregular foreground of local stars. Dwarf ellipticals have ratios of minor to major axes ranging from 1 to about 0.5, with 0.7 a fairly typical value. Their distances are between about 80–250 kpc. From the mass–luminosity ratios of their stars they appear to have masses of 10^5–$10^7 M_{\odot}$ and central densities of about 10^{-4}–$5 \times 10^{-3} M_{\odot} \text{pc}^{-3}$. These are very low for galaxies. Their tidal radii, in the field of our galaxy, range from 0.5–7 kpc. These properties lead to timescales for dynamical relaxation by gentle two-body encounters of about 10^{12}–10^{14} yr, much longer than the age of the Universe. This makes them clear candidates for the effects of violent relaxation, and they should be examined carefully from that point of view.

With this brief summary of the sizes, shapes, masses, and distributions of spherical gravitating systems, we turn to the physical processes which govern their internal evolution.

43

Gravithermal instabilities

Hitherto in Part III, we have considered the properties of stationary relaxed stellar systems. We know such systems cannot be realistic because they are not stable. Their most obvious instability is the evaporation of high energy stars. These carry net energy away from the system; remaining stars swarm inside a deeper gravitational well and cluster more strongly together.

Even if a cluster did not evaporate, something must clearly be wrong with the assumption that it could be stable. The fact that, in a Newtonian system, more and more kinetic energy can be produced by continued gravitational concentration suggests that instability may occur wholly internally through a redistribution of the stars. In a spherical system we would expect the center to become denser, losing energy by ejecting high energy stars into the outer parts, and forming a steeper gravitational well.

A simple physical argument suggests how this instability arises. Systems satisfying the virial theorem (9.30) have $2T + W = 0$ or since the total energy $E = T + W$, they satisfy $E - - T$. Although systems, or parts of systems, may not satisfy the virial theorem exactly, they are close enough that this relation describes the essential part of the physics needed to understand gravithermal instability. The kinetic energy of random motions T is proportional to the temperature. (A useful confusion of nomenclature?) Thus decreasing the energy of a gravitationally bound system, increases its temperature. These systems have negative specific heat. We should therefore not be too surprised if they seem to behave rather strangely compared with ordinary gases.

Thermal energy tends to diffuse from hotter regions to cooler regions as dynamical friction increases the velocities of the more slowly moving stars at the expense of the faster ones. So kinetic energy flows along the thermal gradient. In ordinary gases this flow leads to stability as it diminishes the thermal gradient. In gravitating systems it does not lead to stability. The regions which lose heat have negative specific heat, so they grow hotter. The thermal gradient increases. More heat flows away from the contracting region and it gets hotter still. The result is thermal runaway.

In a spherical cluster, heat flows from the hot center to the cooler outer regions. To reverse this thermal runaway it would be necessary to heat the outer parts until they become hotter than the core. Two effects prevent this from happening in a freely evolving cluster. First, the dynamical relaxation time $\tau_R \propto v^3/\rho$ from (2.9). As long as v^3/ρ increases with radius, τ_R will be longer in the outer part than in the core. This is the case, for example, in an isothermal sphere where $v^3 \approx$ constant and $\rho \propto r^{-2}$ in the outer parts, from (40.21). Therefore high energy stars become less thermalized as they move into the outer parts. Instead of increasing the temperature of these outer regions they increase the anisotropy of the velocity distribution there.

The second effect which inhibits heating of the outer part is its high specific heat. A large change of energy only produces a small change of temperature. The specific heat varies throughout the cluster because some parts of it are bound more strongly than others. The strength of gravitational binding in a spherical shell of stars can be judged from its contribution to the total gravitational energy (even though this energy cannot be localized in a particular part of the cluster). If a spherical shell of mass $dM(r)$ is brought from infinity onto a mass $M(r)$ already present at r, it acquires a potential energy with respect to infinity of $-GM(r)\,dM(r)/r$. Building up the entire sphere this way, we see that its total gravitational energy is

$$W = -G \int_0^R \frac{M(r)\,dM(r)}{r}. \qquad (43.1)$$

Outer shells of a given mass dM will contribute relatively less to the binding energy if $M(r)$ increases more slowly than r. For an isothermal sphere $M(r) \propto r$ in the outer parts, so this is the limiting case. For a real finite system ρ must eventually decrease faster than r^{-2} and the outer parts are more weakly bound than the inner parts. Being more weakly bound, a given addition of heat causes it to expand more than it would expand the same mass in the core. Greater expansion leads to greater adiabatic cooling. The net result is that a given heat input raises the temperature of an outer shell less than it would an inner shell of the same mass. Thus the outer shells have a greater specific heat. This acts to increase the thermal gradient, again promoting instability.

Gravithermal instability leads to a redistribution of the cluster both in configuration space and in velocity space. A central core forms, becoming smaller and denser as it releases energy to the outer parts. Around the core, a halo grows. The timescale for significant redistribution cannot be found from simple instability theory; a kinetic theory is needed. Nevertheless, since the processes is essentially controlled by diffusion, we expect τ_R to play the dominant role. In Section 42 we saw that this can be fairly long for globular clusters. The evolution therefore is not catastrophic (i.e., governed by the dynamical crossing timescale), although this process has sometimes been called the 'gravothermal catastrophe'. Eventually so few stars remain in the core that the diffusion description no longer applies and individual orbits must be followed.

These physical arguments apply to a wide range of realistic clusters, and have been explored in several types of N-body simulations. The next two sections describe the dynamical evolution in more detail. The rest of this section discusses the stability of one of the models of Section 40 – the isothermal sphere. These have been studied extensively because, as already mentioned, they come fairly close to describing the cores of actual systems. Therefore it is interesting to analyze their stability, as an illustrative example.

A pure isothermal sphere stretches to infinity and has an infinite mass, so it is not really suitable for a stability analysis. To keep the essential physics, but provide a suitable mathematical model, we truncate the cluster by surrounding it with a spherical wall. We can (mentally) construct various types of wall. One type is completely rigid and impenetrable; nothing gets through; all stars are reflected at the boundary. This describes an isolated cluster of fixed mass, energy, and volume – a member of a microcanonical ensemble. Less restrictive walls are available. They may be thermally conducting, in equilibrium with a thermal reservoir of constant temperature. In this case energy is exchanged. Such clusters form a canonical ensemble. If stars can also be exchanged with a reservoir, we have a grand canonical ensemble. Another alternative is a rubber wall, in pressure equilibrium with a reservoir. Each of the models describes a different physical situation and their stability properties can be quite different.

Consider, as a first example, whether the equilibrium of a truncated isothermal sphere is stable. The sphere is surrounded by a rubber wall transmitting the pressure which would be produced by the rest of the matter if it were there. It is also in temperature equilibrium with a thermal reservoir. Although this example is more suitable to the internal instability and fragmentation of an isothermal gas cloud which forms a star, than to an isolated globular cluster, it illustrates the main point simply. The point is that global thermodynamics determines stability (Bonnor, 1956, 1958).

Global gravitational thermodynamics is different from the local gravitational thermodynamics studied in Part II. There, an equation of state, modified by gravity, held over local averages of an infinite system. Here, a 'perfect gas' equation of state

$$P = \frac{kT\rho}{m} \tag{43.2}$$

is assumed to hold locally at each point in equilibrium. This is reasonable because near equilibrium the virial theorem holds to a good approximation and the distribution is nearly Maxwellian. Correlations which would modify (43.2) are negligible. At the boundary of a finite volume a pressure must be exerted to keep the equilibrium. This pressure depends on the total gravitational energy of the volume. It can be added to the equation of state at the boundary. This produces a global equation of state which depends on the size and shape of the volume.

To see this we start with an identity, valid at the boundary (subscript zero)

$$P_0 V_0 - N_0 k T_0 = P_0 \int_0^{V_0} dV - \frac{kT_0}{m} \int_0^{V_0} \rho \, dV. \tag{43.3}$$

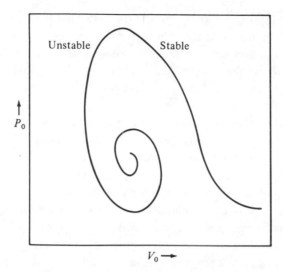

Fig. 45. Schematic illustration of $P_0(V_0)$ for a sequence of isothermal spheres of given N_0 and T_0 but different central densities.

Integrating by parts and substituting (43.2) gives

$$P_0 V_0 - N_0 k T_0 = \frac{kT_0}{m} \int V \mathrm{d}\rho = \frac{4}{3}\pi \frac{kT_0}{m} \int_0^{r_0} r^3 \frac{\mathrm{d}\rho}{\mathrm{d}r} \mathrm{d}r. \qquad (43.4)$$

The last equality supposes spherical symmetry for simplicity. If gravity were not important the equilibrium state would be one of constant density, there would be no boundaries, and the perfect gas law would apply everywhere. That would bring us back to the 'Jeans swindle' described in Section 21. Since finite equilibrium systems have density gradients, the global equation of state (43.4) departs from the perfect gas relation.

To complete the description, we obtain $\mathrm{d}\rho/\mathrm{d}r$ from the hydrostatic equation (40.1) coupled with Poisson's equation (40.2). For the isothermal sphere, of either stars or gas, these lead to (40.18). In this case no closed form for the global equation of state $P_0(V_0)$ seems to exist and it must be found numerically.

Having found $P_0(V_0)$, the nature of the stability follows by examining the derivative $\partial P_0/\partial V_0)_{N_0, T_0}$. If this quantity is negative, then a fluctuation which increases the volume will decrease the pressure, and one which decreases volume increases pressure, so the system is stable. On the other hand, if $\partial P_0/\partial V_0)_{N_0, T_0} > 0$, pushing the wall in causes the internal pressure to decrease so the outside pressure pushes the wall in farther and produces an unstable collapse. At the point where the slope changes a Taylor series expansion shows that $\mathrm{d}P = \frac{1}{2}(\partial^2 P/\partial V^2)_{N_0, T_0}(\mathrm{d}V)^2 + \cdots$, so if the second derivative is negative there, the situation is unstable to contraction.

Figure 45 schematically illustrates the global equation of state for truncated isothermal spheres. This is a sequence of systems which each contain the same total number of particles N_0, at temperature T_0, but have different central densities. (It is not one sphere truncated at different radii.) If N_0 particles are put into a large volume, then their mutual gravitational binding becomes weaker and $P_0 V_0 \to N_0 k T_0$. These isothermal spheres are stable. As the N_0 particles are put into smaller and smaller volumes their gravitational energy becomes more and more important until at the peak of the curve instability sets in and systems become unstable. Numerical calculations (Bonnor, 1956) show this instability sets in at a critical radius

$$r_c = 1.8 \left(\frac{kT}{mG\rho_c} \right)^{1/2}, \tag{43.5}$$

where ρ_c is the central density. For comparison, this is twice the Jeans length in a sphere of uniform density. The ratio of mass to radius at the critical point is $M(r_c)/r_c = 2.4 kT/mG$. Spheres along the sequence between the peak of the $P_0(V_0)$ relation and its singular point at the center of the spiral are larger than r_c, so they are internally unstable. Thus there is no stable equilibrium configuration beyond the peak.

Similarly, if one examines the global equation of state for a single truncated isothermal sphere, then one finds that at r_c the second derivative $(\partial^2 P/\partial V^2)_{N_0, T_0} < 0$. Thus such a sphere is unstable to general fluctuations if its size exceeds r_c. These are two aspects of the same physical phenomenon: gravitational energy has overwhelmed the thermal energy which can be generated for hydrostatic pressure support.

The idea of global thermodynamics can also be applied to thermally isolated systems of given mass, total energy and volume (Antonov, 1962; Lynden-Bell & Wood, 1968; Thirring, 1970; Horwitz & Katz, 1978). These are more closely related to isolated globular clusters, members of a microcanonical ensemble. The internal gravithermal stability of a globular cluster is determined by its global thermodynamics.

Consider now, as our second example, an isothermal distribution isolated inside an impenetrable, non-conducting spherical shell which reflects those stars that hit it. All stars have the same mass. Under what conditions can such systems be stable? Again, we just describe the basic physical principles since the analysis must be carried out numerically. The stability of a finite isolated gravitating system depends on the relation between its total energy and its total entropy, both global quantities. Systems of the same total energy may have different entropies depending on their size and internal distribution. If the entropy has a local maximum for a given energy, then that configuration is stable to small fluctuations (but possibly metastable to large changes). If the entropy does not have a local maximum, then the system of a given energy can redistribute itself internally to increase its entropy, and the situation is unstable.

Thus we need to calculate these thermodynamic quantities for an isothermal system in its isolating box. The total energy may be found using the virial theorem of Section 9. Now, however, we must include the surface pressure at the boundary. Going back to equation (9.7) we see that the surface pressure term, which was previously dropped, is just $3P_0 V_0$. The scalar virial theorem now takes the form

$$2\mathcal{T} + W = 3P_0 V_0, \tag{43.6}$$

where $\mathcal{T}$ is now the total kinetic energy and W the total potential energy. The total energy $E = \mathcal{T} + W$. From equipartition, $\mathcal{T} = 3NkT/2$ where T is the temperature. So the total energy is

$$E = 3P_0 V_0 - \mathcal{T} = 3P_0 V_0 - \frac{3}{2}NkT, \tag{43.7}$$

and the potential energy is

$$W = 3P_0 V_0 - 3NkT. \tag{43.8}$$

Notice the similarity with (43.4) in that the global thermodynamics would reduce to that of a local perfect gas if gravity were absent. The entropy is found by integrating

$$S = -k \int f \ln f \, \mathrm{d}^3 \mathbf{x} \mathrm{d}^3 \mathbf{v} \tag{43.9}$$

over the entire cluster. The single-particle distribution f is the Maxwellian including the gravitational potential energy. The isothermal sphere relations of Section 40 can then be used to express E, M, and S as functions of the radius R of the system for any equilibrium isothermal distribution. In this way it is found that local entropy maxima do not exist for a given total energy if the radius of the spherical box is

$$R > 0.335 \frac{GM^2}{(-E)}. \tag{43.10}$$

If the radius of the box is too large the density contrast between the center and the edge becomes too great. Then the outer stars cannot absorb the energy transported from the center and the system becomes unstable. It tends toward a more and more pronounced core–halo structure. The critical density ratio corresponding to (43.10) is $\rho_{\text{center}}/\rho_0 = 709$, originally found by Antonov. At this density contrast, or greater, the center essentially does not know that the outer part of the cluster exists. Replacing the halo by a vacuum would not change the development of the center significantly. The thermodynamic instability arguments mentioned here can also be formulated more rigorously in terms of Poincaré's theory of linear series of equilibria (Lynden-Bell & Wood, 1968; Katz, 1978; 1979).

Although these general thermodynamic arguments show instability and suggest core–halo evolution, they do not provide any detailed information about further dynamical development. In principle, we might think it possible that a minor change in the density or velocity distribution could quench the instability. After all, we have only examined the stability of one very restricted equilibrium state – that of isothermality. Other quite different sequences of equilibrium states (polytropes would be a simple example) could also exist. If one sequence is unstable the system might latch onto another stable sequence. On the other hand, it might never become stable again. The core might itself develop a halo and a smaller core, which in turn develops a halo and still smaller core, etc.

Some answers to these questions are provided by the interplay between simple theoretical models and computer N-body experiments on globular clusters of stars. We turn to these in the next section.

44

Self-similar transport

For my part, I travel not to go anywhere, but to go.
I travel for travel's sake. The great affair is to move.

R.L. Stevenson

Like Robert Louis Stevenson, the great affair of a star in a cluster is to move. When gravithermal instability sets in we want to understand how the distribution of stars evolves. Some of the earliest N-body simulations with 50 or more stars (e.g., Aarseth, 1963) already showed a clear tendency for stellar systems to form three distinct regions. At the center there is a nearly isothermal core. Its radius shrinks and its mass decreases as the system evolves. Its average density, however, increases. The stars which remain in the core bind more tightly together. Beyond the core is the halo. It contains most of the stars which diffuse from the core. Computer models of spherical systems which use the equations of stellar hydrodynamics (see Section 8) show that the halo develops a power law density distribution when all stars have the same mass (Larson, 1970). More detailed models, for isotropic velocities, which use the Fokker–Planck equation (Sections 4 and 5) confirm this power law structure (e.g., Cohn, 1980). The density in the halo $\rho \propto r^{-\alpha}$, where $2.1 \lesssim \alpha \lesssim 2.4$, depending on details of the model. At the outer edge of the halo an evaporation zone develops. Here the density departs from its power law and plummets to zero.

The Fokker–Planck models also show that this tripartite structure – core, halo, evaporation zone – develops as a result of gravithermal instability. Isolated clusters, in contrast to those in fixed boxes, are better characterized by the quantity

$$x \equiv 3\frac{[\phi(0) - E(r)]}{v_m^2(0)} = \frac{\phi(0) - E(r)}{kT} \tag{44.1}$$

than by the ratio of the central density to the edge density. Physically, x is the scaled energy $E(r)$ for a unit mass star at radius r. The gravitational potential and velocity dispersion at the center are $\phi(0)$ and $v_m^2(0)$. Thus the normalized Maxwellian distribution is just $f_M(x) = \exp(x)$. As the cluster evolves, $\phi(0)$ and $v_m^2(0)$ change. The scaled escape energy $x(E = 0) \equiv x_{esc}$, increases (negatively) roughly linearly from about -5 to about -10 over about ten relaxation times. During this period a sequence of nearly isothermal models described by $f_M(x)$ (see Section 40) represents the development quite well. When x_{esc} reaches about -10, however, a sudden redistribution occurs. In less than one relaxation timescale, x_{esc} increases (negatively) by about 50%. Then its increase markedly slows. This core–halo readjustment occurs at the value of x_{esc} where the isothermal sequence has an entropy maximum for a given total energy and mass. Near this value of x_{esc}, the corresponding Fokker–Planck models also show a significant total entropy increase to a new maximum after their structural transition.

We can now see why the density distribution in the halo follows a power law. The remaining isothermal core keeps shrinking, but the cluster develops along a different sequence than before the gravithermal instability. Before, it went along an isothermal sequence; now it follows a core–halo sequence. The core shrinks; it produces a halo; the new core shrinks; it produces a new halo, and so on. Throughout each step of this process there is only one fundamental length scale characterizing the physics. That is $L = GM^2/(-E)$ and this characterizes the entire core since M and E are its total mass and energy. The radius of the core is a multiple of L at any given time: $r_c = r_c(t) = a(t)L$. Since each part of the halo is produced by a previous shrinking of the core, this suggests that the halo should have the same structure throughout. At this stage, we must assume that the halo does not itself evolve rapidly compared to the core, and since τ_R is longer in the halo this seems reasonable. Thus there is no characteristic length scale within the halo and $\rho(\lambda r)/\rho(r)$ is independent of r. We saw in the beginning of Section 32 that this restriction requires $\rho(r)$ to be a power law. This type of structure is called self-similar, or homologous. Its scale may change but its functional form remains the same.

We continue to follow the implications of this similarity solution (Lynden-Bell & Eggleton, 1980) since it describes the computer models quite well, at least for isotropic velocities. (Effects of including anisotropy are an important area for further study.) Moreover, this solution also provides much insight into the development of the core.

In a similarity solution, all length scales can be made dimensionless by writing them in terms of the characteristic length, which we take to be the radius of the core

$$r_* = \frac{r}{r_c(t)}. \tag{44.2}$$

Other variables are also the product of a time dependent factor which describes their evolution and an r_* dependent factor which describes their unchanging spatial form. Thus the density in the halo is

$$\rho(r, t) = \rho_c(t)\rho_*(r_*), \tag{44.3}$$

where ρ_c is the density at the center of the halo, i.e., the edge of the core.

Outside the core's edge we make the idealized assumption that the relaxation time is so long that the halo does not know it is eating into the core. The density throughout the halo is then constant in time and from (44.2)–(44.3)

$$\dot{\rho}(r, t) = 0 = \dot{\rho}_c\rho_* - \frac{\dot{r}_c}{r_c}\rho_c r_*\rho'_*, \tag{44.4}$$

where the prime denotes differentiation with respect to r_*. Dividing through by

$\rho_* \rho_c \dot{r}_c / r_c$, we see that the independent variables separate, so each term must be a constant, denoted by $-\alpha$. Solving the simple differential equations which result shows that

$$\rho_* = A r_*^{-\alpha} \tag{44.5}$$

in the halo and

$$\rho_c \propto r_c^{-\alpha} \tag{44.6}$$

at the boundary. This last relation is independent of r, so it also holds in (44.3) throughout the halo.

Insofar as the core remains isothermal it retains its homology structure (different from that of the halo, see Section 40.1). Therefore particular properties of the core, as it evolves, are always the same dimensionless numbers times its evolving characteristic properties. For example, its mass is always proportional to $\rho_c r_c^3$, with the same constant of proportionality since its structure remains constant. Thus we have for the core

$$M_c \propto \rho_c r_c^3 \propto r_c^{3-\alpha}, \tag{44.7}$$

$$E_c \propto \frac{-GM_c^2}{r_c} \propto -r_c^{5-2\alpha} \propto -M_c^{(5-2\alpha)/(3-\alpha)}, \tag{44.8}$$

$$v_c^2 \propto \frac{GM_c}{r_c} \propto r_c^{2-\alpha}. \tag{44.9}$$

Already with these results we can make a reasonable guess for the value of α. Suppose the core evolves through a series of quasi-static configurations, each of which nearly satisfies the virial theorem. Evolution occurs because stars of higher total energy leave the core and, as described earlier, the core grows hotter. Therefore its velocity dispersion increases as it shrinks and from (44.9) the value of $\alpha > 2$. The limiting case $\alpha = 2$ would give an isothermal density gradient in the halo and energy could flow back into the core, but we know this is not the case. Equation (44.8) tells us that if the absolute value of the core's energy is to decrease, then $\alpha < 2.5$, so it seems to be confined to the range $2 < \alpha < 2.5$. If we were to argue, more speculatively, that a given decrease of r_c acts to maximize the fractional increases in both the energy E_c and the temperature $\propto v_c^2$, then we would put $\alpha \approx 2.25$, midway in its range. This agrees with the Fokker–Planck models to within a couple of percent.

To estimate the time dependence of $r_c(t)$ in this homology solution we recall that τ_R defines the characteristic timescale (Equation (2.9)). Fractional changes of all the quantities describing a homologous system occur on this same timescale. Writing this specifically for the core radius,

$$-\frac{1}{r_c}\frac{dr_c}{dt} \propto \frac{1}{\tau_R} \propto \frac{\rho_c}{v_c^3}. \tag{44.10}$$

Here we have used (2.9), neglecting the unimportant time dependence of $\log N$ as the core loses stars (an approximation which breaks down, along with all the other approximations in this regime, when $N \lesssim 10^2$). From (44.6)–(44.9) we see that using

(44.10) for fractional derivatives of the other physical quantities only changes the constant of proportionality. This is because, in a similarity solution all the dimensionless quantities like $\tau_R \dot{r}/r$, which do not depend on any scales of the system, are constant.

Substituting (44.6) and (44.9) into (44.10) and integrating gives (for $\alpha \neq 6$)

$$r_c \propto (t_0 - t)^{2/(6-\alpha)}. \tag{44.11}$$

The other relations (44.6)–(44.9) therefore have the time dependences

$$\rho_c \propto (t_0 - t)^{-2\alpha/(6-\alpha)}, \tag{44.12}$$

$$M_c \propto (t_0 - t)^{(6-2\alpha)/(6-\alpha)}, \tag{44.13}$$

$$E_c \propto (t_0 - t)^{2(5-2\alpha)/(6-\alpha)}, \tag{44.14}$$

$$v_c^2 \propto (t_0 - t)^{2(2-\alpha)/(6-\alpha)}. \tag{44.15}$$

Formally as $t \to t_0$, the core becomes a vanishing region of very high density but no mass or total energy. In practice we know that our assumptions break down and the few-body problem takes over. The homology solution shows the direction in which the evolution is heading.

During this homology phase the cluster evolution is so regular and smooth that we are led to ask if there is any analogy between it and the evolution of a star. Both the cluster and the star evolve through a sequence of quasi-static states. The star's evolution is driven by transfer of heat and radiation from its core to the outside vacuum. The cluster's evolution is driven by transfer of hot stars from its core into its halo. Consequent gravitational contraction provides the basic heat source in both cases, although in stars the thermonuclear reactions are a dominant amplifier.

Approximate models of cluster evolution can indeed be constructed by analogy with stellar evolution (Hachisu, *et al.*, 1978; Lynden-Bell & Eggleton, 1980; Sugimoto *et al.*, 1981). This approach starts with the equations of stellar hydrodynamics which are moments of the collisionless Boltzmann equation (Section 8). First divide the cluster into shells of mass having a given radial distance from the center. Since these shells are very close to hydrostatic equilibrium the first two terms of (8.20) are zero and the hydrostatic equation

$$\frac{\partial P}{\partial r} = nF_r = -\frac{GM(r)\rho}{r^2} \tag{44.16}$$

in spherical coordinates applies to a given shell. The mass within the shell follows from mass conservation which has a simple form in these Lagrangian coordinates

$$\frac{\partial M(r)}{\partial r} = 4\pi r^2 \rho. \tag{44.17}$$

The isotropic local pressure comes from the 'perfect gas' equation of state

$$P = \frac{kT\rho}{m} = \frac{\rho v^2}{3}, \tag{44.18}$$

where v^2 is the mean square velocity in three dimensions. These are three equations for the four variables P, M, ρ and T. The crux of the problem is to complete this set of equations.

We already noticed a more basic form of this difficulty in the discussion following (8.20). There I remarked that the next higher moment involves a transport equation for energy, and near equilibrium this transport equation can be constructed to a good approximation from physical arguments. Energy flow is represented by a flux F of energy per unit area per second. In the elementary (phenomenological) theory of heat transfer this flux is proportional to the temperature gradient: $F = -K\partial T/\partial r$ (see Equation (4.2)). The minus sign shows that heat flows down the thermal gradient and K is called the thermal conductivity. The total energy leaving a shell each second is its luminosity $L = 4\pi r^2 F$, so we may write

$$\frac{L}{4\pi r^2} = -K\frac{\partial T}{\partial r}. \tag{44.19}$$

This provides an equation for T at the expense of introducing L and K.

Energy conservation enables us to complete the equations for the thermodynamic variables, leaving only K to be specified. The energy conservation equation is the first law of thermodynamics (29.10). Since we are following a mass shell with constant $N($ and $M)$, no chemical work is done in changing its volume, so (29.10) becomes

$$\mathrm{d}U = \mathrm{d}Q - P\mathrm{d}V. \tag{44.20}$$

In a quasi-statically evolving system

$$\frac{\mathrm{d}U}{\mathrm{d}t} = \frac{\mathrm{d}Q}{\mathrm{d}t} - P\frac{\mathrm{d}V}{\mathrm{d}t}. \tag{44.21}$$

Now the energy flow out of the shell is its luminosity, so $\mathrm{d}L = \mathrm{d}U/\mathrm{d}t$. The heat leaving a shell of constant mass is related to the temperature change

$$\mathrm{d}Q = -N_{\text{shell}}\mathrm{d}(\tfrac{3}{2}kT) = -M_{\text{shell}}\mathrm{d}\left(\frac{3kT}{2m}\right)$$
$$= -4\pi r^2 \rho\,\mathrm{d}r\,\mathrm{d}(\tfrac{1}{2}v^2). \tag{44.22}$$

Since the constant mass of the shell is $M_{\text{shell}} = \rho\mathrm{d}V$, the mechanical work needed to change the shell's volume at constant pressure is $P\mathrm{d}V = PM_{\text{shell}}/\rho = P4\pi r^2\rho\mathrm{d}r/\rho$. Thus (44.21) takes the form

$$\frac{\partial L}{\partial r} = -4\pi r^2 \rho\left[\left(\frac{\partial}{\partial t}\right)\frac{v^2}{2} + P\left(\frac{\partial}{\partial t}\right)\frac{1}{\rho}\right]_M \tag{44.23}$$

in these Lagrangian coordinates following a particular mass shell. Equations (44.16)–(44.19) and (44.23) with the rate of nuclear energy generation in the shell added to the terms in square brackets are also the basic equations of stellar structure.

Finally we must deal with the thermal conductivity K. This is not so trivial because its form hides all the details of the transport processes. Specifying K

effectively determines all the moment equations of Section 8 in one swoop. K is a phenomenological quantity designed to express concisely our ignorance of the way orbits evolve. In order to maintain a closed set of equations the conductivity may only depend on P, ρ, M, L and T(or v). In the interior of a star there are three main contributors to heat transport: convection, radiation, and conduction. In a cluster of stars there is nothing quite like any of these processes! Convection would resemble coherent large scale streaming motions of stars. These may arise in clusters as a result of tidal interaction, but do not seem very probable in isolated, relaxed spherical clusters. No one has yet examined either the sky or N-body simulations very carefully for convection of stars. Radiation and conduction both have similar representations as diffusion processes. The main difference between photon and particle transfer in this regard is the different forms of K.

Inside a star the jiggling of atomic orbits which causes heat to diffuse is a local affair. The mean free path of an atom for large angle scattering is small compared to the size of the star. This is not true for stellar systems, as we saw in Section 2. Thus there is no guarantee that a local 'stellar dynamical conductivity' is very meaningful. Nevertheless, it is tempting to try such a description, and it does indeed turn out to be quite useful. The basic reason it works so well is that the heat flux equation (44.19) is phenomenological in nature, so we can adjust K in accordance with our physical intuition of what should happen.

With this attitude we notice that the dimensions of K are those of Boltzmann's constant times an inverse timescale and an inverse lengthscale: $K \propto \tau^{-1} R^{-1}$. The lengthscale comes from the temperature gradient and the radius of a mass shell. In a self-similar cluster these lengths must both be related to the one characteristic length in the cluster: $R \approx GM/v^2$. The timescale is the one on which higher energy stars diffuse out of the core as they gain energy from many small angle scatterings. This must be the relaxation time τ_R of (2.9). Therefore we have

$$K = \frac{C\rho}{v}, \tag{44.24}$$

where the constant C absorbs the very slowly varying $\log N$ factor in τ_R. Apart from the dimensionless constant C (whose exact value is not important for many problems) we now have a complete set of equations to describe a spherical cluster of stars.

To find self-similar solutions for such a cluster use (44.2), (44.3), and similar equations for the other variables M, v, and L. For any function $M_*(M, t)$ we have

$$\left(\frac{\partial}{\partial t}\right)_{M_*} = \left(\frac{\partial}{\partial t}\right)_M + \left(\frac{\partial M}{\partial t}\right)_{M_*} \left(\frac{\partial}{\partial M}\right)_t \tag{44.25}$$

to apply to the derivatives in (44.23). Substituting these expressions into the basic structural equations ((44.16)–(44.19), (44.23)) shows that they separate into terms dependent on either t or on r_*. This leaves a set of total differential equations containing the constants of separation. For given boundary conditions, this set has solutions only for certain values of the separation constants, rather like eigenvalues. The density has a power law form whose exponent depends on the separation constants which solve the system. Thus these are easily related to α in (44.5). A numerical analysis (Lynden-Bell & Eggleton, 1980) gives the result $\alpha = 2.21$, in good agreement with Cohn's Fokker–Planck simulation mentioned earlier. This value of α may then be used to find the time dependences in (44.11)–(44.15). Somewhat different conductivities and assumptions may modify the value of α by a few per cent.

This general theoretical approach seems to be very useful for describing the long period during which an initially isothermal stellar cluster develops its halo. Whether it describes the actual clusters of our Universe remains to be determined. Relaxation times in the halos of some globular clusters may exceed the Hubble time, and they may retain much memory of the conditions of their formation. Might not the halos of such clusters, possibly born of violent relaxation, differ in their structure from halos left behind by a contracting isothermal core? Here is another clue to their origin, if only we could read it.

The reading of clues is made more difficult by a variety of other processes which affect the development of realistic clusters. Evaporation of stars from the outer parts influences the halo. Not all stars in the cluster have the same mass, and groups of different mass will interact and evolve differently. Large mass motions in clusters, either from segregating groups of stars or from gas will alter the orbits of other cluster members. Binaries can form and function as a new source of dynamical energy in the cluster. In the next sections we begin an elementary exploration of these added features.

45

Evaporation and escape

Fetter and chain, Dungeon of stone,
All are in vain – Prisoner's flown!
Spite of ye all, he is free – he is free!

Gilbert and Sullivan

As clusters evolve, some of their members gain enough kinetic energy to escape from the system. Total energy is conserved, so any escape requires a redistribution of the remaining stars. If this redistribution makes subsequent escape easier the instability can grow. Eventually only a tight binary may remain to soak up the gravitational potential energy of the original system which may be unrecognizably dispersed. Finally, if the binary loses most of its mass to gravitational radiation there may be little to show for a once grand ball of stars.

Stating the criterion for a star to escape is easy (40.56)

$$E_{esc}(r) = \tfrac{1}{2}mv_{esc}^2 - m\phi(r) > 0. \tag{45.1}$$

Calculating the rate at which escapers carry mass and energy away from the bulk of the system is difficult. Only very approximate theories are available. Computer experiments show they give reasonable order-of-magnitude results, occasionally accurate to factors of two or three without too much adjustment. The reason for this difficulty is the range and complexity of energy transport processes leading to escape. Some of these processes in isolated clusters are: (a) few-body interactions among unbound stars, (b) ejection of field stars through interaction with binaries, (c) gravitational slingshot ejection of a member of a bound triple or few-body system, (d) interchange of a field star with a member of a binary causing the former binary member to escape, (e) violent relaxation during the formation of the system, (f) scattering of a halo orbit by the condensed nucleus of the system, (g) diffusion to high velocities caused by many small velocity exchanges. In realistic clusters, additional possibilities are: (h) tidal forces by other clusters or the galaxy, (i) rotation of the cluster interacting with tidal forces, (j) velocity perturbations from massive gas clouds within the cluster or passing through it, (k) supernova explosion of one member of a binary or triple group, (l) gravitational shock as the cluster passes through inhomogeneities (e.g., the plane) of its galaxy. All these processes may occur to different degrees in different parts of the cluster with different values of v_{esc}. Thus the difficulty.

For isolated clusters, only their last-mentioned process, velocity diffusion, seems amenable to a simple theory. Even here, we shall soon find an ambiguity. Sections 2 and 3 described how the cumulative effects of many distant gravitational encounters cause the velocity distribution of a cluster to relax to a nearly Maxwellian form, and Section 40 described the resulting quasi-equilibrium models of clusters. A feature of these models was the escape of stars on the high energy tail of the velocity distribution. This process, by analogy with gaseous systems, is called evaporation. We now ask how the cluster reacts to evaporation by changing its structure.

First we need to estimate the number of stars which evaporate. This seems easy. Just calculate $v_{esc}(r)$ and integrate the velocity distribution from v_{esc} to infinity at each position in the cluster. To simplify the problem even further for an analytic illustration, imagine a nearly homogeneous, uniform cluster. It will have some average mean square escape velocity $\langle v_{esc}^2 \rangle$ related to the average mean square velocity by a factor γ of order unity

$$\langle v_{esc}^2 \rangle = \gamma^2 \langle v^2 \rangle. \tag{45.2}$$

One way often used to estimate γ is to employ the virial theorem in the form $\langle v^2 \rangle = GM/2R_h = \langle \phi \rangle/2$, where R_h is the radius containing half the mass of the cluster. This radius can at least be defined precisely, unlike the total radius of a highly attenuated cluster. The factor two depends weakly on the detailed structure of the cluster. Then from (45.1), $\langle v_{esc}^2 \rangle = 2\langle \phi \rangle$ and $\gamma = 2$.

Integrating over a three-dimensional Maxwellian velocity distribution then shows that a fraction 7.4×10^{-3} of all stars have velocities more than twice the root mean square velocity. Relaxation replenishes this tail in a timescale $\sim \tau_R$. Since τ_R also varies throughout the system, we can take it to have an effective average value equal to $\tau_R(r = R_h)$. This introduces another uncertain factor of order unity, depending on the cluster's structure. Supposing that stars with $v^2 > \langle v_{esc}^2 \rangle$ actually do escape, we may write

$$\frac{1}{N}\frac{dN}{dt} = -\frac{\varepsilon}{\tau_R}, \tag{45.3}$$

where ε is of order 10^{-2}.

Further thought shows that this apparently simple theory is really rather ambiguous. In a cluster of given size, the relaxation timescale $\tau_R \propto v^3/n$ for a star of velocity v. As the velocity of the star increases towards v_{esc}, two things happen. First the probability per unit of real time that a distant encounter will increase the star's velocity further diminishes rapidly. Each further encounter becomes less effective. Second, there are also fewer and fewer encounters because the higher energy star spends more and more of its time in orbits farther from the bulk of the cluster. The net result is that a star approaches escape velocity more and more slowly by this process. Indeed it may never escape at all. It remains in a halo of high energy (i.e., nearly zero energy) stars around the cluster.

Many of the processes listed earlier may now operate on these halo stars, causing them to escape. However, as far as the core itself is concerned, stars evaporated into the halo have effectively escaped since they return only rarely to perturb the core. Continuing with our illustrative model we may suppose the evaporating stars carry no net energy away from the core since they are just below or at the escape velocity. Then if the core evolves through a series of quasi-stationary states in virial equilibrium, its total energy and its gravitational potential energy each remain approximately constant. Therefore $R \propto N^2$ and $\tau_R \propto v^3/n \propto v^3 R^3/N \propto N^{3/2} R^3 / R^{3/2} N \propto N^{7/2}$. With this relation we can integrate (45.3) and obtain

$$\frac{N}{N_0} = \left(1 - \frac{t}{t_{evap}}\right)^{2/7}, \tag{45.4}$$

where the subscript 0 denotes the initial value at $t_0 = 0$. The timescale for the core to evaporate completely by these diffusion processes is

$$t_{evap} = \frac{2}{7\varepsilon} \tau_{R0}. \tag{45.5}$$

It is roughly an order of magnitude longer than the initial relaxation time.

Using the proportionality relations for R, ρ, and $\langle v^2 \rangle$, we readily see that these quantities vary as the factor in parentheses in (45.4) raised to the powers $\frac{4}{7}$, $\frac{-10}{7}$, and $\frac{-2}{7}$, respectively. This result is just what we would expect from the homology relations (44.11)–(44.15) in the special case $\alpha = 2.5$ for which $E_c = $ constant. The structure of a simple evaporative halo follows from the same basic dimensional considerations which describe gravithermal contraction. They are both aspects of the same fundamental gravitational many-body instability, but differ in their timescales and boundary conditions.

This more kinetic view of halo formation gives us a better understanding of the timescale t_{evap} through the estimate of ε. Computer experiments which simulate isolated systems with large N (e.g., Spitzer & Thuan, 1972) can be parameterized by an effective value of ε. In these experiments, about two stars remained in the halo for every one that actually escaped. Counting both results as evaporation typically gave $\varepsilon \approx 1 \times 10^{-2}$, while for the escaping stars alone, $\varepsilon \approx 3 \times 10^{-3}$. In the early phases of cluster evolution, evaporation and escape drive the core's contraction. Later, the density contrast within the cluster increases and its internal gravithermal instability becomes more dominant.

Many stars of the halo may be removed by tidal forces from the galaxy or a nearby cluster. External tidal forces on a star become more important as the star moves farther from the center of its own cluster. There comes a point where the tidal force exceeds the force binding the star to its cluster and from (1.4) we we see this occurs at a radius

$$R_t \approx l(M_{clust}/M_{ext})^{1/3}, \tag{45.6}$$

called the tidal radius or Roche limit. It is important to remember that this is a static criterion, taking no account of either the star's orbit in the cluster or the cluster's orbit in the external field. For example, as a cluster orbits the galaxy, the external field may change on about the same timescale as for a star to orbit around, or be removed from, the outer halo. As a result, the halo star approximately participates in a restricted three-body problem: a test particle orbiting around two moving massive particles. Orbital resonances, which depend on fine details of the forces, may prevent the star from escaping, even though its distance exceeds $\sim R_t$. This would increase the effective value of R_t.

Anisotropy in the velocity distribution of halo stars also changes the effective tidal limit. This is especially important if the cluster rotates. Stars with orbital angular momentum in the same direction as the cluster's orbital angular momentum can escape more easily than stars on retrograde orbits. Stars on direct orbits do not have to be 'turned around', a time-consuming process. Therefore tidal forces themselves contribute to the growth of anisotropy in the halo. Clearly, the tidal limit must be worked out for each individual case, although (45.6) gives the general result to order of magnitude.

Some stars leave isolated clusters with velocities much greater than v_{esc}. These escapes nearly always result from a strong encounter with other stars. The idealized theory of these escapes, even for a single close encounter (Hénon, 1969), is very complicated. Its analysis depends strongly on the mass distribution of the stars. As a result, most of our knowledge about strong escapers comes from direct N-body computer experiments on clusters of a few hundred stars (Aarseth, 1974). Several 'rules of thumb' emerge as rough guides from these experiments:

1. Defining a fractional escape rate per crossing time L_i for a group of N_i stars of mass m_i by

$$\frac{\Delta N_i(m_i)}{N_i(m_i)} = L_i \frac{t}{t_{cr}}, \tag{45.7}$$

the numerical experiments give values of $L_i \approx 10^{-3}$. This may vary by a factor ~ 2 depending on the structure and initial conditions of the cluster.

2. Most escapes result from close encounters, including encounters with a temporarily bound pair.

3. The fractional escape rate L_i does not seem to depend significantly on mass m_i except for very heavy stars which escape less frequently.

4. The excess escape energy relative to the mean energy $v_\infty^2/\langle v^2 \rangle$ usually is between 0.1 and 1.0, with 0.5 being a rough average for a typical ejection.

5. Roughly 5% of the total mass escapes after about 30 initial crossing times in systems with a range of masses. For equi-mass systems, the total escaping mass is about $\frac{2}{3}$ this value.

6. Relatively few ($\lesssim 10\%$) escapes occur during an initial violent collapse of the system. Many occur after the first few crossing times, and are associated with an initial burst of binary formation.

7. Most high energy transitions which produce escape originate throughout the core of the cluster. In some cases, a massive central binary is particularly important in flinging stars out of the system.

It should be emphasized that these rules of thumb are derived from systems of several hundred members. Close encounters may be less important for systems with $N \gtrsim 10^5$, say. Unfortunately there does not seem to be a well-tested theory which describes the transition between small and large systems. One subtle difficulty is that mass segregation occurs. A central subsystem of massive stars may cause large clusters to retain many of the features of small ones, especially with regard to escape.

46

Mass segregation and equipartition

Mass segregation was one of the early important results to emerge from computer N-body simulations of clusters with a few dozen members. The heavier stars would gradually settle towards the center, increasing their negative binding energy. Lighter stars would preferentially populate the halo, with reduced binding energy. Later, direct integrations using many hundreds of stars showed the same tendency, as did models which integrated the Fokker–Planck equation for many thousands of stars.

We would expect this behavior from the basic properties of gentle relaxation by two-body encounters. Equation (14.65) shows that the timescale for dynamical friction to significantly decrease the energy of a massive star of mass M_0 is less than the relaxation timescale τ_R for lighter stars of mass m(moving with the same velocity) by a factor m/M_0. As massive stars in the outer regions of a cluster lose energy to the lighter ones, they fall toward the center and increase their velocity. The massive stars continue to lose the kinetic energy they gain by falling, and continue to fall. The lighter stars, on the other hand, increase their average total energy and move into the halo. As light stars rise through the system, their velocity decreases, altering the local relaxation time for remaining massive stars. The net relaxation for any individual star will therefore involve an average of τ_R over its orbit.

Can there be an end to mass segregation; will the system ever reach an equilibrium? For such an equilibrium the massive stars must gain energy from the fluctuating gravitational field as fast as they lose it to dynamical friction. Two conditions will then be satisfied: mechanical equilibrium determined by the scalar virial theorem (9.30), $2\langle T \rangle + \langle W \rangle = 0$, and thermal equilibrium determined by equipartition of energy among components of different mass m_α,

$$m_\alpha \langle v_\alpha^2 \rangle = 3kT. \tag{46.1}$$

All mass species must have the same temperature, so there is no energy exchange among the different species. Therefore our original question can be answered by asking whether equipartition and the virial theorem are mutually consistent for gravitating systems.

Although we previously developed the virial theorem for systems whose members all had the same mass, we can easily modify the result for a spectrum of masses. Contracting the virial potential energy tensor (9.9) gives the gravitational potential energy of the system

$$W = -\frac{G}{2} \int \int \rho(\mathbf{x})\rho(\mathbf{x}')\frac{1}{|\mathbf{x} - \mathbf{x}'|}\mathrm{d}\mathbf{x}'\mathrm{d}\mathbf{x}. \tag{46.2}$$

Thus is just a more detailed form of (43.1), which provides a physical interpretation. (For ease of notation we drop the brackets which denote time averaging.) Now if the number of stars per unit volume with mass m_α is $n(\mathbf{x}, m_\alpha)$, then

$$\rho(\mathbf{x}) = \int m_\alpha n(\mathbf{x}, m_\alpha)\mathrm{d}m_\alpha. \tag{46.3}$$

Discrete mass distributions can be dealt with by using delta functions in the mass spectrum. Thus the potential energy becomes

$$W = -\frac{G}{2} \int_\mathbf{x} \int_{\mathbf{x}'} \int_{m_\alpha} \int_{m_\beta} \frac{1}{|\mathbf{x} - \mathbf{x}'|} m_\alpha m_\beta n(\mathbf{x}, m_\alpha)n(\mathbf{x}', m_\beta)\mathrm{d}m_\beta \mathrm{d}m_\alpha \mathrm{d}\mathbf{x}'\mathrm{d}\mathbf{x}, \tag{46.4}$$

which we note is symmetric in m_α and m_β, as well as in $\mathbf{x}$ and $\mathbf{x}'$.

To combine virial equilibrium with equipartition we need to know how the velocity dispersion of each mass species is related to its potential energy. Suppose we focus on a particular mass range or species, calling it m_γ. Let

$$n(\mathbf{x}, m_\alpha) = n(\mathbf{x}, m_{\alpha \neq \gamma}) + n(\mathbf{x}, m_\gamma), \tag{46.5}$$

where

$$n(\mathbf{x}, m_\gamma) = n_\gamma(\mathbf{x})\,\delta(m_\alpha - m_\gamma) \tag{46.6}$$

and similar equations with $\mathbf{x}'$ replacing $\mathbf{x}$ and β replacing α. Thus the integral in (46.4) breaks up into the sum of four integrals, each of which involves one of the four products

$$n(\mathbf{x}, m_{\alpha \neq \gamma})n(\mathbf{x}', m_{\beta \neq \gamma}), \tag{46.7a}$$

$$n(\mathbf{x}, m_{\alpha \neq \gamma})n(\mathbf{x}', m_\gamma), \tag{46.7b}$$

$$n(\mathbf{x}', m_{\beta \neq \gamma})n(\mathbf{x}, m_\gamma), \tag{46.7c}$$

$$n(\mathbf{x}, m_\gamma)n(\mathbf{x}', m_\gamma). \tag{46.7d}$$

Equation (9.5) shows that the interior integrals over m_β and $\mathbf{x}'$ in (46.4) give the gravitational potential $\phi(\mathbf{x})$. We may recall from elementary mechanics that inside a spherical system (or a system composed of homogeneous ellipsoidal shells for that matter) there is no net contribution to the force from shells with $\mathbf{x}' > \mathbf{x}$. This is the force which makes its way into the virial theorem through (9.1). Therefore we do not need to include the constant potential for $\mathbf{x}' > \mathbf{x}$ and we have

$$\phi(\mathbf{x}) = \frac{GM(r)}{r}. \tag{46.8}$$

With this decomposition, the first term of (46.4) is

$$W(\alpha \neq \gamma, \beta \neq \gamma) = -\frac{G}{2}\int_{m_\alpha}\int_r m_\alpha n(\mathbf{r}, m_{\alpha \neq \gamma})\frac{M_{\neq \gamma}(r)}{r}\mathrm{d}\mathbf{r}\mathrm{d}m_\alpha. \qquad (46.9)$$

This is the potential energy obtained from 'building up' the cluster by bringing shells containing all stars, except those of mass m_γ, from infinity into the potential field of the previously added stars (also excluding m_γ). (Each star is included in both $n(\mathbf{r})$ and $M(\mathbf{r})$, but this is compensated by the factor $\frac{1}{2}$.) So this is the potential energy of all masses except γ interacting with all other masses except γ. Similarly, the second term

$$W(\alpha \neq \gamma, \gamma) = -\frac{G}{2}\int_{m_\alpha}\int_r m_\alpha n(\mathbf{r}, m_{\alpha \neq \gamma})\frac{M_{=\gamma}(r)}{r}\mathrm{d}\mathbf{r}\mathrm{d}m_\alpha \qquad (46.10)$$

is the potential energy due to bringing the $m_{\alpha \neq \gamma}$ stars into the cluster in the potential field of the m_γ stars. The third term

$$W(\gamma, \alpha \neq \gamma) = -\frac{G}{2}\int_r m_\gamma n_\gamma(r)\frac{M_{\alpha \neq \gamma}(r)}{r}\mathrm{d}r \qquad (46.11)$$

is the potential energy due to bringing the m_γ stars into the potential field of the other stars. (Remember β is just a dummy index which can be replaced by α after the integrations are performed.) Finally, the fourth term

$$W(\gamma, \gamma) = -\frac{G}{2}\int_r m_\gamma n_\gamma(r)\frac{M_{=\gamma}(r)}{r}\mathrm{d}r \qquad (46.12)$$

is the energy of the γ stars in their own field.

Splitting the kinetic energy into two parts also according to (46.5), the virial theorem takes the form

$$\int_{m_\alpha}\int_r m_\alpha n(r, m_{\alpha \neq \gamma})v_\alpha^2 \mathrm{d}m_\alpha \mathrm{d}\mathbf{r} + m_\gamma N_\gamma v_\gamma^2 + W(\alpha \neq \gamma, \beta \neq \gamma)$$

$$+ W(\alpha \neq \gamma, \gamma) + W(\gamma, \alpha \neq \gamma) + W(\gamma, \gamma) = 0. \qquad (46.13)$$

If the virial theorem is to apply separately to each mass species, we see that we should identify twice the kinetic energy of each star with the potential energy it would acquire by being dropped from infinity into the cluster containing all the other stars of whatever mass. Thus we have

$$m_\gamma N_\gamma v_\gamma^2 = -W(\gamma, \alpha \neq \gamma) - W(\gamma, \gamma) \qquad (46.14)$$

and similarly we can identify twice the kinetic energy contributed by all the other masses with $-W(\alpha \neq \gamma, \beta \neq \gamma) - W(\alpha \neq \gamma, \gamma)$.

Next, we turn to the requirement of equipartition. From the discussion around Equation (45.2), it is clear that stars with mass less than about $\frac{1}{4}$ the average mass of their neighbors cannot be in equipartition. They would escape. So we will just be concerned with equipartition among the remaining stars. Equipartition, expressed by (46.1), means that the right hand side of (46.14) divided by N_γ is a constant, independent of γ. It is now convenient to combine the right hand terms of (46.14) by writing

$$W_\gamma \equiv - W(\gamma, \alpha \neq \gamma) - W(\gamma, \gamma)$$

$$= 2\pi G \int_0^{m_0} \int_0^{R_0} \frac{m_\gamma n(r, m_\gamma)}{r} mN(r, m) r^2 \, dr \, dm$$

$$= \frac{G}{2} \int_0^{m_0} \int_0^{R_0} \frac{m_\gamma}{r} \frac{\partial N(r, m_\gamma)}{\partial r} mN(r, m) \, dr \, dm. \tag{46.15}$$

Here $N(r, m)$ is the total number of stars of mass m within the radius r. It satisfies

$$\frac{\partial N(r, m)}{\partial r} = 4\pi r^2 n(r, m) \tag{46.16}$$

for each mass m. The integration in (46.15) is over all stars of mass m, including m_γ, between zero and the maximum mass m_0 (there is no longer need for the general mass subscripts α or β). The cluster extends to a radius R_0.

To simplify notation we can make W_γ dimensionless by using

$$x \equiv r/R_0 \text{ and } y \equiv m/m_0 \tag{46.17}$$

and noticing that $N_\gamma \equiv N(1, y_\gamma)$. Then the condition for equipartition becomes

$$0 = \frac{\partial}{\partial y_\gamma} \left(\frac{W_\gamma(x, y, y_\gamma)}{N(1, y_\gamma)} \right) = \int_0^1 \int_0^1 \frac{y}{x} \frac{N(x, y)}{N(1, y_\gamma)} \frac{\partial^2}{\partial x \partial y_\gamma} \left\{ \frac{y_\gamma N(x, y_\gamma)}{N(1, y_\gamma)} \right\} dx \, dy \tag{46.18}$$

after some minor rearrangement of the derivatives. This is a basic result. It is the necessary and sufficient condition for global equipartition.

We may be interested in applying a condition which is necessary, but not sufficient, to see if a particular mass distribution can possibly be in equipartition. A simple necessary condition follows immediately from the integrand of (46.18). The factors before the second derivative are always positive. So if the integral is to be zero, the second derivative must be positive for some values of x and negative for others, and thus pass through zero for at least one value of x. At this value x_0 the quantity in braces must be independent of y_γ. Performing the differentiation with respect to x and using (46.16) gives

$$n(x_0, y_\gamma) = C(x_0) \frac{N(1, y_\gamma)}{y_\gamma} \tag{46.19}$$

for all values of y_γ where $C(x_0)$ is a function which depends just on x_0. Thus there must exist at least one value of x where the mass spectrum of the local number density is related to the total number spectrum of masses by (46.19). In other words, somewhere in the cluster the local mass spectrum must be proportional to the global number spectrum. The lighter a star's mass the more of them must be present at x_0 relative to their total number. Physically, this prevents the heavier stars from becoming so isolated that they form a self-gravitating subsystem of their own which independently evolves away from equipartition with the lighter stars.

A sufficient condition for equipartition is that

$$\frac{\partial^2}{\partial x \partial y_\gamma}\left\{\frac{y_\gamma N(x, y_\gamma)}{N(1, y_\gamma)}\right\} = 0 \tag{46.20}$$

everywhere. This is also the necessary condition for a state which may be called 'local equipartition' to apply within each thin shell of stars in the cluster. To see whether any reasonable distribution function can satisfy these conditions, we integrate (46.20). It is a simple hyperbolic differential equation whose characteristics are lines parallel to the x and y axes. To find the general solution for $N(x, y_\gamma)$ we may specify its values along two intersecting characteristics which bound the domain of interest. For this problem, it is natural to specify $N(x, 1)$ and $N(1, y_\gamma)$: the total number of the most massive stars within any radius and the total number spectrum of masses for the whole cluster. The general solution of (46.20) is that the quantity in braces is equal to a function of x plus a function of y_γ. Applying these boundary conditions then gives

$$N(x, y_\gamma) = \frac{N(1, y_\gamma)}{y_\gamma}\left[\frac{N(x, 1)}{N(1, 1)} + y_\gamma - 1\right] \tag{46.21}$$

for all values of y_γ. (We could have dropped the γ subscript in (46.19)–(46.21) but have kept it to avoid possible confusion.)

Do reasonable distributions satisfy this constraint? No. For as $x \to 0$ the fraction of most massive stars $N(x, 1)/N(1, 1) \to 0$ also. Then since $0 \le y_\gamma \le 1$ there is a range of masses for which $N(x, y_\gamma) < 0$, which is unphysical (to say the least). Therefore we conclude that local equipartition cannot occur throughout spherical gravitating systems. (It may occur temporarily among some mass species in parts of the system.)

By examining particular models of clusters in detail it is possible to find more specific criteria for equipartition. For example, the first discussion of this criterion (Spitzer, 1969) considered a cluster containing stars of just two masses m_1 and m_2 for which (1) $m_2 > m_1$; (2) $M_2 \equiv m_2 N_2 \ll M_1$; and (3) the heavy stars are already sufficiently concentrated at the center so that the stars of mass m_1 which affect their potential energy can be represented by a distribution $M_1(r) = 4\pi\rho_{01}r^3/3$, where ρ_{01} is their central density. Then equipartition can not be satisfied if $M_2/M_1 > \beta(m_1/m_2)^{3/2}$, where β is a number less than one. The value of β depends somewhat on the density and velocity distributions and is about 0.16 for $m_2 \gg m_1$.

When this equipartition criterion is violated the self-attraction of the heavier stars is so great that they form a high temperature subsystem in the core of the cluster. The velocity dispersion needed to keep this subsystem in virial equilibrium corresponds to a temperature higher than the equipartition temperature with the light stars. So a thermal gradient is established and light stars take more energy out from the core, causing the heavy subsystem to contract still further.

In general it appears that gravitating clusters do not reach the nirvana of equipartition. Such clusters, however, may still find states of mechanical equilibrium which do not require isothermal distributions (see Merritt, 1981; Katz & Taff, 1983). These are usually unstable and remain to be explored further.

If a cluster is not in thermal equilibrium it is natural to ask on what timescale the disequilibrium evolves significantly. Mass segregation is the manifestation of this evolution. Since energy exchange is essentially a diffusion process we again expect this timescale to be of the order τ_R. Remember that τ_R will be different for different masses, being proportional to $1/m_y$. Computer simulations show this is indeed the case for systems with two or three mass species. These are described approximately by phenomenological equations of the form (Spitzer, 1975)

$$\frac{dE_1}{dt} = \frac{\Delta T_{12}}{\tau_R(1,2)}. \tag{46.22}$$

Here E_1 is the mean total energy of the heavier stars, ΔT_{12} is the temperature (kinetic energy) difference between the two mass species, and $\tau_R(1,2)$ is their energy exchange timescale (approximately equal to the relaxation time (14.65)). If there are many mass species, the individual energy exchange times are not so clear cut. For these cases, direct N-body integrations show pronounced mass segregation over several dynamical crossing times, but for these experiments N is only a few hundred and $\tau_{cross} \approx \tau_R$. Considerable work remains to find and verify the detailed forms of (46.22) for general systems. However, it is already clear that clusters containing a range of masses evolve considerably more rapidly than clusters containing identical masses (Inagaki & Saslaw, 1985).

47

Orbit segregation

Up and down, and in and out,
Here and there, and round about

Gilbert and Sullivan

Any process, such as mass segregation, mass loss, or core–halo instability which redistributes the density of a gravitating system will also produce orbit segregation. Orbit segregation is caused primarily by changes in the mean gravitational field. It affects orbits according to their eccentricity.

In a globular cluster, for example, the timescale for mass segregation to redistribute density falls between the dynamical crossing timescale and the relaxation timescale for stars of average mass. Therefore orbits of average or light stars undergo secular changes governed by the slowly changing mean field. (We are isolating changes in the mean field which, averaged over orbits, lead to orbit segregation, so we now ignore close encounters and dynamical friction for the test stars.) Suppose, for clarity, we compare the extremes of eccentricity: circular and radial orbits. Let the cluster's density increase toward the center, as usual. If the cluster were in stationary equilibrium, stars in circular orbits would just go around and around, and those in radial orbits would just go in and out through the center, in equilibrium with the mean field. Stars with intermediate eccentricities, but constant angular momentum, would generally follow open orbits with constant amplitude.

Now suppose part of the cluster begins to contract slowly compared with the timescale for free fall. The contraction will not generally be homologous but will enhance the central concentration. Stars moving in circular orbits around the edge of the cluster do not notice any change, since the mean field is spherical. So these orbits stay put. Stars in interior circular orbits notice an increasing central mass. They increase their eccentricity and spiral inward. Next consider a star on a radial orbit plunging toward the center. As it falls in from its maximum amplitude it gains an additional acceleration from the interior mass, which is moving inward. It therefore overshoots the center with more than equilibrium velocity. As it moves out it loses this additional velocity. But during the outward journey it also loses more velocity since it is further decelerated by the additional mass moving in. The net result is a loss of kinetic energy and a decrease in the amplitude of the orbit. Thus radial orbits whose initial amplitude is at the edge of the cluster will be segregated from circular orbits there, since they sample a different time dependent field.

Segregation of other orbits depends on details of the density distribution. Orbits of intermediate eccentricity will generally show more complicated intermediate behavior, especially since their eccentricity will itself change with time.

To put this physical result on a more quantitative foundation we develop the evolution of radial orbits which pass through the center of a general non-linear spherical density distribution. The density distribution changes slowly compared with the orbital period.

Let the cluster have an initial radius R_0, fiducial density ρ_0, and a radial position coordinate $x = r/R_0$. Its density distribution may be expanded as

$$\rho(x) = \rho_0 \sum_{n=-1}^{\infty} \alpha_n(t) x^n, \tag{47.1}$$

so the mass interior to x is

$$M(x) = 4\pi R_0^3 \rho_0 \sum_{n=-1}^{\infty} \frac{\alpha_n(t)}{n+3} x^{n+3}. \tag{47.2}$$

The coefficients $\alpha_n(t)$ describing the structure of the cluster depend on time. By letting these structure coefficients absorb any time dependence in R_0 and ρ_0, we may regard these latter two quantities as constants. Radial orbits are governed by the equation of motion

$$\frac{d^2 x}{d\tau^2} = - \sum_{n=-1}^{\infty} \frac{\alpha_n}{n+3} x^{n+1}, \tag{47.3}$$

with the usual normalized time

$$\tau \equiv (4\pi G \rho_0)^{1/2} t. \tag{47.4}$$

Now the structural coefficients α_n vary on a timescale which is slow compared with the orbital period. We can represent this by letting α_n depend on a 'slow time' ζ, whose change is of a slower order than τ, i.e.,

$$\zeta = \varepsilon\tau. \tag{47.5}$$

So $\alpha_n = \alpha_n(\varepsilon\tau) = \alpha_n(\zeta)$. In general the position of an object in its orbit will have some components which vary on the slow time ζ and others which vary on a faster timescale, which we call η. For example, ζ could apply to a secular change in the amplitude of an orbit which is approximately periodic on the faster time η. Therefore we can write the position in the form

$$x(\zeta, \eta) = x_0(\zeta, \eta) + \varepsilon x_1(\zeta, \eta) + O(\varepsilon^2). \tag{47.6}$$

The fast time η will be of the same order as τ, but will not, in general, be equal to τ. Therefore we can suppose there is a differential relation between them with the form

$$\frac{\partial \eta}{\partial \tau} = h(\zeta), \tag{47.7}$$

where $h(\zeta)$ is a positive function of order unity. Since we are dealing with orbits which remain bound as τ becomes large, $h(\zeta)$ will be subject to this important asymptotic restriction.

Substituting (44.5)–(44.7) into (47.3) and equating separately the coefficients of $O(1)$ and those of $O(\varepsilon)$ gives

$$h^2(\zeta)\frac{\partial^2 x_0}{\partial \eta^2} + Q(\zeta, x_0) = 0 \tag{47.8}$$

and

$$h^2(\zeta)\frac{\partial^2 x_1}{\partial \eta^2} + \frac{\partial Q(\zeta, x_0)}{\partial x_0}x_1 = -2h(\zeta)\frac{\partial^2 x_0}{\partial \eta \partial \zeta} - \frac{\partial h}{\partial \zeta}\frac{\partial x_0}{\partial \eta}, \tag{47.9}$$

where

$$Q(\zeta, x_0) \equiv \sum_{n=-1}^{\infty} \frac{\alpha_n(\zeta)}{n+3} x_0^{n+1}. \tag{47.10}$$

To lowest order the orbit will be periodic; variations will be of higher order. Therefore we look for periodic solutions of $x_0(\zeta, \eta)$ and they will have the general form

$$x_0(\zeta, \eta) = \sum_{k=0}^{N} A_k(\zeta) \cos k\eta. \tag{47.11}$$

(The phase of the orbit can be chosen so that we do not need the sine terms.) Substituting this form into the equation of motion $d^2x/d\tau^2 = -Q(\zeta, x)$, writing Q as its Fourier cosine series, equating coefficients of the same harmonics, and keeping terms up to the Nth order, shows that

$$-h^2 k^2 A_k + Q_k = 0, \quad k = 0, 1, 2, \ldots, N, \tag{47.12}$$

where

$$Q_k = \frac{2}{\pi}\int_0^{\pi} Q[\zeta, x_0(\zeta, \eta)] \cos k\eta \, d\eta. \tag{47.13}$$

Thus we have a set of $N+1$ equations (47.12) for the $N+2$ unknowns $A_0, A_1, \ldots, A_N, h$. An additional constraint, as mentioned before, is provided by the requirement that the solution be periodic as $\tau \to \infty$. In order that no secular terms appear in the solution of (47.9) for x_1, so that x_1/x_0 remains bounded for all time, the inhomogeneous part of (47.9) must be orthogonal to a solution of the homogeneous part. This prevents the forcing term from destroying the periodic solution. Now $x_1 = \partial x_0/\partial \eta$ is a solution of the homogeneous part of (47.9), as may readily be seen by differentiating (47.8). Therefore, the orthogonality relation is

$$\int_0^{\pi}\left(2h\frac{\partial^2 x_0}{\partial \eta \partial \zeta} + \frac{\partial h}{\partial \zeta}\frac{\partial x_0}{\partial \eta}\right)\frac{\partial x_0}{\partial \eta}d\eta = \frac{d}{d\zeta}\left[h\int_0^{\pi}\left(\frac{\partial x_0}{\partial \eta}\right)^2 d\eta\right] = 0, \tag{47.14}$$

which combines with (47.11) to give

$$h(\zeta)\sum_{k=0}^{N} k^2 A_k^2(\zeta) = C. \tag{47.15}$$

The constant C characterizes the energy of the orbit. The other integration constant of the equation of motion characterizes the phase of the orbit, and has been chosen so that we need only use the Fourier cosine components in (47.11).

Equations (47.12) and (47.15) provide the general solution for radial orbits in a spherical system whose density slowly changes. Note that it is not necessary to actually solve for x_1, so long as it is bounded, since it contributes a negligible correction of order ε to x_0. The solution is usually even simpler because most periodic orbits are dominated by a single frequency, so that $N = 1$. Moreover, for orbits having amplitude $A_1 = A$ at $\eta = 0$ in a density distribution where $\alpha_n = 0$ for all odd n, and $N = 1$, the solution using (47.10)–(47.13) and Wallis' formula for the integrals further simplifies to

$$\frac{2}{\sqrt{\pi}} \sum_{n=0,2,4,\dots} \frac{\alpha_n(\zeta)}{n+3} \frac{\Gamma[(n+3)/2]}{\Gamma[(n+4)/2]} A^{n+4}(\zeta) = C^2 \tag{47.16}$$

and

$$h(\zeta)A^2(\zeta) = C, \tag{47.17}$$

where Γ is the standard gamma function. So specifying C for the orbit and the time development for the density coefficients, the solution is reduced to an algebraic equation of order $(n+4)/2$.

In many cases, determining the orbit segregation is even easier because we do not have to solve this algebraic equation explicitly. As an illustrative example, consider a density distribution $\rho = \rho_0(\alpha_0 - \alpha_2 x^2)$ with all the other $\alpha_n = 0$. This is similar to the core of an isothermal sphere ((40.7), (40.20)), with the edge of the core at $x = (\alpha_0/\alpha_2)^{1/2}$. Alternatively, it may be thought of as representing an entire (unstable) cluster. Its total mass is

$$M = \tfrac{8}{15}\pi R_0^3 \rho_0 \alpha_0^{5/2} \alpha_2^{-3/2}. \tag{47.18}$$

If the total mass remains constant, or if it changes in some given manner, Equation (47.18) relates the change of α_0 to the change of α_2. The evolution of the orbit amplitude $A(\zeta)$ is given by (47.16) for $n = 0, 2$, which then becomes a cubic.

We need not actually solve this cubic to determine the orbit segregation. Nor do we need to know C. Differentiating the cubic with respect to ζ and using (47.18) for constant total mass gives

$$\left.\frac{d\ln A}{d\ln \alpha_2}\right|_{\text{rad}} = -\frac{3}{2}\frac{(4-3y)}{(40-27y)}, \tag{47.19}$$

where

$$y(\zeta) \equiv \frac{A^2(\zeta)\alpha_2(\zeta)}{\alpha_0(\zeta)} \equiv \frac{A^2(\zeta)}{\lambda^2(\zeta)} \leq 1 \tag{47.20}$$

since the ratio $(\alpha_0/\alpha_2)^{1/2}$ is effectively the scale length of the quadratic density decrease. Thus an increase of α_2 decreases the scale length of the cluster and also decreases the amplitude of radial orbits, but not homologously.

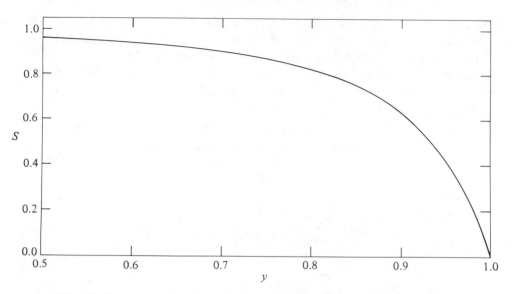

Fig. 46. The segregation index S for radial and circular orbits in a slowly changing quadratic density distribution. Orbit segregation occurs for $S(y) < 1$ where $y^{1/2}$ is the ratio of the orbit's amplitude to the radius of the cluster.

The change of circular orbits is easier to calculate since the product of the radius with the mass inside the orbit is a standard adiabatic invariant (e.g., Landau & Lifshitz, 1960). Adiabatic invariants change only in order ε^2 or higher, a special case of the more general separation of timescales used in this section. From (47.2) we then have

$$\left.\frac{\mathrm{d}\ln A}{\mathrm{d}\ln \alpha_2}\right|_{\mathrm{circ}} = -\frac{3}{2}\frac{(1-y)}{(10-9y)}. \tag{47.21}$$

The reader may like to show that this result also follows directly from the conservation of angular momentum. Again there is a non-homologous variation of the radius of the orbit with changing density distribution. The outermost circular orbits at $y = 1$ do not change since they do not notice the changing internal distribution.

Using (47.19) and (47.21), we can compare the change of radial and circular orbits at the same (or different) radii in the same (or a similar) cluster. For this comparison, we define a general segregation index as

$$S \equiv \frac{\mathrm{d}\ln A|_{\mathrm{circ}}}{\mathrm{d}\ln A|_{\mathrm{rad}}}. \tag{47.22}$$

When $S < 1$, radial orbits move in faster than circular orbits.

As an example, let us compare circular and radial orbits of the same amplitude in the same cluster. Then from (47.19)–(47.21)

$$S = \frac{\mathrm{d}A|_{\mathrm{circ}}}{\mathrm{d}A|_{\mathrm{rad}}} = \frac{(40-27y)(1-y)}{(4-3y)(10-9y)}. \tag{47.23}$$

Figure 46 illustrates this function. The segregation here depends only on the square of the ratio of the amplitude to the scale length. As $y \to 1$, $S \to 0$, and radial orbits move in faster. As $y \to 0$ the contraction becomes homologous, $S \to 1$ and both orbits move in at the same rate, as we would expect from the physical arguments introducing this section.

In general, we would expect that the more eccentric orbits move inward more rapidly than the less eccentric ones. Overall contraction of a cluster will tend to increase the eccentricities of interior orbits ($y < 1$) as the mean field becomes more concentrated. This occurs even if the collapse is homologous and the density uniform (Lecar, 1966). The central force averaged over each orbit increases slowly with time, causing the orbit to spiral inward with increasing eccentricity. This additional effect enhances the segregation of orbit amplitudes as a function of their eccentricity.

Thus orbit segregation caused by a changing mean field contributes, along with gentle relaxation, close encounters, gravithermal instability, evaporation and mass segregation, to the redistribution of orbits in a system of galaxies or stars.

48

Binary formation and cluster evolution

Binaries in clusters act as energy sources and sinks. As sinks of gravitational potential energy, they can absorb more than one-half the total binding energy of a cluster. As corresponding sources of kinetic energy they heat the cluster.

How fast will binaries form? How important can they be? An analytic theory of these processes, even in the approximations for which it is available, is quite complex. N-body simulations are sometimes the only way to answer detailed questions, especially for a broad spectrum of masses. Nevertheless, many useful results are available, and this section describes some. Following discussions of how binaries form by interactions of single stars, we consider how they can affect a cluster's evolution. From a more general point of view, these can be regarded as examples of energy transfer between different levels of hierarchial structure in a cluster.

48.1. Formation by few-body interactions

Binaries may be born directly when stars or galaxies form. We do not really understand how this occurs, so all we can do is add such initial binaries into the cluster in an *ad hoc* way, and then examine their development. Even if there are no initial binaries, however, the fluctuating gravitational field may chance to bring a few stars so close together, with such low relative velocities, that they remain temporarily bound. Usually these subclusters lead a fleeting existence, for later fluctuations may inject enough kinetic energy to make them decay. Occasionally, during this process, the vagaries of complicated few-body orbits may leave a binary bound so strongly that it resists further disruption.

Clearly, the binding energy of this residual binary is its most important property. From the elementary solution to Kepler's problem, (e.g., Landau & Lifshitz, 1960), we know that the binding energy, $E < 0$, is related to the elliptical orbit by

$$|E| = \frac{Gm_1 m_2}{2a} = \frac{Gm_1 m_2}{2p}(1 - e^2) \qquad (48.1)$$

and

$$|E| = \frac{L^2}{2mb^2} = \frac{L^2}{2mp^2}(1 - e^2). \tag{48.2}$$

Here a and b are the major and minor semiaxes of the ellipse and $m = m_1 m_2 / (m_1 + m_2)$ is the reduced mass. Moreover, $L = mr^2 \dot{\phi}$ is the constant total angular momentum of the orbit whose eccentricity is e and whose path is given by

$$\frac{p}{r} = 1 + e \cos \phi, \tag{48.3}$$

with $p \equiv L^2 / Gm_1 m_2 m$ being one-half the latus rectum.

It is most useful to compare the binding energy of a binary with the random thermal energy of field stars in its neighborhood. To give some perspective to the orders of magnitudes involved, consider binaries of four characteristic length scales. First, and largest, there is the scale R of the system itself. If such a binary with radius $a \approx R$ could survive, then from (48.1) and the virial theorem applied to the whole cluster (with all stars about the same mass) we see that the ratio of that binary's binding energy to the whole cluster's binding energy is of order N^{-2}. Therefore the ratio, $|E|/kT$, of the binding energy in the binary to the kinetic energy of a typical star is of order N^{-1}. For the second scale length, suppose the binary orbit is about the size of the average star separation: $a \approx RN^{-1/3}$. Then $|E|/kT \approx N^{-2/3}$. Third, there is the case when $|E|/kT \approx 1$. This is the most important comparison. We will return to it after mentioning the fourth length scale for a binary whose internal energy is about equal to the entire binding energy of the cluster. For such a maximally bound binary, $a/R \approx N^{-2}$. In a typical globular cluster this would be a contact binary.

The case $|E|/kT \approx 3$ is especially important because it is the dividing line between hard and soft binaries. Hard binaries, for which $|E|/kT \gg 3$, tend to lose orbital kinetic energy to their single neighbors. As a result they contract and become even harder. Soft binaries, with $|E|/kT \ll 3$, on the other hand, will generally gain orbital kinetic energy from their neighbors and become less bound.

The basic reason for this effect is that dynamical friction tends towards equipartition of kinetic energy in a system where all masses are the same. To understand the effect of changes in the binding energy of binaries it helps to recall the negative specific heat of gravitating systems (see section 43). Removing internal kinetic energy from hard binaries causes them to contract and heat up. The single stars which thus gain energy spend more time on average in the outer parts of the cluster. Therefore the cluster tends to expand and cool. Stars remaining near the binary are, on average, cooler than those which gained kinetic energy from it. Thus the ratio $|E|/3kT$, using the local temperature, increases and the hard binary hardens. For soft binaries, the inverse process occurs. The hard get harder and the soft get softer.

Since the criterion for hardness is so fundamental it is interesting to interpret it slightly differently. For $|E| \approx 3kT$ and, again using the virial theorem for the cluster, we see that the binary radius and the size of the cluster are related by

$$aN \approx 2R. \tag{48.4}$$

In other words, if all the stars were in binaries of transitional hardness, and their orbits were all laid end-to-end, they would extend approximately across the cluster. Taking the one-dimensional radius of the three-dimensional cluster, and stringing all the stars along it at equal spacings (retaining their thermal energies), would give nearest neighbors the transitional binary energy. We have already met this criterion before because a is also approximately the separation which produces a large-angle scattering by two approaching field stars ((2.2)). We will again meet this criterion when we consider the conditions for colliding stars to explode in very dense stellar systems.

Almost by definition the total binding energy contained in soft binaries cannot contribute much to the overall dynamics of a cluster. Hard binaries, however, may be a very strong influence. So we next ask how many hard binaries form by few-body interactions in a cluster.

To form a hard binary this way requires at least a strong triple encounter, so the third body can carry away the excess energy and momentum. Now the number of two-body encounters per unit volume per second is $n^2 \sigma v$, so the number which occurs throughout the whole cluster during one relaxation time τ_R is $n^2 \sigma v \tau_R (4\pi R^3/3) = n\sigma v \tau_R N$. The probability of finding a third body within this same cross section volume is $\sim N(a/R)^3 \sim N\sigma^{3/2} R^{-3}$. Therefore the total number of three-body encounters, each within a volume $\sim \sigma^{3/2}$, throughout the whole cluster is $\sim n\sigma^{5/2} v \tau_R N^2 R^{-3}$ during one relaxation time. If we assume that each triple encounter within a radius a given by (48.4) yields a hard binary, then using $\sigma \sim \pi a^2$ and τ_R from (2.9), along with the virial theorem in the more accurate form $v^2 \approx GM/2R$ for a finite system, we find the number of hard binaries which form during a time t to be of order (see Spitzer & Hart, 1971).

$$N_{\rm hb} \approx \frac{1.5}{N \ln (N/2)} \frac{t}{\tau_R}. \tag{48.5}$$

A much more precise discussion which also includes the hard binaries formed from the hardening of soft binaries (Heggie, 1980) gives the rate of hard binary formation per unit volume roughly as

$$\frac{\mathrm{d}n_{\rm hb}}{\mathrm{d}t} \approx 10^2 \frac{G^5 m^5 n^3}{\langle v^2 \rangle^{9/2}}$$

$$\approx 2 \times 10^{-13} \left(\frac{n}{10^4 \,{\rm pc}^{-3}} \right)^3 \left(\frac{m}{m_\odot} \right)^5 \left(\frac{100 \,{\rm km}^2 \,{\rm s}^{-2}}{\langle v^2 \rangle} \right)^{9/2} {\rm pc}^{-3} \,{\rm yr}^{-1}. \tag{48.6}$$

Note the strong dependence on n^3, m^5 and v^{-9}. Considering the uncertainties in both (48.5) and (48.6), they are in reasonable agreement. The semimajor axis of the 'softest hard' binaries, i.e., those with $|E| = 3kT$, is

$$a \approx 4 \left(\frac{m}{m_\odot} \right) \left(\frac{100 \,{\rm km}^2 \,{\rm s}^{-2}}{\langle v^2 \rangle} \right) {\rm AU} \tag{48.7}$$

in handy Astronomical Units. (One AU is the distance from the earth to the sun, 1.5×10^{13} cm or about 200 solar radii.) We should remember that this is a rather approximate limit in clusters with a range of masses, since they cannot usually be characterized by a single equipartition temperature (Section 46). Nevertheless, it serves as a convenient reference scale, at least until the theory of multimass systems is understood better.

The analysis leading to (48.5) has been relatively straightforward. But as soon as we try to apply it to realistic clusters a difficulty arises: What is the effective value of N? If the cluster were homogeneous and all its stars were of equal mass, then the principles underlying (48.5) would apply and N would be the total number of stars. In this case, for $N \gtrsim 10^3$ the number of hard binaries would be negligible even after 10 or 100 τ_R. But realistic models of clusters, as we have seen, are usually very inhomogeneous and τ_R may be orders of magnitude shorter in the center than in the outer regions. This would decrease the effective value of N appreciably. Physically, the central core would behave like a relatively isolated system of fewer stars and it would form hard binaries at a more rapid rate. Moreover, the substantial mass segregation of a realistic cluster would enhance this effect considerably.

Sure enough, direct N-body integrations of clusters with several hundred members typically develop hard binaries after several tens of crossing times, τ_c (see (2.11)), or ~ 10 relaxation times τ_R for equal masses. If there is a smooth mass spectrum with $m_{max}/m_{min} \sim 10$, a hard binary develops roughly ten times faster. This binary almost always contains the two most massive bodies. Therefore the effective value of N in (48.5) may decrease as the cluster evolves and become much less than the total number of stars. This probably remains essentially true for clusters with many more stars, but direct N-body simulations have not yet explored this in detail.

A straightforward way to estimate the effects of inhomogeneity and mass segregation would be to integrate (48.6) over the cluster using local averages for n, m and $\langle n^2 \rangle$ at each evolutionary timestep, and integrate the results over the cluster's history. The binary development that (48.6) predicts could then be compared with actual results in the N-body simulation. There are some analytic estimates based on this approach, (Heggie, 1975b), but much remains to be done.

Once binaries form they evolve by exchanging energy with field stars, and occasionally with other binaries. Heggie (1975a) gives a very detailed analysis, with an extensive bibliography, of the evolution of both hard and soft binaries. In each case the two main mechanisms for evolution are the cumulative effects of many distant random encounters, and the sharp effects of a few close encounters. The relative contribution of close and distant encounters depends on details of the velocity distributions of the field stars and the orbital parameters of the binaries. Typically, when averaged over the whole range of these properties, the close and distant encounters contribute about equally to energy changes in the binary.

For distant encounters with hard binaries it should not surprise us that the timescale for significant energy changes in the binary is of the order of the diffusion relaxation time τ_R. For close encounters this may not be so obvious but it follows readily from the following physical argument. Since a single close encounter changes the energy by a large fraction it only takes one or at most several such encounters to alter the binary significantly. This cross section for a typical binary whose hardness is of order unity is $\sigma \sim a^2 \sim R^2 N^{-2}$ from (48.4). Therefore the collision time $(n\sigma v)^{-1}$ is of order $NR/v \sim N\tau_{\mathrm{cross}}$, which is roughly τ_{R}. With the rate of change of E we can use (48.1)–(48.3), to see how the orbital parameters of the binary evolve, also on this same timescale. The result of this evolution, as mentioned earlier, is a tendency for soft binaries to break up and for hard binaries to become tighter. If a cluster has a roughly Maxwellian distribution, the soft binaries tend toward a stationary equilibrium distribution while hard binaries tend to accumulate. Over very long timescales this causes hard binaries to 'precipitate' into the center. We shall soon see how the presence of even one hard binary can affect the appearance of a system dramatically.

48.2. Formation by stellar dissipation – tidal energy transfer

Go, measure earth, weigh air, and state the tides.

Alexander Pope

A completely different way to produce binaries involves energy transfer from a star's orbit into its internal structure. Only two stars need interact in this process. Tidal dissipation of orbital energy has long been studied in relation to the earth–moon system, and essentially the same physical principles apply to stellar systems.

Two extended deformable gravitating bodies raise tides on each other as they pass by. The effects of these tides depend on the ratios of three timescales. First there is the time τ_{tide} for the tide to rise, or subsequently to relax. This depends on the effective viscosity of the star. Second, there is the time separating successive encounters τ_{sep}. Third, there is the time of the encounter τ_{enc} during which the stars are near their closest separation. If the internal viscosity is very small, then $\tau_{\mathrm{tide}} \ll \tau_{\mathrm{enc}}$ and the internal structure of the star is always in equilibrium with the external perturbing force. As a result, very little energy is dissipated within the star during the encounter. If the internal viscosity is large, then $\tau_{\mathrm{tide}} > \tau_{\mathrm{enc}}$ and the star responds sluggishly to the external force. The star's response lags behind the current direction and magnitude of the force. In the limiting case when $\tau_{\mathrm{tide}} \gg \tau_{\mathrm{enc}}$ the star feels the external perturbation as an impulse. This impulse distorts the stars, and after it is over they are left oscillating. Gradually viscosity damps these oscillations and their energy is eventually radiated as heat.

Since the total internal plus orbital energy of the stars is constant (the existence of stars other than the interacting pair is irrelevant here) the energy radiated away must have been lost from the orbits. After the encounter the mutual binding energy of the pair is greater than it was before. Alternatively, from the point of view of the orbits this change has been caused by the altered gravitational attraction of the tidally deformed stars. Tides enhance the gravitational multipole moments of the star which react back on the orbit. If the internal structure of the star does not respond instantaneously to the encounter, then its effects during approach will not cancel symmetrically with its effects during departure.

When $\tau_{sep} \gg \tau_{tide} > \tau_{enc}$, the star undergoes just one perturbation at any time, and the total energy removed from orbits of successive encounters adds linearly (neglecting higher order residual modifications to the star's structure, resulting, for example, from changes in the nuclear reactions or mixing produced by the oscillations). If this condition is not met resonances may be set up which change the evolution dramatically. These nonlinear resonances are quite unexplored for stellar systems. Some of their weak effects have been examined in binary stars and in the spin-orbit coupling of the earth and moon.

We can readily find the criterion for tidal dissipation in a close passage to produce capture. At closest approach, i.e. periastron, the stars will have a relative velocity $v_{peri} \approx [2G(m_1 + m_2)/r_{12}]^{1/2}$. For separations of a few stellar radii, this velocity is several hundred $km\,s^{-1}$. The velocity dispersion of stars in the cluster v_{rms} is typically $\sim 10\,km\,s^{-1}$ for a globular cluster. So in order to form a binary, it is necessary to dissipate a fraction $\sim (v_{rms}/v_{peri})^2$ of the maximum orbital kinetic energy. In globular clusters this is typically just a few tenths of 1%. In dense galactic nuclei, however, tidal dissipation ceases to be important after v_{rms} reaches $\sim 10^3\,km\,s^{-1}$.

Simple calculations will show us how to estimate the tidal dissipation both for impulsive and for slow encounters. We can do this without needing to know the viscous details of the dissipation mechanism. First, for the impulsive case, suppose a star of mass m moves quickly by another star whose mass is M, on an approximately straight orbit (Figure 47). Although the tides work on both stars we can safely neglect the second order effect of each star's distortion on the other star.

Focus attention on a small volume element in the star M of Figure 47. The z and z' axes are out of the diagram perpendicular to the velocity of star m, and the x–y and x'–y' planes contain the velocity vector of star m. The x' axis always points toward m. In the rotating primed coordinate system, the impulsive tidal acceleration components on this element are,

$$F_{y'} = -\frac{Gmy'}{R^3}, \tag{48.8}$$

$$F_{z'} = -\frac{Gmz'}{R^3}, \tag{48.9}$$

and

$$F_{x'} = \frac{2Gmx'}{R^3}. \tag{48.10}$$

Notice that the correction to the distance R in the denominators has been neglected in deriving $F_{y'}$ and $F_{z'}$ where it would just give a second order term. However, for $F_{x'}$ it is critical because we need to subtract the center of mass acceleration to obtain the differential tidal effect along x'.

From the geometry we find the standard relation between the rotating and fixed

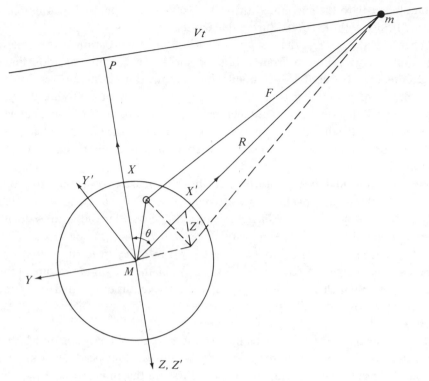

Fig. 47. The geometry of an impulsive tidal encounter.

coordinate systems

$$x' = \frac{xP}{R} + \frac{yS}{R}, \tag{48.11}$$

$$y' = -\frac{xS}{R} + \frac{yP}{R}, \tag{48.12}$$

$$z' = z, \tag{48.13}$$

with

$$R^2 = P^2 + S^2 \tag{48.14}$$

and

$$S = Vt \tag{48.15}$$

since $V^2 \gg GM/P$ is assumed to be constant. Transforming (48.8)–(48.10) to the fixed coordinates and letting v denote the perturbed velocity of the mass element in M, we see that its equations of motion are

$$\frac{dv_x}{dS} = \frac{Gm}{VR^3}\left(2x - \frac{3xS^2}{R^2} + \frac{3yPS}{R^2}\right), \tag{48.16}$$

$$\frac{dv_y}{dS} = \frac{Gm}{VR^3}\left(2y - \frac{3yP^2}{R^2} + \frac{3xPS}{R^2}\right), \tag{48.17}$$

$$\frac{dv_z}{dS} = -\frac{Gmz}{VR^3}. \tag{48.18}$$

To find the tidal velocity produced by the impulse we substitute (48.14) and integrate over the entire orbit, $-\infty \leq S \leq \infty$. The result is

$$v_x = \frac{2Gmx}{P^2 V}, \tag{48.19}$$

$$v_y = 0, \tag{48.20}$$

$$v_z = -\frac{2Gmz}{P^2 V}. \tag{48.21}$$

The total energy transferred to the star's interior follows by integrating the kinetic energy density $\frac{1}{2}[\rho(v_x^2 + v_z^2)]$ of the perturbation over the entire star

$$\Delta T = \frac{1}{3} M \left(\frac{2Gm}{P^2 V}\right)^2 r_*^2 \tag{48.22}$$

where r_*^2 is the star's mean square radius. If the density is constant, for example, r_* is $(\frac{3}{5})^{1/2}$ of the star's radius. In an actual star, ΔT will be somewhat less since the dense central regions have smaller perturbations. To get a feeling for this energy transfer we may compare it with the total gravitational energy of the star

$$\frac{\Delta T}{Gm^2/r_*} = \frac{4}{3} \frac{GM}{r_* V^2} \left(\frac{r_*}{P}\right)^4. \tag{48.23}$$

Thus, for a grazing encounter, the star's energy increases by a fraction $\sim V_{esc}^2/V^2$, and the result depends sensitively on the distance of closest approach. This ratio of energies is less than unity in the impulsive case, so such encounters can neither disrupt the stars nor form bound pairs. However, they can be a significant sink of orbital energy in some clusters, as the reader may see by forming the ratio $\Delta T/T$ with T the star's kinetic energy.

We may now continue by examining in more detail how the star is perturbed. This will also provide some insight into the harder problem of slow encounters. As the stars pass by, many internal modes become excited. When these perturbations are linear they do not interact with each other and they can be examined independently. Standard texts on mathematical physics, or potential theory, show that such perturbations may generally be represented by ordinary Legendre polynomials if there is azimuthal symmetry, and by spherical harmonics if there is not. During an impulsive encounter, most of the tidal energy transfer will go into the fundamental mode. This elongates the star in the direction of closest approach, but maintains its symmetry around the axis of elongation.

To determine the approximate amplitude of the induced pulsation we therefore assume that this fundamental mode dominates, and we relate its energy and amplitude. If θ is measured from the line of centers of the stars (see Figure 47), r_0 is the equilibrium radius of a shell under consideration, and $k(t)$ is a dimensionless amplitude of displacement, then the shape of the distorted star may be written as

$$r = r_0 \left[1 + \frac{k}{2}(3\cos^2\theta - 1) \right].$$ (48.24)

The quantity multiplying k is the Legendre polynomial $P_2(\cos\theta)$ which describes the fundamental mode. For small perturbations the perturbed volume element involving $d(r - r_0)$ is therefore

$$d\tau = \frac{k}{2}(3\cos^2\theta - 1)r_0^2 \sin\theta \, dr_0 \, d\theta \, d\phi.$$ (48.25)

The total kinetic energy of the perturbation is

$$\Delta T = \frac{1}{2}\int \rho \dot{r}^2 r_0^2 \sin\theta \, dr_0 \, d\theta \, d\phi = I_1 \dot{k}^2,$$ (48.26)

which defines I_1, using (48.24). For a uniform column displaced above the equilibrium surface by an amount $\Delta r = r - r_0$, the perturbation in potential energy density is $GM(r_0)\rho\Delta r/2r_0^2$. Multiplying this by the volume element $d\tau$ and integrating over the star gives the total potential energy of the perturbation

$$\Delta W = I_2 k^2,$$ (48.27)

which defines I_2. For constant density, $I_1 = 3MR_*^2/50$ and $I_2 = 3GM^2/50R_*$. From these potential and kinetic energies of the perturbation we may form its Lagrangian $L = \Delta T - \Delta W$ and obtain its equation of motion

$$I_1 \ddot{k} + I_2 k = 0.$$ (48.28)

This is just a simple harmonic oscillation, as befits a linear perturbation.

The solution of (48.28) is

$$k = A\cos(\omega t + \delta),$$ (48.29)

with frequency $\omega = (I_2/I_1)^{1/2}$, amplitude A and phase δ. The total energy of the oscillation is just its maximum kinetic energy and (48.26) gives

$$\Delta T = I_1 A^2 \omega^2 = I_2 A^2.$$ (48.30)

Equating this to the energy transferred in the encounter (48.22) gives the maximum perturbation

$$A = \left.\frac{\Delta r}{r}\right|_{\max} = r_* \left(\frac{2Gm}{P^2 V}\right)\left(\frac{M}{3I_2}\right)^{1/2} \approx \frac{2 \times 10^3}{\alpha^2 V(\mathrm{km\,s}^{-1})}.$$ (48.31)

The last expression is for approximately uniform stars of solar dimensions and $\alpha \equiv P/r_*$. This will only be valid for $A \lesssim 0.1$. Physically we expect close, slow encounters to excite higher modes of oscillation, as well as the fundamental mode, and then the approximations leading to (48.31) break down.

We now turn to the case of slow encounters. An important result, which we can take over from the previous analysis, is that the total energy in any linear mode is proportional to the square of its amplitude. Reversing our previous procedure we can estimate the maximum tidal amplitude induced by slow encounters, and use this to give a rough value for the energy transfer.

If a tidal acceleration $\sim GmR_*/P^3$ acts for a time τ, the resulting displacement is $\Delta r \sim GmR_*\tau^2/P^3$. To get maximum tides, P should be a grazing collision for which $\tau^2 \sim R_*^3/GM$ if both stars have about the same mass. Under these conditions the amplitude is

$$A = \frac{\Delta r}{r} \approx \frac{m}{M}\left(\frac{R_*}{P}\right)^3. \tag{48.32}$$

A more distant encounter will not lead to such a good resonance because the timescale for the encounter is much greater than the fundamental timescale for the star to oscillate. Less resonant encounters therefore reduce A below the value of (48.32), for $P \gtrsim 2r_*$.

If the perturbations remain linear, then from (48.30) and (48.32) the energy transfer is

$$\Delta T \approx I_2 \left(\frac{m}{M}\right)^2\left(\frac{R_*}{P}\right)^6 \approx \frac{3}{50}\frac{Gm^2}{R_*}\left(\frac{R_*}{P}\right)^6, \tag{48.33}$$

where the last expression is for an approximately uniform star. If the perturbation is very large and non-linear, the energy is still altered by an amount $\propto \Delta r \, d\tau \propto A^2$, as discussed before (48.27). Although the numerical coefficient in (48.33) is now increased, its exact value requires a more detailed analysis. The main result is that the energy transfer decreases extremely rapidly as the impact parameter P increases. Notice that forcing the impulsive approximation (48.23) to the slow limit by setting $V^2 \approx GM/P$ would not give the correct dependence (48.33) on R_*/P. This is because a slow encounter is a much more sensitive resonance process than an impulsive encounter.

It is these slow encounters which can dissipate enough orbital energy to form a binary. Equation (48.33) shows that the binding criterion discussed at the beginning of this section, that $\Delta T/T \gtrsim (v_{\mathrm{rms}}/v_{\mathrm{pcri}})^2$, leads to

$$P \lesssim P_{\mathrm{bin}} \approx \left(\frac{GM}{R_* v_{\mathrm{rms}}^2}\right)^{1/6} R_* \approx \left(\frac{v_{\mathrm{esc}}}{v_{\mathrm{rms}}}\right)^{1/3} R_* \tag{48.34}$$

for roughly equal mass stars. Here v_{esc} is the escape velocity at the star's surface. The resonance is so highly tuned that $1 \lesssim P_{\mathrm{bin}}/R_* \lesssim 3$ for a wide range of conditions, the actual value depending on details of the model.

To determine the cross section for capture we must now distinguish between the original impact parameter P and the distance of closest approach R_{min} just before capture. Although these are the same for a straight line orbit, they differ for curved orbits leading to bound states, as discussed in Section 17. The same analysis of energy and momentum conservation which gave (17.6) now provides the relation

$$\frac{P^2}{R_{\mathrm{min}}^2} = \frac{2G(M+m)}{R_{\mathrm{min}}v^2} + 1, \tag{48.35}$$

where we may usually neglect the second term on the right hand side and v is the relative velocity. Therefore the capture cross section is

$$\sigma_{\text{cap}} = \pi P^2 \approx 2\pi G(M + m)\frac{R_{\min}}{v^2}. \tag{48.36}$$

The number of captures per second per unit volume occurring between masses m and M is $\dot{n}_{\text{cap}} = n(M)n(m)\sigma_{\text{cap}}v$. To find the total number of tidal binaries which form, we should integrate $\dot{n}_{\text{cap}}$ over the spatial and velocity distributions of the system. In keeping with our current level of approximation, however, we may simply replace v by a typical value, $\sim v_{\text{rms}}$, and consider a uniform cluster of radius R_c with all stars of mass m. Then the fractional rate of binary formation becomes

$$\frac{\dot{N}_{\text{tid}}}{N} \approx \frac{R_{\min}}{R_c}\frac{1}{\tau_c} \sim \left(\frac{v_{\text{rms}}}{v_{\text{esc}}}\right)^2 \frac{1}{N\tau_c} \propto \left(\frac{v_{\text{rms}}}{v_{\text{esc}}}\right)^2 \frac{1}{\tau_R}. \tag{48.37}$$

Here we have applied the virial theorem, the relation (48.34) with $R_{\min} \approx P_{\text{bin}} \approx R_*$, and the relation (2.11) between the cluster's crossing time and its relaxation time. The essential result is that in one crossing time a fraction $\approx R_*/R_c$ of all stars are tidally captured. A more detailed calculation (Press & Teukolsky, 1977) which sums over all the distorted modes of an $n = 3$ polytrope and integrates over a Maxwellian velocity distribution gives a capture rate per unit volume for stars of identical mass and radius

$$\dot{n}_{\text{tid}} = 2 \times 10^{-8} \left(\frac{R_*}{R_\odot}\right)^{0.9} \left(\frac{m}{m_\odot}\right)^{1.1} \left(\frac{100\,\text{km}^2\,\text{s}^{-2}}{\langle v^2 \rangle}\right)^{0.6} \left(\frac{n}{10^4\,\text{pc}^{-3}}\right)^2 \text{pc}^{-3}\,\text{yr}^{-1}. \tag{48.38}$$

Comparison with (48.6) directly shows the conditions (such as in globular clusters) where tidal capture dominates three-body capture.

When most binaries are first formed by either tidal or three-body capture their orbits will be quite eccentric. This is because the stars' original paths were hyperbolic. Having formed, subsequent tidal dissipation will cause the binary to harden. Equation (48.2) shows that for a given angular momentum (which remains constant) the orbit with greatest binding energy is a circle: $e = 0$. So as the binary hardens it becomes more circular. During each orbit soon after formation, a fraction $\sim (v_{\text{rms}}/v_{\text{peri}})^2$ of the energy is typically lost, as described near the beginning of this subsection (also compare 48.37). Therefore the orbit circularizes after roughly $(v_{\text{peri}}/v_{\text{rms}})^2$ revolutions. These tidal processes apply to the capture and merging of galaxies as well as stars. In the case of galaxies, tidal dissipation may accentuate the effects of dynamical friction during a merger. Here the dissipation occurs not through the radiation of heat, but through the ejection of high velocity stars into a halo or beyond.

48.3. Effects on cluster evolution

Binaries may strongly influence both the gravitational evolution and the astronomical evolution of a cluster. One of their milder astronomical effects is to modify the relation between colors and magnitudes of stars. Hard binaries, especially those involving compact white dwarfs, neutron stars, or black holes, may also become strong X-ray and γ-ray sources. Their radiation, often changing quickly in time, results from complicated patterns of mass flow and accretion in the binary system. On a larger scale, binaries whose components are supermassive objects or black holes with $M \gtrsim 10^6 m_\odot$ may eject smaller stars from the nucleus of a galaxy. Supermassive binaries may themselves be broken up or ejected, wreaking havoc on the stellar distribution in their path and leaving behind a trail of radiating debris.

Gravitational evolution, however, is our main theme here. Both Equation (48.6) and (48.38) show that high densities and mass segregation favor binary formation. These properties characterize the later stages of cluster evolution, after a central core forms. Additional binaries may remain from the initial state, produced by the poorly understood processes of multiple star formation. Through encounters with single stars, hard binaries inject kinetic energy into the system. (Soft binaries, which are less important, absorb some kinetic energy, but then break up to replace much of that energy in the system.) The most effective hard binaries in this regard are the less tightly bound ones. They have larger interaction cross sections, and do not usually transfer so much energy to the single star that it escapes completely.

Details of how binaries scatter single stars will be discussed in Section 49. Fortunately, a little pure thought tells us the most important results for cluster evolution without needing these details. If the single star's closest approach is more than two or three times the binary semimajor axis, then we would expect the binary to interact like a single object. Its effective mass would be the total mass of the binary, but its monopole moment would dominate the scattering. On the other hand, if the single star comes within one or two binary radii, a large-angle scattering occurs which significantly alters the energy of both the star and the binary. These will dominate energy transfer from the internal energy of the binary to the system.

Let us suppose (consistent with three-body experiments) that an average large-angle scattering roughly alters the binary energy by an amount

$$\Delta E \approx \tfrac{1}{2}|E| = \frac{Gm^2}{4a} \tag{48.39}$$

in an equi-mass system. This implies a velocity for the departing body of about $20(m/m_\odot)^{1/2}(a_{\mathrm{AU}})^{-1/2}\,\mathrm{km\,s^{-1}}$. The single stars in a unit volume then gain energy at a rate

$$\dot E \approx (\Delta E) n_{\mathrm{sing}} n_{\mathrm{bin}} \sigma v. \tag{48.40}$$

The cross section is $\sigma \approx \pi a^2$, where a is given by (48.4), and the velocity dispersion v is again obtained from the virial theorem. For simplicity, let the system be homogeneous with $N_{\mathrm{bin}} < N$. Combining these results gives the total rate of transfer from binary binding energy into kinetic energy throughout the system

$$\frac{\dot E_{\mathrm{bin}}}{E} \approx \frac{3}{2}\frac{N_{\mathrm{bin}}}{N}\frac{1}{N\tau_{\mathrm{c}}} \approx 10^{-2}\frac{N_{\mathrm{bin}}}{N\ln(N/2)\tau_{\mathrm{R}}}. \tag{48.41}$$

The last equality uses the relation (2.11) between crossing time and relaxation time (including the more exact factor of $\frac{1}{2}$ in the potential energy for a uniform spherical system). When applying (48.41) to realistic scattering over a distribution of binary sizes in an inhomogeneous model cluster, it is important to remember that the numerical coefficient may be several times larger or smaller than the value here.

Tidal dissipation from binary formation heats the stars but cools the stellar system. We expect this cooling to be less than the heating by hard binary scattering for two reasons. First, hard binaries are hard to destroy. Since they accumulate there are likely to be more present at a typical time than there are tidally forming binaries. Second, the energy input of each hard scattering is greater than the energy lost, $\sim \frac{1}{2}mv^2$, in each tidal capture. From (48.38) we have

$$\frac{\dot{E}_{\text{tid}}}{E} = \frac{\dot{N}_{\text{tid}}}{N} = \frac{\dot{n}_{\text{tid}}}{n} = 9 \times 10^{-5} \left(\frac{r_*}{r_\odot}\right)^{0.9} \left(\frac{m}{m_\odot}\right)^{-0.9} \left(\frac{v}{10\,\text{km s}^{-1}}\right)^{1.8} \frac{1}{\tau_{\text{R}} \ln(N/2)}$$

(48.42)

in terms of the relaxation time (2.9). For typical globular cluster parameters, gains from hard binary scattering exceed tidal losses if more than $\sim 1\%$ of all stars are hard binaries. Whether this is true in any particular model may depend critically on the initial binary distribution.

How large do $\dot{E}_{\text{bin}}$ and $\dot{E}_{\text{tid}}$ have to be to change the homology developments described in Sections 44 and 45 significantly? When evolution proceeds along a quasi-static virial sequence the radius changes according to

$$\frac{1}{R}\frac{dR}{dt} = \frac{2}{N}\frac{dN}{dt} - \frac{1}{E}\frac{dE}{dt}.$$

(48.43)

In the simple evaporative cluster (45.4), $dE/dt = 0$ and $\dot{N}/N \approx -\varepsilon/\tau_{\text{R}}$, where $\varepsilon \approx 10^{-2}$. So this cluster would be strongly affected if $\dot{E}/E$ were greater than about $10^{-2}/\tau_{\text{R}}$. For this to happen, (48.41) shows that a large fraction of all stars must be hard binaries! Similarly (48.42) shows that tidal energy loss is unlikely to be important except in extreme circumstances (which may sometimes occur in particular models). Comparison with the more general homology evolution of cluster cores (44.8) also shows that large numbers of binaries are needed to have a strong overall effect on these cores. More detailed models (e.g., Inagaki, 1984) reach this same basic conclusion.

Even though these binaries do not alter the global dynamics of the cluster they can have an important effect on the very central parts. This is especially true if mass segregation causes heavy binaries to accumulate near the center. By ejecting single stars from the center, even a very small number of binaries (just one!) can decrease the average density there and dramatically alter the core's structure. This result is almost always found in direct N-body simulations, but is hard to treat analytically because it is so irregular, depending on detailed orbits. It is also possible for several binaries to interact among themselves at the center. In the next section we describe some statistical results, mainly experimental, for the detailed interactions of these binaries with single stars.

49

Slingshot

And certain stars shot
Madly from their spheres.

Shakespeare

The simplest examples of the gravitational slingshot are a binary which scatters a single star and the breakup of an initially bound three-body system. Resulting recoil may eject both the single body and the binary from the system. If galactic nuclei contain supermassive objects which form binaries, these processes may have especially dramatic astronomical consequences (Saslaw, Valtonen & Aarseth, 1974). They are also important in ordinary star clusters with $N \lesssim 10^2$, and in the very center of richer clusters.

Performing large numbers of three-body experiments shows how scattering behaves. There is an amusing contrast with high energy physics. Our (inexpensive) gravitational accelerator is a computer. We experiment not to find the microscopic law describing the interaction – we've known that since Newton – but to understand realistic applications of this law. Even our relatively low energy particles, typical stars moving at $\sim 10^2 \, \mathrm{km \, s^{-1}}$, have kinetic energies exceeding $\sim 10^{50} \, \mathrm{GeV}$!

Although there are a great many exact mathematical results on the analytical dynamics of three-body systems, they usually apply only under very restricted circumstances. Normally they are asymptotic or perturbation calculations (see Arnold, 1978; Pars, 1965). These are not sufficient to understand the rich range of important phenomena occurring in realistic physical scattering, so we must resort to numerical simulations and approximate non-linear analyses (see Heggie, 1975(*a*, *b*), reference in Section 48).

Basic input into a three-body slingshot experiment is a binary and a third approaching body. The final results, after complex interactions which jumble all the orbits together, can take several forms. The simplest is a *flyby*, when the incoming body leaves immediately after the interaction. *Resonance* occurs when the incoming body is captured by the binary, forming a close triple system which breaks up rapidly. *Exchange* occurs when one of the original binary components is immediately ejected and the other forms a new binary with the incoming body. *Ionization* occurs when the incoming body unbinds the binary and all three objects fly apart.

Which one of these results actually occurs depends on the precise initial conditions. A complete description of a three-body configuration requires 21 parameters: three positions, three velocities, and a mass for each body. Of the total, three positions and three velocities just represent the center of mass, so in a frame where the origin moves with the center of mass 'only' 15 parameters are needed. Various combinations of orbital properties may be used to represent these parameters. Not all these properties are equally important, and some can be scaled to unity. Moreover, most basic physical results involve averages over some orbital parameters. Therefore we need not go into details of the geometry here; they can be found elsewhere (e.g., Saslaw, Valtonen & Aarseth, 1974).

Early extensive experiments (Saslaw, Valtonen & Aarseth, 1974; Valtonen, 1975) established several general results. For cases with total negative energy, ionization cannot occur. The relative probability of flyby, resonance and exchange depends primarily on the pericenter distance of the nearly parabolic orbit of the incoming body. It depends secondarily on the relative mass of the incoming body. It also has a tertiary dependence on the initial eccentricity of the binary and on the inclination of its orbit to that of the incoming body. If the incoming pericenter distance is $\gtrsim 3$, in units of the initial binary semimajor axis, there is a nearly 100% probability of a flyby. At smaller distances, exchanges and resonances dominate. Lowering the incoming body's mass decreases the probability of exchange. The exchanges are then replaced by flybys at large distances and by resonances at small pericenter distances. If the incoming body has the smallest mass, the probability of exchange is almost negligible. Increasing the energy of the incoming body increases the probability of flybys at the expense of resonances. For all orbits in a plane there are more exchanges and resonances at larger distances if the binary has high initial eccentricity. For most orbits, a resonance or exchange is more probable at large distances if the orbits are direct rather than retrograde.

In the case of a flyby, the most interesting quantity is the energy gained or lost by the passing body. This decreases rapidly with pericenter distance, usually becoming negligible at distances $\gtrsim 4$. At shorter distances it depends strongly on the relative phase of the binary and incoming orbits. This phase dependence becomes especially pronounced for high eccentricity, for low inclination and for small pericenter distance, as we would expect. The phase dependence does not alter much with mass ratio until the incoming mass exceeds about twice the total mass of the binary, when exchanges become more common. Typical fractional velocity changes for pericenter distances between about 1 and 2.5, range from about 80% to 10%, if all masses are the same order of magnitude. Decreasing the relative mass of the incoming body, also decreases its velocity change, since it distorts the binary's orbit less strongly.

At close approaches, strong resonances between orbits lead to temporary capture and ejection. In order for such resonances to occur, the energy of the binary must be closely matched with the energy of the incoming body. This is so the incoming body has a good chance to move slowly relative to one of the binary components, and deflect it significantly. At incoming velocities much higher than the binary's internal velocity, the resonance cross section decreases rapidly. This suggests expanding the velocity dependence of the cross section in powers of E_{bin}/E_{inc}. Detailed analyses (Heggie, 1975(a)) show this is the case, with the first term dominating: $\sigma \approx \pi a^2 v_{bin}^2/v_{inc}^2$. The velocity distribution of the escaping body usually has a broad peak at $v_{esc}^2 \approx 0.5|E_{tot}|/m$ with m the average mass of the escaping body. For high total angular momentum, however, v_{esc} is less by a factor ~ 2. There is usually a high energy tail in this velocity distribution, extending to two or three times v_{esc}. Parameters of the experiments other than m and L do not seem to affect the escape velocity strongly.

The distribution of escape velocities of the single body of mass m can be expressed in a simple form due to Heggie which fits a wide range of cases (Saslaw *et al.*, 1974)

$$f(v)\mathrm{d}v = 3.5\frac{mv}{z}\left(1 + 0.5\frac{mv^2}{z}\right)^{-4.5}\mathrm{d}v. \qquad (49.1)$$

The mass m can be associated with the particle which is most likely to be ejected, i.e., a weighted average of the lightest and second lightest mass, and z is the absolute value of the total energy. This distribution works well both for initially close systems, and loosely bound systems, including resonance cases. It is essentially a statistical result of significant orbital mixing. The process of ejection does not require a close encounter. Often the escape is preceded by a triangular configuration with all bodies roughly equidistant.

Resonances also cause exchanges; usually the least massive body escapes. Its typical squared escape velocity is also of the order of magnitude $|E_{tot}|/2m$. If all three masses are the same, and the original semimajor binary axis is a, then the N-body experiments show that the exchange cross section is given by

$$\sigma_{exch} \approx \frac{80\pi G^3 m^3}{3av^6} \qquad (49.2)$$

for incoming velocity v greater than about twice the velocity for which the total energy of the whole system is zero (Hut & Bahcall, 1983; Hut, 1983).

If the energy of the incoming body is so high that the total energy of the system is positive, it can ionize the binary. This also involves a resonance cross section which at high velocities decreases as (Hut, 1983)

$$\sigma_{ion} \approx \frac{40}{3}\pi a G\frac{m_{inc}^2}{m_{bin}}\frac{1}{v^2}. \qquad (49.3)$$

Like σ_{exch}, this cross section is an average over orbital parameters. Higher order approximations show that it also depends somewhat on the eccentricity of the original binary.

In addition to binary scattering, close three-body systems may also condense directly from fragmenting gas clouds. The final properties of these systems are very similar to those resulting from resonance capture. Typically after about 10^2 initial (or binary) crossing times the lightest body usually escapes. Escape can take two or three times longer, however, if all three masses are the same. Very occasionally, a more stable form of the restricted three-body problem arises with the light mass circling around a distant tight binary. After escape, momentum recoil ejects the binary in the opposite direction. In general, a low angular momentum for the total system tends to result in a high eccentricity for the remaining binary.

When escape occurs from a three-body system the angular distribution of the ejected single body and the binary is strongly peaked in the plane of the total angular momentum. This occurs for nearly all orbital parameters. It is also interesting to dress the slingshot by surrounding each body with a disk of test particles to see how they fare during the ejection (Lin & Saslaw, 1977). This may apply to the accretion disks of compact objects. Two of the main results which emerge from N-body experiments are that retrograde accretion disks survive the ejection much more easily than directly rotating disks, and that survival is also easier if the ejected mass is about equal to the total binary mass. Both these results agree with our intuition.

Stars scattered by slingshot processes in the nucleus of a galaxy may form a halo around the galaxy. If the binary contains two compact supermassive objects and it slings out ordinary stars, this method of halo formation may be quite efficient. Could it explain some of the structure of giant elliptical galaxies which are strong radio sources (see Saito, 1977)?

The internal dynamics of a star flung out by a binary involve a race between escape and tidal destruction. Which wins? Applying the analysis of (1.4) to a star of mass m and radius d, which approaches a binary of mass M, shows that tidal disruption begins at a distance $l \approx d(M/m)^{1/3}$. At this distance, the timescale for the star's disruption, $(G\rho_*)^{-1/2}$, is also about equal to its orbital escape time, $(GM/l^3)^{-1/2}$. So the star must not go closer if it expects to survive. The maximum orbital energy (per unit mass) that the binary could impart to the star is of order GM/l. Therefore, in terms of the escape velocity at the star's surface, v_{esc}, the maximum slingshot velocity the star could acquire without tidally disrupting is of order

$$v_{max} \approx v_{esc} \left(\frac{M}{m} \right)^{1/3}. \tag{49.4}$$

For greater velocities the slingshot would enucleate the star, leaving its core to escape as a remnant.

50

Role of a central singularity

There are too many stars in some places and not enough in others, but that can be remedied presently, no doubt.

Mark Twain

One can imagine the fun Mark Twain would have had with the concept of black holes and their influence. Unfortunately, he died in 1910, six years before K. Schwarzschild discovered these singular solutions of general relativity. Although this book is concerned with Newtonian systems, in which the ratio GM/Rc^2 of gravitational energy to rest mass energy is very small, the presence of a central singularity which destroys or absorbs stars can have a significant effect on the surrounding Newtonian dynamics. Actually a black hole is just one example of such a singularity. Others are a supermassive star or spinar, and even just a region of such high density that stars disrupt inside it by physically colliding (Section 52).

We have already seen (Section 40.2) how a massive gravitating point modifies the distribution of surrounding stars when this distribution obeys a simple polytropic equation of state. The large point mass induces a central density cusp, in contrast to the normal flat distribution. We did not approach the central mass very closely in Section 40.2 because it destroys the polytropic behavior. Here we venture further in to see how the stellar orbits are distorted.

Let us suppose that the mass of the hole M_h is much greater than the mass of any individual star m_*, but much less than the total mass of all stars in the cluster $m_* N$. The hole is characterized by a destruction radius, r_D, from which stars never return. This radius, to order of magnitude is the larger of either the Schwarzschild radius

$$r_s = \frac{2GM}{c^2} = 3 \times 10^5 (M_h/m_\odot) \, \text{cm} \tag{50.1}$$

or the tidal breakup radius

$$r_T \approx 1.5 \times 10^{11} \left(\frac{M_h}{m_\odot}\right)^{1/3} \left(\frac{m_\odot}{m_*}\right)^{1/3} \frac{r_*}{r_\odot} \, \text{cm}. \tag{50.2}$$

(Recall $r_\odot \approx 7 \times 10^{10} \, \text{cm}$.)

The Schwarzschild radius characterizes simple non-rotating black holes. Texts on general relativity show that a particle's energy and angular momentum determine how closely it can approach a black hole before becoming unstable and falling in.

For a circular orbit this is $3r_s$, a fairly typical value. The tidal 'Roche' radius follows from the analysis at the end of Section 49 for stars whose structure is similar to the sun. As section 48 discusses, stars with impact parameters $\sim 2r_T$ may be captured by tidal dissipation, and eventually spiral into r_s, if $r_s < r_T$. Notice that for $M_h \gtrsim 3 \times 10^8 m_\odot$ the Schwarzschild radius exceeds the tidal radius. Therefore very massive black holes are likely to swallow whole stars. These may release less tidal gas to radiate during slow accretion. People have speculated (e.g., Hills, 1975) that this might be important for energy sources in quasars and active galactic nuclei.

The destruction region r_D is so small that only a few stars are buzzing around it at any given time. To determine their fate it becomes necessary to follow complicated individual orbits. The relative velocities of these stars are usually so high that they do not influence each other very much, unless they come so close they collide. This high velocity regime occurs within a region where the velocities of the stars, $\sim (GM_h/r)^{1/2}$, are about equal to the escape velocity from a star's surface, i.e.

$$r_{coll} \approx 7 \times 10^{10} \left(\frac{M_h}{m_\odot}\right) \frac{r_*}{r_\odot} \frac{m_\odot}{m_*} \, \text{cm}. \qquad (50.3)$$

Under most circumstances $r_s < r_T < r_{coll}$.

Farther from the singularity than r_{coll}, enough stars are swarming about that it is reasonable to estimate their statistical distribution. These stars may or may not be bound to the singularity. Even if they are unbound, however, the singularity strongly influences their motions out to a distance (2.1)

$$r_h \approx \frac{GM_h}{v_c^2} \approx 2 \times 10^{16} \left(\frac{M_h}{10^3 m_\odot}\right) \left(\frac{n_c}{10^5 \, \text{pc}^{-3}}\right)^{-1} \left(\frac{r_c}{1 \, \text{pc}}\right)^{-2} \left(\frac{m_\odot}{m_*}\right) \text{cm}. \quad (50.4)$$

Here the virial theorem is used to estimate the average velocity dispersion v_c^2 in the cluster of radius r_c and average density n_c.

Within r_h, stars move on roughly Keplerian orbits with velocities $v \approx (GM_h/r)^{-1/2}$. The energy and, more importantly, the angular momentum of these orbits diffuse slowly by two body encounters on the relaxation timescale τ_R (2.9). There is no guarantee that these orbits are isotropically distributed. The collapsing matter which formed the singularity could have segregated orbits as discussed in Section 46. Anisotropic orbits may remain from an inhomogeneous collapse, since collapse occurs on a dynamical crossing time if it is close to free-fall, and this is much shorter than τ_R. Anisotropy is also produced by the absorption–destruction process which depopulates low angular momentum orbits.

Models which incorporate these anisotropies rapidly become very complicated. So, for a rough overview, it is easiest to assume that most of the important physics can be described by an isotropic distribution function. From Section 39 we recall that it will depend only on the energy: $f = f(E)$ with $E = v^2/2 - GM_h/r$. There are no fully self-consistent derivations of $f(E)$ in these circumstances, but it has been conjectured that a simple power law

$$f(r,v) \propto E^p \qquad (50.5)$$

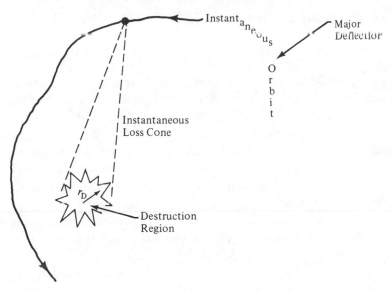

Fig. 48. Schematic illustration of a star orbiting a singularity in a cluster.

would be a reasonable approximation, and this is supported by numerical experiments (see Shapiro & Marchant, 1978, and references in it). For this power law, with a velocity dispersion $v \propto r^{-1/2}$, the density distribution is

$$n(r) = \int f \, \mathrm{d}^3 v \propto r^{-(p+3/2)}. \tag{50.6}$$

To determine p we may specify the overall outflow of kinetic energy from r_D. As some stars fall more deeply into the gravitational well around r_D they transfer part of their kinetic energy to the surrounding stars, as discussed in Sections 43–6. This occurs on the relaxation time τ_R, during which the stars receiving kinetic energy drift outward. The energy drifting through any spherical shell at radius r is $\sim n(r) \, v^2(r) \, r^2 v_\mathrm{drift} \approx n(r) \, v^2(r) \, r^3 / \tau_\mathrm{R}(r)$. Assuming that this energy flow is independent of radius (Bahcall & Wolf, 1976), and substituting (2.9) for τ_R with $v \propto r^{-1/2}$ gives $n(r) \propto r^{-7/4}$, in reasonable agreement with Fokker–Planck simulations. This means that the outward flow of energy is almost completely controlled by the inner boundary condition at r_D with no additional sources or sinks of energy in the cluster.

Such a solution cannot apply everywhere, however. The angular momentum of a star anywhere in the cluster may happen to diffuse to such a low value that it plunges directly into r_D. This leads to the idea of a 'loss cone' (Frank & Rees, 1976), illustrated in Figure 48. Writing the specific energy of a star in terms of its radial and tangential velocities and specific angular momentum $J = r v_\mathrm{t}$, we have

$$E = \tfrac{1}{2} v_r^2 + \frac{J^2}{2r^2} - \frac{GM_\mathrm{h}}{r}. \tag{50.7}$$

For a closest approach just outside r_D, the star's radial velocity is zero, implying that the minimum angular momentum for avoiding the loss cone is

$$J_{\min}(E) = \left[2 \left(E + \frac{GM_{\mathrm{h}}}{r_{\mathrm{D}}} \right) \right]^{1/2} r_{\mathrm{D}}. \qquad (50.8)$$

Since stars whose angular momentum is less than $J_{\min}$ are destroyed, unless a subsequent deflection flicks them out of the loss cone, the distribution function will be significantly depleted for $J < J_{\min}$. This means that the distribution function cannot generally be isotropic, and so $f = f(E, J)$. The caveat in the first sentence is important. It determines the radius where loss-cone depletion becomes prominent. Angular momentum is transferred by diffusion on a timescale $\tau_{\mathrm{R}} \propto v^3/n \propto r^{1/4}$ for the run of velocity and density discussed earlier as an example. The timescale for capture by the destruction region is $\tau_{\mathrm{cross}} \propto r/v \propto r^{3/2}$. Thus there will be a critical radius where these two timescales are equal. Within this radius, there will not be enough time for momentum diffusion to change the star's orbit before it is destroyed. Outside this radius, stars may hope to survive.

To find the actual value of this critical radius, it is necessary to know the detailed structure and orbit interactions in a model cluster. It is worth noting that the $r^{-7/4}$ density distribution implies that the critical radius is less than r_{h}, so that the gravitational potential of the singularity dominates the velocity field in the region of loss-cone capture. Since τ_{R} depends fairly strongly on radial position, especially near the critical radius, and stars of different angular momenta spend different amounts of time at different radial positions, it is necessary to average τ_{R} over the orbits. This also means that the critical radius will depend to some extent on the way in which the central singularity formed, and the orbits it consequently left behind. Significant orbit segregation (Section 47) may result. Some of these transient properties may damp out if the distribution tends toward a steady state on the τ_{R} timescale. But by then many of the astronomically interesting effects of the singularity may also be over.

Most of these astronomical effects, especially the radiative ones, are governed by the rate of mass infall. Again this depends on details of the orbits close to r_{D}. If we simply assume that all stars from a basically isotropic velocity distribution which reach r_{D} are in fact absorbed, then we can easily estimate the swallowing rate to order of magnitude. Amusingly, the essential analysis of Section 17 applies equally to a black hole swallowing stars as to a star swallowing dust grains. The second term of (17.6) dominates the cross section, so we have

$$\frac{\mathrm{d}m}{\mathrm{d}t} \approx \rho_* \sigma v \approx 10^{-15} \frac{\rho_*}{(m_\odot/\mathrm{pc}^3)} \frac{r_{\mathrm{D}}}{r_\odot} \frac{M_{\mathrm{h}}}{m_\odot} \frac{1}{v_*(\mathrm{km\,s}^{-1})} \, m_\odot \, \mathrm{yr}^{-1}. \qquad (50.9)$$

Here v_* is an average root mean square velocity of the stars at $r > r_{\mathrm{h}}$, far from the singularity. Moreover ρ_* represents an average mass density of stars, and will need to be determined for specific models, as will v_*. The details of (50.9) will also have to be modified for the relative numbers and location of bound and unbound orbits. Nevertheless, since $10^{-15} \, m_\odot \, \mathrm{yr}^{-1}$ corresponds to a rest mass energy loss rate of

$\sim 6 \times 10^{31}\,\mathrm{erg\,s^{-1}}$, it is clear that quasar-like energies are easily possible for the conditions in galactic nuclei. On the smaller scale of globular clusters this process has also been suggested as a source of high energy activity.

In addition to radiation from decomposing stars as they fall into the singularity, it may be possible to observe some direct dynamical effects on the surrounding velocity disperion. This requires high spatial and spectral resolution. Merely examining the distribution of light in the nucleus is inadequate, because these regions may well have mass–luminosity ratios which change rapidly in space and time. Synchrotron, Compton, free–free and free–bound emission, as well as starlight, may all combine to reproduce a confusing pattern. Within this pattern it may be hard to ferret out a velocity dispersion of stars which varies as $v \propto r^{-1/2}$ around a massive singularity.

The relation between dynamics and luminosity – especially starlight – may also be obscured by anisotropy in the velocity dispersion. This anisotropy may remain from the recent formation of the singularity, possibly with a companion. It may also be promulgated by large scale density inhomogeneities, strongly correlated stellar orbits, or continuous infall into the nucleus. In other words, it does not necessarily vanish after a relaxation timescale τ_R. When orbits are not highly correlated the effects of anisotropy can be gleaned from the moments (8.19) of the collisionless Boltzmann equation (with suitable closure approximations). Depending on their form, Equations (8.14) and (8.15) show that velocity anisotropy can either steepen or reduce a density gradient.

This result perhaps appears more transparent when (8.15) is rewritten in spherical polar coordinates

$$\frac{\mathrm{d}(n\overline{v_r^2})}{\mathrm{d}r} = -n\frac{GM(r)}{r^2} - n\left(\frac{2\overline{v_r^2} - \overline{v_T^2}}{r}\right). \tag{50.10}$$

Only if motions are isotropic (or specially contrived) does the tangential mean square velocity equal twice the radial value. To solve (50.10) of course requires additional assumptions, as discussed in Section 8. Even so, we see that if radial velocities become more dominant with increasing radius, then the radial density gradient decreases faster than for the isotropic case. This, in turn, alters the luminosity pattern.

The overall view of a galactic nucleus where a compact massive object or black hole breaks stars apart may have been described by Wallace Stevens:

> First one beam, then another, then
> A thousand are radiant in the sky.
> Each is both star and orb; and day
> Is the riches of their atmosphere.

51

Role of a distributed background

The effects of an extended background on a cluster have more in common with those of a central point mass than might first seem to meet the eye. In both cases the stellar orbits are dominated by an external mass distribution, not by their own self-consistent gravitational field. One result is that the basic physical processes discussed in Part I must be averaged over the particular distribution of orbits for each system.

Observational evidence for a distributed background comes from several sources. Rotation curves of galaxies, particularly of giant spirals measured with 21 cm emission from neutral hydrogen clouds, sometimes appear to be flat far beyond the optical image of the galaxy. This non-Keplerian relation between distance and velocity may indicate a halo surrounding the galaxy. Its total mass is very uncertain. To produce a significant dynamical effect it would have to be at least comparable with the stellar mass of the galaxy, and it could be much larger. The nature of this halo is unknown. Within galaxies, the orbits of globular clusters are dominated by the distributed stellar background, and possibly by the halo as well.

On a larger scale, Sections 41 and 42 already described the 'missing mass' problem for clusters of galaxies. The resolution of the problem may involve an unseen mass distributed throughout the cluster. The mass of very hot intracluster gas found in X-ray emission is usually an order of magnitude too small. Various suggestions such as gas lost from protogalaxies as they formed, or subsequently stripped from galaxies by their ram pressure against initial surrounding gas, or massive neutrinos and other hypothetical elementary particles abounding from the early Universe are possible. Many such models have been worked out in detail, but which, if any, are correct is again uncertain (see also Section 35.1).

Whatever the origins of distributed backgrounds, their dynamical effects on spherical systems can be important. The most straightforward of these effects is modification of the tidal limit. Figure 49 illustrates this for a spherical subsystem (e.g., globular cluster or a galaxy) inside a spherical background (e.g., a galaxy or a cluster of galaxies). The distance **r** to an arbitrary element whose distance from the center of the subsystem is **s** is clearly related to the distance of the center of the

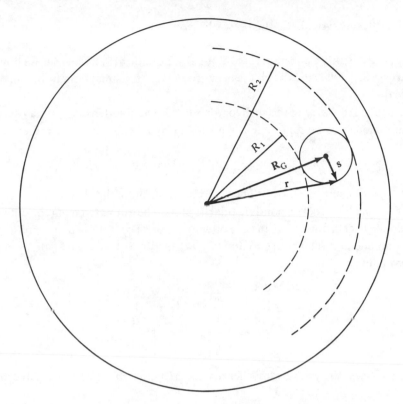

Fig. 49. The geometry of a spherical subsystem interacting tidally with a spherically distributed background.

subsystem by $\mathbf{r} = \mathbf{R}_G + \mathbf{s}$. The background force on an element at $\mathbf{r}$ has three contributions: the force from matter more central than the subsystem, the force from a shell between R_1 and r which overlaps the subsystem, and a correction term to exclude the background within the subsystem (since it counts as part of the subsystem's own mass). This last term amounts to saying that the background contains a sphere of negative mass within the subsystem. There is no net force on the element from the spherical shell beyond r. Therefore the total background force is

$$\mathbf{F}_b = -\frac{GM_b(R_1)}{r^3}\mathbf{r} - \frac{G\mathbf{r}}{r^3}\int_{R_1}^{r} 4\pi x^2 \rho_{\text{shell}}(x)\mathrm{d}x + \frac{GM_b(s)}{s^3}\mathbf{s}, \qquad (51.1)$$

where $M_b(R_1)$ is the background mass within R_1 and $M_b(s)$ is the background mass in the sphere within s.

At this stage physical intuition should shout that there are two obvious limiting cases. If the subsystem is at the center of the background, then $\mathbf{F}_b = 0$ from the symmetry of the situation. Equation (51.1) shows this immediately, for then $R_G = 0$, $\mathbf{r} = \mathbf{s}$, $R_1 = 0$, $M_b(R_1) = 0$, and the last two terms cancel out. The second limiting case is that $F_b = $ constant for a constant background density, and there is no tidal force on the subsystem. In this case the component of the force produced by $M_b(s) \propto s^3$ along $\mathbf{r}$ plus the contribution of the shell cancel the force along $\mathbf{r}$ from $M_b(R_1)$, leaving only a constant force along $\mathbf{R}_G$. Equation (51.1) also shows how this special case depends critically on the background and the subsystem both remaining spherical

To calculate the force of the shell for an inhomogeneous density, we may suppose it is thin enough to be represented reasonably by a linear density gradient

$$\rho_{\text{shell}} = \frac{1}{(R_2 - R_1)} [\rho_1(R_2 - r) + \rho_2(r - R_1)], \qquad (51.2)$$

with $\rho_1 = \rho(R_1)$ and $\rho_2 = \rho(R_2)$. Then the contribution of the shell to $\mathbf{F}_b$ in (51.1) is easily evaluated. The corresponding tidal force may be estimated from the difference between the background force in the $\mathbf{r}$ direction along $\mathbf{R}_G$ at the center of mass of the subsystem and at s. This is approximately $F_{\text{tid}} \approx (dF_b/dr) \, \Delta r = s \, dF_b/dr$. From the last two equations

$$s^{-1}F_{\text{tid}} = \frac{2GM_b(R_1)}{r^3} - \frac{4}{3}\pi G \frac{(\rho_1 R_2 - \rho_2 R_1)}{(R_2 - R_1)}\left[1 + \frac{2R_1^3}{r^3}\right]$$

$$+ \frac{\pi G(\rho_1 - \rho_2)}{(R_2 - R_1)}\left[2r + \frac{2R_1^4}{r^3}\right] + \frac{2}{3}\pi G(\rho_1 + \rho_2). \qquad (51.3)$$

In the last term, the background density for $M_b(s)$ has been approximated by its average value $\frac{1}{2}(\rho_1 + \rho_2)$.

This formula becomes more transparent if we suppose that the subsystem is far from the center of the background, so that $r \approx R_G \approx R_1 \approx R_2$, but $R_1 = R_G - s$ and $R_2 = R_G + s$ so $R_2 - R_1 = 2s$. Then

$$s^{-1}F_{\text{tid}} = \frac{2GM_b(R_1)}{R_1^3} - \frac{4}{3}\pi G(\rho_1 + \rho_2). \qquad (51.4)$$

The right hand side is independent of s, and its first term is just the standard tidal force of the background within R_1. The second term, representing the shell, is equivalent to the tidal effect of a background of negative mass within the subsystem. This applies whether the contribution of the background mass is small or large relative to the subsystem.

An estimate of the tidal limit (without worrying about details of individual orbits) follows from the value of s at which $F_{\text{tid}} = GM_{\text{substructure}}(s)/s^2$. The effect of the shell between R_2 and R_1 is to increase the tidal radius relative to the case where the tidal force is completely exterior to the galaxy. We can see this geometrically because the dominant tidal force from the shell on points at R_1 and R_2 along $\mathbf{R}_G$ comes from the nearby matter outside the galaxy, and has a component pulling toward the galaxy's center. The ratio of the two terms on the right hand side of (51.4) is the ratio of the average density inside R_1 to the average density of the shell. As this ratio tends to unity, the tidal radius becomes infinite, agreeing with the previous basic result for uniform density. Other things being equal, which is not usually the case, tidal limits may act as probes of the total mass distribution in clusters, including any invisible background matter.

Other effects of a distributed background also modify the evolution of finite systems. Orbit segregation, produced by a changing background, has already been discussed in Section 47. Mass segregation, equipartition, and dynamical friction can be significantly altered by the different mean field orbits of different backgrounds. These areas have not been much studied.

52

Physical stellar collisions

The stars may dissolve, and the fountain of light
May sink into ne'er ending chaos and night.

P.B. Shelley

At the very beginning of this book, our first calculation was to find the condition for which stars are far enough apart to act as point masses. Now, we have reached the opposite extreme. In the nuclei of galaxies, the density of stars may be so great that they collide bodily. What happens then?

For a typical star of radius r, the mean time between collisions is

$$t_c = \frac{1}{n\sigma V} = \frac{9.7 \times 10^{21} R^{7/2}(\text{pc})}{N^{3/2}(m^{1/2}r^2/m_\odot^{1/2}r_\odot^2)(1 + 8.8 \times 10^7 R(\text{pc})r_\odot/(Nr))}\text{yr.} \quad (52.1)$$

To derive this relation the virial theorem is used in the form $\langle v^2 \rangle = GNm/2R$ for an average polytropic distribution (see Section 54.1) with a root mean square radius R. The mean square velocity, $\langle v^2 \rangle$ is one-half the squared relative velocity, V^2 at large separation. Moreover, the geometrical cross section has been increased by the factor $(1 + 2Gm/rV^2)$ to account for the gravitational attraction (see 17.6 and 50.9), ignoring tidal deformation. Table 6 provides a feel for these collision times by comparing them with the average relaxation time τ_R from (2.9) and the root mean square velocity dispersion in spherical systems consisting of solar size stars.

The great dividing line in the theory of stellar collisions is between collisions which can disrupt stars completely, and those which cannot. The requirement for a disruptive head-on collision can be estimated easily. Most of the kinetic energy is converted into thermal energy, and substantial liberation of gas can occur if the total energy is positive. If each star is a polytrope of index n, then using its potential energy derived (by the reader) in Section 54.1, the criterion for complete dissolution is

$$\frac{1}{2}m\left(\frac{V}{2}\right)^2 - \frac{3}{2(5-n)}\frac{Gm^2}{r} > 0. \quad (52.2)$$

(Remember that the star's binding energy is half its potential energy.) For two solar-type stars ($n = 3$), this implies $V \gtrsim 1500\,\text{km s}^{-1}$. This result is approximate since it does not include the effects of shocks, the excess energy carried off by the gas liberated with more than escape velocity, or the time dependent gravitational field.

A solar type star smashing head-on into its mirror image was explored in a two-

Table 6. *Properties of compact, spherical star systems*

No. of stars	Radius, R(pc)	0.1	1	10	100
$N = 10^6$	Stellar velocity $\langle v^2 \rangle^{1/2}$ (km s^{-1})	150	50	15	5
	Relaxation time, τ_R (yr)	5.3×10^5	1.7×10^7	5.3×10^8	1.7×10^{10}
	Collision time, t_c (yr)	3.2×10^8	1.1×10^{11}	3.5×10^{13}	1.1×10^{16}
N = 10^8	Stellar velocity	1500	500	150	50
	Relaxation time	3.9×10^6	1.2×10^8	3.9×10^9	1.2×10^{11}
	Collision time	2.8×10^6	5.2×10^9	3.1×10^{12}	1.1×10^{15}
N = 10^{10}	Stellar velocity	15 000	5000	1500	500
	Relaxation time	3.1×10^7	9.9×10^8	3.1×10^{10}	9.9×10^{11}
	Collision time	3.1×10^3	9.7×10^6	2.8×10^{10}	8.5×10^{13}

dimensional numerical hydrodynamic simulation (Seidel & Cameron, 1972). As the encounter proceeds, the stars become squashed and a sheet of gas is heated, then ejected in the plane perpendicular to the initial relative velocity. Simultaneously a recoil shock forms in the outer layers and ejects gas from the backs of the stars, an effect which is relatively more important at low collision velocities. Experiments for distant relative velocities of zero, 1000 and 2000 km s^{-1} showed that about 5%, 18%, and 60% respectively, of the stars was liberated. Thus (52.2) provides quite a good criterion for disruption, despite the simplicity of its derivation.

Rewriting the disruption criterion (52.2) in a simpler form using the virial theorem for the cluster, we find

$$\frac{V^2}{Gm/r} \approx \frac{N}{R/r} \approx 1. \tag{52.3}$$

Therefore collisions begin to be highly disruptive when the density of stars is so high that they will just be touching if lined up along a radius of the cluster.

If the disruption criterion is not satisfied by a large margin, most of the gas still interacts supersonically relative to the local sound speed. Shocks convert much of the kinetic energy of stellar motion into thermal energy which is then radiated away. Thus the collision is highly inelastic, and much of the stars' material will coalesce. The distended, newly formed object pulsates for some time, and then settles down to a well-defined star. During the collision, the temperature and density are not great enough to generate an important amount of energy by thermonuclear reactions (Spitzer & Saslaw, 1966; Mathis, 1967).

Only a small fraction of collisions are head-on, and the rest must be handled gingerly by approximate methods. Their detailed hydrodynamics is very complex, and not well understood. Simple models (Spitzer & Saslaw, 1966; Colgate, 1967; Sanders, 1970), have provided some insight, at least into the questions to ask. In these models, the two stars are divided into long rectangular tubes of gas parallel to their relative velocity. Each tube collides only with its geometric counterpart, and their changes are not coupled to the rest of the star. The collision converts the kinetic energy of the star's motion into heat, conserving linear momentum. This thermal energy is divided between the two mass tubes in proportion to their kinetic energy before impact (relative to their own center of mass frame). If the thermal energy of a mass element is greater than its binding energy to the star to which it is most strongly bound, the mass element is assumed to escape.

Clearly the accuracy of these assumptions is very uncertain, especially for low velocity collisions which may transfer substantial momentum perpendicular to the relative velocity of the gas tubes. For a head-on collision at low velocites no mass would be lost on this approximation; therefore we would expect the approximation to become worse as the impact distance p becomes smaller. The general shape of the curve of mass loss as a function of p would then be zero for $p = 0$ and for $p > 2r$, with a maximum in between. As an example, Sanders' computations for two suns colliding at $V = 62 \, \mathrm{km \, s^{-1}}$ give a total maximum mass loss of ~ 0.04 solar masses at $p \approx 0.4$ solar radii.

What is the condition that the two stars coalesce? If the first collision converts enough orbital energy into heat, the stars will become bound. Successive collisions will then reduce the orbit's semimajor axis until most of the mass merges permanently. This occurs if the kinetic energy of motion which is converted into heat exceeds the orbital kinetic energy of the two stars at infinity. Dynamical and tidal binding discussed in Section 48 may assist the merger. Without detailed computations we do not know how much energy is irreversibly lost, i.e., how inelastic the collision is. Assuming some degree of inelasticity, one can compute how much thermal energy is produced in each colliding mass tube, add up the total for all tubes, and see if this is enough to coalesce the stars.

In applying this procedure to stars of different mass it is important to know their internal density distributions. More massive stars will generally have lower average density than less massive ones. Their density distributions dominate the question of whether two colliding stars of greatly different mass can convert enough orbital energy and momentum into heat so that they coalesce. Colgate first suggested that stars of $m \gtrsim 50 \, m_\odot$ would not coalesce with the more numerous field stars of $\sim 1 \, m_\odot$, but would simply have holes punched through them. Thus there would be an upper limit to the mass which could form by such coalescence. However, Sanders' computations showed that this result is very sensitive to the model used, and coalescence may still occur, but at lower velocities. This remains an important question which is not well understood.

Stars are very nutritious. A large star can increase its longevity by swallowing a smaller one for two main reasons. First, of course, there is the added hydrogen fuel. Second, more hydrogen of the massive star is mixed throughout the core, extending the main sequence lifetime. Thus whether a massive star evolves into a supernova depends on the ratio of its coalescence-mixing timescale to its main sequence lifetime in a given state. If the core of a coalesced star mixes faster than it burns, it may be possible to build up extremely massive stars in the center of the stellar system.

Having been swallowed, a star must be digested, then absorbed. The additional heat created in the collision distends the coalesced star beyond the normal size for its total mass. The bloated object pulsates awhile and eventually settles down to mechanical equilibrium after several relaxation times of order $(G\rho)^{-1/2}$ (about half an hour for the sun). To reach thermal equilibrium requires the Kelvin–Helmholtz radiative diffusion timescale (about 3×10^7 yr for the sun). This may be compared to the collision timescales in Table 6.

Our knowledge of the structure and evolution of coalesced stars is very meager. This remains an important, but difficult, problem for our understanding of dense stellar systems.

Stellar disruption dominates at impact velocities much greater than the escape velocity, for off-center collisions. Using the simple model mentioned earlier, Spitzer & Saslaw computed the amount of gas liberated by two identical main sequence stars for a variety of impact parameters and relative velocities. Averaging these results over all distances of closest approach, weighing each distance by its geometrical cross section proportional to $p\,dp$, gives the total mass loss Δm in units of the stellar mass m as a function of the ratio V_{esc}/V of the escape velocity from the surface of one of the two stars to the relative velocity at infinity for the two stars. Typical approximate values are $\Delta m/m = (0.04, 0.06, 0.1, 0.2, 2)$ for $V_{esc}/V = (0.4, 0.2, 0.1, 0.05, 0)$. There is a rapid rise in $\Delta m/m$ as $V_{esc}/V \to 0$ and substantial disruption by non-central collisions becomes important.

Another important question which arises in disruptive collisions is the amount of energy which is converted into relativistic particles by shocks. De Young (1968) has computed a numerical hydrodynamic simulation of a head-on collision of two suns with a relative velocity of 5000 km s^{-1}. Typically, shocks convert about 3% of the total collision energy into relativistic electrons with energies $\gtrsim 0.5$ MeV. This is not very much, but the efficiency could be greatly increased if the stars have even a weak magnetic field which is then compressed during the collision.

Stellar collisions, along with the normal mass loss from a star's own evolution, are an important source of hot gas in the cluster. Some of this gas may escape from the cluster. The rest may fall to the center and be partially reprocessed into new stars. The next sections gives a brief description of several star–gas interactions, to indicate the directions along which this story may be completed. A stellar collision may be a star's way of making another star, or perhaps a quasar.

53

More star–gas interactions

And all dishevelled wandering stars.

W.B. Yeats

53.1. Galactic winds

Stellar collisions, novae, supernovae, planetary nebulae, stellar winds and flares all dump gas into a star cluster or galaxy. How much of this gas cools and remains in the cluster, and how much escapes is a fundamental question. The answer is difficult because it depends completely on details of the mass loss process and on the subsequent interactions of gas lost from different stars. Clouds may collide and the heat generated by shocks may cause evaporation. Wandering stars may pass through the clouds, heating them as described in Section 17. The results are clearly very model dependent, so we will just consider a very simple example to illustrate the nature of galactic winds. (See Mathews & Baker, 1971, for a detailed analysis.)

A necessary condition for an atom or ion to escape from the cluster is that its energy exceed roughly twice the average energy per atom of the stars' random motions. (The collisional mean free path determines sufficient escape conditions.) If we can average very crudely over the cluster, then supposing that the star and gas motions are in rough energy equipartition (per particle) gives an equivalent escape temperature $T_{esc} \approx \mu v_*^2/3k \approx 2 \times 10^7 (\mu/0.5)(v_*/1000 \, \mathrm{km \, s^{-1}})^2$ where μ is the mean particle mass in proton units. From Table 6 we see that, in typical systems where the stellar collision time is much less than a Hubble time, gas must be heated to $\sim 5 \times 10^7$ K. Such collisions provide nearly this energy to much of the gas, so the resulting high energy tail of the distribution may easily escape the cluster.

The evolution of the gas may be described by modifying the fluid Equations (15.2)–(15.4). These equations apply at every point and, clearly, conditions will be vastly different from place to place throughout the cluster. It would be possible to add any desired distribution of sources of hot gas and sinks of cold gas to these equations and see how they affect the overall flow. More simply, we could imagine the fluid equations to apply to averages over larger volumes of the system than usual, or over an ensemble of macroscopically similar systems. Then to represent the injection of gas we add a source term to the continuity equation. It may have any space and time dependence, but it is particularly reasonable to make it proportional to the stellar density. In spherical coordinates with v the radial velocity

$$\frac{\partial \rho}{\partial t} + \frac{1}{r^2}\frac{\partial}{\partial r}(r^2 \rho v) = a\rho_*. \tag{53.1}$$

The gas density is $\rho(r, t)$ and $a(t)$ is the rate at which stars lose mass.

The equation of momentum conservation is also modified because the injected gas will not generally have the same average momentum as its surrounding flow. Again this depends on detailed local interactions, but it may be represented in an average sense for the simple case where the injected gas has no net velocity. Then if local interactions act quickly to give this new gas the same velocity as the surrounding flow, an equivalent amount of momentum must be lost from the general flow. Therefore (15.3) becomes

$$\rho\frac{\partial v}{\partial t} + \rho v\frac{\partial v}{\partial r} = -\frac{\partial P}{\partial r} - \frac{GM(r)\rho}{r^2} - a\rho_* v. \tag{53.2}$$

The gravitational term is the spherical solution of Poisson's equation (15.4) and includes all the mass interior to r, whether from stars, gas, a central massive object, or a distributed background.

Pressure and density may be related, to close the set of equations, either by the polytropic equation of state (40.4) or by the perfect gas equation. The latter case introduces temperature, so closure requires an additional equation – the first law of thermodynamics (29.10). Here details of the atomic heating and cooling mechanics of the gas enter, as well as energy input from the sources and those adiabatic changes caused just by density changes. Expressing the thermal energy change in terms of the local density, temperature, pressure and bulk velocity finally closes the description.

Pursuit of galactic winds would rapidly take us out of this book. Here we shall go just far enough to see how the essential effect, that of a distributed source, changes matters relative to a wind from a single source such as a star. The first major approximation is to consider a steady state wind, setting the time derivatives in (53.1) and (53.2) equal to zero. Physically this is unreasonable for active galaxies, although it is all right for fairly quiescent stars like the sun. Nevertheless, changes in mass injection processes, and thermal and other wind instabilities are not understood. So specific detailed models may not really be more accurate than simple ones. The second major approximation is to suppose that the wind is isothermal. This allows us to dispense with the whole radiative transport input into the equation of state. It is not so unreasonable as the first approximation because of the combination of high plasma conductivity and energy deposition averaged throughout the system.

Thus the perfect gas equation of state, $P = kT\rho/\mu m_H$ with $\mu = \frac{1}{2}$ for a predominantly hydrogen plasma, and constant T, defines our simple example. It is useful to scale the velocity and radius by writing (in solar wind notation)

$$\xi \equiv \frac{r}{r_0}, \psi \equiv \frac{v^2}{U^2}, U^2 = \frac{2kT}{m_H}. \tag{53.3}$$

Here r_0 is some fiducial radius at the base of the wind (where a may be taken to be zero); it can be related to a central massive object in models which have one. The speed U is the most probable (i.e., $(\frac{2}{3})^{1/2}$ times the root mean square) speed of protons in a Maxwell–Boltzmann distribution. To allow some flexibility we let the density of stars have the form

$$\rho_*(\xi) = \rho_*(1)\xi^{-\beta}, \tag{53.4}$$

although many other forms could easily be used.

These assumptions and notation give a simple differential equation for the radial dependence of the wind velocity. First integrate (53.1), writing the constant of integration in terms of the gas density $\rho_g(1)$ and velocity v_0 at r_0. Substitute this, along with the equation of state, to eliminate the density and pressure in (53.2). (As an exercise the reader can see what difference a temperature gradient would make.) The result for $\rho_g(1)/\rho_*(1) \ll 1$ is

$$\left(1 - \frac{1}{\psi}\right)\frac{d\psi}{d\xi} = \frac{4[1 + (\beta - 1)f/2]}{(1+f)}\frac{1}{\xi} - \frac{2GM(\xi=1)}{r_0 U^2}\frac{1}{\xi^2} - \frac{2ar_0}{(1+f)v_0}\frac{\rho_*(\xi)}{\rho_g(1)}\xi^2\psi$$

$$- \frac{2G[M(\xi) - M(1)]}{r_0 U^2}\frac{1}{\xi^2}, \tag{53.5}$$

where

$$f(\xi) \equiv \frac{a\rho_*(1)}{(3 - \beta)\rho_g(1)}\frac{r_0}{v_0}\xi^{3-\beta}. \tag{53.6}$$

The physical meaning of $f(\xi)$ is apparent from (53.6). Apart from the radial dependence it is essentially the ratio of the travel time for the wind across r_0 to the time for mass injection to double the gas density there. (The radial dependence may be rewritten using the total masses within r_0 and r.)

In the limit of no mass injection, $a = f = 0$ and the first three terms of (53.5) give the basic equation of solar wind theory (Parker, 1963). The last term shows the gravitational effect of the distributed mass. For $\psi < 1$ it makes ψ increase more rapidly with radius, while for $\psi > 1$ it causes ψ to decrease more rapidly. The value $\psi = 1$, or $v = U$, is clearly a critical point of the equation. Since the sound speed $(\gamma P/\rho)^{1/2} = (5kT/3\,\mu m_H)^{1/2}$ is approximately equal to U, this is sometimes called the 'sonic point', although it is not necessarily the same. Thus the distributed gravitational field tends to keep the flow velocity near the critical velocity; its importance depends on the detailed run of $M(\xi)$.

It is straightforward to integrate (53.5) analytically or numerically and build a wide range of models. Some of their general features, however, emerge directly from the form of the equation:

1. There is usually a critical point at which $\psi = 1$ or $d\psi/d\xi = 0$. Its location is found by setting the right hand side of (53.5) equal to zero, and it depends mainly on the mass distribution and the strength of the sources. At this critical point the solution has a dual nature, and the boundary conditions select the relevant flow. For

example, if the second term on the right hand side dominates at small radii, then if $\psi < 1$ there, $d\psi/d\xi > 0$ so the velocity will increase with radial distance. It can continue increasing at the critical point, and eventually become supersonic, only if the $\psi = 1$ branch applies there. Along the $d\psi/d\xi = 0$ branch with $\psi < 1$, the velocity decreases at large radii where the right hand side becomes positive, and the flow does not become supersonic.

2. If the thermal velocity U is sufficiently small the negative gravitational terms dominate the right hand side and, for $\psi > 1$, the velocity decreases with radius. For $\psi < 1$ the velocity increases with radius. This tends to hold the streaming velocity about equal to the velocity dispersion of the gas and prevent escape, as discussed at the beginning of this section. If U is very large the mass injection rate dominates the position of the critical point.

3. The exact value of β, indicating the stellar density concentration, is not too important unless $f \gtrsim 1$. For high mass injection into a nearly uniform cluster, the nature of the wind is altered. If the right hand side is always negative there is no critical radius and the outflow velocity decreases for $\psi > 1$ and is finally brought to rest by the increasing gravitational force at larger radii.

4. Behavior of the wind is also significantly altered if the central mass within r_0 always dominates the gravitational field.

5. This approach is readily extended to include sinks as well as sources of gas. The presence of sinks can describe accretion flows. It is possible to have inner regions of infall and outer regions of outflow, separated by a stagnation radius where the velocity is zero. Some cases of pure infall which become supersonic require the existence of shocks.

Realistic winds may become thermally unstable; regions cool and form denser clouds. Whether the wind's momentum continues to drive clouds out, or whether they lose pressure support and fall to the center of the cluster, depends on details of each model.

53.2. Central disks and star formation

Gas falling to the center of a cluster may accumulate there, at least temporarily. Its overall shape depends importantly on its net angular momentum. Gas lost from individual stars in rapidly rotating clusters will retain most of its angular momentum. Gas lost from stellar collisions in the nuclei of galaxies may have less specific angular momentum than the average star. This is because stars with opposite orbital angular momentum have a somewhat greater probability of colliding than stars moving in the same direction.

Suppose we wish to make a simple order-of-magnitude estimate for the size of the resulting disk. Let gas retain its net angular momentum after being lost from stars. Its specific angular momentum (per unit mass) is $J = Rv_\theta$, where R and v_θ are a characteristic radius (such as the root mean square) and rotational velocity for the cluster. Define a dimensionless parameter ε which compares J with the initial radius of the gas (which is the same as the radius of the cluster R_0) and with the initial root mean square velocity dispersion of the stars v_0

$$\varepsilon \equiv \frac{J}{R_0 v_0} = \frac{R v_\theta}{R_0 v_0} = \frac{v_{\theta 0}}{v_0} = \frac{R_d v_{\theta d}}{R_0 v_0}. \tag{53.7}$$

Here the subscript 'd' refers to values achieved by the gas when it forms a disk.

We know v_0 from the virial theorem, and have $v_{\theta d}$ for the disk from the approximate rotational force balance

$$v_{\theta d}^2 \approx \frac{G m N_d}{R_d}. \tag{53.8}$$

Here N_d is the number of stars within a radius R_d of the cluster's center. So from the last two equations we obtain

$$\frac{R_d}{R_0} \approx \varepsilon^2 \frac{N}{N_d}. \tag{53.9}$$

If gas cools and contracts much faster than the stellar dynamical evolution, we may roughly suppose $N_d \approx N R_d^3 / R^3$ and find

$$\frac{R_d}{R_0} \approx \varepsilon^{1/2}. \tag{53.10}$$

Note that this is little more than a dimensional argument. It neglects the shape of the disk, the exact factors in the virial theorem, and the detailed form of the density distribution. To some extent, these approximations balance out, but only detailed models can provide an exact result. The main point is that fairly modest amounts of rotation, say $\varepsilon = 0.1$, can strongly inhibit the contraction of a gaseous disk perpendicular to its angular momentum.

Parallel to its angular momentum, the disk is supported by gas pressure, magnetic pressure, and turbulence. The vertical scale height of the disk can therefore be estimated by combining (15.3) and (15.4) for the case of hydrostatic equilibrium

$$\frac{d}{dz}\left(\frac{1}{\rho}\frac{dP}{dz}\right) = \nabla^2 \phi = -4\pi G \rho_{tot}. \tag{53.11}$$

Here ρ_{tot} is the total density and ρ is just the gas density. If only gas pressure is important, $P = \rho c^2$ with c the sound speed from the perfect gas equation of state. Moreover if stars dominate the gravitational force, then the dynamical crossing time is $\tau_c = (4\pi G \rho_{tot})^{-1/2} = R/v_0$. Setting the gradient operator equal to the inverse scale height of the gas z_g^{-1} we see that

$$\frac{z_g}{R} \approx \frac{c}{v_0}. \tag{53.12}$$

Thus the disk may be very thin compared to the radius of the stellar system, even for hot ionized clouds. If there is also magnetic or turbulent pressure support, the approximate relation (53.12) still holds with c^2 replaced by the equivalent energy per particle for the total pressure. This disk is very idealized and it is probably unstable, possibly fragmenting into stars.

Of star formation in the cores of clusters, virtually nothing is known. Indeed we are barely beginning to understand this process for more normal conditions. How its rate and resulting mass spectrum depend on the density and chemical composition, remain for the future to tell.

53.3. Embedded stars

Normally a star radiates into a void. But what would happen to a star surrounded by a cloud so dense as to be opaque? Would the star just add the cloud to its atmosphere? Or would it become unstable and try to blow the cloud away?

These questions pertain to stars in the cores of galactic nuclei as more and more gas falls to the center. To describe how a star evolves when it is embedded in a hot gas or radiation field, we need to join the stellar structure equations discussed in Section 44 with the stellar wind equations for its outer atmosphere. These stellar wind equations are just the special case of (53.1) and (53.2) which has all the gravitating mass within the spherical boundary of the star and the mass source term $a = 0$. Thus the evolution of an embedded star is (tenuously) related to the problem of galactic winds.

The main parameter in the join between a star's interior structure and its outer wind is the temperature of the surrounding radiation field. Radiation is taken to be in thermal equilibrium with the outflowing gas. If there is no surrounding matter – just radiation – then the critical stellar wind solution (see the discussion following (53.6)) with an indefinitely increasing velocity applies. For this case the radiation temperature determines a unique self-consistent rate of mass loss for the star. Detailed numerical models (Kato, 1980) show that this mass loss rate depends very sensitively on the external temperature. If this temperature is about 100 times the effective surface temperature of the star, then the star evaporates on roughly a Kelvin–Helmholtz timescale $\sim Gm^2/rL$, where m, r, and L are the star's mass radius and luminosity. For a solar mass star, this takes about 3×10^7 yr.

More realistically, a star is surrounded by dense gas, as well as by radiation. Then the subcritical (subsonic) wind solution, which reaches zero velocity at a finite density, applies. It may be truncated at a value of velocity or density which matches onto the surrounding gas cloud. This truncation parameter, along with the temperature, then provides the boundary conditions which determine the mass loss rate. It is significantly reduced compared to a pure radiation boundary, and may even turn into an accretion flow if the cloud density is high enough. If, even more realistically, the star moves rapidly through the cloud, strong shocks (Section 17.1) may modify its mass loss.

Whether conditions in quasars or galaxy nuclei are sufficiently extreme to dissolve stars is a fascinating problem for the future. This is perhaps the ultimate star–gas interaction, counterpoised with the original condensation of stars out of gas.

54

Problems and extensions

54.1. Potential energy of a polytrope

From the discussions in Section 46, or Equation (43.1), show that the negative gravitational potential energy of a polytrope of index n (Section 40), total radius R and mass M is

$$W = \frac{3}{5-n} \frac{GM^2}{R}.$$ (1)

The value $n = 0$ gives the result for a sphere of uniform density. Larger values of n indicate more centrally condensed spheres with greater potential energy. At first sight, the value of W for $n = 5$ is infinite. This would seem to contradict the result of Section 18.2. The escape from this contradiction is that the radius for $n = 5$ is infinite and $(5 - n)R$ tends to a finite limit.

54.2. Virial mass discrepancy for a system which loses mass

One possible explanation for the discrepancy between the virial mass and the luminosity mass in a cluster of galaxies is that the cluster is expanding. This expansion may be caused by mass loss from galaxies. Some mechanisms for this mass loss are ejection from galactic nuclei, gravitational radiation from forming black holes, and emission of massive high energy particles, but there are others as well. Suppose that the mass loss is of the form $\dot{M} = -M/\tau_1$ where τ_1 is the relevant timescale. Show for $\tau_1 > \tau_c$, where τ_c is the dynamical crossing timescale, that the expansion is adiabatic and the virial theorem is satisfied, i.e., the system evolves along a quasi-static sequence. On the other hand, show that if $\tau_1 < \tau_c$ there is an apparent virial mass discrepancy given by

$$\frac{M_v}{M} = \left(1 + \frac{t - t_0}{\tau_{co}}\right) \exp\left(\frac{t - t_0}{\tau_1}\right),$$ (1)

where subscript '0' refers to the initial time. (Ikeuchi, *Publ. Astron. Soc. Japan* **31**, 169, 1979). This discrepancy can be large.

54.3. Effect of angular momentum on the contraction of a stellar system

In sections 44 and 45 we examined the contraction of the core of a non-rotating stellar system. In Section 53.2 we saw how rotation inhibits the contraction of a gaseous disk. Adapt the analysis of Section 53.2 to a rotating stellar system which evolves by evaporation without changing its total energy or specific angular momentum. Show that it changes into a flattened disk whose radius is approximately

$$\frac{R_d}{R_0} = \varepsilon^{4/3}. \tag{1}$$

Therefore angular momentum need not inhibit the contraction of a stellar system as much as it does a gaseous disk. Examine this for other forms of contraction. (Spitzer & Saslaw, *Ap. J.*, **143**, 400, 1966).

54.4. Calculation of slingshot results

Many important results of three-body slingshot experiments can be understood on the basis of a semianalytical statistical theory. It accounts for such properties as the eccentricity distribution and angular momentum direction of the remaining binary, as well as the mass, velocity and solid-angle distributions of the escaping particle.

The fundamental approximation of the theory is that after a slingshot interaction occurs, the probability of finding a given result is proportional to the phase space volume available for that result. Further assumptions are that the total energy and angular momentum are the only relevant conserved quantities, and that the slingshot interaction occurs in a bounded volume of configuration space. With these assumptions, show that the total phase space volume available is

$$\sigma = \int \delta(E_B + E_s - E_0)\delta(\mathbf{J}_B + \mathbf{J}_s - \mathbf{J}_0)\mathrm{d}\mathbf{r}\,\mathrm{d}\mathbf{p}\,\mathrm{d}\mathbf{r}_s\mathrm{d}\mathbf{p}_s. \tag{1}$$

Here the total energy is

$$E_0 = \tfrac{1}{2}\mu\dot{r}^2 + \frac{1}{2}m\dot{r}_s^2 - \frac{Gm_am_b}{|\mathbf{r}|} \tag{2}$$

and the total angular momentum is

$$\mathbf{J}_0 = \mu(\mathbf{r} \times \dot{\mathbf{r}}) + m(\mathbf{r}_s \times \dot{\mathbf{r}}_s), \tag{3}$$

with

$$\mu = \frac{m_am_b}{m_B}, \quad m = \frac{m_sm_B}{M}. \tag{4}$$

The subscripts *a*, *b*, *B* and *s* denote particles *a* and *b*, the total binary, and the single escaper. The total mass is *M*, and **r** is the relative position of the binary components, and $\mathbf{r}_s$ is the position of the escaper relative to the binary's center of mass. The vector δ functions are products of their scalar components.

Since the probability of finding a given range of final configurations is proportional to the fraction of phase space volume occupied by those configurations, it is necessary to solve the 12-dimensional integral of Equation (1). The domains of these integrals can become very complicated because the energy and angular momentum limits depend on each other. But it can be done with a combination of analytical and numerical Monte Carlo techniques. The results agree reasonably well with the three-body experiments described in Section 49 (Nash & Monaghan, *MNRAS*, **184**, 119, 1978).

55

Bibliography

Section 37:

Gunn, J.E. & Gott, J.R. III, 1972. 'On the infall of matter into clusters of galaxies and some effects on their evolution', *Ap. J.*, **176**, 1.

Roxburgh, I.W. & Saffman, P.G., 1965. 'The growth of condensations in a Newtonian model of the steady-state universe', *MNRAS*, **129**, 181.

Sandage, A.R., 1963. 'The ability of the 200-inch telescope to discriminate between selected world models', *Ap. J.*, **133**, 355.

Section 38:

Aarseth, S.J. & Lecar, M., 1975. 'Computer simulations of stellar systems', *Ann. Rev. Astron. and Astrophys.*, **13**, 1.

Aarseth, S.J. and Saslaw, W.C., 1972. 'Virial mass determinations of bound and unstable groups of galaxies', *Ap. J.*, **172**, 17.

Hénon, M., 1964. 'L'evolution initiale d'un amas spherique', *Annals. Astrophys*, **27**, 83.

Kadomtsev, B.B. & Pogutse, O.P., 1970. 'Collisionless relaxation in systems with Coulomb interactions', *Phys. Rev. Lett.*, **25**, 1155.

King, I.R., 1962. 'The structure of star clusters. I. An empirical density law', *A. J.*, **67**, 471.

Lynden-Bell, D., 1967. 'Statistical mechanics of violent relaxation in stellar systems', *MNRAS*, **136**, 101.

Saslaw, W.C., 1969. 'Gravithermodynamics. II. Generalized statistical mechanics of violent agitation', *MNRAS*, **143**, 437.

Saslaw, W.C., 1970. 'Violent relaxation in galaxies and stellar systems', *MNRAS*, **150**, 299.

Shu, F.H., 1978. 'On the statistical mechanics of violent relaxation', *Ap. J.*, **225**, 83.

Zwicky, F., 1960. 'The age of large globular clusters of galaxies', *PASP*, **72**, 365.

Section 40:

Bailey, M.E. & MacDonald, J., 1981. 'A comparison between velocity dispersion profiles of de Vaucouleurs and King galaxy models', *MNRAS*, **194**, 195.

Chandrasekhar, S., 1939. *An Introduction to the Study of Stellar Structure* (Chicago: Chicago UP).

Huntley, J.M. & Saslaw, W.C., 1975. 'The distribution of stars in galactic nuclei: loaded polytropes', *Ap. J.*, **199**, 328.

Kent, S.M. & Gunn, J.E., 1982. 'The dynamics of rich clusters of galaxies. I. The Coma cluster', *A. J.*, **87**, 945.

King, I.R., 1965. 'The structure of star clusters. II. Steady-state velocity distributions', *A. J.*, **70**, 376.

King, I.R., 1966. 'The structure of star clusters. III. Some simple dynamical models', *A. J.*, **71**, 64.

Spitzer, L. & Härm, R., 1958. 'Evaporation of stars from isolated clusters'. *Ap. J.*, **127**, 544.

Section 41:

Aarseth, S.J. & Saslaw, W.C., 1972., 1972. 'Virial mass determinations of bound and unstable groups of galaxies', *Ap. J.*, **172**, 17.

Smith, H., Jr, 1982. 'The potential-estimation method of cluster mass determinations', *Ap. J.*, **259**, 423.

Section 42.1:

Bahcall, N.A., 1977. 'Clusters of galaxies', *Ann. Rev. Astron. and Astrophys.*, **15**, 505.

Bahcall, N.A., 1980. 'Clusters of galaxies', *Highlights of Astronomy*, **5**, 699.

Beers, T.C., Geller, M.J. & Huchra, J.P., 1982. 'Galaxy clusters with multiple components. I. The dynamics of Abell 98', *Ap. J.*, **257**, 23.

Bucknell, M.J., Godwin, J.G. & Peach, J.V., 1979. 'Studies of rich clusters of galaxies–V. Photometry and luminosity functions for eight clusters', *MNRAS*, **188**, 579.

Capelato, H.V., Gerbal, D., Mathez, G., Mazura, A., Salvador-Sole, E. & Sol, H., 1980. 'On the luminosity-segregation in rich clusters. Application to Coma'. *Ap. J.*, **241**, 521.

Dekel, A. & Shaham, J., 1980. 'Secondary peak in clusters of galaxies – a clue to their formation', *Astron. and Astrophys.*, **85**, 154.

Dressler, A., 1981. 'The dynamics of the cluster of galaxies A2029', *Ap. J.*, **243**, 26.

Faber, S.M. & Gallagher, J.S., 1979. 'Masses and mass of light ratios of galaxies', *Ann. Rev. Astron. and Astrophys.*, **17**, 135.

Field, G.B. & Saslaw, W.C., 1971. 'Groups of galaxies: Hidden mass or quick disintegration?' *Ap. J.*, **170**, 199.

Godwin, J.G. & Peach, J.V., 1979. 'Brightest members of clusters of galaxies', *Nature*, **277**, 364.

Hartwick, F.D.A., 1981. 'Groups of spiral galaxies around the Coma cluster and upper limits to its mass', *Ap. J.*, **248**, 423.

Klinkhamer, F.R., 1980. 'Groups of galaxies with large crossing times', *Astron. and Astrophys.*, **91**, 365.

Quintana, H. & Havlen, R.J., 1979. 'A detailed photometric and structural study of the southern cluster of galaxies CA 0340–538', *Astron. and Astrophys.*, **79**, 70.

Quintana, H., 1979. 'Core radii and mass segregation in clusters of galaxies', *A. J.*, **84**, 15.

Rose, J.A. & Graham, J.A., 1979. 'Mass-to-light ratios of two compact groups of galaxies', *Ap. J.*, **231**, 320.

Struble, M.F., 1979. 'Velocity dispersion profiles of clusters of galaxies', *A. J.*, **84**, 27.

Struble, M.F. & Rood, H.J., 1981. 'Binary-galaxy-rich clusters of galaxies', *Ap. J.*, **251**, 471.

Thompson, L.A. & Gregory, S.A., 1980. 'The luminosity function of the Coma cluster galaxies', *Ap. J.*, **242**, 1.

Vorontsov-Vel'yaminov, B.A., Dostal', V.A. & Metlov, V.G., 1980. 'The most compact nest of galaxies VV 644'. *Sov. Astron. Lett.*, **6**, 217.

Section 42.2:

Dressler, A., 1980. 'The spectrum of the central luminosity spike in M87'. *Ap. J. Lett.*, **240**, L11.

Faber, S.M. 1980. 'The structure of galactic nuclei: recent observations'. *Highlights of Astronomy*, **5**, 135.

Geller, M.S., Turner, E.L. & Bruno, M.S., 1979. 'Night-to-night variations in the optical continuum of NGC 1275', *Ap. J. Lett.*, **230**, L141.

Jones, D.L., Sramek, R.A. & Terzian, Y., 1981. 'VLBI Observations of galactic nuclei', *Ap. J.*, **246**, 28.

Petrosyan, A.R., Saakyan, K.A. & Khachikyan, Eh.E., 1979. 'Galaxies with ultraviolet continuum that have double and multiple nuclei. II.' *Astrophys.*, **14**, 36.

Saslaw, W.C., 1974. 'Theory of galactic nuclei', in *IAU Symp.*, 58, J.R. Shakeshaft (ed.) (Dordrecht: D. Reidel), pp. 305–34.

Tonry, J.L. & Davis, M. 1981. 'Velocity dispersion of elliptical and S0 galaxies. I Data and mass-to-light ratios', *Ap. J.*, **246**, 666.

Tremaine, S. & Ostriker, J.P., 1982. 'The dynamics of the nucleus of M31', *Ap. J.*, **256**, 435.

Section 42.3:

Burbidge, E.M. & Sandage, A.R., 1958. 'Properties of two intergalactic globular clusters', *Ap. J.*, **127**, 527.

Hanes, D., 1977. 'The luminosity distribution of globular clusters in the Virgo cluster of galaxies', *Mem. RAS*, **84**, 45.

Hanes, D. & Madore, B. (eds.), 1980. *Globular Clusters* (Cambridge UP).

Hodge, P.W., 1966. 'Radii, orbital properties, and relaxation times of dwarf elliptical galaxies', *Ap. J.*, **144**, 869.

Madore, B.F. & Arp, H.C., 1979. 'Three new faint star clusters', *Ap. J. Lett.*, **277**, L103.

Sandage, A.R. & Hartwick, F.D.A., 1977. 'Remote halo globular cluster Palomar 5'. *A. J.*, **82**, 459.

Section 43:

Antonov, V.A., 1962. 'Most probable phase distribution in spherical stellar systems and conditions of its existence', *Journ. Leningrad Univ.*, **7**, Issue 2.

Bonnor, W.B., 1956. 'Boyle's law and gravitational instability', *MNRAS*, **116**, 351.

Bonnor, W.B., 1958. 'Stability of polytropic gas spheres', *MNRAS*, **118**, 523.

Horwitz, G. & Katz, J., 1978. 'Steepest descent technique and stellar equilibrium statistical mechanics. III. Stability of various ensembles', *Ap. J.*, **222**, 941.

Katz, J., 1978. 'On the number of unstable modes of an equilibrium', *MNRAS*, **183**, 765.

Katz, J. 1979. "On the number of unstable modes of an equilibrium – II', *MNRAS*, **189**, 817.

Lynden-Bell, D. & Wood, R., 1968. 'The gravo-thermal catastrophe in isothermal spheres and the onset of red-giant structure for stellar systems', *MNRAS*, **138**, 495.
Thirring, W. 1970. 'Systems with negative specific heat', *Z. Physick*, **235**, 339.

Section 44:

Aarseth, S.J., 1963. 'Dynamical evolution of stars in clusters. I, *MNRAS*, **126**, 233.
Cohn, H., 1980. 'Late core collapse in star clusters and the gravothermal instability', *Ap. J.*, **242**, 765.
Hachisu, I., Nakada, Y., Nomoto, K. & Sugimoto, D., 1978. 'Gravothermal catastrophe of finite amlitude', *Prog. Theor. Phys.*, **60**, 393.
Larson, R.B., 1970. 'A method for computing the evolution of star clusters', *MNRAS*, **147**, 323.
Lynden-Bell, D. & Eggleton, P.P., 1980. 'On the consequences of the gravothermal catastrophe', *MNRAS*, **191**, 483.
Sugimoto, D., Eriguchi, Y. & Hachisu, I., 1981. 'Gravothermal aspects in the evolution of the stars and the universe', *Prog. Theor. Phys. Supp.*, **70**, 154.

Section 45:

Aarseth, S.J., 1974. 'Dynamical evolution of simulated star clusters. I. Isolated models', *Astron. and Astrophys.*, **35**, 237.
Hénon, M., 1969. 'Rates of escape from isolated clusters with an arbitary mass distribution', *Astron. and Astrophys.*, **2**, 151.
Spitzer, L., Jr & Thuan, T.X., 1972. 'Random gravitational encounters and the evolution of spherical systems. IV. Isolated systems of identical stars', *Ap. J.*, **175**, 31.

Section 46:

Inagaki, S. & Saslaw, W.C. 1985. "Equipartition in multi-component gravitational systems', *Ap. J.* May 15.
Katz, J. & Taff, L.G., 1983. 'Stability limits for 'Isothermal' cores in globular cluster models: two component systems', *Ap. J.*, **264**, 476.
Merritt, D., 1981. 'Two component systems in thermal and dynamical equilibrium', *A. J.*, **86**, 318.
Spitzer, L., Jr, 1969. 'Equiparition and the formation of compact nuclei in spherical stellar systems', *Ap. J.*, **158**, L139.
Spitzer, L., Jr, 1975. 'Dynamical theory of spherical stellar systems with large *N*', in *IAU Symp.*, 69, A. Hayli (ed.) (Dordrecht: D. Reidel), p. 3.

Section 47:

Lecar, M., 1966. 'Stellar orbits in a time-varying gravitational field', *A.J.*, **71**, 706.
Saslaw, W.C., 1977. 'Orbit segregation in evolving galaxies and clusters of galaxies', *Ap. J.*, **216**, 690.

Section 48:

Fabian, A.C., Pringle, J.E. & Rees, M.J., 1975. 'Tidal capture formation of binary systems and X-ray sources in globular clusters', *MNRAS*, **172**, 150.

Heggie, D.C., 1975*a*. 'Binary evolution in stellar dynamics', *MNRAS*, **173**, 729.

Heggie, D.C., 1975*b*. 'The dynamical evolution of binaries in clusters', in *IAU Symposium*, 69, *Dynamics of Stellar Systems*, A. Hayli (ed.) (Dordrecht: Holland), p. 73.

Heggie, D.C., 1980. 'Dynamical theory of binaries in clusters', in *Globular Clusters*, D. Hanes & B. Madore (eds.) (Cambridge UP), p. 281.

Inagaki, S., 1984. 'The effects of binaries on the evolution of globular clusters', *MNRAS*, **206**, 149.

Landau, L. & Lifshitz, E.M., 1960. *Mechanics* (London: Pergamon).

Press, W.H. & Teukolsky, S.A., 1977. 'On formation of close binaries by two-body tidal capture', *Ap. J.*, **213**, 183.

Spitzer, L., Jr & Hart, M.H., 1971. 'Random gravitational encounters and the evolution of spherical systems. I. Method', *Ap. J.*, **164**, 399.

Section 49:

Arnold, V.I., 1978. *Mathematical Methods of Classical Mechanics* (New York: Springer-Verlag).

Hut, P. & Bahcall, J.N., 1983. 'Binary-single star scattering. I. Numerical experiments for equal masses', *Ap. J.*, **268**, 319.

Hut, P., 1983. 'Binary-single star scattering. II. Analytic approximations for high velocity', *Ap. J.*, **268**, 342.

Lin, D.N.C. & Saslaw, W.C., 1977. 'The dressed slingshot and the symmetry of double radio galaxies', *Ap. J.*, **217**, 958.

Pars, L.A., 1965. *A Treatise on Analytical Dynamics* (London: Heinemann).

Saito, T., 1977. 'Massive compact objects in elliptical galaxy and their dynamical relation to the halo formation', *PASJ*, **29**, 421.

Saslaw, W.C., Valtonen, M.J. & Aarseth, S.J., 1974. 'The gravitational slingshot and the structure of extragalactic radio sources', *Ap. J.*, **190**, 253.

Valtonen, M.J., 1975. 'Statistics of three-body experiments: probability of escape and capture', *Mem. RAS*, **80**, 77.

Section 50:

Bahcall, J.N. & Wolf, R.A., 1976. 'Star distribution around a massive black hole in a globular cluster', *Ap. J.*, **209**, 214.

Frank, J. & Rees, M.J., 1976. 'Effects of massive central black holes on dense stellar systems', *MNRAS*, **176**, 633.

Hills, J.G., 1975. 'Possible power source of Seyfert galaxies and QSOs', *Nature*, **254**, 295.

Shapiro, S.L. & Marchant, A.B., 1978. 'Star clusters containing massive, central black holes: Monte Carlo simulations in two-dimensional phase space', *Ap. J.*, **225**, 603.

Section 52:

Colgate, S.A., 1967. 'Stellar coalescence and the multiple supernova interpretation of quasi-stellar sources', *Ap. J.* **150**, 163.

De Young, D.S., 1968. 'Two-dimensional stellar collisions', *Ap. J.*, **153**, 633.

Mathis, J.S., 1967. 'Nuclear reactions during stellar collisions', *Ap. J.*, **147**, 1050.

Sanders, R.H., 1970. 'The effects of stellar collisions in dense stellar systems', *Ap. J.*, **162**, 791.

Seidel, F.G.P. & Cameron, A.G.W., 1972. 'A numerical hydrodynamic study of coalescence in head-on collisions of identical stars' *Ap. and Space Sci.*, **15**, 44.
Spitzer, L., Jr & Saslaw, W.C., 1966. 'On the evolution of galactic nuclei', *Ap. J.*, **143**, 400.

Section 53:

Mathews, W.G. & Baker, J.C., 1971. 'Galactic winds', *Ap. J.*, **170**, 241.
Parker, E.N., 1963. *Interplanetary Dynamical Processes* (New York: Interscience).
Kato, M. 1980. 'Structure and mass loss rate of stars immersed in hot radiation field', *Prog. Theor. Phys.*, **64**, 847.

PART IV

Finite flattened systems – galaxies

As we move from the idealizations of mathematical physics to the messy realities of astronomical systems, the very nature of our understanding is transformed. From basic principles we move to models, and from general models to special examples. Observed examples seldom do more than whisper the approximations we need to decode them. All we have left for guidance is a jumble of parametrized ideas, and a hope of relations among them. We understand through the inspired guess, or by accident.

56

Observed dynamics of galaxies

examine this region of short distances and definite
places

W. H. Auden

History has shown that the Universe is complex. The more closely we examine astronomical systems the more branches of physical theory we need to understand their behavior and origin, and the more we must specialize the general theory of the last three parts to apply to particular systems. Passing through this book we have successively examined systems of decreasing symmetry. In Part II we dropped the homogeneity that dominated Part I. In Part III we dropped the infinite boundaries that dominated Part II. Now, in Part IV, we are about to drop the spherical symmetry that dominated Part III. As a result there is less to say of a general nature. To compensate, there arises a richness of detail which can often be compared with observations. Although such comparisons are not within the scope of this book, a sketch of some basic observed dynamical properties of non-spherical systems helps give a useful perspective.

In forming impressions of galactic structure we must be especially careful of selection effects. These influence what we observe in subtle ways. Rates of extragalactic information flow are slow, and the natural desire to get interesting results from long nights spent observing often subconsciously affects the choice of objects and places to examine. Versatile and sensitive detectors continue to improve this situation.

On a more fundamental level the brightness of the night sky limits our vision. Looking at the darkest parts of our galaxy, toward the galactic pole, there is an average diffuse illumination of ~ 23 visual magnitudes per square arcsecond. This comes just from faint stars in our galaxy and from other faint galaxies; there are additional contributions from earth's atmosphere and the sun's reflections off interplanetary dust. Photographic techniques can detect signals at a level of several percent of the night sky noise. Sensitive photometric or spectroscopic techniques can use filters to find faint objects at frequencies where the night sky is darker than average. But these are very difficult procedures and it helps to already know what one is seeking. As a result, it is hard to find objects or parts of objects with very low surface brightness. What we see may be only the tip of an iceberg, and we have no idea what lies in store underneath.

At the opposite extreme from low surface brightness are the objects of such small angular size that they look like stars. Life is too short to examine every pinpoint of light on the sky. So quasars continued to slip through our astronomical nets until the 1960s. Nearly 150 years before we were warned against such selection effects by John Herschel the English astronomer. Meeting with seven contemporaries at a London tavern in January 1820 he founded the Royal Astronomical Society. After the meeting, he prepared a summary of the motives and objects of the society. There was one sentence present in the original draft of his summary but not in the final one. Herschel wrote: 'Yet it is possible that some bodies, of a nature altogether new, and whose discovery may tend in future to disclose the most important secrets in the system of the Universe, may be concealed under the appearance of very minute single stars no way distinguishable from others of less interesting character, but by the test of careful and often repeated observations.' We do not know what caused Herschel to delete this percipient remark from his final version. But it is clear that he was not letting mere appearance confine his imagination!

Mere appearance, nonetheless, is responsible for the major method of galaxy classification. This is all right if appearances are correlated with useful physical properties; fortunately this has more or less turned out to be true. On the basis of photographs taken in the blue part of the spectrum, galaxies are broadly classified as elliptical, spiral, or irregular. Spirals are further subdivided into those with and without central structures like bars (see Sandage, 1961; de Vaucouleurs, 1959). Any galaxy too obdurate to fall into these categories is known as peculiar. Its peculiarities may include split arms, connected companions, surrounding rings, jets, loops, and large internal disturbances (Arp, 1966).

Figure 50 shows schematic examples of a useful classification scheme for normal galaxies. Within the spiral category, subdivisions depend mainly on the prominence and openness of the arms. Our own galaxy appears to be an Sb or Sc type. Within the elliptical category, subdivisions depend mainly on axis ratio (e.g., 0.4 for E6 galaxies). Correlations among these properties have led to other more complex classification schemes which emphasize the degree of central concentration of light, although this may vary greatly with color. An example is the cD type, a massive elliptical galaxy which may be more than 100 kpc in diameter and whose light mostly comes from its center. Often, cDs are the brightest members of a cluster of galaxies (e.g., Thuan & Romanishin, 1981). The big question about any classification scheme is whether it represents an evolutionary sequence, or whether its features result mainly from the initial conditions of star and galaxy formation. The answers, so far, are unknown.

Physical properties of galaxies vary with their form. Consider first their sizes. At once we run into the problem of finding a useful definition of size. A surface brightness limit seems reasonable, but taking it too high gives only the nucleus; taking it too low runs into trouble with the night sky background. Color again makes a difference. So does inclination to the line of sight, which alters the isophotes. And why restrict the definition to optical? Radio emission at 21 cm from neutral

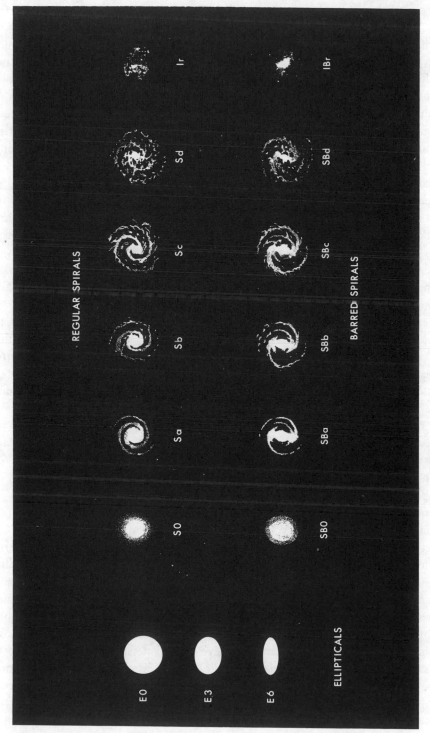

Fig. 50. Composite of Hubble and de Vaucouleurs galaxy classification systems shown in schematic form. (From Roberts (1972).)

hydrogen or synchrotron and X-ray emission from very hot gas also define sizes, usually different again. Clearly, the manner of measuring size must always be specified. A large amount of optical diameter information is available for thousands of bright nearby galaxies (de Vaucouleurs, *et al.*, 1976). These are based on a limiting isophotal surface brightness of $25^{m}.0$ arc sec^{-2}, about 10% of the night sky brightness. Such optical diameters range from about 1 kpc for dwarf galaxies to about 50 kpc for giant spirals. These may be lower limits if low luminosity halos are present. In the radio, some galaxies are associated with neutral hydrogen out to distances of ~ 100 kpc and others eject relativistic electrons over scales of several megaparsecs.

Shapes of galaxies are more complex than the schematic diagrams of Figure 50 suggest. While there are some simple two-armed spirals with quite regular structure, most spirals have several arms and bits of arms. They also contain extensive dark streaks and patches where dust absorbs their starlight. Nor does a whole spiral galaxy always lie in a plane. Many are the warps, bends and twists of real systems. If these uncoordinated features dominate completely, the galaxy is classed as irregular at the end of the sequence. Most irregular galaxies are small and some have glimmerings of spiral arms; there is a continuum of shapes.

Whether galaxy shapes extend continuously from the spirals, through the SO, to the elliptical classes is not so clear. The SO's are a rather miscellaneous bunch of disk galaxies with great central bulges, sometimes having faint spiral or bar structure. Appearance of a bar may depend on the color of the observation.

Elliptical galaxies tend to have more regular shapes, although when examined closely they too have distortions. Often their isophotes show changes in eccentricity and orientation from the center to the outer regions. These are affected by projection, but show that galaxies are more than just simple oblate or prolate ellipsoids. We will pursue their possible triaxial nature further in Section 63. Changing ellipticity may be ether intrinsic, or produced by tidal interactions of companions.

One of the useful aspects of the simple shape-classification scheme is its correlation with spectral properties of stars. A star's spectrum and total luminosity indicate its age and chemical composition. On this basis the stars of a galaxy form two main populations – with many gradations between them. Population I contains metal rich stars, both young and old ($\sim 10^{8} - 10^{9}$ yr), closely associated with interstellar gas. Their orbits are mostly circular, lying in the disks of spiral galaxies with low velocity dispersion (roughly 20–60 km s^{-1}). Population II contains very old ($\sim 10^{10}$ yr) metal poor stars whose positions are unrelated to interstellar gas, whose orbits are inclined and eccentric, and whose velocity dispersion is high (roughly 150–200 km s^{-1}). They occupy the halos of spirals and the globular clusters. As one moves along the Hubble sequence from the ellipticals through more and more open spirals to the irregulars, the relative importance of Population I stars tends to increase. Along this sequence the galaxies' average light becomes bluer; they contain more neutral hydrogen and ionized gas.

The fairly smooth luminosity profiles of elliptical galaxies are often fit to an empirical relation of the form (Reynolds, 1913; Hubble, 1930)

$$\frac{I}{I_0} = \left(\frac{r}{a} + 1\right)^{-2},$$ (56.1)

or of the form (de Vaucouleurs, 1948)

$$\log\left(\frac{I}{I_0}\right) = -3.33\left[\left(\frac{r}{a}\right)^{1/4} - 1\right].$$ (56.2)

Here I is the surface brightness, I_0 is a normalization (at the radius a containing half the total light in 56.2), and a is a scale length. These relations often fit reasonably well over ranges of a few magnitudes in surface brightness and ranges of ~ 10 or 100 in radius, but there are always exceptions. There is not yet any physical basis for these relations. They usually provide a better fit than Fokker–Planck models (Section 40), which do have a physical basis, using a constant mass–luminosity ratio. This last assumption is rather dubious.

Spiral galaxies have much more complicated luminosity distributions, but many attempts have been made to uncover common structures for them (reviewed, for example in Kormendy, 1982). There is some evidence for an average exponential decrease of disk surface brightness with radius. Irregularities, contributions of a central bulge, and selection effects make this result quite uncertain, however. It looks as though a multiparameter fit to several underlying components may be the best that can be done for the integrated light. Future measurements in different colors – or even for separate groups of stars with different velocity dispersions – are likely to tell a clearer story. It is also natural to ask whether the central bulges of spirals have properties similar to elliptical galaxies. Much work is being done on these questions, but the answers have not yet settled down.

Masses of galaxies range from $\sim 10^8 M_\odot$ for dwarfs, to $\sim 10^{11} M_\odot$ for large spirals such as the Milky Way and Andromeda galaxies, to $\sim 5 \times 10^{12} M_\odot$ for giant ellipticals such as M87. These masses are estimated in several ways. The virial theorem can be applied to the velocity dispersion within individual galaxies. Orbits of binary galaxies may be used to estimate the mass of their members. There may be a mass–luminosity relation for galaxies, but this is much less clear than for stars. Also, galaxy masses may be estimated from models of their rotation curves. All these methods, however, are rather uncertain.

As might be expected from their differing shapes, spirals and ellipticals have quite different forms of rotation. Not only do ellipticals rotate less rapidly than spirals, many do not even rotate enough for their shapes to reflect centrifugal balance. Astronomers were quite surprised to learn this in the late 1970s.

Galaxy rotation is a strong clue both to formation and to appearance. Most early observations were analyzed in terms of a galaxy's 'rotation curve'. This is the rotation velocity, $v(r)$, averaged over angle at a given projected radial distance from the center. For spirals which can be assumed to present a circular face-on

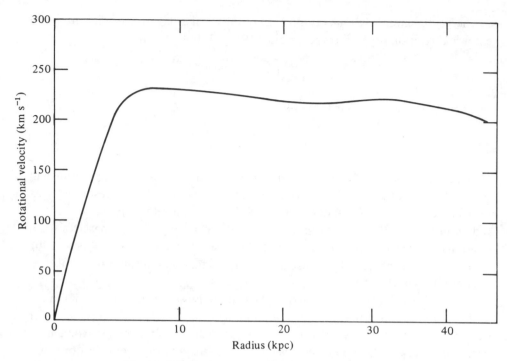

Fig. 51. Schematic rotation curve of a spiral galaxy.

appearance the observed inclination is used to estimate the deprojected rotation curve. With increasing information it becomes clearer that averaged rotation curves are often not representative, and it is better to work with the general 'rotation field' whenever possible. This is most easily done with 21 cm radio observations of neutral hydrogen, but is also possible with optical Fabry–Perot interferometry of bright galaxies (Rubin, 1982).

Figure 51 shows a schematic rotation curve for a spiral galaxy. The central region is roughly in solid-body rotation. At a distance somewhere between about 0.5 and 5 kpc the curve turns over and remains fairly flat. Measurements at large distances are difficult since few ionized patches or neutral hydrogen clouds are detected. Angular averaging becomes increasingly coarser. Some rotation curves show signs of a $v \propto r^{-1/2}$ Keplerian fall-off, but most remain flat to observational limits. It is rather remarkable that although the radius, R, of the solid body part and its angular velocity, ω, each range over a factor ~ 10 among spirals, the maximum rotational velocity, $v_{max} = R\omega$, is narrowly confined to $\sim 200 \pm 50 \,\mathrm{km\,s^{-1}}$ in most cases (Saslaw, 1970). This is a clue to galaxy formation.

Ellipticals, one might think, become flatter as they rotate faster. This seems natural. It is also wrong. Historically it was just another example, like Herschel's, of how astrophysical expectation is limited by simplistic thinking. Indeed we had a clue, at least in retrospect, from the long relaxation times τ_R of elliptical galaxies. If such galaxies formed by the merging of smaller systems, violent relaxation would not be complete and the resulting distribution function could remain anisotropic for a Hubble time. Rotational velocities of bright ellipticals are often (Kormondy, 198?) only $\frac{1}{3}$ or $\frac{2}{3}$ the value needed to support an oblate spheroid with isotropic velocities (although faint ellipticals and bulges of spirals may rotate sufficiently fast). Combined with observations of twisted isophotes it is likely that many ellipticals are less relaxed than they superficially appear to be.

Bar galaxies contain central flattened structures, typically with ratios of long to short axes of about 5 to 1. They may be triaxial. Along the minor axis the luminosity profile is usually steep, but along the major axis it is quite flat until reaching the edge. Within bars the stellar velocity dispersion is considerably greater than in the surrounding disk. Thus bars are probably significant density enhancements. It is not clear whether the orbits of stars in a bar are generally confined to the bar or whether a bar can also be a substantial density wave passing through the disk. Gas and stars are often found to stream along bars, suggesting bars are fairly long lived.

Many, perhaps most, spirals have bars, but ellipticals never (so far) have been found with a bar. Evidently, rotation is critical for producing a bar. In turn, the coherent bar mode reacts back on the disk structure of the galaxy, quite possibly setting off or maintaining the spiral arms.

Lens-like structures with nearly constant surface brightness and elliptical shape appear in some barred galaxies between the disk and the bulge. Their dynamical importance is not understood. Also associated with bars, and perhaps produced by them, are rings around the inner or outer parts of some galaxies. Much variety is found in the texture of galaxies. We should not assume this texture is governed solely by gravity; gas processes and star formation are also important.

Disturbances critical to the shape of a galaxy may also be induced by nearby companions. Galaxies in groups are sometimes strongly affected by a close passage. But those with constant attendants are perturbed all the time. The great Whirlpool Nebula (NGC 5194/5195, Sandage, 1961) is a spectacular example of a satellite stirring up spiral structure.

57

Kinematics of motion

Into this Universe, and *Why* not knowing
Nor *Whence*, like water willy-nilly flowing

Omar Khayyam

57.1 General kinematics

The title of Section 57 is redundant. Kinematics *is* motion. Pure motion. Three properties – rotation, translation, and strain – suffice to describe local motions of flows without telling how their motions arose. This result is often called Helmholtz's theorem, although it goes back further to Stokes and before him to Cauchy (see Lamb, 1932). Forces are absent from 'kinematic' descriptions; their realm is 'dynamics'.

Motion can be described by examining changes in the position or the velocity of a particle (or fluid element). In either case we are dealing with the way a vector field alters over small distances. Generally, a velocity description is more useful for loosely coupled systems of particles, and a positional description is better for elastic bodies. Since velocities give the rate of change of position their analysis gives the rate of rotation or strain.

Consider, at any given time, a region located at $\mathbf{x}$ where the average velocity of stars is $\mathbf{u}(\mathbf{x})$. A nearby region at $\mathbf{x} + \mathbf{r}$ contains stars whose average velocity is $\mathbf{u} + \delta\mathbf{u}$. A Taylor series expansion in Cartesian coordinates shows, to first order in $\mathbf{r}$, that

$$\delta u_i = r_j \frac{\partial u_i}{\partial x_j}. \tag{57.1}$$

Here we adopt the usual convention of summing over all indices appearing twice in a term. Any translational velocity common to the entire system is subtracted out from this description. Now we can apply the same little trick we used in Section 9 to write the second order tensor $\partial u_i / \partial x_j$ as the sum of a symmetric and an antisymmetric tensor. These are found, as before, by exchanging indices

$$e_{ij} = \frac{1}{2} \left(\frac{\partial u_i}{\partial x_j} + \frac{\partial u_j}{\partial x_i} \right) \tag{57.2}$$

and

$$\xi_{ij} = \frac{1}{2} \left(\frac{\partial u_i}{\partial x_j} - \frac{\partial u_j}{\partial x_i} \right), \tag{57.3}$$

so that

$$\delta u_i = r_j e_{ij} + r_j \xi_{ij}. \tag{57.4}$$

The symmetric tensor e_{ij}, like all symmetric tensors, can be diagonalized by rotating its coordinate basis. Then, in the frame where the coordinate axes are the principal axes of the tensor,

$$e'_{ij} = \frac{\partial r_k}{\partial r'_i} \frac{\partial r_m}{\partial r'_j} e_{km}, \tag{57.5}$$

only the diagonal elements ($i = j$) of e'_{ij} are non-zero. Since the sum of the diagonal components of a tensor is invariant to a change of coordinate basis,

$$e'_{ii} = e_{ii} = \frac{\partial u_i}{\partial x_i} = \nabla \cdot \mathbf{u}. \tag{57.6}$$

Physically it is reasonable that the divergence of a flow does not depend on the coordinate system.

A typical component e'_{11}, say, measures the amount by which the velocity in the r'_1 principal axis direction changes along this axis. A line of stars near $\mathbf{x}$ parallel to the r'_1 axis is being stretched at a rate e'_{11}. Lines of stars along the other principal axes are being stretched out at rates e'_{22} and e'_{33}. Thus e'_{ij} converts an initially spherical distribution of stars near $\mathbf{x}$ into an ellipsoid, since generally $e'_{11} \neq e'_{22} \neq e'_{33}$. This describes a pure straining motion, which is why e_{ij} is called the rate of strain tensor, and e'_{11}, e'_{22}, e'_{33} are its three principal rates of strain. In an arbitrary Cartesian reference frame the component e_{ij} would deform a rectangular distribution in the i, j plane into a parallelogram and displace it from its original position. In general the principal axes of e_{ij}, like e_{ij} itself, will vary throughout the system and depend on details of the flow.

A simple calculation shows the explicit relation between the rate of strain tensor and the changing separation of stars. Two nearby stars are separated by a distance δs where

$$(\delta s)^2 = r_i r_i. \tag{57.7}$$

Their separation changes at a rate

$$\frac{D(\delta s)^2}{Dt} = 2r_i \frac{Dr_i}{Dt} = 2r_i \delta u_i = 2r_i r_j \frac{\partial u_i}{\partial x_j} = 2r_i r_j e_{ij}. \tag{57.8}$$

Thus if all the symmetric combinations e_{ij} of the velocity gradients are zero there is no relative motion of the two stars. If $r_2 = r_3 = 0$ along the principal axes, then the rate of relative separation along the x_1 axis is

$$\frac{1}{\delta s} \frac{D(\delta s)}{Dt} = \frac{1}{r_1} \frac{Dr_1}{Dt} = \frac{\partial u_1}{\partial x_1} = e_{11}. \tag{57.9}$$

More generally, the relative rate of change of a local volume, δV, is

$$\frac{1}{\delta V} \frac{D(\delta V' - \delta V)}{Dt} = e_{ii}. \tag{57.10}$$

This is the rate at which the flow diverges. Similarly, strain itself is just the relative change of a length scale.

We can easily rewrite e_{ij} in yet another physically instructive way. Adding and subtracting an average of the diagonal terms multiplied by δ_{ij} gives the identity

$$e_{ij} = (e_{ij} - \tfrac{1}{3}\delta_{ij}e_{kk}) + \tfrac{1}{3}\delta_{ij}e_{kk}. \tag{57.11}$$

The sum of diagonal terms for the expression in parentheses is zero since $\delta_{ii} = 3$. So it represents a pure shear of the local velocity field, changing its shape without any overall expansion (or compression). The last term gives a pure expansion. Any rate of strain can be decomposed into a rate of shear and an isotropic rate of expansion.

When analyzing fairly flat rotating galaxies it is often helpful to use the rate of strain tensor in cylindrical coordinates r, ϕ, z. It can be worked out similarly to the Cartesian case, but the algebra is more complicated and I just list the components for reference

$$e_{rr} = \frac{\partial u_r}{\partial r}, \quad e_{\phi\phi} = \frac{1}{r}\frac{\partial u_\phi}{\partial \phi} + \frac{u_r}{r}, \quad e_{zz} = \frac{\partial u_z}{\partial z},$$

$$2e_{\phi z} = \frac{1}{r}\frac{\partial u_z}{\partial \phi} + \frac{\partial u_\phi}{\partial z}, \quad 2e_{rz} = \frac{\partial u_r}{\partial z} + \frac{\partial u_z}{\partial r},$$

$$2e_{r\phi} = \frac{\partial u_\phi}{\partial r} - \frac{u_\phi}{r} + \frac{1}{r}\frac{\partial u_r}{\partial \phi}. \tag{57.12}$$

For a more spheroidal or irregular galaxy it is useful to have the rate of strain tensor in spherical coordinates, r, θ, ϕ (where ϕ is the azimuthal angle around the $\theta = 0$ axis)

$$e_{rr} = \frac{\partial u_r}{\partial r}, \quad e_{\theta\theta} = \frac{1}{r}\frac{\partial u_\theta}{\partial \theta} + \frac{u_r}{r}, \quad e_{\phi\phi} = \frac{1}{r\sin\theta}\frac{\partial u_\phi}{\partial \phi} + \frac{u_\theta}{r}\cot\theta + \frac{u_r}{r},$$

$$2e_{\theta\phi} = \frac{1}{r}\left(\frac{\partial u_\phi}{\partial \theta} - u_\phi\cot\theta\right) + \frac{1}{r\sin\theta}\frac{\partial u_\theta}{\partial \phi}, \quad 2e_{r\theta} = \frac{\partial u_\theta}{\partial r} - \frac{u_\theta}{r} + \frac{1}{r}\frac{\partial u_r}{\partial \theta},$$

$$2e_{\phi r} = \frac{1}{r\sin\theta}\frac{\partial u_r}{\partial \phi} + \frac{\partial u_\phi}{\partial r} - \frac{u_\phi}{r}. \tag{57.13}$$

Recalling Section 9 leads us to suspect that the antisymmetric contribution $r_j\xi_{ij}$ to the local velocity field δu_i represents rotation. This is essentially correct, but not quite complete. Rotation is represented most directly by angular velocity, which is a vector. Since the antisymmetric tensor ξ_{ij} has only three independent components we may obtain a vector, Ω_i, from it by writing

$$\xi_{ij} = -\tfrac{1}{2}\,\varepsilon_{ijk}\Omega_k. \tag{57.14}$$

Here ε_{ijk} is the alternating tensor whose value is zero if any indices are equal. When all the indices are different, its value is $+1$ if the indices are in cyclic order, and -1 if they are not. We are not to know in advance if Ω_k is the angular velocity since (57.14) is a general functional form. To find its physical interpretation, we note that

$$\delta u_i^{(\text{anti})} = r_j\xi_{ij} = -\tfrac{1}{2}\,\varepsilon_{ijk}r_j\Omega_k \tag{57.15}$$

is the ith component of the vector $\frac{1}{2}\boldsymbol{\Omega} \times \mathbf{r}$. Thus a solid-body rotation with angular velocity $\frac{1}{2}\boldsymbol{\Omega}$ around a point at distance $\mathbf{r}$ produces a velocity $\delta\mathbf{u}^{(\text{anti})}$. This is a clue to the meaning of $\boldsymbol{\Omega}$, but not yet the whole story.

Generally all the stars are not revolving in step like a solid body about some common axis, so $\boldsymbol{\Omega}$ will vary from place to place. Comparing (57.14) with (57.3) we see from its components that

$$\boldsymbol{\Omega} = \boldsymbol{\nabla} \times \mathbf{u}. \tag{57.16}$$

As in fluid dynamics, $\boldsymbol{\Omega}$ is called the local vorticity of the flow; systems with $\boldsymbol{\Omega} = 0$ are irrotational. The factor $\frac{1}{2}$ was put into (57.14) just so it would not appear in (57.16), in case you were wondering about it.

A related physical interpretation of vorticity follows by integrating it over the area, $\mathrm{d}A$, of a small circular disk having radius a, unit normal $\mathbf{n}$ and an element of circumference $\mathbf{dr}$

$$\int (\boldsymbol{\nabla} \times \mathbf{u})\cdot\mathbf{n}\,\mathrm{d}A = \oint \mathbf{u}\cdot\mathbf{dr} \approx \pi a^2 (\boldsymbol{\nabla} \times \mathbf{u})\cdot\mathbf{n}. \tag{57.17}$$

The second step follows from Stokes' theorem and the third step from assuming $\boldsymbol{\nabla} \times \mathbf{u}$ to be nearly constant over the small area. Thus $\boldsymbol{\Omega}\cdot\mathbf{n}$ is an average of the tangential component of velocity, $\mathbf{u}\cdot\mathbf{dr}$, over the area of the disk. This generalizes the idea of local angular velocity for a system whose motions also include strain deformations. For the special case where the motion is a completely rigid rotation with angular velocity ω, then $\mathbf{u} = \omega \times \mathbf{r}$ and the vorticity $\boldsymbol{\Omega} = \boldsymbol{\nabla} \times (\omega \times \mathbf{r}) = 2\omega$ is just twice the angular velocity everywhere.

We would not want to confuse local vorticity with the motion of a group of stars around a closed path. Vorticity may be present even if the stars move along straight lines. An example is shear flow with $u_1 = u_1(x_2)$, $u_2 = u_3 = 0$ in Cartesian coordinates. Then $\Omega_1 = \Omega_2 = 0$ and $\Omega_3 = -\partial u_1/\partial x_2$, producing a net vorticity. If one inserted a gigantic paddle along the x_2 axis into this stream of stars, the velocity shear would make the paddle rotate. On the other hand, a group of stars can move around in a closed path without any local vorticity, as illustrated in Figure 52.

The net angular momentum of stellar motions in a spherical volume whose center is at x is

$$L_i = \int \varepsilon_{ijk} r_j \left(u_k + r_m \frac{\partial u_k}{\partial x_m} \right) \rho \,\mathrm{d}V. \tag{57.18}$$

The volume is so small that u_k and $\partial u_k/\partial x_m$ are effectively constant within it. Thus the first term is zero and

$$L_i = \varepsilon_{ijk} \frac{\partial u_k}{\partial x_m} \int r_j r_m \rho \,\mathrm{d}V = \tfrac{1}{2} \varepsilon_{ijk} \frac{\partial u_k}{\partial x_m} I\delta_{jm} = \tfrac{1}{2}\Omega_i I, \tag{57.19}$$

where I is the moment of inertia of the sphere along any diameter. So the spherical volume acts like a solid body whose angular velocity is $\boldsymbol{\Omega}/2$. Other shapes do not usually behave so simply because their principal axes of strain do not coincide with

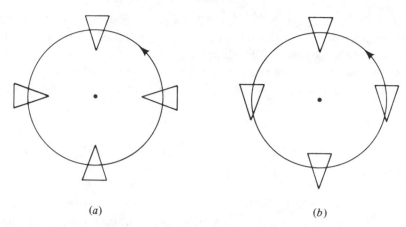

(a)   (b)

Fig. 52. Motion of a triangular group of stars (a) in solid-body rotation and
(b) circulating without local rotation or vorticity.

the principal axes of the shape. However, (57.18) is a useful general expression for local angular momentum.

It is possible to repeat these kinematic analyses incorporating second and higher order terms in (57.1). This would lead to spatial curvature in the rate of strain. Expansion of a spherical volume, for example, would not produce an ellipsoid along the principal axes of strain, but rather a skewed figure. The distance at which second order terms become important must be determined for each flow pattern. In our own galaxy it is thought to be about two kiloparsecs from the sun's position.

57.2 Motions in galactic coordinates

When examined in detail, most galaxies appear to have quite complicated distributions of stars and stellar motions, as Section 56 briefly sketched. To describe these motions around any local observer, we transform the kinematic equations into the galactic coordinates suitable for such an observer. Here we will not make the usual simplifying assumptions that the motion is in a plane, or symmetric about a plane, or that only circular velocities are to be found. Full generality is useful to see just how restrictive the simplifying assumptions can be, and also to see how to describe the complexities of real galaxies.

Consider an observer located at 0, not necessarily in the plane of the galaxy, illustrated in Figure 53. We will need just two coordinate systems: a Cartesian system centered on the observer, and a moving system centered on the star he observes. Let $\hat{\mathbf{z}}$ be a unit axis toward the zenith and $\hat{\mathbf{r}}$ along the direction $\mathbf{r}$ to the star. To form an orthonormal coordinate system on the celestial sphere at the star's location, first define $\mathbf{l} = \hat{\mathbf{z}} \times \hat{\mathbf{r}}$ which is tangent to the circle of latitude. Its magnitude is $l = \sin(\pi/2 - b) = \cos b$, since $\hat{\mathbf{z}}$ and $\hat{\mathbf{r}}$ are unit vectors. So to normalize $\mathbf{l}$ we have $\hat{\mathbf{l}} = \mathbf{l} \sec b$. This gives us $\hat{\mathbf{r}}$ and $\hat{\mathbf{l}}$ at the star. To get the third basis vector orthogonal to these two, we take their cross product: $\hat{\mathbf{b}} = \hat{\mathbf{r}} \times \hat{\mathbf{l}} = \sec b \,\hat{\mathbf{r}} \times \mathbf{l} = \sec b \,\hat{\mathbf{r}} \times$

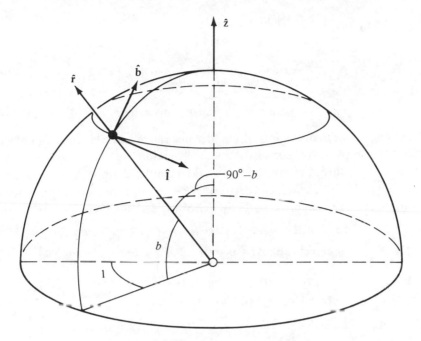

Fig. 53. The coordinate systems used by a galactic observer.

$(\hat{\mathbf{z}} \times \hat{\mathbf{r}}) = \sec b \, [\hat{\mathbf{z}} - (\hat{\mathbf{r}} \cdot \hat{\mathbf{z}})\hat{\mathbf{r}}] = \sec b \, \hat{\mathbf{z}} - \tan b \, \hat{\mathbf{r}}$ using the vector identity $\mathbf{a} \times (\mathbf{b} \times \mathbf{c}) = (\mathbf{a} \cdot \mathbf{c}) \, \mathbf{b} - (\mathbf{a} \cdot \mathbf{b}) \, \mathbf{c}$. Thus the orthonormal coordinates at any observed star are

$$\hat{\mathbf{r}} = r^{-1}(x\hat{\mathbf{i}} + y\hat{\mathbf{j}} + z\hat{\mathbf{z}}), \quad \hat{\mathbf{l}} = \hat{\mathbf{z}} \times \hat{\mathbf{r}} \sec b = \left(-\frac{y}{r}\hat{\mathbf{i}} + \frac{x}{r}\hat{\mathbf{j}} \right) \sec b.$$

$$\hat{\mathbf{b}} = \sec b \, \hat{\mathbf{z}} - \tan b \, \hat{\mathbf{r}}. \tag{57.20}$$

The three components of velocity change δu_i given by (57.1) form the Cartesian differential velocity vector

$$\delta \mathbf{u} = \frac{\partial u_1}{\partial x_j} x_j \hat{\mathbf{i}} + \frac{\partial u_2}{\partial x_j} x_j \hat{\mathbf{j}} + \frac{\partial u_3}{\partial x_j} x_j \hat{\mathbf{z}}, \tag{57.21}$$

where $(x, y, z) = (x_1, x_2, x_3)$ since the observer is located at the origin. (If the observer is not at the origin, we need another linear transformation which adds to the algebra but not to the physics.)

To find the radial and tangential velocity components on the sky, we project $\delta \mathbf{u}$ onto the basis (57.20)

$$v_r = \delta \mathbf{u} \cdot \hat{\mathbf{r}} = r^{-1} \left[\frac{\partial u_1}{\partial x_1} x_1^2 + \frac{\partial u_2}{\partial x_2} x_2^2 + \frac{\partial u_3}{\partial x_3} x_3^2 \right.$$

$$\left. + \left(\frac{\partial u_1}{\partial x_2} + \frac{\partial u_2}{\partial x_1} \right) x_1 x_2 + \left(\frac{\partial u_2}{\partial x_3} + \frac{\partial u_3}{\partial x_2} \right) x_2 x_3 + \left(\frac{\partial u_3}{\partial x_1} + \frac{\partial u_1}{\partial x_3} \right) x_3 x_1 \right], \tag{57.22}$$

$$v_1 = \delta\mathbf{u}\cdot\hat{\mathbf{l}} = r^{-1}\sec b\left[\frac{\partial u_2}{\partial x_1}x_1^2 - \frac{\partial u_1}{\partial x_2}x_2^2\right.$$

$$\left. + \left(\frac{\partial u_2}{\partial x_2} - \frac{\partial u_1}{\partial x_1}\right)x_1 x_2 + \left(\frac{\partial u_2}{\partial x_3}x_1 - \frac{\partial u_1}{\partial x_3}x_2\right)x_3\right],$$

(57.23)

$$v_b = \delta\mathbf{u}\cdot\hat{\mathbf{b}} = -v_r\tan b + \frac{\partial u_3}{\partial x_j}x_j\sec b.$$

(57.24)

Spherical, rather than cylindrical, coordinates can give the best account of all the bumps and lumps in realistic distributions. Therefore we set $x = r\cos b\cos l$, $y = r\cos b\sin l$, $z = r\sin b$ and use the half-angle formulas $2\sin\alpha\cos\alpha = \sin 2\alpha$ and $2\cos^2\alpha = 1 + \cos 2\alpha$. The velocity derivatives can be converted directly into the rate of strain tensor (57.2) and the vorticity tensor (57.3). This gives the basic result

$$v_r = [e_{12}\sin 2l + \tfrac{1}{2}(e_{11} - e_{22})\cos 2l + \tfrac{1}{2}(e_{11} + e_{22})]r\cos^2 b$$
$$+ e_{13}r\sin 2b\cos l + e_{23}r\sin 2b\sin l + \tfrac{1}{2}e_{33}r(1 - \cos 2b),$$

(57.25)

$$v_l = [e_{12}\cos 2l - \tfrac{1}{2}(e_{11} - e_{22})\sin 2l + \xi_{21}]r\cos b$$
$$+ [(e_{23} + \xi_{23})\cos l - (e_{13} + \xi_{13})\sin l]r\sin b,$$

(57.26)

$$v_b = -v_r\tan b + (e_{31} + \xi_{31})r\cos l + (e_{32} + \xi_{32})r\sin l + e_{33}r\tan b.$$

(57.27)

It would be possible to write the first two terms of (57.25) and (57.26) more compactly by using the addition formulas for $\sin(\alpha + \beta)$ and $\cos(\alpha + \beta)$, but that would tend to obscure the underlying strains and rotations.

Now we can see the consequences of the usual simplifying assumptions made when applying (57.25)–(57.27) to our galaxy. First, suppose we observe only motions in the plane of the galaxy where we are located. Then v_r retains just the terms in brackets, v_l loses its terms multiplying $\sin b$, and v_b loses its first and fourth terms. Next, suppose we observe outside the galactic plane, but the motions are symmetric about the plane. Then among terms with 'index 3' only $\partial u_3/\partial x_1$ and $\partial u_3/\partial x_2$ survive, and these would also vanish if the motions were parallel to the plane.

As a third example, suppose there were no z motions at all and the motions in the plane were completely circular. Then $e_{11} + e_{22}$, which is the divergence of the flow in the plane, would also vanish. This is often called the 'K term' in galactic structure. The only remaining terms would be the e_{12}, ξ_{21}, and $(e_{11} - e_{22})$ terms. The strain $e_{12} = e_{r\phi}$ is then given by (57.12) with $\partial u_r/\partial\phi = 0$, and $-e_{12}$ is called 'Oort's constant A' after the Dutch astronomer who pioneered the kinematic study of our galaxy's rotation. The vorticity term ξ_{21} is Oort's constant

$$B \equiv -\frac{1}{2}\left(\frac{\partial u_\phi}{\partial r} + \frac{u_\phi}{r}\right)$$

(57.28)

in this special case. Another interpretation of B will emerge after Equation (59.16). The $(e_{11} - e_{22})$ term, which could be formally combined with the e_{12} term as mentioned earlier, generates ellipticity from the different rates of strain along the principal axes. This term depends on the observer's longitude relative to the principal axes.

Much work has gone into estimating the values of A and B for our own galaxy (see Mihalas & Binney, 1981 for a review of these observations.) Values of $A = 15 \, \mathrm{km \, s^{-1} \, kpc^{-1}}$ and $B = -10 \, \mathrm{km \, s^{-1} \, kpc^{-1}}$, along with related values of 10 kpc for the sun's distance from the galactic center and $250 \, \mathrm{km \, s^{-1}}$ for the local rotational velocity around the galaxy seem generally acceptable. However, it is important to bear in mind that most studies of galactic rotation assume circular, planar velocities from the start. Since we are near the galactic plane, b is small and we are not well situated to see the effects of large streaming velocities in the z direction. There may also be irregular local expansion or contraction in the plane, which we have seen modifies A and B, as well as introduces the K term. Formal errors in these values do not usually consider the possible errors of the underlying assumptions on which they are based. A change of these basic assumptions may alter our picture of the local flow of stars considerably.

Except for a temporary lack of information, there is no particular reason to confine kinematic analyses to our own galaxy. When enough redshifts and distances are determined for nearby galaxies, we should know how our local cluster behaves. All kinematic studies, however, are preludes to understanding dynamics.

58

Transfer of angular momentum

'If everybody minded their own business', the
Duchess said, in a hoarse growl, 'the world would
go round a deal faster than it does.'

Lewis Carroll

'Tis torque that makes the world go round, to paraphrase another favorite saying of
the Duchess. Torques are angular forces, and they bring us back to the realm of
dynamics. The main goal of dynamics is to motivate kinematics. Dynamics provides
a reason, in terms of equations of motion with their initial and boundary conditions,
for the existence of any particular pattern of flow.

Spin is one of the fundamental properties of matter on all scales from molecules to
clusters of galaxies. (Elementary particles are supposed to have spin, but not in the
same sense as classical objects.) The Universe itself may possess global rotation,
although observations suggest its rotational timescale is longer than the timescale
for its Hubble expansion. For any region we can divide the gravitational timescale
by the angular velocity timescale to obtain the spin ratio

$$\sigma \equiv \frac{\omega}{(G\rho)^{1/2}}. \tag{58.1}$$

A simple example is a test particle in a stable circular orbit of radius r around a
massive body. Gravitational and centripetal forces balance (i.e., the virial theorem
applies) giving $\sigma = 2 \left(\frac{1}{3}\pi\right)^{1/2}$. Here ρ is the density obtained by distributing the
massive body throughout the spherical volume of radius r; when r is the surface of
the body, then ρ is its average density. Thus, regions where the spin is not large
enough to make gravitationally bound material fly apart will generally have
$0 < \sigma \lesssim 2$. Numerical factors are not included in the definition (58.1) because they
will depend strongly on the local geometry.

To generalize the spin ratio for a local region where strain is also present, we
simply replace ω by half the vorticity, $\frac{1}{2}\Omega_k$, from (57.14)–(57.17) or by $\frac{1}{2}\Omega$ when
directional information is not needed. Therefore σ will be constant throughout the
whole system if it has uniform density and is in solid-body rotation.

We do not know the origin of spin for large scale objects such as galaxies and
clusters of galaxies. On smaller scales such as stars and planets it seems reasonable
that spin arose from initial irregularities of density, which produced torques, and
from initial irregularities of velocity, which added spin directly. We can believe these

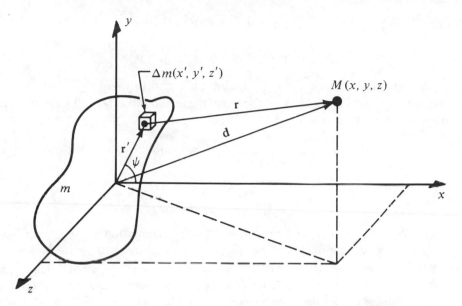

Fig. 54. The geometry of a point mass M exerting torque on an extended body of mass m.

initial states were irregular because we observe them quite generally where stars are now forming.

Conditions for galaxy formation, however, are much more obscure. We have not seen galaxies forming. We can only ask whether gravitational torques alone were sufficient to endow galaxies with their observed spin, or whether initial vorticity was also required. If we could answer this question we would gain great insight into processes of galaxy formation, even if they were very complex.

Whether or not there was initial vorticity, nothing escapes from the force of the tides.

Earlier, in Sections 48 and 51, we examined tidal interactions of spherical systems. They were too symmetric to produce a net torque, and only energy was transferred. Now that we have dropped the condition of spherical symmetry, we can see how tidal torques transfer angular momentum from the orbits of two bodies to their internal rotation. As usual, we will consider a simple problem illustrating the essential physics which can then be applied to more complex situations.

Suppose there are two bodies, of which one is a point mass but the other has an irregular structure. The point mass exerts a force which differs in different regions of the extended body; see Figure 54. The positional moment of this force is the external torque

$$\mathbf{N} = \sum_i \mathbf{r}_i \times \mathbf{F}_i, \tag{58.2}$$

where the subscript refers to the ith particle or mass element in the extended body. For a continuous distribution, the sum over masses, involving $\mathbf{F}_i = \mathbf{F}(m_i)$, is replaced by an integral of the density, using $\mathbf{F}(\rho)$, over the volume of the body (see (58.8)). The external torque induces a net angular momentum since

$$\mathbf{N} = \sum_i \mathbf{r}_i \times \mathbf{F}_i = \sum_i (\mathbf{r}_i \times \dot{\mathbf{p}}_i) = \sum_i \frac{d}{dt}(\mathbf{r}_i \times \mathbf{p}_i) = \frac{d\mathbf{L}}{dt} \tag{58.3}$$

(recalling that the internal torques cancel in pairs and $\mathbf{p}_i = m\dot{\mathbf{r}}_i$).

Basic mechanics texts show that the angular momentum is related to the inertia tensor (note $(x_1, x_2, x_3) = (x, y, z)$),

$$I_{jk} = \int_V \rho(r)(r^2\delta_{jk} - x_j x_k)dV, \tag{58.4}$$

and the angular velocity ω by

$$L_x = I_{xx}\omega_x + I_{xy}\omega_y + I_{xz}\omega_z \tag{58.5}$$

for the x component of angular momentum; cyclic permutation $x \to y, y \to z, z \to x$, gives the other two components. The diagonal components of I_{jk} are the 'moments of inertia' and the off-diagonal components are the 'products of inertia'. The x, y, z axes we will use here are fixed in space. If, however, we were to use axes fixed in the body of the galaxy, then the angular momentum relative to these body axes would change at a rate given by

$$\left(\frac{d\mathbf{L}}{dt}\right)_{body} + \omega \times \mathbf{L} = \mathbf{N}, \tag{58.6}$$

and substituting (58.5) into (58.6) would give Euler's equations.

To calculate the net torque which M exerts on the galaxy, we place the origin of the x, y, z coordinate system at the galaxy's center of mass. The torque on the small element of mass $dm = \rho(\mathbf{r}')dV'$ at $\mathbf{r}'$ is

$$d\mathbf{N} = \frac{GM\rho dV'}{r^3}\mathbf{r}' \times \mathbf{r}. \tag{58.7}$$

Now it is easiest to work in Cartesian coordinates with just one component of angular momentum, say L_z, and obtain the others by cyclic permutation. Thus integrating $d\mathbf{N}$ over the volume of the extended body and using (58.3) shows that

$$\frac{dL_z}{dt} = GM \int \frac{\rho(\mathbf{r}')}{r^3}(x'y - y'x)dx'dy'dz'. \tag{58.8}$$

This simplifies considerably when $r' < d$ for all r'. From the geometry of Figure 54 we have $r^2 = d^2 + r'^2 - 2\mathbf{d}\cdot\mathbf{r}'$ so that when M is well outside the extended body

$$r^{-3} = d^{-3}\left[1 - \frac{3}{2}\frac{r'^2}{d^2} + 3\frac{\mathbf{d}\cdot\mathbf{r}'}{d^2} + \cdots\right]. \tag{58.9}$$

Substituting (58.9) into (58.8), recalling that the center of mass is at the origin, and integrating, we find

$$\frac{dL_z}{dt} = \frac{3GMxy}{d^5}\int \rho(x'^2 - y'^2)dx'dy'dz' = \frac{3GMxy}{d^5}(I_{yy} - I_{xx}). \tag{58.10}$$

The other components L_x and L_y follow similarly, or by cyclic permutation.

It is also straightforward, but a bit more messy, to do this calculation in spherical coordinates, using the multipole expansion for the field of the extended body (an exercise for the reader). Physically, it is then more obvious that the monopole term only affects the motion of the center of mass and does not contribute to the torque. The dipole term vanishes since the center of mass is at the origin. This leaves the quadrupole moment as the lowest order contribution to the torque, and the result is equivalent to (58.10).

Angular momentum gained by the extended body must come from a change in the angular momentum of its orbit with the mass M. Thus it is possible to invert the problem to see how the orbit changes. The most common example is the orbit of a satellite of mass M around a non-spherical planet. The analogous calculation for the gravitational potential, ϕ, of the planet then leads to Mac Cullagh's formula

$$\phi = -\frac{GMm}{r} + \frac{GM}{2r^3}[3I_r - (I_x + I_y + I_z)]. \tag{58.11}$$

Here I_r is the planet's moment of inertia around the $\mathbf{r}$ axis. The set of terms in brackets leads to the precession of bound orbits by the torque of a non-spherical planet.

To apply (58.10) it is important to remember that all the quantities on the right hand side may depend significantly on time, as the orbit and internal structure of the body change. If, however, these changes are slow compared to changes in the spin ratio (58.1) we may approximate the quantities on the right hand side by their average values. Using (58.5) for L_z in the case when the $I_{zz}\omega_z$ term dominates, we may write (58.10) as

$$\frac{d\omega_z}{dt} = (G\rho)^{1/2}\frac{d\sigma_z}{dt} \simeq -\frac{3GM\langle\eta_z\rangle}{d^3}, \tag{58.12}$$

where

$$\eta_z \equiv \frac{xy}{d^2}\frac{(I_{yy} - I_{xx})}{I_{zz}}. \tag{58.13}$$

The first factors of η_z are the direction cosines, and these are just slightly less than unity for $r' \ll d$. Generally the second factor will also be of order unity unless the mass distribution is nearly spherical. For example, along the principal axes of a uniform density ellipsoid having mass m and semiaxes R_x, R_y and R_z,

$$I_{xx} = \frac{m}{5}(R_y^2 + R_z^2), \tag{58.14}$$

with cyclic permutation for the other moments of inertia. So

$$\eta_z = \frac{xy}{d^2}\frac{R_x^2 - R_y^2}{R_x^2 + R_y^2}, \tag{58.15}$$

and if $R_y = R_x/2$, then $\eta_z = 3\ xy/5d^2$. Under many circumstances, therefore, we would expect η_z to be of order unity. Similar relations hold for the other components of angular velocity and usually one of them, say ω_z, will dominate.

The spin ratio, in (58.12), increases as

$$\frac{d\sigma}{dt} \approx 4\pi \langle \eta \rangle \frac{\tau_m}{\tau_M^2}. \tag{58.16}$$

Generally $\tau = (G\rho)^{-1/2}$ and, in particular, τ_m gives the average dynamical timescale within the extended body denoted by m. Similarly, τ_M gives the dynamical time-scale for a test particle in circular orbit of radius d around M; it is the timescale τ which would result if the mass M were distributed uniformly over a spherical volume of radius d. Therefore a necessary criterion for time averaging the right hand side of (58.10) is that

$$4\pi \langle \eta \rangle \frac{\tau_m}{\tau_M} = 4\pi \langle \eta \rangle \left(\frac{M R^3}{m\, d^3} \right)^{1/2} > 1, \tag{58.17}$$

where $R^3 = 3m/4\pi\bar{\rho}_m$. A sufficient condition, which can always be checked in retrospect, is that σ has increased faster than the unaveraged form of the right hand side of (58.16) has changed. After σ increases from zero to ~ 1 the tidally induced rotation plays a major role in the force balance which determines the internal equilibrium structure. Thus we may define a characteristic growth timescale for σ as

$$\tau_\sigma = \frac{1}{4\pi \langle \eta \rangle} \left(\frac{m\, d^3}{M R^3} \right)^{1/2} \tau_M. \tag{58.18}$$

Hoyle (1949) was the first to calculate tidal transfer of angular momentum among protogalaxies, and he concluded that it would be significant. We can readily see this from (58.18) because τ_M is approximately the dynamical timescale of the smoothed out density in the region of the protogalaxies. This is also about equal to the Hubble expansion timescale of the Einstein–Friedman cosmologies (see Equation (21.31)). Most of the angular momentum is transferred when the protogalaxies are close together with a significant density contrast. Under optimum conditions, $\langle \eta \rangle \approx m/M \approx 1$ and $d \lesssim 3R$, giving $\tau_\sigma \approx \tau_{\text{Hubble}}$. Since $\dot{\sigma} \propto d^{-3}$, most angular momentum transfer will be caused by the nearest large protogalaxy, although summing the random contributions of more distant protogalaxies may increase $\dot{\sigma}$ by a factor of order unity. Another contribution may arise from the large scale inhomogeneity of a protocluster of protogalaxies. Clearly, more precise results will depend on detailed models of galaxy formation, and particularly on the changing density contrast of protogalaxies as they separate. For examples of more detailed calculations, see Harrison (1971), Efstathiou & Jones (1980), and their many references. After the galaxies form and cluster, close chance encounters can transfer more than the average amount of angular momentum. (Could our galaxy and Andromeda be examples?) Although it is possible to make these estimates of total angular momentum transfer, its resulting distribution within the galaxy is poorly understood.

We still do not know whether tidal torques alone can account for the observed angular momentum of galaxies. In reaching astrophysical conclusions it is important not to let the deficiencies – or even the successes – of particular models cloud our judgment. Many realistic effects are difficult to model, except by detailed simulations which then become so specific that they may not apply to individual observations. For example, in a realistic situation, matter which is spun up on its way to becoming a galaxy may not ultimately join the galaxy for which it initially seems destined. The same strong tides which induce spin also mix up the material lines of flow. Under these complex conditions, results like Kelvin's theorem for conservation of vorticity (see texts on fluid mechanics) do not apply. If there are gaseous shocks or interpenetration of star streams, Kelvin's theorem breaks down on a fine grained scale. If there is exchange of matter among protogalaxies, Kelvin's theorem breaks down on the coarse grained scale which corresponds to the present observations of the angular momentum of galaxies.

Tidal torques are far from the only process of angular momentum transfer. The others are even less well understood. Perhaps the most straightforward of these are actual mergers of galaxies or protogalaxies. Suppose N subgalaxies, each of mass m and average random velocity v, merge to form a galaxy of radius R. Then we would expect the average net angular momentum to be of order $L \approx NmRv/\sqrt{N} = mRv\sqrt{N}$. The internal angular momentum of each subunit will also contribute. If these subunits are in virial equilibrium, each can have a maximum internal angular momentum of order $L_i = (GM^3 r)^{1/2}$. Their total contribution may then be as high as $\sim L_i N$ if the spins strongly tend to align, or $\sim L_i \sqrt{N}$ if they do not, and they could have any direction relative to the orbits.

If protogalaxies are primarily gaseous, forming in turbulent regions, then the turbulent viscosity can also transfer angular momentum. Here I repeat the refrain that there is little basic understanding of the physics of turbulent viscosity, so it is only possible to construct phenomenological models (e.g., Saslaw, 1971) as illustrations. Magnetic fields, which may be effective even if the gas is predominantly neutral, would also introduce additional mechanisms for angular momentum transfer.

All momentum transfer processes – tidal torques, merging, turbulent viscosity, magnetic fields – apply to galaxies which form either by accumulating subgalaxies or by fragmenting from a larger cloud. Whether protogalaxies formed first and then clustered, or whether protoclusters formed first and then fragmented into galaxies, is still an overriding cosmogonical question. If we could discover definitive differences in the distributions of angular momenta which result from the various clustering and fragmentation pictures of galaxy formation, it would be a great step forward.

59

Rotation curves and galaxy mass

Here's fine revolution, an we had the trick to see't.

Shakespeare

To compare theories of angular momentum transfer with properties of observed galaxies we must know how observed rotation curves are related to a galaxy's internal distribution of mass and angular momentum. There is a direct relation only if the balance between gravity and centrifugal force dominates the galaxy's internal motions. This is almost certainly the case in the plane of a flat spiral galaxy. But in the outer warps and central bulges of spirals, as well as in ellipticals and irregulars, the situation is less clear. In these more distended objects, unrelaxed streaming and random motions of stars, and pressures and magnetic forces in gas, may contribute significantly to the observed velocities. Models may, of course, take all these effects into account, but unless they can be determined independently they will magnify the uncertainties.

Here we consider just the gravitational effects, for they do give a qualitative understanding of the typical rotation curve illustrated in Figure 51. So far, these are the best determined results and they deserve – or at any rate will get – a section to themselves. To see how gravity determines the overall shape of spiral galaxy rotation curves we start with the balance between gravitational and centrifugal acceleration in the plane of rotation

$$\frac{v^2(r)}{r} = -\frac{\partial \phi}{\partial r}\bigg|_{z=0}, \tag{59.1}$$

where $v(r)$ is the rotational velocity (denoted by u_ϕ in 57.28) and the gravitational potential ϕ follows from Poisson's equation (15.4)

$$\nabla^2_\phi = -4\pi G \rho. \tag{59.2}$$

Here the positive gradient of the potential is the force per unit mass. So if gravity is the only force the density distribution completely determines the rotation curve.

The simplest solution is for test particles revolving around a point of mass M, whence $\phi = GM/r$ and $v^2 = GM/r$. This circular Keplerian motion also follows from the virial theorem, as mentioned at the end of Section 9. Far enough from any mass distribution for its multipole moments to be inconsequential, the orbits should be Keplerian. Applying this formula to the observed $v(r)$ within a galaxy gives an order of magnitude estimate of the mass within that radius, and thus a rough limit to the total mass of the galaxy.

The next most simple solution is for test particles revolving within a spherically distributed mass. The non-constant part of the potential which determines the force is $\phi = GM(r')/r$ and $v^2(r') = GM(r')/r'$. If the density is uniform, $M(r') \propto r'^3$ and $v \propto r'$, which is solid-body rotation. Any non-singular density distribution can be expanded in a Taylor series starting at the center. Since the first term in this series is a constant (the central density), it should not surprise us that galaxies have a region of solid-body rotation near their center, at least when local irregularities are averaged out. What might surprise us is the extent of this region. In spiral galaxies it is reassuring to know that it is roughly co-extensive with the central bulge. To understand the dynamical origin of a central bulge is a harder problem. Farther from the center, a spherical distribution dominated by $\rho \propto r^{-2}$ will give a region of constant rotational velocity, also similar to what is often observed. This density distribution is characteristic of an isothermal sphere at large radius (Section 40), suggesting, to some astronomers, evidence for massive spherical halos around galaxies. Finite, non-spherical distributions can also have a constant rotational velocity (as we shall soon see), so isothermal halos do not provide a unique interpretation.

Although points and spheres contain the essence of the physical problem, more elaborate models are easily constructed for distributions with rotational symmetry, such as ellipsoids and spheroids. Their potentials can be worked out from Poisson's equation, or looked up in MacMillan (1958). Looking up results in texts may be easier, but working them out is better style. The general result that finite disks lead to flatter rotation curves than idealized infinite disks, other things being equal, is readily understood physically. The forces on a circular orbit in an infinite disk are produced by the inner mass, the centrifugal force, and the outer mass. Remove the outer mass and the orbit loses a component of its outward force. To remain in equilibrium it must speed up and make an increased centrifugal force compensate for this loss.

To make more realistic models of galaxy rotation we must allow for many components in the density distribution. There are several ways to do this. One is to represent a galaxy by a concentric sequence of thin spheroidal shells, each having constant density (these are called homeoids). The main virtue of this approach (Burbidge, *et al.*, 1959) is that within an homeoid there is no net gravitational force; the force at a given radius depends only on the mass within that radius. This simplifies the integral form of (59.2). If $U(r,a)$ is the potential of a homogeneous spheroid of unit density and semimajor axis a then

$$\phi(r) = -\int_0^r \rho(a)\frac{dU}{da}\,da. \tag{59.3}$$

We neglect the exterior mass which contributes to the potential, but not to its gradient. The densities $\rho(a)$ of different shells may differ. The integral is over all the mass within r. Substituting (59.3) into (59.1) gives an integral equation for the rotational velocity

$$v^2(r) = r \int_0^r \rho(a) \frac{\partial^2 U(r,a)}{\partial r \partial a} \, da. \tag{59.4}$$

Now the procedure is to expand both sides in a power series in r. The coefficients on the left hand side are selected to fit the observed rotation curve. Since $U(r,a)$ is a known function, when the right hand side is expanded only the coefficients of the density expansion are unknown. These can then be determined by equating coefficients of the same power of r on both sides. This gives $\rho(r)$.

In practice, significant problems befuddle this approach. First, forcing a galaxy to fit homeoids introduces errors whose size is unknown. Basic structures with any shape could equally well be used to build up the actual density, although they would complicate the integral equation. Second, the results will depend on the eccentricity of the spheroids. This can only be guessed from the galaxy's axial ratio and general structure; it cannot be determined self-consistently. Third, a polynomial does not always fit the observed velocity curve very well. When polynomials of different orders are tried the density often appears to converge slowly and irregularly. Fourth, even if the results are reasonable in the bulk of the galaxy, they cannot be extrapolated accurately to large radii. Most of the angular momentum resides at large radii. So although the inner mass may be well determined – as it often is – the total angular momentum is very uncertain.

An alternative approach exploits the large axial ratio of spiral galaxies. If we write out Poisson's equation (59.2) in cylindrical coordinates (r, θ, z),

$$\frac{\partial^2 \phi}{\partial r^2} + \frac{1}{r} \frac{\partial \phi}{\partial r} + \frac{1}{r^2} \frac{\partial^2 \phi}{\partial \theta^2} + \frac{\partial^2 \phi}{\partial z^2} = -4\pi G \rho. \tag{59.5}$$

For axial symmetry, ρ does not depend on θ so the third term vanishes. The homogeneous form of this equation, with $\rho = 0$, is just Laplace's equation and is solved as usual by separating variables. Substituting $\phi(r,z) = R(r) Z(z)$ into (59.5) with $\rho = 0$, and dividing through by RZ gives one set of terms which depends only on z and another which depends only on r. Because a change of one variable cannot affect the terms in the other variable, each set of terms must be a constant in order to maintain the equality. Denoting this separation constant by k^2 we get the two ordinary differential equations

$$\frac{d^2 Z}{dz^2} - k^2 Z = 0 \tag{59.6}$$

and

$$\frac{d^2 R}{dr^2} + \frac{1}{r} \frac{dR}{dr} + k^2 R = 0. \tag{59.7}$$

A solution of (59.6) which is symmetric around the plane of the galaxy is

$$Z(z) = e^{-k|z|}. \tag{59.8}$$

Equation (59.7) is an elementary form of Bessel's equation and its solutions which remain finite at the origin are Bessel functions of order zero, $J_0(kr)$. (Higher order Bessel functions are needed if azimuthal symmetry breaks down.) Thus the eigenfunctions of the homogeneous equation are $e^{-k|z|}J_0(kr)$ for the different eigenvalues k.

These simple eigenfunctions will also give solutions of the inhomogeneous axially symmetric Poisson equation for special density distributions in which the mass per unit surface area of a thin disk is proportional to $J_0(kr)$. A more general axisymmetric surface density can be expressed as a Bessel integral over $J_0(kr)\,k\,dk$ (analogous to a Fourier integral over $\exp(ikr)\,dk$), and (59.5) gives its potential. Equation (59.1) then relates the density distribution to the velocity profile.

This procedure has been carried through for the case of thin disks (Toomre, 1963). The surface density, $\mu(r)$, of the disk is related to its rotation curve by

$$\mu(r) = 2\int_0^\infty \rho dz = \frac{1}{2\pi G}\int_{s=0}^\infty \frac{dv^2(s)}{ds}\int_{k=0}^\infty J_0(rk)J_0(sk)dkds. \qquad (59.9)$$

Here s is a dummy radius variable and the inner integral may also be expressed in terms of a hypergeometric function or of the first complete elliptic integral. An especially simple example is the velocity squared profile

$$v_0^2(r) = c_0^2(1 + r^2/a^2)^{-1/2}, \qquad (59.10)$$

to which corresponds the surface density

$$\mu_0(r) = \frac{c_0^2}{2\pi G}\left[\frac{1}{r} - (a^2 + r^2)^{-1/2}\right], \qquad (59.11)$$

where c_0 and a are arbitrary constants. Since (59.9) shows that $\mu(r)$ and $v^2(r)$ are linearly related, any sum, difference, or derivative of (59.10) with respect to the parameter a has a similar sum, difference or derivative for its corresponding surface density. In this way Toomre builds up classes of models such as

$$v_n^2(r) = c_n^2\left\{-\frac{\partial}{\partial a^2}\right\}^{n-1}\left[\frac{r^2}{a}(a^2 + r^2)^{-3/2}\right] \qquad (59.12)$$

and

$$\mu_n(r) = \frac{c_n^2}{2\pi G}\left\{-\frac{\partial}{\partial a^2}\right\}^{n-1}[(a^2 + r^2)^{-3/2}]. \qquad (59.13)$$

These are useful schematic representations of observed rotation curves with their inferred surface densities. Examples of further analyses along these lines have been worked out by Nordsiek (1973, with a following paper applying the results to observations). The more complicated problem of non-circular motions is not well understood.

Finding the surface density of a galaxy is really just a first step toward understanding the origin of its rotation. The next step is to find its distribution of angular momentum. In Section 58, I glossed over the detailed internal distribution of angular momentum which tidal torques produce because very little is known about this, or about subsequent momentum transfer. Now, however, we are in a position to see where the simplest picture of the collapse of a spinning protogalaxy, with detailed conservation of angular momentum, would lead (Mestel, 1963; Crampin & Hoyle, 1964).

With the surface density available, it is easy to find the angular momentum distribution in principle. In practice, it is harder because the errors of the rotation curve enter both the surface density and the velocity in the angular momentum. Again consider a thin disk and let $h(r) = rv(r)$ be the angular momentum per unit mass at a distance r from the rotation axis. The amount of mass in an annulus between r and $r + dr$ is

$$dm(r) = 2\pi\mu(r)rdr. \tag{59.14}$$

From

$$dh = \left(v + r\frac{dv}{dr}\right)dr \tag{59.15}$$

we have that the amount of matter with specific angular momentum between h and $h + dh$ is

$$dm(h) = 2\pi\mu(r)\left(\frac{v}{r} + \frac{dv}{dr}\right)^{-1}dh = -\frac{\pi\mu(r)}{B(r)}dh, \tag{59.16}$$

where $B(r)$ is Oort's 'constant' from (57.28). Thus B may also be interpreted, apart from a numerical factor, as the Jacobian of the transformation between area and specific angular momentum in the case of circular rotation. The radial distance can be expressed in terms of h from $h = r\,v(r)$ as long as h increases monotonically with r. Then $dm(h)$ can be integrated over all h to get the total angular momentum of a galaxy. For the disk of our own galaxy, angular momentum estimates are in the area of $10^{74}\,\mathrm{g\,cm^2\,s^{-1}}$, with a corresponding mass of about 10^{11} solar masses.

Now consider a spheroidal protogalaxy spun up by gravitational torques, as described in the last section. The equation of a spheroid in cylindrical coordinates is $a^2z^2 + b^2r^2 = a^2b^2$, where a and b are the major and minor semiaxes. As a toy model, suppose the entire system has a uniform density ρ in solid body rotation. Then $h = r^2\omega$ or

$$dh = 2\omega rdr. \tag{59.17}$$

To find the amount of matter between r and $r + dr$, where the angular momentum is between $h + dh$, we note that the volume enclosed by the spheroid at a distance r from the z axis is

$$V(r) = \int_0^{2\pi}\int_0^r 2\int_0^{b(a^2 - r^2)^{1/2}/a} sdzdsd\theta = 4\pi\frac{b}{a}\int_0^r s(a^2 - s^2)^{1/2}\,ds. \tag{59.18}$$

This last integral is trivial, but we do not even have to do it because we are interested in the shell between r and $r + dr$ which is given by dV/ds. Therefore

$$dm(h) = \frac{4\pi b\rho}{a}(a^2 - r^2)^{1/2}rdr = \frac{2\pi b\rho}{\omega}\left(1 - \frac{h}{a^2\omega}\right)^{1/2}dh, \tag{59.19}$$

using (59.17).

This protogalaxy, not being in centrifugal (or pressure) equilibrium, must be contracting. The simplest way for it to contract is for each material element to conserve its angular momentum: no turbulent mixing, no internal torques. The spheroid contracts, maintaining the distribution (59.19) until it forms a thin disk in centrifugal balance (with gas pressure or random stellar velocities giving a finite vertical dimension). Although both v and r will change during contraction, $h = h' = rv(r) = r'v'(r')$ remains the same, with primes denoting post-contraction quantities. After contraction, (59.16) will hold with primed quantities. But since $dm(h) = dm'(h')$ we can equate (59.16) and (59.19), now dropping the primes to keep notation simple. Moreover, the semimajor axis a of the spheroid contracts to the radius R of the equilibrium disk. The result, normalized to the central surface density, is

$$\mu(r) = \mu(0)\left(\frac{v}{r} + \frac{dv}{dr}\right)_{r=0}^{-1} \left(\frac{v}{r} + \frac{dv}{dr}\right)\left(1 - \frac{r^2}{R^2}\right)^{1/2}. \tag{59.20}$$

This formula relates the surface density to the rotation curve in a disk formed by contracting a spheroid of initially uniform density, and angular velocity, and conserving momentum in detail. The special case where the final disk is also in solid-body rotation gives the simple surface density $\mu(r) = \mu_0(1 - r^2/R^2)^{1/2}$ over the whole disk, see Mestel (1963).

To see if real galaxies could have formed in this simple way we could use their observed rotation curve to calculate $\mu(r)$ from the general formula (59.9), and then calculate $\mu(r)$ for the special cosmogony of (59.20). If they agree, then the simple cosmogony is at least consistent with observations, although not necessarily demanded by them. Other more complicated models can be constructed similarly and tested against the observations.

We can use the general principles of this section to derive interesting order-of-magnitude relations between a spiral galaxy's total mass, peak rotation velocity, luminosity, and total angular momentum. These are little more than dimensional results, but they provide a useful feel for what to expect. Here we will suppose that all the physical quantities represent average or characteristic values for a given galaxy. The total angular momentum of a spiral galaxy is $H \approx MRv$ and in centrifugal balance $GM \approx Rv^2$, so that

$$H \approx \frac{GM^2}{v} \approx \frac{R^2v^3}{G} \approx \frac{R^5\omega^3}{G}. \tag{59.21}$$

From (59.13) and (59.20) we see that reasonable surface density relations may be expanded in the form $\mu = \mu_0(1 - r^2/R^2 + \cdots)$. Therefore in the central part of the spiral disk, out to some characteristic length scale R, the characteristic mass is $M \approx \pi\mu_0 R^2$. Combining this with centrifugal balance shows that

$$v^2 \approx \pi G\mu_0 R \approx G(\pi\mu_0 M)^{1/2}. \tag{59.22}$$

Therefore the angular momentum

$$H \approx \left(\frac{G^2}{\pi\mu_0}\right)^{1/4} M^{7/4} \propto R^{7/2}. \tag{59.23}$$

These relations can be written in terms of the central (and perhaps the dominant) luminosity by making the additional assumption that $L \propto M$. This means that the characteristic lengthscale for a galaxy's central surface luminosity to change is the same as the lengthscale for its surface density to change. In other words, its mass–luminosity ratio does not change much near the center. Then $v^4 \propto M \propto L$. Furthermore, if we repeat the arguments of the last paragraph, substituting the virial theorem with velocity dispersion σ for the requirement of centrifugal balance, and supposing $L \propto M \propto R^2$, we arrive at the relation

$$\sigma \propto L^{1/4}. \tag{59.24}$$

In order to generalize this result to a set of galaxies it is necessary that the constant of proportionality (involving the mass–luminosity ratio, the central surface density, and the several lengthscales) be common to all those galaxies. There is some somewhat ambiguous evidence that that may be the case.

60

Orbits and third integrals

If you ask the special function
Of our never-ceasing motion,
We reply, without compunction,
That we haven't any notion!

Gilbert and Sullivan

Few orbits in smooth axisymmetric galaxies are exactly circular, and no orbits in unsymmetric force fields are regular. Some orbits, like the denizens of Arcadia, move ergodically hither and thither, described by no special function. Others, more constrained, are partly predictable.

In this section we explore some general properties of orbits in the smooth force field of a flattened galaxy. After a galaxy has formed, the two-body relaxation timescale τ_R usually becomes very long compared to the dynamical crossing time τ_c or even to the Hubble time. Therefore, the collisionless Boltzmann equation provides a good description of the internal dynamics. We saw from (7.15) that this description is equivalent to knowing the orbits of stars in the smooth mean field. Thus it is possible to invert the approach of Section 7 by starting from the orbits and building up a self-consistent distribution function (see Section 63). Orbits in unsymmetric systems are naturally very complicated. Finding them usually requires extensive numerical integrations which are highly specific to individual models. Some types of orbits, however, are common to a reasonable range of idealized models, so we may consider them here.

After circular orbits, the next simplest are motions across the symmetry plane $z = 0$. If these orbits do not stray far above or below the plane, they respond primarily to the local density $\rho(r,z)$ rather than to the overall galactic field. To obtain the general equations of motion for a star in a rotating axisymmetric galaxy, we start with the kinetic energy in cylindrical coordinates for a star of unit mass

$$T = \tfrac{1}{2}(\dot{r}^2 + r^2\dot{\theta}^2 + \dot{z}^2). \tag{60.1}$$

Thus the canonical momenta

$$p_k = \frac{\partial T(\dot{q}_k)}{\partial \dot{q}_k} \tag{60.2}$$

are

$$p_r = \dot{r}, \quad p_\theta = r^2\dot{\theta}, \quad p_z = \dot{z}. \tag{60.3}$$

The Hamiltonian (with negative potential energy, $-\phi$, per unit mass) is

$$H(\mathbf{p},\mathbf{q}) = \tfrac{1}{2}(p_r^2 + r^{-2}p_\theta^2 + p_z^2) - \phi(r,\theta,z). \tag{60.4}$$

Substituting (60.3) and (60.4) into the canonical equations (7.16) gives the equations of motion in non-rotating coordinates

$$\ddot{r} = r\dot{\theta}^2 + \frac{\partial \phi}{\partial r}$$ (60.5)

(which generalizes 59.1),

$$r\ddot{\theta} = -2\dot{r}\dot{\theta} + \frac{1}{r}\frac{\partial \phi}{\partial \theta}$$ (60.6)

(which includes the Coriolis force),

$$\ddot{z} = \frac{\partial \phi}{\partial z},$$ (60.7)

where $\partial \phi / \partial \theta = 0$ for axisymmetry. Incidentally, substituting the canonical formulae (60.3) and (60.4) into (7.17) is an easy way to find Boltzmann's equation in cylindrical geometry.

Immediately we notice that vertical motion decouples from the other velocities, since it depends only on $\phi(r,z)$. This is because there are no centrifugal forces in the z direction. Physical arguments readily give the form of the force term, F_z, on the right hand side of (60.7) for stars close to the galactic plane. Expanding F_z in powers of z, the constant term must be zero to provide equilibrium at $z = 0$. The next term, proportional to $-z$, which stabilizes the equilibrium, presumably dominates and gives standard harmonic oscillations. To determine their frequency more precisely, we use Poisson's equation (59.5). Throughout most of a highly flattened disk – except near the edge – the radial gradient of the potential is much less than its vertical gradient. Its angular gradient vanishes completely for axisymmetry. Therefore (59.5) may be approximated by

$$\frac{\partial^2 \phi}{\partial z^2} = -4\pi G\rho.$$ (60.8)

If the density is symmetric around the $z = 0$ plane, it has the form $\rho = \rho_0 - \text{const } z^2 + \cdots$. Therefore, to lowest order in z the force is

$$\frac{\partial \phi}{\partial z} = -4\pi G\rho_0 z.$$ (60.9)

Substituting (60.9) into (60.7) and solving the resulting harmonic oscillator equation shows that

$$z = z_0 \cos\left[\alpha(t - t_\alpha)\right],$$ (60.10)

where the angular frequency

$$\alpha = (4\pi G\rho_0)^{1/2}$$ (60.11)

and t_α is the initial time. This simple oscillation occurs only if r is much less than the radius of the smooth disk and z is also much less than the vertical scale height of the density. Otherwise, detailed models of ρ and ϕ are necessary (see Section 66 for a useful example.) Moreover, detailed models are also necessary if the density distribution near the plane is so irregular that it destabilizes orbits there; warped disks are examples.

If the gravitational field is fairly symmetric around $z = 0$ we can use (60.9) to estimate the local density ρ_0 near the plane from observation. To do this we need an additional equation for the force. The simplest estimate follows for a model where the components of the disk have motions described by the collisionless Boltzmann equation (Section 7 and 8). Then the momentum transport equation (8.20) further simplifies to

$$\frac{\partial \phi}{\partial z} = \frac{1}{\rho} \frac{\partial P}{\partial z} = \langle v^2 \rangle \frac{\partial \ln \rho}{\partial z} \tag{60.12}$$

for isotropic pressure under steady state conditions with $P = \rho \langle v^2 \rangle$ (from 8.18) and velocity dispersion independent of position. Again, only the vertical gradient is important.

Clearly there are already many simplifications in the discussion of this problem. Even more are needed. The density gradient in (60.12) cannot be determined directly from observations because we may not be able to observe all the constituents of the disk. So an additional set of simplifying assumptions is invoked. These suppose that a given type of constituent (say stars of a given spectral class or gas clouds with certain properties) can be considered as test particles which trace the mean force field of all the constituents. The given constituent must neither interact with the other constituents nor show correlations remaining from earlier formation. It must be well mixed throughout the disk. If this is the case its gradient at any value of z gives an estimate of the force in (60.12) and the average density ρ_0 near the plane from (60.9).

Applying this procedure to our own galaxy gives a rather uncertain result. The review volume of Blaauw & Schmidt (1965, see especially the chapters by Oort and by Schmidt with regard to the last and present sections) summarizes many of the details. The average density obtained in the region of the sun is $\rho_0 \approx 10^{-23}\,\mathrm{g\,cm^{-3}}$. About half this value is found by adding up the luminosity of all the observed stars and using their independently estimated mass–luminosity ratios. The difference (sometimes called the 'Oort limit') represents the maximum amount of dark matter or interstellar gas which could be present locally. Considering all the theoretical simplifications which go into these analyses, and their observational difficulties, the results could easily be wrong by at least a factor of two. Much remains to be learned about the local structure of our galaxy.

Turning next to orbits in the plane we generally must solve (60.5) and (60.6) for specific mass models. For angular symmetry (60.6) is just our old friend angular momentum conservation: $h = r^2 \dot{\theta} = \text{constant}$. If the orbits are primarily circular, we return to the models described in Section 59. We can now go a step further and see how these circular orbits respond to small perturbations. Writing the departure from an arbitrary orbit as

$$r = r_0 + r_1, \quad \theta = \theta_0 + \theta_1 \tag{60.13}$$

and substituting into (60.5)

$$\ddot{r}_0 + \ddot{r}_1 = \frac{h^2}{(r_0 + r_1)^3} + \frac{\partial}{\partial r}\phi(r_0 + r_1). \tag{60.14}$$

The orbit is perturbed at constant angular momentum, although this could be generalized for systems with slight external torques. The unperturbed orbit $r_0(t)$ could be anything stable, but circles are simplest. Expanding $\phi(r)$ in a Taylor series around r_0, and subtracting the unperturbed orbit, the linear approximation to (60.14) becomes

$$\ddot{r}_1 = -\kappa^2 r_1, \tag{60.15}$$

where

$$\kappa^2 = -\left[\frac{3}{r_0}\left(\frac{\partial\phi}{\partial r}\right) + \left(\frac{\partial^2\phi}{\partial r^2}\right)\right]_{r_0}. \tag{60.16}$$

The solution again has the form

$$r_1 = r_1(t_\kappa)\cos\kappa(t - t_\kappa), \tag{60.17}$$

giving harmonic oscillations for $\kappa^2 > 0$. The frequency κ is called the epicyclic frequency. When $\kappa^2 < 0$ the perturbation r_1 increases as $\cosh\kappa(t - t_\kappa)$ and the orbit is unstable. It diverges from its initial form at time t_κ until higher order terms begin to either damp or accelerate this divergence.

To gain some feeling for force laws which are compatible with stable circular orbits, consider the simple potential

$$\phi = ar^{-m} + b, \tag{60.18}$$

where m, a, and b are constants. An attractive force $\partial\phi/\partial r$ has $ma > 0$ and it is easy to check that the requirement $\kappa^2 > 0$ for stability implies $m < 2$. Thus forces which fall off more steeply than r^{-3} are too weak to prevent perturbations from destabilizing circular orbits. As an exercise, the reader can extend this analysis to find the most effective ways of destroying non-circular orbits.

The perturbation (60.13) can also affect the angular coordinate of an orbit. At some given time the angular position of the star can be displaced relative to its unperturbed value, and its velocity and radius will alter accordingly. For a circular orbit whose angular momentum remains constant, these perturbations must be related to first order by

$$h = r_0^2\dot{\theta}_0 = r^2\dot{\theta} = (r_0 + r_1)^2\frac{\mathrm{d}}{\mathrm{d}t}(\theta_0 + \theta_1) = r_0^2\dot{\theta}_0 + 2r_0\dot{\theta}_0 r_1 + r_0^2\dot{\theta}_1 \tag{60.19}$$

or

$$\dot{\theta}_1 = -\frac{2\dot{\theta}_0}{r_0}r_1 = -\frac{2\omega}{r_0}r_1. \tag{60.20}$$

Substituting (60.17) and integrating then gives

$$\theta_1 = -\frac{2\omega r_1(t_0)}{r_0\kappa}\sin\kappa(t - t_0). \tag{60.21}$$

Notice that, for a positive initial radial perturbation, θ_1 decreases with time, so the

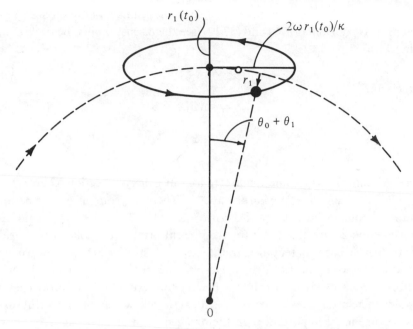

Fig. 55. An epicycle of frequency κ in a rotating frame associated with a perturbed circular orbit of frequency ω. The unfilled circle would be the star's unperturbed position; the big dot is its actual position.

perturbation is moving backward relative to ω. This is nothing more than the statement that the angular velocity must decrease as r increases to conserve angular momentum. Similarly, after half an epicycle, r_1 is negative and θ_1 positive, so the perturbation moves faster than ω. Thus the perturbed angular velocity has the opposite sense to the angular velocity of the basic orbit, i.e., the motion is retrograde. The perturbation traces out an ellipse with axes $r_1(t_0)$ and $2\omega r_1(t_0)/\kappa$ in the rotating frame, as illustrated in Figure 55. Such motion is called an epicycle.

The epicyclic frequency is clearly a basic property of simple orbits. Circular Keplerian orbits with $\phi = GM/r$ have $\kappa^2 = GM/r^3 = \omega^2$. These are the most centrally condensed potentials and have rotational velocity $v \propto r^{-1/2}$. Flat rotation curves, $v = $ constant, are characterized by $\phi = -c \ln r$ where c is a positive constant and thus $\kappa^2 = 2c/r^2 = 2\omega^2$. Solid-body rotation curves, $v \propto r$, are characterized by $\phi = -cr^2/2$ and for these $\kappa^2 = 4c = 4\omega^2$. All this illustrates a fairly wide range of possibilities. Relations between epicyclic frequencies and rotational frequencies are especially important for understanding collective perturbations of orbits leading to spiral structure in galaxies.

Sometimes it is useful to relate the epicyclic frequency to Oort's constants A and B for circular orbits. First, substituting (60.5) in the form of (59.1) into (60.16), we find using $v = r\omega$ that

$$\kappa^2 = 4\omega^2 \left(1 + \frac{r}{2\omega} \frac{\mathrm{d}\omega}{\mathrm{d}r} \right) = \frac{1}{r^3} \frac{\mathrm{d}}{\mathrm{d}r} (r^2\omega)^2. \tag{60.22}$$

Thus the epicyclic frequency is closely related to the gradient of the squared angular momentum. Next, from (57.28) and (57.12) (see the discussion just before 57.28), we obtain

$$\omega = A - B \tag{60.23}$$

and therefore

$$\kappa^2 = -4B(A - B) \tag{60.24}$$

or

$$\left(\frac{\kappa}{\omega}\right)^2 = -\frac{4B}{A - B}. \tag{60.25}$$

Using the local values of A and B adopted for our galaxy in Section 57 gives $\omega = 25$ km s^{-1} kpc^{-1} and $\kappa = 31.6$ km s^{-1} kpc^{-1}. These correspond to a period of revolution around the galaxy of $2\pi r/r\omega = 2\pi/(A - B) = 2.5 \times 10^8$ yr and to an epicyclic period $2\pi/\kappa = 2 \times 10^8$ yr for the present orbit of the solar neighborhood.

So far we have been content to examine simple illustrative orbits. Now we turn to face the complexities of real orbits. Despite the peculiarities of any individual trajectory there are classes of orbits with very interesting general properties. Our viewpoint therefore moves closer to statistical mechanics, and the distributions of orbits become more important than the detailed form of each one.

Consider a simplified version of phase space: the phase plane. A phase plane is just a two-dimensional cut across phase space, which includes any two orthogonal axes. We will discuss two of the simplest and most informative phase planes here: the z–r plane which characterizes orbits in and through axisymmetric disks, and the y–$\dot{y}$ plane (usually called 'the phase plane' for ordinary differential equations) which characterizes the position and velocity in any one dimension of an orbit. These phase planes are sometimes called 'surfaces of section' because they are sections of phase space through particular values of the remaining coordinates. Additional types of phase diagrams whose axes are functions of the phase space coordinates are also informative for particular systems, and the reader might think some out as examples.

Figure 56 shows how the z–r phase plane represents several general classes of orbits. The zero velocity curve is the locus of limiting coordinates for any bound orbit of a given total energy in a given potential. At the limit of a bound orbit its velocity vanishes. So the zero velocity curve is found by setting the total energy of the orbit

$$E = T + \phi, \tag{60.26}$$

equal to just the potential energy: $E = \phi(r, z)$. For example, epicyclic orbits which move through the plane and are represented by (60.10) and (60.17) have constant energy

$$E = \tfrac{1}{2}\dot{z}^2 + \tfrac{1}{2}\dot{r}_1^2 + \tfrac{1}{2}\alpha^2 z^2 + \tfrac{1}{2}\kappa^2 r_1^2. \tag{60.27}$$

Thus their zero velocity curve is the ellipse $\alpha^2 z^2 + \kappa^2 r_1^2 = 2E$. Ellipses are zero velocity curves for any two-dimensional harmonic oscillator.

A given value of E isolates the bound orbits within the system – i.e., those within

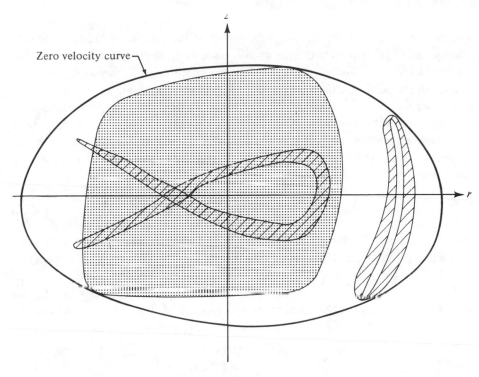

Fig. 56. Schematic boundaries of different classes of orbits in the z–r phase plane. Stars can move round and round throughout the areas which are dotted (box orbits) or shaded (tube and shell orbits).

the zero velocity curve – from all other possible orbits through the system. This is a feature of an isolating integral, and we already saw in Section 39 that energy is such an integral. If energy were the only isolating integral, then orbits could go anywhere within the zero velocity curve, eventually filling its area completely. Numerical integrations of orbits in special potentials, however, show that they are often confined to subregions within the zero velocity curve, at least for very long times. Characteristic shapes for these subregions are rough boxes, long tubes and doubly connected shells. These are all illustrated in Figure 56.

Box orbits follow complicated trajectories, sweeping from one side of the dotted area to the other, sometimes reaching the boundaries of the box. Orbits with the same energy, but different initial positions and velocities, will cover the whole area of the box. Tube orbits follow long loops within a narrow confined region, shaded in the diagram. For a strictly periodic stable closed orbit, the tube would contract to a line. Shell orbits move up and down through the plane with a constant sense of rotation. The diagram shows shell orbits to the right of the box orbit but they could occur anywhere. If the central forbidden region of the shell becomes vanishingly small the result is a box orbit; an infinitesimally thin shell is also a line orbit.

What confines orbits around a disk to these picturesque subregions? Why, another isolating integral of course. Section 39 showed that a stationary collisionless system can have five isolating integrals. For disks the extra integral, additional to energy and angular momentum, is usually called the 'third integral' in studies of galactic structure. Whether it exists as an isolating integral depends on both the energy of the orbit and on the detailed shape of the potential.

Let us consider some examples. The simplest is a closed periodic orbit. In this case, the third integral is the orbit itself, which is constant in time. Any vector whose amplitude depends only on constants of the orbit and whose direction (say from the origin to the point of closest approach) is an invariant of the orbit can represent an integral of the motion. For instance, the Laplace–Runge–Lenz vector

$$\mathbf{A} = \mathbf{p} \times \mathbf{L} - \frac{Gm^3}{r}\mathbf{r} \qquad (60.28)$$

is the third integral for Keplerian orbits (48.3), and represents their eccentricity e and perihelion direction (see Goldstein, 1980). The dot product $\mathbf{A} \cdot \mathbf{r}$ is the orbit itself and $|\mathbf{A}| = Gm^3 e$. Tube orbits tend to be found around closed periodic orbits.

For the second example, we turn again to the harmonic epicycles (60.10) and (60.17) in z and r_1. Since the forces in these two coordinates are not coupled, i.e., the potential $\phi(r_1, z)$ is separable, the motions in z and r_1 are independent of each other. Therefore the energies in both dimensions are separately conserved

$$E_z = \tfrac{1}{2}\dot{z}^2 + \tfrac{1}{2}\alpha^2 z^2 \qquad (60.29)$$

and

$$E_{r_1} = \tfrac{1}{2}\dot{r}_1^2 + \tfrac{1}{2}\kappa^2 r_1^2. \qquad (60.30)$$

We have already used the total energy (60.27) to find the zero velocity ellipse. But now we see that there are actually two independent isolating energy integrals in the set E_z, E_{r_1} and E. Taking these to be E_z and E_{r_1}, shows that the orbits are further confined to a box with rectangular boundaries $r_1 = \pm(2E_{r_1})^{1/2}/\kappa$ and $z = \pm(2E_z)^{1/2}/\alpha$ inscribed within the ellipse. In general, separable potentials lead to box-type constraints, and approximately separable potentials (whose interaction terms are relatively small for long times) lead to approximately box-like constraints, perhaps with curved boundaries as illustrated in Figure 56.

Is there a third integral for harmonic epicycles? We just saw from the Kepler example that a closed periodic orbit is itself a third integral, so we ask whether harmonic epicycles can follow such orbits. Eliminating the time from (60.10) and (60.17) gives the orbit

$$I_3 \equiv t_\alpha - t_\kappa = \frac{1}{\kappa}\cos^{-1}\left(\frac{r_1}{r_{10}}\right) - \frac{1}{\alpha}\cos^{-1}\left(\frac{z}{z_0}\right), \qquad (60.31)$$

with $r_{10} \equiv r_1(t_\kappa)$. Solving for r_1 with particular values for z and the third integral I_3 gives

$$r_1 = r_{10}\cos\left[\kappa I_3 + \frac{\kappa}{\alpha}\mathrm{Cos}^{-1}\left(\frac{z}{z_0}\right) + \frac{\kappa}{\alpha}2n\pi\right], \qquad (60.32)$$

where n is any integer and Cos^{-1} is the principal value. A given height z occurs only for a finite number of radii r_1 if the ratio κ/α is rational. Then the orbit will be periodic and further constrained to just one line within the box. Rational values of κ/α are thus the condition for I_3 to be an isolating integral. If, on the other hand, the ratio of epicyclic to angular frequency is irrational, then I_3 is infinitely multiple valued and does not constrain the orbit. The star can move anywhere within the box, eventually coming arbitrarily close to any value permitted by E_{r_1} and E_z. Since there are many fewer rational than irrational fractions (the rationals being a lower order of infinity), the existence of an exact third isolating integral is a rather rare and special circumstance. It reflects the local and global symmetries and separability of the potential.

In a realistic potential, the energy and angular momentum are usually the only obvious isolating integrals. The angular momentum affects the zero velocity curve through its appearance in the 'centrifugal' contribution to the potential energy in (60.5). For two-dimensional orbits in a four-dimensional phase space $(r, z, \dot{r}, \dot{z})$, the energy and angular momentum already provide two relations among r, z, $\dot{r}$ and $\dot{z}$, thus restricting the orbit to a two-parameter surface. A third isolating integral would restrict the (periodic) orbit to a one-parameter surface, a line. A fourth isolating integral would restrict the orbit to a point, which is hardly an orbit at all! So there can be at most three isolating integrals. This is a special case of the general situation described in Section 39 which may have five isolating integrals. In three-dimensional orbits, four of these integrals are already accounted for by energy and angular momentum, so again there can be, at most, one additional isolating integral in stationary collisionless systems.

How can we tell if there is a third isolating integral for the mass distribution of our galaxy? Observations may provide a clue. Recall from Section 39 that if energy and angular momentum are the only isolating integrals of an axisymmetric system, its distribution function must have the form

$$f = f(E, L_z) = f(v_r^2 + v_\theta^2 + v_z^2 - 2\phi, rv_\theta). \tag{60.33}$$

This has several straightforward observational implications, although their observation is not straightforward. First, the velocity dispersions of stars in the r and z directions should be equal, $\sigma_{rr} = \sigma_{zz}$. Second there should be no net radial outflow or escape from the plane since $\bar{v}_r = \bar{v}_z = 0$. Third, there should be no correlation between the radial and angular velocities, $\sigma_{r\theta} = 0$. None of these statements appear to hold for stars near the sun. One explanation may be the existence of a third isolating integral. But this is not a unique explanation. Another possibility is that the local mass distribution is not axisymmetric; large star clumps or gas clouds may be distorting the flow. A further possibility is that the local flow is not stationary, either because of global instabilities in the galaxy, or because some stars still reflect the initial conditions of their formation, or for some other reason.

Since the direct observational evidence for a third integral is rather ambiguous, much work has gone into inverting the problem by asking whether observationally

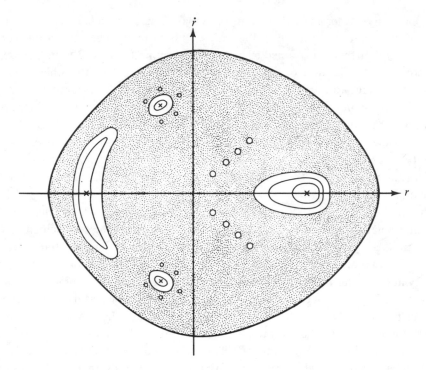

Fig. 57. Schematic representations of different classes of orbits in the r–$\dot{r}$ phase
plane.

reasonable models of the galactic potential have a third isolating integral. Here, as in most problems of gravitating systems, both analytical calculations and numerical orbit computations have been necessary to provide adequate insight. Contopoulos (1963) has summarized much of the early analytical work. New mathematical techniques have stimulated many recent results (see *Astronomy and Astrophysics Abstracts*), but the analytical problem remains a difficult one.

Numerical integrations illustrate the full complexity of orbits and third integrals in a model galaxy (e.g., Aarseth, 1966). Low energy orbits have a stronger tendency to have isolating third integrals than high energy orbits, perhaps because the low energy orbits respond mainly to a local, more symmetric part of the potential. For a given energy the existence of an isolating third integral depends sensitively on the initial position and velocity of the star. In particular, orbits which have such low initial angular momentum that they pass close to the galaxy's core (typically within ∼ 1 kpc from the center), can be scattered strongly by the core. This can convert radial motion into motion perpendicular to the plane after only one or two crossings, which typically take a few hundred million years. A strong scattering center destroys isolating integrals. Slight variations in the initial orbit produce great differences in the scattered orbit. As a result the scattered orbits lose the memory of their initial state. We cannot learn much about the early structure of our galaxy from the orbits of stars which now happen to be nearby but once were strongly scattered.

Let us leave the r–z plane and take another slice across phase space. Energy, $E(r, z, \dot{r}, \dot{z})$, is always an isolating integral, providing one relation between the four coordinates. Any three coordinates may be considered independent and used to describe a trajectory. Suppose we use r, z and $\dot{r}$ and look at the points where orbits of the same energy but different initial conditions pierce the plane $z = 0$. (For simplicity we continue with axisymmetric systems.) Every passage of a star through $z = 0$ is characterized by a value of r and $\dot{r}$. Figure 57 schematically illustrates various rights of passage.

Simple stable periodic orbits always cross the $z = 0$ plane at the same place and with the same radial velocity. In Figure 57, invariant points denoted by '·×' represent these line orbits. Each × marks the crossing of one trajectory. Around each invariant point are closed curves. Each closed curve marks all the crossing points of a single tube orbit. Many tube orbits may surround an invariant point, or the invariant point may stand alone. Chains of closed curves with no invariant points lie like islands near tube orbits, or unrelated to other structures. Each chain of islands marks the crossing points of a single quasi-stable orbit. This orbit crosses the $z = 0$ plane by jumping discontinuously from island to island. Finally, all the sprinkled dots represent a single unstable orbit moving ergodically throughout the permitted region.

Islands in an ergodic sea may occur with many arrangements on smaller and smaller scales as the resolution of the phase plane increases. Beautiful patterns emerge which depend sensitively on energy and the form of the potential (Hénon & Heiles, 1964; Mandelbrot, 1982). These patterns indicate the percentage of orbits which have a third isolating integral relative to those remaining ergodic. Generally, as the energy increases for a given potential, ergodicity spreads over more and more of the phase plane.

61

Axisymmetric and bar instabilities

Worlds on worlds are rolling ever
From creation to decay,
Like the bubbles on a river
Sparkling, bursting, borne away.

P. B. Shelley

Having calculated examples of disks in rotational equilibrium (e.g., Section 59) we may begin to wonder if they are stable. If we dent a disk here, or speed its stellar orbits up there, will the disk gradually recover its original form, or will it slide into a lower energy state with quite a different configuration? These questions are much more difficult than may appear at first sight, especially for realistic cases. For purely stellar disks, difficulties arise from the strong gravitational coupling between stellar positions and velocities. For gaseous disks (not discussed here), complications include radiative transfer, shocks, and star formation. Realistic disks, of course, contain a multitude of star–gas interactions.

Instabilities may either be global, involving changes which are coherent over large parts of the system, or local, mainly involving particular orbits with resonances confined to small regions. The borderline between these distinctions is necessarily rather vague. Global instabilities may be triggered by local instabilities. In the complexity of galactic structure, non-linear feedback effects between different types of instabilities are often important.

Two main techniques are used for sorting out instabilities. The first is linear mode analysis. The six-dimensional distribution function (or its integral the density distribution for gaseous systems) is perturbed infinitesimally, and the linearized Boltzmann equations of motion tell if these perturbations grow. This is essentially the same technique used in Sections 15, 16, and 21, for example. Its application to finite flattened systems is more complicated because only a small class of perturbations will preserve any initial symmetry. Perturbations that destroy symmetry, either in position or velocity space, are usually more important and more difficult to follow. A variant of this technique perturbs averages of the equations of motion, such as the tensor virial equations (Section 9). This can be useful for discovering global instabilities and developing average energy criteria for changes of global configurations.

The second main technique uses computer N-body simulations to explore the evolution of unstable systems. This has the great advantage of being able to follow a wide range of initial perturbations well into their non-linear development. As computers become faster and their memories increase, these experiments become more and more realistic. Dangerous approximations such as coarse grids, small number of particles, artificially softened gravitational forces, or two-dimensional confinement are all becoming less of a problem. Errors can also be removed by examining the same basic problem with several computational methods (see Section 26).

Unfortunately, the overlap between analytic techniques and computer experiments is rather small. Linear analysis is usually clearer and more informative as far as it goes, but it seldom goes far enough. Most interesting developments occur in the non-linear regime; perturbation theory can only tell us to stay tuned for something exciting to follow. On the other hand, we often need many computer experiments to get enough insight even to ask the right non-linear questions. Their answers appear in a conceptual framework which is too thin. A good feeling for the historical flavor of these difficulties may be found in Toomre's (1977) excellent review.

Studies of bar and spiral instabilities, and associated observations of galactic structure, have produced a vast and rapidly changing literature which deserves a vast treatise of its own. This section, and the next, merely describe some of the dynamical principles basic to the general discussion. The subject is reviewed every couple of years in published symposia of the International Astronomical Union and elsewhere.

Let us start with the simplest disk instability, one which remains completely axisymmetric. If we squeeze the disk uniformly around its circumference, as indicated by Figure 58, we push its outermost stars from R into $R - \delta R$. Whether these orbits return or become unstable depends on the relative changes of gravitational, centrifugal, and pressure forces (per unit mass). To order of magnitude (mainly neglecting geometrical multipole factors) the gravitational force of the disk is $F_d \approx - GM_d/r^2$. Its change, evaluated at $r = R$, is $\delta F_d \approx 2GM_d\,\delta R/R^3$ since M_d is conserved. With little extra trouble we can allow for the possibilities that the disk contains a significant central concentration and also contains an independent spherical component, so we might as well add these effects.

The change of gravitational force contributed by the central mass has the same form as that of the disk to the approximation we are considering: $\delta F_0 \approx 2GM_0\delta R/R^3$. The contribution of the spherical component, which may represent the inner part of a massive halo, depends on its detailed run of density $\rho(r)$. For a simple power law $\rho = \rho_0/(r_0/r)^n$, the resulting change in gravitational force is $\delta F_h = (n - 1)\,GM_h\delta R/R^3$, where M_h is the mass of the spherical halo within R. Note that, for $n < 1$, the gravitational force decreases as the orbit moves in because the potential is sufficiently non-singular that it resembles a harmonic well. The values $n = 0$ for a uniform halo and $n = 2$ approximating an isothermal distribution (Section 40) have special interest.

Centrifugal force keeps the orbits in equilibrium when there are no random motions. These are known as cold disks. We know from Part I that cold disks of finite thickness would not be self-consistent states because phase mixing and gentle two-

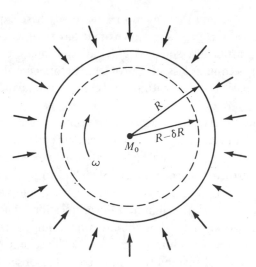

Fig. 58. Geometry of the simplest compression mode of a disk.

body relaxation would eventually generate significant thermal motions. This is even true for thin disks. But if the relaxation time is much longer than the rotation period, and if there are no large inhomogeneities to scatter the stars, a cold disk is a fair idealization. The change of centrifugal force $F_c = r\omega^2 = r^4\omega^2/r^3$ is $\delta F_c = -3\omega^2\delta R$ since angular momentum $r^2\omega$ is conserved in axisymmetric perturbations. (Imagine an impulsive central force perturbing each star.)

If the total perturbed gravitational force pulls the stars at R in toward the center more strongly than the perturbed centrifugal force pushes them out (δR being negative in our example), the net perturbation is unstable. Therefore, the approximate criterion for instability of axisymmetric modes is

$$R^3 \lesssim \frac{2G}{3\omega^2}\left[M_0 + M_d + \frac{(n-1)}{2}M_h \right]. \tag{61.1}$$

Notice that the addition of a central mass M_0, for a given angular velocity, makes larger radii unstable. In this approximation it does not matter how a given total mass is divided between M_0 and M_d, but it will matter if the shape of the disk is included in F_d. Adding a halo may tend either to stabilize or destabilize perturbations of a given scale, depending on the halo's density distribution and how much it increases ω. Writing (61.1) in terms of the average surface density $\sigma = M_d/\pi R^2$, the epicyclic frequency $\kappa(r)$ (60.22), or Oort's constants (60.23) for $M_0 = M_h = 0$, gives

$$R \lesssim \frac{2\pi G\sigma}{3(A-B)^2}, \tag{61.2}$$

which is more directly related to observational quantities. The important general conclusion is that rotation tends to stabilize the large scale perturbations.

Thermal pressure, on the other hand, tends to stabilize small scale perturbations.

We found this already for Jeans' instability (Section 15). Might it be possible, then, by adding random velocities to the stars, to stabilize perturbations of all scales in a rotating disk? We saw that Jeans' instability could be understood roughly as an excess of gravitational energy over thermal energy in a perturbed region. Similar reasoning, including rotational energy, gives an approximate stability criterion for a warm disk. Random motions thicken the disk. With a scale height z and radius R, its gravitational energy (per unit mass) is $\sim 4\pi G\rho/(R^{-2} + z^{-2})$ from Poisson's equation (59.5). The rotational energy is $\sim R^2\omega^2$ and the thermal energy is $\sim \langle V^2 \rangle$. For instability the gravitational energy of the perturbation should exceed its combined rotational and thermal energy

$$\frac{4\pi G\rho}{(R^{-2} + z^{-2})} \gtrsim R^2\omega^2 + \langle V^2 \rangle. \tag{61.3}$$

A detailed normal-mode analysis of an infinite homogeneous isothermal gas with uniform rotation, just as in Section 15.2 but adding centrifugal force, gives a dispersion relation showing that waves with $\mathbf{k}\cdot\boldsymbol{\omega} = 0$ are unstable if

$$4\pi G\rho > k^2c^2 + 4\omega^2 \tag{61.4}$$

(an exercise for the reader). Except for the numerical coefficients this is essentially the limiting case of (61.3) as $z \to \infty$. So a physical energy argument again gives the fundamental result.

The instability criterion (61.3) shows that if either ω^2 or $\langle V^2 \rangle$ is very large, all scales become stable; i.e., there is an overlap between the small scales stabilized by pressure and the large scales stabilized by rotation. Moreover, for large values of R the finite thickness z limits the gravitational energy of the perturbation, making it easier for rotation or pressure to provide stability. These results are illustrated in Figure 59 which schematically plots both sides of (61.3). For ω less than a critical value, of order $(4\pi G\rho)^{1/2}$, the curves pass through each other and perturbations with scale lengths between R_1 and R_2 are unstable.

Now the reader can see quite readily that for a gravitating system in centrifugal or pressure equilibrium, each side of (61.3) has the same order of magnitude. This means that stability of the equilibrium to a particular mode (even for isotropic velocities) will depend crucially on the exact numerical coefficients. These coefficients are determined mainly by details of the equilibrium geometry, the perturbation geometry, and the velocity distribution function. Their analysis can be quite complex even for simple examples of homogeneous gas disks (e.g., Goldreich & Lynden-Bell, 1965) and thin stellar disks (e.g., Toomre, 1964). As might be expected under such circumstances, it is usually possible to find some unstable modes, particularly if we relax the requirement of axisymmetry.

Although there are local inhomogeneities which are not axisymmetric, the simplest non-axisymmetric perturbations are global modes. The circular mode we just discussed is a limiting case which is, in a sense, both local and global. Local, because examining any small part is equivalent to examining the whole circular

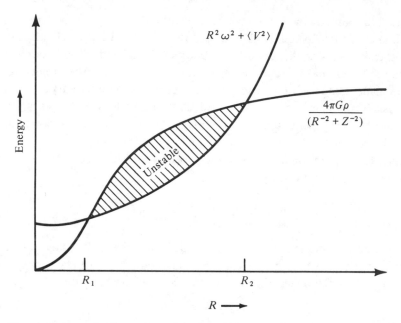

Fig. 59. Schematic illustration of the instability criterion of Equation (61.3).

mode; global because the mode encompasses the entire system. In non-axisymmetric modes the global features are more dominant because their local aspects change from place to place in a coherent way throughout the entire system.

Bars are usually the most prominent non-axisymmetric instability. When the perturbation amplitude of a bar is linear it is often represented by the $m = 2$ mode of $e^{im\phi}$. Highly non-linear bars may be represented by isolated ellipsoids or triaxial configurations. All these representations are highly idealized; real bars are seldom precisely defined.

Conditions under which bars form are not well understood. Numerical N-body simulations (e.g., Sellwood, 1981) show that a thin cold disk of stars, initially in centrifugal equilibrium, can deform over a few rotation periods. It develops a massive bar which slowly rotates through the remaining disk. The bar itself is a much more stable structure; streaming orbits dominate its internal motions (and make its observed rotation curve very dependent on projection). Complicated momentum transfers between bar and disk may induce intermittent spiral patterns. The size of the bar is governed largely by the initial rotation curve and mass distribution of the disk. Roughly speaking, the bar extends over the solid-body part of the rotation curve.

Figure 60 (kindly provided by Dr. J.A. Sellwood for this book) illustrates the development of a bar in a 50 000-body simulation representing a disk with a bulge component. Each star has the same mass, and the initial surface density of the disk is

$$\mu(r) = \frac{fM}{2\pi a^2}\left(1 + \frac{r^2}{a^2}\right)^{-3/2}, \tag{61.5}$$

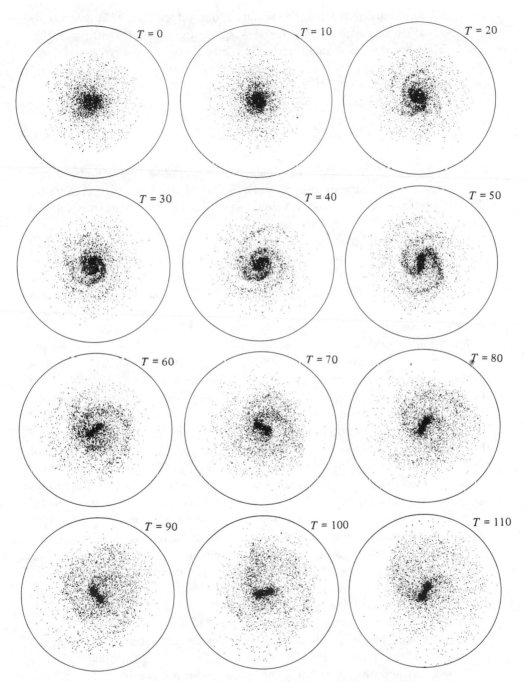

Fig. 60. A computer experiment illustrating the formation of global instabilities in rotating disks. (Courtesy of Dr. J.A. Sellwood.)

which is the $n = 1$ case of the distribution (59.13). Here f is the fraction, 0.7, of the total mass M which is in the disk. The initial circular velocity is (59.12), modified for artificially softened gravity $\phi(r) = -Gm(r^2 + \varepsilon^2)^{-1/2}$ by replacing a with $a + \varepsilon$. Here $\varepsilon = 0.2a$ (see Section 22). A small random velocity averaging $3.36\,QG\mu(r)/\kappa(r)$ with $Q = 1$ is added to inhibit local axisymmetric instability (Toomre, 1964). Again $\kappa(r)$ is the epicyclic frequency (60.22).

The bulge component, rigid and unresponsive, has a spherical density

$$\rho(r) = \frac{3M(1-f)}{4\pi b^3}\left(1 + \frac{r^2}{b^2}\right)^{-5/2}. \tag{61.6}$$

Its scale length $b = 0.2a$, and 90% of its mass lies within radius $4b$. The unit of time is $T = (GM/a^3)^{-1/2}$. The numerical technique (Sellwood, 1981) Fourier analyses the gravitational forces over a two-dimensional mesh. Ten per cent of the stars, selected randomly, are plotted in the diagram.

Evidently instabilities set in after several dynamical periods. Inhomogeneities shear into prominent, but evanescent, spiral patterns. All the while, random motions are growing; the disk heats up. Heating eventually destroys most of the spiral, leaving its loosely wound center as a fairly stable bar-like relic. The bar has roughly the scale of the original bulge. It is easy to imagine shocks and star formation amplifying these patterns so they closely resemble observed galaxies.

Hot initial states discourage bar formation, as we would expect from the discussion of axisymmetric instability. The ratio, $T_{\rm rot}/|W|$, of the kinetic energy of rotation to the gravitational energy provides a rough description of the 'coldness' of the disk. Equation (9.16) shows that this ratio measures the importance of kinetic energy of average motion relative to the thermal pressure in the global virial theorem. Early N-body simulations and idealized linear mode analyses suggested that the empirical relation $T_{\rm rot}/|W| \gtrsim 0.15$ might provide a sufficient (but not necessary) criterion for global bar instability (Kalnajs, 1972; Ostriker & Peebles, 1973). Linear perturbation of the tensor virial equations also provides insight into this criterion (Vandervoort, 1982). Unfortunately, the criterion is not very general and a variety of disk systems have been found which are stable even if $T_{\rm rot}/|W| \gtrsim 0.15$ (see discussions in Athanassoula, 1983 and references therein). The early analyses and N-body simulations were not very adventurous in their choice of initial conditions, usually sticking with smooth uniformly rotating disks. But details such as differential rotation, anisotropic velocities and mass distribution of the disk are very important for stability.

A quite different way to stabilize disks is to embed them in a halo of roughly equal or greater mass. The smooth, relatively unresponsive gravitational field from the part of the halo within the outer radius of the disk inhibits global instabilities. It is an important, but unsettled, question whether such halos are present in flat galaxies without bars.

Figure 60 shows that the formation and maintenance of a bar may be closely connected with the streaming and shearing motions throughout the disk as a whole.

Once a massive bar forms, its tidal field dominates the orbits and momentum transfer to nearby stars. Orbital resonances determine regions of stability and density enhancement. So it is not a very good approximation to regard the bar as an isolated object. Nevertheless such isolated models do provide useful insights. To construct these models, one starts with the collisionless Boltzmann equation (7.2) for a steady state bar. In a frame rotating with angular velocity $-\omega\hat{z}$ we must include the centrifugal and Coriolis forces, so (7.2) becomes

$$\mathbf{v}\cdot\frac{\partial f}{\partial \mathbf{r}} + (\nabla\phi + \omega^2\mathbf{R} + 2\omega\hat{z}\times\mathbf{v})\cdot\frac{\partial f}{\partial \mathbf{v}} = 0, \qquad (61.7)$$

with $\mathbf{R} \equiv (x, y, 0)$. Poisson's equation has the form

$$\nabla^2\phi = -4\pi G\rho = -4\pi G \int f\,\mathrm{d}^3\mathbf{v}. \qquad (61.8)$$

Very few self-consistent solutions of these equations are known. One is for an ellipsoid of uniform density whose centrifugal and gravitational forces balance along the major axis (Freeman, 1966). The ellipsoid rotates with a uniform angular velocity $-\omega\hat{z}$. It has no velocity dispersion σ_{xx} along the major axis, but σ_{yy} and σ_{zz} are non-zero so the pressure is very anisotropic. Within the ellipsoid are very powerful streaming motions, so strong that for some axis ratios the total angular momentum is in the $+\hat{z}$ direction, opposite to the angular velocity. The mean rotational velocity may also have the opposite sense to ω in certain regions of the ellipsoid. Coriolis forces are very important in regions of fast circulation. In a companion paper, Freeman develops a solution for a uniformly rotating elliptical disk with surface density of the form $\mu(r) = \mu_0(1-\eta)^{1/2}$ for $\eta \leq 1$, and $\mu = 0$ for $\eta \geq 1$, where $\eta = (x/a)^2 + (y/b)^2$ and axis $a > b$. This has velocity dispersions in both directions in the plane. Neither of these models is centrally condensed enough to be realistic, but they show how complexities develop at very early stages of the analysis.

Between the rather intricate mathematical analyses of Boltzmann's equation and detailed orbits on one side, and the non-linear computer simulations on the other, is a large gap. This gap may be largely responsible for our slow progress in understanding gravitating disks. New physical approaches and conceptual structures are needed to fill the gap. Their relation to more fundamental approaches may at first be obscure, but they can equally well be tested against computer simulations and eventually against observations.

Many paths lead into the maze of galactic structure. Some stop after short explorations; others lead us farther, but at the end of each is a closed wall. Reaching these walls, we may try to bore through them, to connect with another path, hoping that by perforating the maze we may succeed when the straightforward approaches are stymied. After many perforations, we may find that the maze was never really there at all.

62

Spiral instabilities

Whirlpools and storms his circling arm invest
With all the might of gravitation blest.

Alexander Pope

62.1. Introduction

Spiral structure is the frosting on the cake of galactic rotation. Like frosting, it is very alluring and has been greatly admired. Self-gravitating stellar disks are very lively objects; many processes contribute to the observed spiral patterns in galaxies. Our knowledge of this subject, greatly extended in the last two decades, still changes quickly. This section just describes some of the basic gravitational principles and questions common to many discussions.

What are the forces that drive spiral structure? To answer 'differential rotation' is true, but not much more informative than the ancient physicians who replied that morphine produces sleep because it contains a 'dormative principle'. Spiral instabilities resulting from differential rotation are familiar in both magnetic and fluid systems, with or without gravitation.

The patterns of rotating magnetic field lines are so suggestive that for many years they were popular as an explanation for spiral structure, either directly or through their influence on star formation. The decline of magnetic explanations occurred when more accurate measurements of the galactic magnetic field strength showed it was significantly less than expected. Even today these measurements are uncertain because the topology of the field is obscure and we are not really sure how to average it over large distances. Current estimates suggest a 'typical' field strength near the galactic plane of about 5×10^{-6} gauss, which would imply an energy density $B^2/8\pi \approx 10^{-12}$ erg cm^{-3}. This is comparable with the energy densities of cosmic rays, starlight, and random gas motions. But it is several orders of magnitude less than the total kinetic energy density of rotation. For a typical total density $\sim 10^{-23}$ g cm^{-3} in the plane, and a rotational velocity ~ 200 km s^{-1}, the rotational energy density is $\sim 10^{-9}$ erg cm^{-3}. The magnetic field would have to be several times greater than observed to perturb the rotational energy by 1%. We cannot conclude from this type of argument (as is sometimes done) that magnetic fields are unimportant. They could still produce local instabilities which might be sheared into spiral patterns. On a larger scale they might provide a 'trigger instability' which would act like a valve releasing much greater pent up gravitational energies. These possibilities are not well understood. Nevertheless, it seems reasonable to see if instabilities in the dominant gravitational energy reservoir can explain spiral patterns.

The main problem, addressed by all theories of spiral structure, is the winding dilemma. When local instabilities produce regions of high density or recent star formation, differential rotation does indeed shear them into the form of a spiral. But, like the sorcerer's apprentice, it continues to work with a vengeance, winding them up tighter and tighter until they merge and the overall spiral appearance fades away. Since spiral arms often extend over regions where the angular velocity differs by 30%, say, we would expect them to become severly distorted after only several rotations. A typical rotation period is $\sim 2 \times 10^8$yr, so the spiral should be destroyed in less than one-tenth the age of the galaxy. Yet nearly all disk galaxies have spiral appearance. Either spiral structure is very long lived, or new arms must form as the old die away.

Perhaps too much emphasis has been laid on this winding dilemma. It tended to force each theoretical approach into either one or the other category, whereas it has now become clear that long time coherence and regeneration are both needed to understand spiral structure. In the early discussions, long time coherence was supposed to be provided by magnetic fields – which now seem inadequate – or by large amounts of matter streaming out from the galaxy center – which are hard to maintain for ten billion years. On the other hand, regenerative theories which depend on shearing local instabilities, find it hard to explain the very high overall symmetry of spirals in many galaxies.

A new path into this maze came by realizing that spiral arms need not contain the same material all the time. The arms could be merely a pattern, a place where stars linger in their orbits. Early discussions, before the mid-1960s, which pioneered this new gravitational explanation were not very successful. They relied mainly on orbit theory, so it was difficult for them to come to grips with collective instabilities. Once again, it was the application of techniques developed for plasma physics which provided new insight into a gravitational problem. Spiral arms could be thought of as density waves in a galactic disk. The collisionless Boltzmann equation would describe them. The winding dilemma, which mostly applies to arms whose material all moves together, would be replaced by the problem of forming and maintaining the density waves. Several waves, interacting together, could then form modes providing a standing pattern of spiral structure. The energy to support these modes would come from galactic rotation, perhaps through the intermediaries of star formation, large density inhomogeneities, turbulence, and orbital resonances. Many of these general hopes of the density wave approach have been developed successfully, although the results were often not those first expected. The theory has provided a standard to show how complex reality actually is, and the idea that spirals are density patterns, while not the whole story, has been exceedingly fruitful.

62.2. Basic properties of patterns

First we should see how to describe patterns and how, once formed, they interact with stars in the disk. In cylindrical coordinates, with azimuthal angle θ, a smooth monotonic relation $r(\theta)$, or its inverse $\theta(r)$, specifies a spiral form. Even if this relation is not very smooth it will give an impression of spirality, especially if averaged. The tangent of the pitch angle, $\imath$, is the ratio of increments along the radial and azimuthal directions:

$$\tan \imath = \frac{dr}{r d\theta} = \left(r \frac{d\theta}{dr} \right)^{-1} = (rk)^{-1}, \qquad (62.1)$$

where

$$k = \frac{d\theta}{dr}. \qquad (62.2)$$

The logarithmic spiral, $r = r_0 \exp(a\theta)$, is an example and (62.1) shows why this is also called the equi-angular spiral. Archimedean spirals, $\theta = kr$, are another simple case.

Generally, the spiral pattern $\theta(r, t)$ is also a function of time. For example, if

$$\theta = \theta_0(r) + \omega(r)t \qquad (62.3)$$

the whole pattern will rotate with a sheared angular speed $\omega(r)$, the case $\omega =$ constant being rigid rotation. The angular speed of the pattern will not usually be the same as the rotation curve for the disk, especially if the pattern is just a low amplitude density wave.

To portray a spiral pattern with more than one arm we need a function which is reasonably periodic in θ. There are many possibilities, of which the trigonometric ones happen to be simplest and best studied. Therefore we represent a perturbed quantity (such as density or velocity) by the real part of the functional form

$$f(r, \theta, t) = f_0(r) \exp \{i(\omega t - m\theta + P(r)\}. \qquad (62.4)$$

The phase (with χ_0 a constant which determines where the spiral starts) is

$$\chi \equiv \chi_0 + \omega t - m\theta + P(r) \qquad (62.5)$$

and the amplitude is $f_0(r)$. Increasing θ by $2\pi/m$ leaves the value of $f(r, \theta, t)$ unchanged, so the spiral has m symmetric arms. In a reference frame rotating with angular speed

$$\omega_p = \frac{\omega}{m} \qquad (62.6)$$

so that $\theta = \omega_p t$, the phase remains constant and the pattern appears stationary. Thus ω_p is the angular pattern speed for the special pattern in rigid rotation. The angle θ is usually taken to increase in the direction of rotation. If r decreases with increasing θ, so that $k < 0$, the spiral is trailing, otherwise it is leading, as shown in Figure 61. Spirals of given phase χ have pitch angle

$$\tan \imath = \frac{m}{r P'(r)} = \frac{1}{r k(r)}, \qquad (62.7)$$

where P' denotes the radial derivative.

Each arm of a spiral pattern may have many winds around the galaxy. To estimate the radial distance λ between arms, we may suppose that all arms have the same

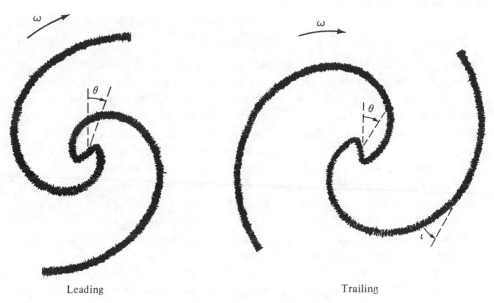

Leading   Trailing

Fig. 61. Schematic illustration of leading and trailing spirals with pitch angle ι.

phase χ. If there is an arm at (θ_0, r_0), then the next arm will be at $(\theta_0 + 2\pi/m, r_0 + \lambda)$, but its phase will be unaltered

$$\chi = -m\theta_0 + P(r_0) = -m\theta_0 - 2\pi + P(r_0 + \lambda). \tag{62.8}$$

Expanding $P(r_0 + \lambda)$ in a Taylor series around r_0 shows that to first order

$$\lambda = \frac{2\pi}{m|k|}, \tag{62.9}$$

so that $mk(r)$ is the radial wavenumber of the pattern.

Once a pattern is introduced into a disk, five new radii become especially important to the basic dynamics. These radii arise for patterns dominated by self-gravity, as well as for density waves of small amplitude, and are common to very wide classes of models. The first two radii are the center and edge of the disk – so obvious as to seem trivial, yet playing a critical role as boundary conditions for the pattern. The center is a singularity where stars' high random velocities destroy any pattern which may penetrate inward. This provides an innermost scalelength where patterns are dissipated. It may represent a bulge in the galaxy. At the edge, the behavior of patterns depends very much on how sharply the gravitational field is truncated. A slow decrease of density may absorb density waves which propagate outward, while a sharp edge may reflect some waves back into the pattern, and after its structure. Processes at these boundaries are not well understood.

The other three radii all occur at related dynamical resonances. As the pattern rotates, there will be special radii at which stars in the pattern, i.e., the density waves, interact strongly with other stars in the disk. The *corotation radius* occurs where the pattern's angular speed is the same as the angular speed of stars around the disk

$$\omega_p = \omega_s(r). \tag{62.10}$$

Many galaxies have nearly flat rotation velocity curves in their outer regions, so $\omega_s(r) \propto r^{-1}$ decreases there. Outside the corotation radius, the pattern speed is therefore faster than ω_s and inside it is slower. Galaxy rotation curves which are not monotonic may have more than one corotation radius if they can support stable global patterns. At the corotation radius it is easy for the density wave to exchange energy and angular momentum with stars in the disk over long periods of time. Many wave–star processes such as Landau damping, resonance capture, reflection, transmission, and stimulated emission interact here in a complicated non-linear way. This is poorly understood, but it may be crucial for maintaining the pattern.

Away from corotation, the difference between the pattern speed and the star's speed can also produce strong resonances. An arm of the spiral, with its associated gravitational perturbation, will move past a disk star at a frequency $|\omega_p - \omega_s|$. If there are m arms in the spiral, the star is perturbed at a frequency $m|\omega_p - \omega_s|$. As we saw from (60.17), stellar orbits are particularly susceptible to perturbations at their epicyclic frequency κ. Continued perturbation at this frequency may occur within the corotation radius where $\omega_p < \omega_s$ and

$$\kappa = -m(\omega_p - \omega_s), \tag{62.11}$$

and outside the corotation radius where $\omega_p > \omega_s$ and

$$\kappa = m(\omega_p - \omega_s). \tag{62.12}$$

The non-linear resonances which develop at these radii are called the inner and outer *Lindblad resonances*, after the Swedish astronomer B. Lindblad who pioneered the early gravitational theories of spiral structure. These resonances do not exist for all rotation curves, and sometimes the outer Lindblad resonance occurs in a region of very low density, depending on the model.

Like the center, edge, and corotation radii of the disk, the pair of Lindblad resonances with their wave–star interactions help determine the fate of self-consistent patterns.

62.3. Self-consistent patterns

Gravitationally self-consistent patterns are the only ones likely to survive for long times in a disk. Arbitrary patterns, imposed by passing tides or by fiat, either decay or grow into other, more stable, forms. To find self-consistent patterns is a difficult task; not many are known. One starts the search by choosing an equilibrium model for the disk, usually based on either the collisionless Boltzmann equation or the equations of stellar hydrodynamics (8.14)–(8.15). If the hydrodynamic equations are chosen they must be truncated somehow. The most straightforward truncation is to suppose the disk to be entirely gaseous, whereupon the equations become those of ordinary gas dynamics. Although this approach is not realistic for galaxies, where the stellar component dominates the gravitational field, it has great advantages of simplicity. It also leads to many important results shared by stellar disks (see Lin & Lau, 1979 for a detailed review; Toomre, 1977 referred to in Section 61; Bertin, 1980).

In particular, gas models show that a spiral pattern may be a quasi-stationary interference mode containing several propagating waves whose forms are similar to (62.4). Leading and trailing waves, moving in and out, may coexist in the mode at any time. Quasi-stationary modes are determined by the transmission, reflection, and absorption of these waves at the five critical radii mentioned earlier. It is at these radii that the basic deficiencies of gas models become especially important. By suppressing the detailed velocity distribution function, the gas models make it impossible to learn how waves really behave at these radii, where non-linear effects are important.

A more basic approach is to use the moment equations of stellar hydrodynamics and truncate them by making assumptions about the stellar velocities. For example, Hunter (1970) has supposed that most stars depart only slightly from circular orbits. The perturbed radial velocities are of order εc_r and the rotational velocities are $r\omega(r) + \varepsilon c_\theta$. By taking moments with respect to powers of εc_θ of the collisionless Boltzmann equation (in cylindrical coordinates), the procedures of Section 8 again give a hierarchy of coupled moments. Now, however, each moment of order n is explicitly smaller than the $(n-1)$th moments by a factor of order ε. Thus the coupling terms in the hierarchy can be dropped consistently. Poisson's equation can be used to close the system self-consistently at the level of third order moments.

Both these approaches rapidly become intractable analytically unless further (but less fundamental) simplifications are made. A major simplification is the short wavelength approximation $kr \gg 1$. From (62.1) we see this implies that the pitch angle $\iota \ll 1$ so the spiral is very tightly wound. Under these conditions, the azimuthal derivatives are much less than the radial derivatives and a WKBJ type approximation applies to the fluid or Boltzmann equations. Such restricted modes provide great insight into the problem, but are seldom applicable directly to observed galaxies.

To sketch the technique for finding self-consistent modes in a little more detail we adopt the approach with fewest fundamental restrictions. This uses the collisionless Boltzmann equation directly (Kalnajs, 1971). If the disturbance caused by the pattern is small, we may write the distribution function and the Hamiltonian in the linearly perturbed forms

$$f = f_0 + f_1, \qquad (62.13)$$

$$H = H_0 + H_1, \qquad (62.14)$$

where $H_1 = \phi_1$ is the potential energy (per unit mass) of the disturbed surface density. Substituting (62.13)–(62.14) into the collisionless Boltzmann equation

(7.17), and setting the zero order equilibrium terms equal to zero, gives the linearized equation

$$\frac{\partial f_1}{\partial t} - \{H_0, f_1\} = \{H_1, f_0\}. \tag{62.15}$$

Similarly, the linearized Poisson equation (7.13) is

$$\nabla^2 \phi_1(\mathbf{r}, t) = -4\pi G \int f_1(\mathbf{r}, \mathbf{v}, t) d\mathbf{v}$$

$$= -4\pi G \delta(z) \int \int f_1(r, \theta, v_r, v_\theta, t) dv_r dv_\theta$$

$$= -4\pi G \sigma_1. \tag{62.16}$$

The second equality in (62.16) applies to infinitesimally thin disks with a perturbed surface density σ_1. This is the usual approximation in spiral structure theory. Allowing motions in the z direction would introduce many interesting new phenomena, especially at density inhomogeneities near resonances, but these have not been much studied.

Now suppose we select a general form of $\phi_1(r, \theta)$ which represents some desirable pattern such as a spiral. From Poisson's equation, this would imply a corresponding surface density σ_1 (select). The system would evolve in response to this disturbance according to (62.15). Since H_0 and f_0 are known from the equilibrium model, and H_1 is given through ϕ_1, we can solve (62.15) in principle for f_1 (response). In general, f_1 (response) will be a complicated functional of ϕ_1 (select) and therefore of σ_1 (select). Integrating f_1 (response) over the velocity coordinates gives its corresponding surface density σ_1 (response), which will be an even more complicated functional of σ_1 (select); we may write it symbolically as σ_1 (response) $= Q[\sigma_1(\text{select})]$. In order to obtain a solution of the combined Poisson and Boltzmann equations we therefore must have the self-consistency condition σ_1 (response) $= \sigma_1$ (select) or

$$\sigma_1(r, \theta, t) = Q[\sigma_1(r, \theta, t)], \tag{62.17}$$

where Q is a complicated integral operator.

Only special forms of σ_1 will satisfy the self-consistency condition. To find the self-consistent spatial forms of σ_1, we can either Fourier analyze f_1 and ϕ_1 (and therefore σ_1) in time, or we can treat (62.17) as an initial value problem. These techniques are related because the solutions of the initial value problem are most easily expressed in terms of the Fourier modes. Indeed, after the transient responses die away it is usually these normal modes that are left. The method for solving the initial value problem is essentially the same as the Laplace transform analysis of Landau damping in Section (16.2). The modes which emerge under different circumstances and assumptions are characterized by their dispersion relations $\omega(k, r)$. These dispersion relations generally become singular at the Lindblad resonances and at corotation, showing that the linear theory breaks down at those radii. The fact that ω depends on r for more general patterns than (62.4) means that in a differentially rotating galaxy ω will change in time. Only exceptional perturbations, carefully adjusted to the rotation curve, have a chance of remaining stationary.

62.4. Sustaining the spiral

One Fourier wave does not make a realistic perturbation, as remarked in the related but much simpler context of Section 15.2. Realistic small amplitude patterns will generally be a superposition of waves, each of which is a solution of (62.17). The peak of this wave packet propagates with a group velocity $V_g = (\partial \omega / \partial k)_r$ as described by (14.51). Therefore the spiral pattern will generally move radially through the galaxy. As it moves into regions of different rotation velocity, its group velocity also changes. Different parts of the pattern may move at different rates and be distorted differently by the differential rotation.

Toomre (1969) first analyzed these group velocities and applied them to sample modes in disk models representing our own galaxy. Tightly wound trailing patterns near the sun would have $V_g \approx -10 \, \text{km s}^{-1}$ and would move radially inward. At this velocity, the pattern would reach the center in $\sim 10^9 \, \text{yr}$. This is just several galactic rotations, and much less than the ages of galaxies, so simple density patterns do not solve the winding dilemma for very long. Other waves also propagate toward or away from corotation, depending mainly on whether their radial wavelength (62.9) is short or long, and whether they are leading or trailing. When the patterns reach the galaxy center, or a Lindblad resonance, they are strongly modified. Thus it is still necessary to find mechanisms to sustain or regenerate the pattern.

Several possibilities have been suggested (reviewed by Lin & Lau, 1979; Toomre 1981): internal local instabilities, internal global instabilities, and external forcing. Internal local instabilities set off by shocks from star formation, or by gravitational perturbations from large gas clouds and globular clusters moving through the plane, may occur anywhere throughout the disk. These may produce short sheared waves and material bits of spirals, but they do not coordinate themselves over large distances. It is especially difficult for them to produce the extraordinary global symmetry of many (although a minority) of the observed spirals. These local instabilities, however, may *result* from shocks made in the gas by gravitational perturbations along a spiral arm. The resulting star formation and illumination of the arm is a very important problem of gas dynamics.

A much more promising active role for local instabilities occurs at the corotation and Lindblad resonances. Radially propagating density waves are not usually absorbed completely at these resonances, but may be transmitted, reflected, and amplified. To see roughly how amplification may occur, imagine a density pattern propagating from the center of the galaxy toward corotation. Within the corotation radius, the angular pattern speed is less than the rotational speed of the disk, so relative to the disk the angular momentum of the pattern is negative. Suppose some waves of the pattern are transmitted through corotation; their relative angular momentum becomes positive. To conserve total angular momentum the waves reflected back from corotation towards the interior gain increased negative angular momentum and are therefore amplified. The increased energy of the reflected waves comes from the differential rotation at the resonance, although the lifetime and other details of the process are not well understood. Variants of these phenomena may occur at the inner and outer Lindblad resonances, and the galactic center.

A consequence of reflection and amplification at resonant radii is that inward and outward traveling waves are present simultaneously throughout much of the disk. The regeneration and combination of these waves may produce a standing spiral mode which is quasi-stationary for much longer than the lifetime of any one propagating wave. Therefore this mechanism requires both instability and feedback. Lack of either (perhaps produced by very long term outward momentum transport and density evolution which removes the Lindblad resonances) would inhibit these global spiral patterns.

Internal global instabilities, such as material bars, may also generate spiral modes. The example of Figure 60 shows that these spirals may not last very long. Their existence depends on feedback and amplification from resonances, on the random velocity dispersion throughout the disk, and on direct gravitational torques. The spiral in turn reacts back on the bar, changing the resonances. No self-consistent stationary states have been found. Nor would any be expected since finite gravitating systems are never thermodynamically stable.

In addition to bars, thicker structures such as lenses may be present. We do not know if they arise during galaxy formation, or if they result from subsequent heating of a disk, perhaps by a bar. Non-axisymmetric instabilities in a lens could stimulate spiral structure. But even symmetric lenses will alter spiral patterns by changing the positions and strengths of the resonances.

Once we admit that galaxies have three dimensions, a variety of global buckling modes can play with the spiral. Many galaxies, including our own, are known to have warps. Although these warps are more readily observed in the gas, they may also be found in the distribution of stars. Regions of the disk where the vertical oscillation frequency α (from (60.11)) is equal to the epicyclic frequency

$$\alpha = \kappa \tag{62.18}$$

are especially prone to instability. Local inhomogeneities there in the density or velocity distribution may be amplified parametrically in the z direction. In axisymmetric systems this resonance occurs around the whole system, leading to a global warp. If axisymmetry does not hold completely, a long arc may form and be stretched out into a three-dimensional spiral warp. A similar forced resonance may arise when the restoring force in the plane changes periodically in time as a density pattern sweeps by

$$\alpha = m(\omega_p - \omega_s). \tag{62.19}$$

Energy is transferred from the pattern into z motions, eventually damping the pattern. The extent of these interactions will depend on the anisotropy of the velocity distribution as well as on its average dispersion and on the amplitude of the density pattern. These resonances are important, although they are not the only ones which can produce warps (e.g., asymmetric halos, external tides, and free wobbles).

External forcing may also be important, at least in some galaxies. A non-symmetric halo or a satellite companion may raise tides which are amplified. These tides may interact with spiral patterns directly, or indirectly through a resulting warp. In galaxies with large differential rotation there are always likely to be regions which resonate strongly with the satellite orbit. In central regions where rotation is more rapid, the 'clock runs faster' and once instabilities arise, they develop more quickly, other things being equal (Toomre, 1981).

Several galaxies with dramatic clear spiral structure do have substantial companions. It is hard to detect smaller companions with low surface brightness, but they may well exist around many more galaxies. Detailed conditions under which satellites generate long lasting spirals are yet to be understood. Even if any one mode is transient, the excitation a satellite provides is continuous.

Spiral structure is really a syndrome of galactic instability. It is many phenomena rolled into one. Here I have emphasized the role of gravitation, just skimming rapidly over its surface. In some places, the theory beneath this surface has been heavily developed, but most places still await exploration.

63

Triaxial and irregular systems

And enterprises of great pitch and moment
With this regard their currents turn awry

Shakespeare

63.1. Introduction – spinning polytropes

In our continued progression toward decreasing symmetry, we now arrive at the most individualistic systems of all. Previously we discussed flattened systems which were mainly thin disks. But those cannot represent the large observed class of elliptical galaxies. Ellipticals show moderate rotation, and as a result they have less global coherence than either spirals or spherical non-rotating systems. Pressure, rotation and internal currents all combine to determine a variety of structures and shapes.

Historically, some of the great eighteenth, nineteenth- and early-twentieth-century mathematicians began to develop this subject by rigorously analyzing certain simple, highly specialized cases. Magnificent though these achievements were – and they often led to important mathematical insights – they also raised the stature of certain models so high that their astronomical importance became greatly exaggerated. All these models were built from uniform, incompressible fluids. The first of these non-spherical self-gravitating models were Maclaurin's (mid-eighteenth-century) spheroids. They have a stately uniform rotation, with no internal motions. Gravity is balanced by incompressibility and rotation. This balance can exist only for certain values of the spin ratio (58.1), nearly equal to unity. Two types of Maclaurin spheroids, with small and large eccentricities, are possible equilibrium figures.

So simple and satisfying seemed Maclaurin's results, that nearly a century went by before Jacobi realized that other equilibrium figures exist. These are ellipsoids rather than spheroids. The Jacobi ellipsoids also rotate uniformly with no internal motions. As the angular momentum of a Maclaurin spheroid increases, it can turn into a Jacobi ellipsoid. The Maclaurin sequence bifurcates at this point. The Jacobi sequence itself bifurcates into the pear-shaped equilibrium figures discovered by Poincaré and studied by Darwin, Liapounov, Jeans, and Cartan.

Allowing internal motions in the ellipsoids adds a new degree of freedom to the problem, introduced by Dirichlet and developed by Dedekind and Riemann. The Dedekind ellipsoids do not rotate at all, but are just supported by fluid circulating along ellipses. Riemann's ellipsoids have both rotation and internal circulation. Chandrasekhar (1969, see the bibliography of the Introduction) has summarized and extended these classical results in great detail.

Astronomical systems are neither incompressible, nor uniform, nor ellipsoidal. So, although the classical models have great intrinsic interest, their application can be misleading. Dropping the basic assumptions of the classical models leads to much richer and more realistic, but also more complicated, systems. To illustrate how these complications arise, we generalize the polytrope models of Section 40.

Polytropes are more accurate representations of spherical stellar systems, where the pressure is isotropic, than of flattened systems. As Section 62.3 mentioned, we need to impose additional assumptions on stellar orbits in flattened systems to close the moment equations of stellar hydrodynamics and obtain the polytrope equations. Since our purposes here are illustrative, we shall take the standard easy way out and suppose we are dealing with a gaseous system. Gaseous systems can illustrate many basic properties of their stellar counterparts, where details of resonances are not important, but they must be applied with some caution.

To add the effects of compression and differential rotation, consider a thick cylindrical disk. The density and potential depend only on r and z, so Poisson's equation (59.5) has the form

$$\frac{1}{r}\frac{\partial}{\partial r}\left(r\frac{\partial \phi}{\partial r}\right) + \frac{\partial^2 \phi}{\partial z^2} = -4\pi G\rho(r, z),\qquad(63.1)$$

which is the first of a series of simplifications. The equation of hydrostatic equilibrium (15.3) has the form

$$\frac{1}{\rho}\nabla P = \nabla \phi - \omega \times (\omega \times \mathbf{r}) - 2\omega \times \mathbf{v}\qquad(63.2)$$

in a rotating system. The second term on the right is the centrifugal force, and the third term is the Coriolis force on an element of gas moving with velocity $\mathbf{v}$ in the rotating frame. Our next simplification is to suppose $\mathbf{v}$ is small compared with the rotational velocity $\omega \times \mathbf{r}$ and drop the Coriolis term. Then hydrostatic equilibrium in the radial and vertical directions requires

$$\frac{1}{\rho}\frac{\partial P}{\partial r} = \frac{\partial \phi}{\partial r} + r\omega^2\qquad(63.3)$$

and

$$\frac{1}{\rho}\frac{\partial P}{\partial z} = \frac{\partial \phi}{\partial z}.\qquad(63.4)$$

There is no reason, in principle, why ω should not also be a function of height, describing a state of vertical shear. But we will further simplify matters by supposing $\omega = \omega(r)$. This appears to be a good approximation for some observed galaxies.

The polytropic equation of state (40.4) completes the description

$$P = K\rho^\gamma,\qquad(63.5)$$

where we have used $\gamma = 1 + 1/n$ instead of the polytrope index n. In a cylindrical system it is easy to explore some possible effects of anisotropic pressure by supposing the constant K to differ in the vertical and radial directions, denoted by K_z and K_r. The value of γ might also be anisotropic, but here we will keep it the same.

Substituting (63.5) into (63.3) and (63.4) and then eliminating the potential in (63.1) gives

$$K_r\gamma\rho^{\gamma-2}\frac{\partial^2\rho}{\partial r^2} + K_r\gamma(\gamma-2)\rho^{\gamma-3}\left(\frac{\partial\rho}{\partial r}\right)^2 + K_r\gamma\frac{\rho^{\gamma-2}}{r}\frac{\partial\rho}{\partial r}$$

$$-\frac{1}{r}\frac{d(r^2\omega^2)}{dr} + K_z\gamma\rho^{\gamma-2}\frac{\partial^2\rho}{\partial z^2} + K_z\gamma(\gamma-2)\rho^{\gamma-3}\left(\frac{\partial\rho}{\partial z}\right)^2 = -4\pi G\rho(r,z). \quad (63.6)$$

This generalizes Emden's equation (40.7) to thick cylindrical rotating polytropes. It simplifies considerably, becoming linear, for $\gamma = 2$. This special case illustrates the effects of compressibility nicely and is close to the expected value for such systems. The resulting equation is separated by writing

$$\rho(r,z) = R(r) + Z(z) \quad (63.7)$$

and yields

$$\frac{d^2Z}{dz^2} + \frac{2\pi GZ}{K_z} = c \quad (63.8)$$

and

$$\frac{d^2R}{dr^2} + \frac{1}{r}\frac{dR}{dr} + \frac{2\pi G}{K_r}R - \frac{1}{2K_r}\frac{1}{r}\frac{d}{dr}(r^2\omega^2) + \frac{K_z}{K_r}c = 0, \quad (63.9)$$

where c is the constant of separation.

The solution of (63.8) satisfying the symmetry condition $\rho(z) = \rho(-z)$ is

$$Z = A\cos\alpha_z z + \frac{c}{\alpha_z^2}, \quad (63.10)$$

where A is a constant of integration. We also write

$$\alpha_z^2 \equiv \frac{2\pi G}{K_z}, \alpha_r^2 \equiv \frac{2\pi G}{K_r}. \quad (63.11)$$

The finite solution at $r = 0$ for the homogeneous part of (63.9) (i.e., its first three terms) is $B J_0(\alpha_r r)$, for it is Bessel's equation of order zero (see (59.7)). Adding a special solution of the full inhomogeneous equation gives the general solution. The inhomogeneous part clearly depends on the form of the rotation curve $\omega(r)$. Two cases are particularly interesting for galactic structure: solid-body rotation $\omega =$ constant, and constant rotational velocity $V_\theta = r\omega =$ constant. In both these cases the special solution is a constant and we may write

$$R = B J_0(\alpha_r r) - \frac{c}{\alpha_z^2} + \frac{\omega^2}{2\pi G} \quad (63.12)$$

for $\omega = $ constant. This formally subsumes the case $V_\theta = $ constant which is obtained by setting $\omega = 0$ in (63.12).

Thus the density distribution is

$$\rho(r, z) = B J_0(\alpha_r r) + A \cos \alpha_z z + \frac{\omega^2}{2\pi G}. \tag{63.13}$$

We see that pressure anisotropy just produces different scale heights in r and z. Having noticed this, we now suppose for further simplicity that the pressure is isotropic and $\alpha_r = \alpha_z \equiv \alpha$. Moreover, we adopt the dimensionless distances $r_* = r\alpha$ and $z_* = z\alpha$ so that

$$\rho(r_*, z_*) = B J_0(r_*) + A \cos z_* + \frac{\omega^2}{2\pi G}. \tag{63.14}$$

The relation between r_* and z_* for a constant value of ρ gives the isodensity contours. To obtain the constants A and B, we may specify ρ at two places, say at the center $\rho(0, 0) = \rho_c$ and at the edge of the minor axis $\rho(0, z_{*e}) = 0$. More closely related to observations, however, we could specify ρ at the edges of both the major and minor axes: $\rho(r_{*e}, 0) = \rho(0, z_{*e}) = 0$. In this latter case for solid body rotation

$$A = -\frac{[1 - J_0(r_{*e})]}{[1 - J_0(r_{*e}) \cos z_{*e}]} \frac{\omega^2}{2\pi G} \equiv -a \frac{\omega^2}{2\pi G} \tag{63.15}$$

and

$$B = -\frac{[1 - \cos z_{*e}]}{[1 - J_0(r_{*e}) \cos z_{*e}]} \frac{\omega^2}{2\pi G} \equiv -b \frac{\omega^2}{2\pi G}, \tag{63.16}$$

where a and b are positive. Substituting A and B into (63.14) gives a particularly simple result for the spin ratio (58.1)

$$2\pi\sigma^{-2} = \frac{2\pi G\rho}{\omega^2} = 1 - b J_0(r_*) - a \cos z_*. \tag{63.17}$$

This measures the ratio of the local gravitational frequency to the rotation frequency needed for equilibrium.

It is also interesting to examine the shape of the edge of the polytrope. The edge is not a cylinder, although it is axially symmetric. To obtain this shape, $z_*(r_*)$, we set $\rho = 0$ in (63.17) and select values of r_{*e} and z_{*e} to determine a and b in (63.15) and (63.16). The resulting arc cos (J_0) profile is illustrated in Figure 62. It may provide a reasonable fit (when projected, see (40.24)) to the 'box-shaped' central bulges observed in some galaxies like NGC 128, NGC 1381, and NGC 4565. The frontispiece contains an extreme example of such a box galaxy, other examples have rather more rounded profiles.

From these illustrations of very simple thick axisymmetric systems, the reader can begin to imagine the added complexities when there is less symmetry, or none at all. Tri-axial systems manage to maintain some global coherence – at least they have overall axes. When their irregularities are not too great, or too localized, the tensor

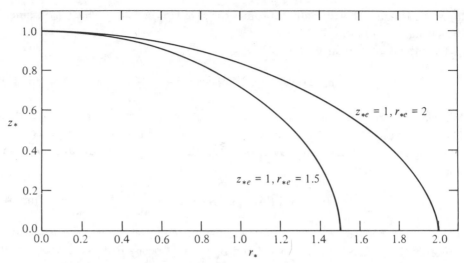

Fig. 62. The arccos (J_0) profile of spinning polytropes with $z_{*e} = 1, r_{*e} = 1.5$ and $z_{*e} = 1, r_{*e} = 2$.

virial equations provide some useful information (Binney, 1978). If the irregularities become greater, self-consistent currents of stars and pressure anisotropies become critical to maintaining a quasi-stable figure. Then the equations of stellar hydrodynamics, which govern polytrope and tensor virial models, are much harder to truncate consistently.

A somewhat more general approach (e.g., Vandervoort, 1980; Petrou, 1983) starts by postulating a plausible form for the stationary one-particle distribution function $f(\mathbf{r}, \mathbf{v})$. We saw from Jeans' theorem (Section 39) that for axisymmetric systems $f = f(E, L_z)$. Near the center of such systems $L_z \to 0$ and $f \to f(E)$ so the velocity distribution there becomes isotropic. Moreover, by applying the analog of (39.11) we see that the net rotation $\bar{v}_z$ depends just on the part of $f(E, L_z)$ which is odd in L_z. This odd part of f does not contribute to the density, so models with the same density can rotate quite differently. Physically one can understand this by standard trick of reversing the sense of rotation in some of the orbits. Thus one can use physical properties of the system to guess the form of $f(E, L_z)$. Then one computes the resulting density distribution and examines its stability. Such stability analyses are seldom easy tasks.

Extending this distribution function approach to triaxial systems is imaginably more difficult. Only for special cases does L_z remain an integral of the motion. We saw in Section 60 that long term stability requires the existence of third integrals of the motion. But we cannot generally find the third integral without knowing the density distribution $\rho(\mathbf{r})$ which determines the smooth orbits, and we cannot find the density distribution from $f(E, L_z, I_3)$ without knowing the third integral. For nearly spherical systems, we can try to break this deadlock by guessing that I_3 depends on the amplitude L of the total angular momentum. Perhaps $I_3 \approx L^2 - L_z^2 = L_\perp^2$ in some simple cases (see (39.10)). The form of the third integral is not unique since any function of E, L_z, and I_3 is also a third integral. However a good guess for I_3 can lead to informative models. Good guesses are harder to make for systems which are neither triaxial nor nearly spherical. For specific insights into more irregular systems we turn to computer N-body experiments.

63.2. Computer experiments

Adventurous computer experiments show that many irregular galaxies can be quasi-stable. A useful amount of quasi-stability is about 10–20 dynamical crossing times, approaching the age of the galaxy. Third integrals may not last forever, but while they operate they can be very effective. Gerhard (1983) computed an amusing example. He found it in an ensemble of galaxy merges. A triaxial remnant of an earlier merger collided head-on with a disk–halo galaxy. The collision was parabolic; the galaxies merged. After several gross oscillations the system settled down. Its inner parts were prolate, its outer ones oblate. The intermediate transition region was triaxial. Although this prolate–triaxial–oblate galaxy had no global axes it was stable for at least 17 crossing times. Velocity anisotropy, changing throughout the galaxy, supported its irregular shape. Rotation was not important, although the central prolate region slowly tumbled about.

Rotation's importance can be measured by the ratio of the rotation velocity to the velocity dispersion. This parameter varies throughout irregular models, and is usually small, say between 0.05 and 0.5. Its value at the center is sometimes compared with observations, but these comparisons are ambiguous: the central velocity dispersion characterizes varied dynamical processes in the nucleus (Sections 38–53) which may be quite independent of the rest of the galaxy. Although comparisons throughout the galaxy are more valuable, they are distorted by projection effects.

Gerhard's example also illustrates these projection effects. They show up especially well in the twisted density contours. These contours can be converted to isophotes by making the simple but dangerous assumption of constant mass–luminosity ratio. The ellipticity of such isophotes depends strongly on the viewing angle of their beholder. Apparent ellipticity for a given model may remain constant with radius, increase, decrease, or do both. Projected position angles of the isophotes' axes also show this varied behavior. Irregular galaxies do not reveal much of their shape from a two-dimensional projection.

Straightforward N-body experiments have the advantage that they show how a model galaxy evolves from its initial formation, and how much of its initial structure remains. If the galaxy reaches a quasi-stable state it is easy to examine its persistent properties self-consistently and in detail. The computation automatically takes any non-classical integrals of motion into account. But the disadvantage of N-body experiments is that we do not know in advance how the result will turn out. So it may not be straightforward to reproduce the observed, or otherwise desired, shapes of galaxies. To answer some questions about the quasi-stability of a given shape, another procedure is useful (Schwarzschild, 1979; Heisler, Merritt & Schwarzschild, 1982; Richstone, 1982).

Back in Section 7 we saw that solving the collisionless Boltzmann equation for systems where only the mean field is important is equivalent to knowing the orbits of all the stars. Therefore, instead of trying to guess a form for $f(\mathbf{r}, \mathbf{v}, t)$ and seeing if it gives a quasi-stable solution to the combined Boltzmann and Poisson equations, we can explore galactic structure by following orbits in a given global potential. If the galaxy as a whole has reached a quasi-stable state, then most stellar orbits should themselves be quasi-stable. For this to happen, the orbits must be constrained by non-classical (e.g., third) integrals. But we do not need to know these integrals explicitly. To learn whether a given shape galaxy can exist quasi-stably we can numerically compute orbits in the given density distribution and see if they really are quasi-stable. This is easier than a full N-body simulation. Computing orbits for about $10^2 - 10^3$ periods should give a good idea of their stability during the age of the galaxy, assuming the mean field does not change substantially (which it may when underlying processes redistribute gas and stars). In $\sim 10^3$ periods the cumulative effects of very high order resonances are not likely to overwhelm the orbits (unlike the stability of solar system orbits where $\sim 10^9$ periods have elapsed) so extreme accuracy is not usually needed.

Which orbits? We would not learn much from an orbit circling round the galaxy at a great distance. Nor would a bunch of orbits be helpful if the stars populating it contribute a gravitational potential different from that of the desired galactic potential. We must use the very orbits which produce the gravitational potential which results from the given density distribution. Then all will be self-consistent. The main problem is to find these relevant orbits.

Numerical techniques are available for finding large classes of approximately self-consistent orbits and using them to judge quasi-stability (see the last three references). Several axisymmetric and triaxial examples suggest that quasi-stability may be quite common in galaxies like those observed. This agrees with impressions from full N-body simulations. Until there are more general results, however, we will have to rely on libraries of special examples.

64

Gravitational shocks

Awaiting the sensation of a short sharp shock

Gilbert and Sullivan

Leaving behind the properties of a single isolated system we now begin to examine what happens when two systems interact. Sections 48.2 and 58 already showed how tides can transfer energy and angular momentum. Now we explore some effects of tidal transfer on the internal structure of stellar systems. Since stellar systems have many more degrees of freedom than the interior of an individual star they may respond to external forces in more varied ways. This section describes the basic response of a spherical cluster which passes quickly through the plane of a flat galaxy. Section 65 describes the distortions of two flat systems as they attempt to move slowly past one another.

'Quick' and 'slow' are relative terms. We saw in Section 47 that what counts is the ratio of the orbital period (or quasi-period) to the timescale for external forces to change. If this ratio is small we can use adiabatic invariants or, more generally, multiple timescale expansions to examine such phenomena as orbit segregation. If this ratio is large we can use the impulsive approximation. A sudden impulsive change of the external field produces a gravitational shock.

It is certainly possible, and probably even common, for a system to contain some orbits which respond smoothly, and others which are shocked, by the same external change. Different parts of the system may respond differently, leading to complicated internal behavior. For example, the halo of a spherical cluster is more easily shocked than its core.

As a globular cluster moves through a galaxy, it takes the brunt of shocks from many sources. Other globular clusters, giant gas clouds, and the mean field of the galaxy itself may all change quickly along the cluster's orbit. Consider, as an illustration, the compressive shock which occurs when a globular cluster crosses the plane of a galaxy (Ostriker, Spitzer & Chevalier, 1972).

This shock is compressive because, from the dynamical viewpoint of a star on a long period orbit, the galactic plane suddenly appears to materialize in the center of the cluster, and then, just as suddenly, vanishes away. The resulting tidal forces momentarily causes the cluster to contract. Then the cluster expands and vibrates. Phase mixing, Landau damping and two-body relaxation gradually redistribute the energy of these vibrations throughout the cluster. When the cluster settles down again it will have a new structure determined by the amount of energy absorbed in the shock. Many successive shocks may transfer enough energy into the cluster to speed up its evaporation and trigger thermodynamic instabilities.

To estimate the energy transfer suppose the cluster's orbit cuts the galactic plane sharply so the gravitational acceleration $g(z)$ depends only on the vertical height, z, above the plane. The vertical velocity of a star in the cluster, relative to the cluster center at z_c, is the differential velocity $v_z = \dot{z} - \dot{z}_c$. The differential acceleration of this star caused by the differential (i.e., tidal) gravitational force is

$$\frac{dv_z}{dt} = g(z) - g(z_c) \approx \frac{dg}{dz}\bigg|_{z_c} \Delta z \qquad (64.1)$$

to first order, where $\Delta z = z - z_c$. The essence of the impulsive approximation is that the cluster star moves very little during the time the cluster takes to pass through the galaxy's plane, so $\Delta z \approx$ constant. This passage time is related to the approximately constant vertical velocity, v_{z_c}, of the cluster by $dz_c = v_{z_c}\, dt$. Therefore, the integral of (64.1) is approximately

$$\Delta v_z \approx \frac{\Delta z}{v_{z_c}} \int_{-g_m}^{g_m} dg(z_c) = \frac{2g_m}{v_{z_c}} \Delta z, \qquad (64.2)$$

where g_m is the maximum acceleration above and below the plane of symmetry.

During the crossing time $t_z = 2\, z_m/v_{z_c}$, from the height z_m, where $g = g_m$, the change in Δz caused by the shock, i.e., $t_z \Delta v_z$, must be less than Δz. So the condition for the impulsive approximation to hold is $2 z_m \Delta v_z \ll v_{z_c} \Delta z$ or

$$\tfrac{1}{2} v_{z_c}^2 \gg 2 z_m g_m. \qquad (64.3)$$

This just says that the work the galaxy does on a cluster star must be much less than the star's initial energy.

The shock changes a star's energy per unit mass by the amount

$$\Delta E = \Delta(\tfrac{1}{2}\mathbf{v}\cdot\mathbf{v}) = \tfrac{1}{2}[2\mathbf{v}\cdot\Delta\mathbf{v} + (\Delta v)^2] \qquad (64.4)$$

to second order in a Taylor series expansion. Multiple crossings of the plane will produce many shocks during the history of a cluster. But since neither the crossings nor the orbits of stars in the cluster are exactly periodic, $\mathbf{v}$ and $\Delta\mathbf{v}$ are generally uncorrelated from crossing to crossing, so $\langle \mathbf{v}\cdot\Delta\mathbf{v}\rangle = 0$ when averaged over many crossings (or over all stars during one crossing). This leaves $\Delta E = (\Delta v)^2/2$. The time between shocks is about half the period P_c of a cluster orbit, giving an approximate rate of energy increase

$$\frac{dE}{dt} = \frac{(\Delta v)^2/2}{P_c/2} = \frac{4g_m^2(\Delta z)^2}{P_c v_{z_c}^2}. \qquad (64.5)$$

To determine a characteristic timescale τ_s for shocks to alter the cluster's structure significantly, we write $\tau_s dE/dt = T$. The average energy per unit mass T of a star in a virially bound cluster is $W/2$. For a polytropic cluster of mass M, radius R, and index *n*, we may use the result of Section 54.1 to obtain

$$\tau_s = \frac{3}{8(5-n)} \frac{GMP_c v_{z_c}^2}{R(\Delta z)^2 g_m^2} \tag{64.6}$$

Increasing the cluster's central concentration (i.e., increasing *n*) makes its stellar orbits harder to disturb. Halo orbits with large average Δz are considerably easier to disturb than those which stay mostly in a core, in accord with our intuition.

For a numerical estimate of τ_s we may use some fairly typical values for globular clusters in a large spiral galaxy (like our own): $n = 3$, $M = 10^6$ solar masses, $R = 10$ pc, $\Delta z = 5$ pc, $P_c = 10^8$ yr, $v_{z_c} = 100$ km s^{-1}, $g_m = 10^{-8}$ cm s^{-2}. These give $\tau_s = 3 \times 10^{11}$ yr, which scales with these quantities as in (64.6). Only in rather extreme cases will τ_s be less than the age of the galaxy. Moreover, τ_s is also usually much greater than the two-body relaxation timescale τ_R (see Section 42.3) which governs evaporation and internal relaxation. Therefore gravitational shocks are likely to dominate only in exceptional conditions. More generally, they will just enhance the direct tidal stripping and Fokker–Planck evaporation of halo stars from the cluster.

From the gravithermal instabilities discussed in Sections 43–5, we would expect that weak shocks which evaporate the halo would also promote shrinking of the core. This is because the core evaporates faster by two-body relaxation in order to repopulate the halo. When its high energy stars are lost, the remaining core forms a more negative gravitational well. Strong shocks, however, which feed substantial energy directly into the core, should have the opposite effect. They heat the core on timescales shorter than the two-body relaxation time, causing the core to expand. Strong shocks still evaporate the halo, but that now plays a minor role. Computer simulations (Spitzer & Chevalier, 1973) show that this behavior does indeed occur. When τ_s is much longer than τ_R, for weak shocking, the core contracts faster than with no shocking. But as τ_s/τ_R becomes smaller, the core stops shrinking, and when τ_s is several times τ_R, the rapid shock heating reverses the core contraction. The exact value of τ_s/τ_R at which this reversal occurs depends on the detailed structure of the cluster and the definition of the core.

65

Passing–merging

I am a part of all that I have met.

Alfred, Lord Tennyson

Some of the grandest encounters in the Universe occur when galaxies pass by one another. In the jetsam of their tidal debris, we often find circumgalactic evidence for close encounters. Sometimes the outer parts of a galaxy pair are distorted into long bridges and tails. We also find large irregular galaxies which have multiple nuclei, perhaps the recent result of a merger.

For a long time, interest in the subject of galaxy merging was quenched by a straightforward $n\sigma v T$ calculation which gave a very small merger probability over the Hubble age T of the Universe. This result does not depend much on the actual value of the Hubble constant. If the observed value of H were larger the volume would decrease, so the observed number density $n \propto H^3$. Supposing the merger cross section to be roughly proportional to the observed visible cross section gives $\sigma \propto H^{-2}$. (Increasing H means the galaxy is closer and has a smaller linear visible diameter.) The Hubble age $T \approx H^{-1}$. Therefore the resulting probability remains small, regardless of the uncertain current value of H.

The apparent conclusiveness of this argument, however, resulted from the unjustified simplicity of its assumptions. Merging encounters of galaxies are not like random collisions in a gas of particles. For one thing, most mergers within groups and clusters of galaxies occur among the more massive galaxies which tend to collect in the center through mass segregation (Section 46). Near the center they follow complicated resonant orbits based on few-body interactions; these increase the chance of a close encounter. Another condition conducive to merging may have been the initial process of galaxy formation. If galaxies did not have large initial random velocities, near neighbors in the Hubble flow would have attracted each other strongly enough to overcome their small initial peculiar velocities. Some encounters would occur soon after galaxy formation, but others would be delayed until the orbits turned around. A third stimulus to galaxy merging is the redistribution of orbits which occurs when two clusters merge. Moreover, the effective cross section for galaxy merging, which may include the effect of a large halo, is not well determined from observations.

About a hundred observed galaxies are good candidates for mergers. But this is only an order-of-magnitude estimate because the signs of merging are often ambiguous and the selection effects in compiling lists are uncertain. We do not yet know the percentages of galaxies of different types which have resulted from mergers. Nor do we know the overall effects of a merger on the relaxed shape, size, and luminosity of the coalesced galaxy. This is more than a dynamical problem. It involves the effects of star formation and supernovae, as well as the production and redistribution of gas and of dust. Here I just summarize some readily calculable, but still incomplete, dynamical aspects. Further discussion may be found in Alladin & Narasimhan's (1982) comprehensive review.

What determines the dynamical results of a galactic encounter? Clearly, the impact parameter is the most important factor. The velocity of the encounter, relative to the escape velocity from the galaxy is next most important. At relatively low velocities, the spins of the galaxies also become important. Their detailed shapes and density distributions usually play a lesser, but still significant role.

Most results in galaxy merging are very specific to detailed models and computer simulations. But there are some simple limiting cases which provide useful insight. At high relative velocities and large impact parameters the impulsive tidal shocks of Section 64 describe most of the energy transfer from the orbits into the galaxies. Examples occur near the centers of rich clusters of galaxies, where the velocity dispersion may be one or two thousand $km\,s^{-1}$, several times the escape velocity from the outer regions of an individual galaxy.

The order-of-magnitude effect of a distant, high velocity, impulsive encounter follows directly from (64.2) and (64.4). If the impact parameter is p, the maximum acceleration $g_m \approx GM/p^2$, and $\Delta z \approx r_{gal}$, then for ΔV approximately perpendicular to V (appropriate to a distant encounter) we obtain

$$\Delta E = \Delta T \approx \tfrac{1}{2} M \left(\frac{2GM}{p^2 V} \right)^2 r_{gal}^2. \tag{65.1}$$

This result is essentially (48.22) because in this limit, where the galaxy remains spherical, the processes are very similar to stellar tidal interactions. The ratio $\Delta E/E$ may be written as

$$\frac{\Delta T}{T} \approx \left(\frac{GM^2}{r_{gal}} \right)^2 (\tfrac{1}{2}MV^2)^{-2} (r_{gal}/p)^4 \approx \left(\frac{V_{esc} r_{gal}}{Vp} \right)^4. \tag{65.2}$$

Not unexpectedly, only a small fraction of the orbital energy is removed in this limit, and merging does not happen.

The next limiting case is an encounter at high velocity and small impact parameter. The impulsive approximation still applies (remember we are neglecting any gaseous interactions which may be visually spectacular but dynamically subordinate), but now $(r_{gal}/p) \approx 1$. Moreover, for a nearly head-on encounter there is the possibility that the tidal acceleration is in the same direction as the velocity, so $\mathbf{V} \cdot \mathbf{\Delta V}$ is not negligible during a single encounter. This would multiply (65.2) by a factor $2V/\Delta V \approx (MV^2)(GM^2/r_{gal})^{-1}(p/r_{gal})^2$. Even with this increase the ratio

$\Delta T/T$ would be of order $(V_{esc}/V)^2$, which is still small. Apparently both small velocities and small impact parameters are required for merging.

Not quite. To draw the conclusion of the last paragraph would be to overlook the most important possibility: bound orbits. The case of small velocity and large separation leads to merging by virtue of the small but cumulative and continuous interaction in a bound orbit. One would hardly expect off-hand that the impulsive treatment would give good results for bound orbits. And yet, even the early tidal analyses (Spitzer, 1958) showed that the impulsive approximation often rises above its origins to give a reasonable order-of-magnitude result at low velocities. The reason is partly because the two main non-impulsive effects tend to work in opposite directions, cancelling each other. On the one hand, tidally escaping high energy stars reduce the galaxy's energy below that of the impulsive approximation. On the other hand, tidal distortions enhance the resonances in the system and increase the energy transferred to the galaxy above that of the impulsive approximation. As a result, the impulsive approximation may often be close to more accurate numerical simulations. A detailed analytical understanding of this fortunate coincidence remains an important problem.

Exploiting this coincidence gives a simple rough estimate of tidal effects on bound orbits (Alladin & Parthasarathy, 1978). For two galaxies with about the same mass in roughly circular orbit, we may substitute the orbital radius, a, and period $\tau_{orb} = \sqrt{2}\,\pi a^{3/2}\,(GM)^{-1/2}$ into (64.6) to obtain an approximate disruption time

$$\tau_d \approx \frac{3\pi}{4(5-n)} \frac{a^{9/2}}{(GM)^{1/2}R^3} \approx \left(\frac{a}{R}\right)^3 \tau_{orb}. \qquad (65.3)$$

If the galaxies have different masses M and M_1 and move in an orbit of semimajor axis a and eccentricity e, then the disruption time for M is approximately (65.3) multiplied by $(1-e^2)^3 M^{3/2}(M+M_1)^{1/2}M_1^{-2}$. Eccentric orbits endanger survival.

As with the tidal formation of binary stars (Section 48.2), the energy gained by the galaxies is lost from their mutual orbit. Eventually the shrinking orbit brings the galaxies into contact. If the galaxies have not disrupted completely their remnants may merge. It is easy to obtain the approximate criterion for disruption versus merging. Both processes involve the same energy transfer, so their relative importance is mainly determined by the ratio of binding energies. If about half the tidal energy goes into each galaxy, then the ratio of galaxy binding energy to orbital binding energy gives

$$\frac{\tau_d}{\tau_{merge}} \approx \frac{6}{(5-n)} \frac{a}{R} \frac{M}{M_1}. \qquad (65.4)$$

This does not take spin or changing geometry into account, nor does it consider nuances of defining τ_d and τ_{merge}, which will all be different for each special case. But it does show that merging is more likely if both galaxies are centrally condensed and have similar masses.

In addition to tidal energy transfer, two other processes may help promote merging. They are both more effective for already bound orbits. The first is the slingshot mechanism and its generalizations. We may think of these as strong resonances which eject stars and groups of stars from the orbiting galaxies. Stars which have been tidally detached from their galaxy are especially likely to fall into these resonances. Details depend quite critically on the exact density and velocity (especially spin) distributions. The upshot is loss of orbital energy.

Dynamical friction is the other process which promotes merging. The end of Section 14.3 has already described an example where dynamical friction causes a single heavy star, initially on a circular orbit, to spiral inward and merge with the galaxy's core. When a satellite galaxy orbits around inside the halo of a more massive galaxy, numerical simulations (e.g., Lin & Tremaine, 1983) show that dynamical friction is fairly well described by the standard formulas. Using the estimate of merging time at the end of Section 14.3 gives

$$\tau_{\text{merge}} \approx 5 \times 10^8 \left(\frac{V_0}{100 \,\text{km s}^{-1}} \right) \left(\frac{r_0}{50 \,\text{kpc}} \right)^2 \left(\frac{10^{11} \,\text{M}_\odot}{M} \right) \text{yr} \qquad (65.5)$$

for dynamical friction (with the logarithmic factor equal to unity). Usually tides will be important as well, and fully self-consistent simulations are needed. Therefore, if massive galaxies have massive haloes, we would not expect to find close satellites near them. Whether mergers actually created massive galaxies cannot be solved by these simple timescale analyses but requires much more knowledge of initial conditions.

Finally, we have the case of passing and merging with both small velocity and small impact parameter. As one would expect by applying this limiting case to equations (65.2) or (65.3), the timescales for passing disrupting, and merging are comparable. Although the analysis behind these equations does not apply, numerical simulations show that they lead to the right qualitative conclusion. This is not surprising since it is basically a dimensional result under extreme, but quasi-adiabatic, physical conditions. Naturally, only self-consistent numerical simulations can yield detailed conclusions.

Early simulations of slow, close merging (e.g., Toomre & Toomre, 1972) treated each galaxy as a disk of non-interacting test particles orbiting a central point mass. Two such galaxies moved along a predetermined orbit, and the test particles responded to both central masses. These simulations were already sufficient to portray the complicated structures emerging with plumes, bridges, ridges, and tails. The results depend strongly on details of initial conditions, as well as on projection. By suitably arranging these circumstances it is possible to simulate many peculiar structures observed (and not yet observed!) in close pairs of galaxies. Much of this structure clearly is tidal. However, galactic winds. radiation pressure, and explosive or compact ejection may also be important in some examples.

Simulations which are more self-consistent (e.g., Gerhard, 1981) show that the stars' self-gravity determines the protrait of merger. The main problem in using self-consistent models is that they are subject to all the internal bar and spiral instabilities discussed in earlier sections. Using a small number of stars, artificially enhances these internal instabilities through the increased gravitational graininess. Large numbers of particles are necessary, even though they require much more computing time. The results of self-consistent experiments emphasize the asymmetry and irregularity of merging and the final remnant. Angular momentum is drastically reshuffled by torques and mass loss. A small fraction of all the stars escape from the galaxies, but they may carry off as much as $\sim 20\text{–}30\%$ of the total angular momentum. The remaining core may be oblate, prolate, triaxial or irregular, with anisotropic velocity dispersions. It all depends on the spins and shapes of the galaxies. The central core may be relaxed, but the outer parts may retain enough memory of the initial conditions that we can hope to recognize them in the observations.

66

Problems and extensions

66.1. Uniform ellipsoids and Maclaurin spheroids

To develop some properties of Maclaurin spheroids, start with a uniform ellipsoid of density ρ bounded by the surface

$$\frac{x^2}{a^2} + \frac{y^2}{b^2} + \frac{z^2}{c^2} = 1, \tag{1}$$

with $a \geq b \geq c$. The gravitational potential within this ellipsoid is

$$\phi(x, y, z) = -\pi G \rho (\alpha x^2 + \beta y^2 + \gamma z^2 - \delta), \tag{2}$$

where

$$\alpha = abc \int_0^\infty \frac{d\lambda}{(a^2 + \lambda)\Delta}, \tag{3a}$$

$$\beta = abc \int_0^\infty \frac{d\lambda}{(b^2 + \lambda)\Delta}, \tag{3b}$$

$$\gamma = abc \int_0^\infty \frac{d\lambda}{(c^2 + \lambda)\Delta}, \tag{3c}$$

$$\delta = abc \int_0^\infty \frac{d\lambda}{\Delta} \tag{3d}$$

and

$$\Delta^2 = (a^2 + \lambda)(b^2 + \lambda)(c^2 + \lambda). \tag{4}$$

Show that the total potential energy of this ellipsoid, integrated over its volume τ, is

$$W = -\tfrac{1}{2} \int_\tau \rho \phi \, d\tau = -\tfrac{8}{15} \pi^2 G \rho^2 a^2 b^2 c^2 \int_0^\infty \frac{d\lambda}{\Delta}. \tag{5}$$

If the ellipsoid is a spheroid with $a = b$ and eccentricity e given by $c^2 = a^2(1 - e^2)$, show that

$$W = -\tfrac{16}{15} \pi^2 G \rho^2 (abc)^{5/3} (1 - e^2)^{1/6} \frac{\sin^{-1} e}{e}. \tag{6}$$

For a configuration rotating with constant angular velocity ω to be in hydrostatic equilibrium, the pressure, gravitational, and centrifugal forces must balance

$$\frac{1}{\rho}\frac{\partial P}{\partial x} = \frac{\partial \phi}{\partial x} + \omega^2 x, \tag{7a}$$

$$\frac{1}{\rho}\frac{\partial P}{\partial y} = \frac{\partial \phi}{\partial y} + \omega^2 y, \tag{7b}$$

$$\frac{1}{\rho}\frac{\partial P}{\partial z} = \frac{\partial \phi}{\partial z}. \tag{7c}$$

Show that Equations (7) have the integral

$$\frac{P}{\rho} = \phi + \tfrac{1}{2}\omega^2(x^2 + y^2) + \text{constant}, \tag{8}$$

where the constant can be chosen to satisfy the boundary condition $P = 0$ on the surface, and therefore $\phi + \omega^2(x^2 + y^2)/2$ is constant on the surface. Substituting Equation (2) into (8), gives the surfaces of equal pressure

$$\left(\alpha - \frac{\omega^2}{2\pi G\rho}\right)x^2 + \left(\beta - \frac{\omega^2}{2\pi G\rho}\right)y^2 + \gamma z^2 = \text{constant}. \tag{9}$$

In order that one of these equilibrium surfaces of equal pressure coincide for some value of ω with the ellipsoid (1), show that the condition

$$(a^2 - b^2)\int_0^\infty \left\{\frac{a^2 b^2}{(a^2 + \lambda)(b^2 + \lambda)} - \frac{c^2}{c^2 + \lambda}\right\}\frac{d\lambda}{\Delta} = 0 \tag{10}$$

must be satisfied. Moreover, the angular velocity is then related to the axes by

$$\frac{\omega^2}{2\pi G\rho} = \frac{a^2\alpha - b^2\beta}{a^2 - b^2} = \frac{b^2\beta - c^2\gamma}{b^2}$$

$$= \frac{ac}{b}(b^2 - c^2)\int_0^\infty \frac{\lambda\, d\lambda}{(b^2 + \lambda)(c^2 + \lambda)\Delta}. \tag{11}$$

This is just the spin ratio (58.1).

There are two solutions of (10). If $a = b$, we get the Maclaurin spheroids. If $a \neq b$, the condition that the integral vanish gives the Jacobi ellipsoids. For the Maclaurin spheroids show that

$$\frac{\omega^2}{2\pi G\rho} = \frac{3 - 2e^2}{e^3}(1 - e^2)^{1/2}\sin^{-1}e - \frac{3}{e^2}(1 - e^2), \tag{12}$$

independent of its size. Letting $r \equiv (abc)^{1/3}$, derive the total angular momentum L and kinetic energy T of the Maclaurin spheroid of mass $M = 4\pi\rho a^2 c/3$

$$\frac{L^2}{GM^3 r} = \frac{6}{25}\frac{\omega^2}{2\pi G\rho}(1 - e^2)^{-2/3} \tag{13}$$

and

$$T = \frac{3}{10}\frac{a^2}{r^2}\frac{\omega^2}{2\pi G\rho}\frac{GM^2}{r}. \tag{14}$$

Show that the maximum angular velocity is given by

$$\frac{\omega^2}{2\pi G\rho} = 0.2247 \tag{15}$$

for $e = 0.9299$ and $a/c = 2.7198$. Also show that the kinetic energy is zero for $e = 0$, rises to a maximum 0.1719 at $e = 0.9912$ and then becomes zero for an infinite disk. (R.A. Lyttleton, 1953, *The Stability of Rotating Liquid Masses*. Cambridge: Cambridge University Press).

66.2. Simple evolution of bars

Suppose the bar structure in a galaxy evolves by losing considerable angular momentum, but negligible mass, from the end of the bar. Idealize this model further by considering a constant mass rotating at the end of a rigid rod. Let the angular momentum change by a small amount keeping the mass and energy constant, but letting the length and angular velocity of the rod adjust. Show that removing angular momentum causes the length to decrease and the angular velocity to increase. Make up other simple models along these lines and see how they react to change.

66.3. Distribution function for a uniformly rotating disk of stars

Consider an equilibrium model of a rotating disk of stars with surface density (in cylindrical coordinates)

$$S(r) = S_0[1 - (r/R)^2]^{1/2} \quad \text{for } r \leq R,$$

$$= 0 \qquad\qquad \text{for } r > R. \tag{1}$$

Show that its gravitational potential has the simple form

$$\phi(r) = \tfrac{1}{2}\Omega_0^2 r^2 \quad \text{for } r \leq R, z = 0, \tag{2}$$

and that the orbits oscillate harmonically in x and y coordinates with frequency Ω_0 related to the central density S_0 and total mass M by

$$\Omega_0^2 = \frac{\pi^2 G S_0}{2R} = \frac{3\pi G M}{4R^3}. \tag{3}$$

Let E and J be the energy and angular momentum, per unit mass of a star, in units where $\Omega_0 = S_0 = R = 1$ and consequently $G = 2\pi^{-2}$. Show that a consistent distribution function $f(\mathbf{x}, \mathbf{v})$ is

$$f(E, J) = [2\pi(1 - \Omega^2)^{1/2}]^{-1}[1 - \Omega^2 - 2(E - \Omega J)]^{-1/2} \tag{4}$$

when the second factor in brackets is positive, and $f = 0$ otherwise. Show that in a coordinate system rotating with angular velocity Ω the velocity distribution is isotropic, so in this sense Ω represents a mean angular velocity. Are these disks stable? (A.J. Kalnajs, 1972, Ap. J., **175**, 63).

474   *Part* IV : *Finite flattened systems*

66.4. Models of thick disks

Many models of flattened galaxies are either infinitesimally thin disks (represented by just a surface density), or chunky ellipsoids. Intermediate models are possible, and one sometimes used is a disk with an exponential or other simple vertical density distribution. It is a disk in the sense that radial gradients are much smaller than vertical gradients. To derive such a structure, we may use (63.6) for a polytropic equation of state (with all the assumptions that implies), and neglect the radial gradients. Using the central density ρ_c in the plane to normalize the density

$$\rho_* = \rho/\rho_c, \tag{1}$$

and normalizing the vertical scale by

$$z_*^2 = \frac{4\pi G}{K\gamma\rho_c^{\gamma-2}} z^2 \tag{2}$$

gives

$$\rho_* \frac{d^2\rho_*}{dz_*^2} + (\gamma - 2)\left(\frac{d\rho_*}{dz_*}\right)^2 = -\rho_*^{4-\gamma}. \tag{3}$$

Show that (3) has a first integral

$$\left(\frac{d\rho_*}{dz_*}\right)^2 = c_1\rho_*^{4-2\gamma} - \frac{2}{\gamma}\rho_*^{4-\gamma}. \tag{4}$$

Solve this for the isothermal case ($\gamma = 1$) with suitable boundary conditions to find

$$\rho_* = \operatorname{sech}^2 \frac{z_*}{\sqrt{2}}, \tag{5}$$

which decreases exponentially at large z_*. For $\gamma = 2$, show that

$$\rho_* = \cos z_* \tag{6}$$

with a boundary at $z_* = \frac{1}{2}\pi$.

Solve for the vertical component of the gravitational force in these disks and use it to examine the properties of large amplitude vertical orbits. Suppose the disks were arbitrarily truncated at a finite radius. How would general tube and box orbits behave? Would the system be stable?

66.5. Origin of the density profiles of elliptical galaxies

There have been many attempts to explain why equation (56.2) provides such a good fit to the surface brightness of elliptical galaxies. Computer simulations of collapsing stellar systems show that one of the important requirements to produce (56.2) is a strong peak of the stellar energy distribution near $E = 0$. Why should this be, and what initial conditions lead to this distribution? (T.S. van Albada, 1982, *MNRAS*, **201**, 939). Such profiles may be observed to result from a galaxy merger (F. Schweizer, 1982, *Ap. J.*, **252**, 455).

66.6. Orbits and adiabatic invariants in a triaxial galaxy

Often it is possible to make considerable analytical progress toward understanding the nature of orbits in triaxial systems by using averaging techniques. If the non-linear terms in the potential change slowly over space, they generally modify the orbits on long timescales, i.e., with low frequencies. Thus spatial changes are related to temporal changes, and techniques related to adiabatic invariants and two-time analyses (such as in Section 47) can be used to develop the orbits. Adiabatic invariants are not discussed at any length in the present book, since it does not emphasize orbit theory. Many standard accounts are available. An excellent introduction is Born's *The Mechanics of the Atom* (1924, reprinted by F. Ungar Publishing Co., New York). Adiabatic invariants, first developed for celestial mechanics, later became useful in atomic structure, and are now important in some areas of galactic structure. Perhaps this is a slightly ironic result of Born's student days when he computed planetary orbits in the Breslau Observatory. The tedium of the task drove him out of astronomy and into physics (see Born, 1969, *Physics in My Generation*, New York: Springer-Verlag), thereby promoting several sciences. For an application of these methods to triaxial galaxies, see de Zeeuw & Merritt, 1983, *Ap. J.*, **267**, 571.

67

Unanswered questions

To follow knowledge, like a sinking star,
Beyond the utmost bound of human thought.

Alfred, Lord Tennyson

Ask a fundamental astronomical question, and the chance of an accurate answer is small. Throughout this account of gravitating systems, nearly each section ends on an incomplete note. We are just beginning to understand the richness and subtlety implicit in a system whose components interact with a force as simple as an inverse square. On all scales, from the sun and its planets to cosmic clusters of galaxies, flock insistent but unanswered questions.

For how long will the solar system be stable? Secular perturbations that grow as the planets follow their courses may eventually end in a resonance that forces ejection. Analytic theory gives us some reassurance, but not a definite answer. Computer simulations do not have the accuracy needed. And if we think of our solar system itself as an analog computer, then the calculation has not yet been done. All we know for certain is that the orbits of planets remaining today have avoided ejection for billions of years. Someday the sun may not rise tomorrow.

The uncertainties of planets orbiting the sun are relatively tame compared to the few-body problem whose masses nearly are equal. More opportunities for resonance flourish. A star hardly knows which way to go. Are computer experiments the only method for predicting the outcome from given initial conditions? How do associations of stars in our galaxy disrupt?

Globular clusters are the largest stellar groups moving within most normal galaxies. Their origin is the most pressing basic question about them. Do they predate galaxy formation and tell us about the earlier Universe, or do they grow within galaxies while they are forming and tell us about early galactic dynamics? Do they do both?

Nuclei of galaxies remain an enigma. Long regarded as an abode of black holes and mysterious forces, they are starting to reveal some of their secrets. How much are they determined by galactic dynamics, and how much do they influence the form of their galaxy?

Galaxies themselves are so closely related to the rest of the Universe that they may hint of an underlying cosmology. No definite answers to basic questions about the origins of galaxies are available yet. Was the Universe always inhomogeneous on the scales and with the amplitudes required for galaxy growth? Or was it initially smooth but subject to subtle instabilities which our searches so far have failed to uncover? What is the minimum information which the early Universe must store if galaxies are to develop? We know many types of initial conditions that suffice for some sort of galaxy formation, but are they necessary and do they really produce the various structures we see today? Were most galaxies formed by coalescing small pieces, or by fragmenting out of much larger clouds? Do dwarf galaxies provide a significant clue?

Once galaxies form, how do they evolve? From where does their angular momentum come? Is it primarily innate or tidal? How is it redistributed as the galaxy ages? What are the properties of galaxies which arise from the wrecks of earlier mergers? How does the external environment alter internal galactic dynamics? What are the effects of the infall of gas? What is the role of dark matter?

Galaxies are gregarious and like to form groups; gravity pulls them together. Does the resulting dynamical dissipation remove all the signs of initial conditions? Are there still clues to cosmogony in the structures of clusters? Beyond the scale of even large clusters like Coma there remain correlations in positions of galaxies. Nor are their velocities entirely random. Can we find more clues to cosmology here?

All these are just representative problems, and when they are solved they will generate more. Their solutions are not to be sought only within realms of gravitational physics, but the ideas and techniques of gravitational physics will play a major role in converting relevant observations into understanding. After great efforts spent on basic astrophysical questions it may seem amazing that we still know so little about them. And yet, it is without doubt even more amazing that we have learned so much.

68

Bibliography

Section 56:

Arp, H.C., 1966. 'Atlas of peculiar galaxies', *Ap. J. Supp.*, **123**, 1.

Disney, M.J., 1976. 'Visibility of galaxies', *Nature*, **263**, 573.

Hubble, E.P., 1930. 'Distribution of luminosity in elliptical nebulae', *Ap. J.*, **71**, 231.

Kormendy, J., 1982. 'Observations of galaxy structure and dynamics', in *Morphology and Dynamics of Galaxies*, L. Martinet & M. Mayor (eds.) (Saverny: Geneva Observatory).

Reynolds, J.H., 1913. 'The light curve of the Andromeda Nebula (NGC 224)', *MNRAS*, **74**, 132.

Roberts, M.S., 1972. 'Galaxies: landmarks of the universe', in *The Emerging Universe*, W.C. Saslaw & K.C. Jacobs (eds.) (Charlottesville: Virginia UP), p. 90.

Rubin, V.C., 1982. 'Systematics of HII rotation curves', in *Internal Kinematics and Dynamics of Galaxies, IAU Symposium No. 100*, E. Athanassoula (ed.) (Dordrecht: Reidel), p. 3 (see also the papers following this one).

Sandage, A., 1961. *The Hubble Atlas of Galaxies.* (Washington: Carnegie Institution).

Saslaw, W.C., 1970. 'On the rotation curves of Sb and Sc galaxies', *Ap. J.*, **160**, 11.

Sersic, J.L., 1982. *Extragalactic Astronomy* (Dordrecht: Reidel).

Thuan, T.X. & Romanishin, W. 1981. 'The structure of giant elliptical galaxies in poor clusters of galaxies', *Ap. J.*, **248**, 439.

de Vaucouleurs, G., 1948. 'Researches on the extragalactic nebulae', *Ann. d'Ap.*, **11**, 247.

de Vaucouleurs, G., 1959. 'Classification and morphology of external galaxies', in *Hand. d. Phys.*, Vol. 53 (Berlin: Springer-Verlag) pp. 275–310.

de Vaucouleurs, G., de Vaucouleurs, A. & Crowin, H.G., Jr, 1976, *Second Reference Catalogue of Bright Galaxies* (Austin: Texas UP).

Section 57:

Lamb, H., 1932. *Hydrodynamics* (Cambridge UP).

Section 58:

Efstathiou, G. & Jones, B.T., 1980. 'Angular momentum and the formation of galaxies by gravitational instability', *Comments on Astrophys.*, **8**, 169.

Harrison, E.R., 1971. 'On the origin of galactic rotation', *MNRAS*, **154**, 167.

Hoyle, F., 1949. 'The origin of the rotations of the galaxies', in *Problems of Cosmical Aerodynamics* (Dayton: Central Air Documents Office), p. 195.

Saslaw, W.C., 1971. 'Rotation curves of turbulent protogalaxies', *MNRAS*, **152**, 341.

Section 59:

Burbidge, E.M., Burbidge, G.R. & Prendergast, K., 1959. 'The rotation and mass of NGC 2146', *Ap. J.*, **130**, 739.

Crampin, D.J. & Hoyle, F., 1964. 'On the angular momentum distribution in the disks of spiral galaxies', *Ap. J.*, **140**, 99.

MacMillan, W.D., 1958. *The Theory of the Potential* (New York: Dover).

Mestel, L., 1963. 'On the galactic law of rotation', *MNRAS*, **126**, 553.

Nordsiek, K.H., 1973. 'The angular momentum of spiral galaxies. I. Methods of rotation curve analysis', *Ap. J.*, **184**, 719.

Toomre, A., 1963. 'On the distribution of matter within highly flattened disks', *Ap. J.*, **138**, 385.

Section 60:

Aarseth, S.J., 1966. 'Third integral of motion for high-velocity stars', *Nature*, **212**, 57.

Blaauw, A. & Schmidt, M., 1965. *Galactic Structure* (Chicago UP).

Contopoulos, G., 1963. 'On the existence of a third integral of motion', *A. J.*, **68**, 1.

Goldstein, H., 1980. *Classical Mechanics* 2nd edn (Reading, Mass.: Addison-Wesley).

Hénon, M. & Heiles, C., 1964. 'The applicability of the third integral of motion: some numerical experiments', *A. J.*, **69**, 73.

Mandelbrot, B., 1982. *The Fractal Geometry of Nature* (San Francisco: W.H. Freeman and Co.).

Section 61:

Athanassoula, E., 1983 (ed.). *Internal Kinematics and Dynamics of Galaxies, IAU Symp. 100* (Dordrecht: Reidel).

Freeman, K.C., 1966. 'Structure and evolution of barred spiral galaxies. II', *MNRAS*, **134**, 1.

Goldreich, P. & Lynden-Bell, D., 1965. 'Gravitational stability of uniformly rotating disks', *MNRAS*, **130**, 97.

Kalnajs, A.J., 1972. 'The equilibria and oscillations of a family of uniformly rotating stellar disks', *Ap. J.*, **175**, 63.

Ostriker, J.P. & Peebles, P.J.E., 1973. 'A numerical study of the stability of flattened galaxies: or, can cold galaxies survive?', *Ap. J.*, **186**, 467.

Sellwood, J.A., 1981. 'Bar instability and rotation curves', *Astron. Astrophys.*, **99**, 362.

Toomre, A., 1964. 'On the gravitational instability of a disk of stars', *Ap. J.*, **139**, 1217.

Toomre, A., 1977. 'Theories of spiral structure', *Ann. Rev. Astron. Astrophys.*, **15**, 437.

Vandervoort, P.O., 1982. 'The dynamical instability of a rotating, axisymmetric galaxy with respect to a deformation into a bar', *Ap. J. Lett.*, **256**, L41.

Section 62:

Bertin, G., 1980. 'On the density wave theory for normal spiral galaxies', *Phys. Rep.*, **61**, 1.

Hunter, C., 1970. 'Stellar hydrodynamic equations for a thin disk galaxy', *Studies in App. Maths.*, **49**, 59.

Kalnajs, A.J., 1971. 'Dynamics of flat galaxies. I', *Ap. J.*, **166**, 275.

Lin, C.C. & Lau, Y.Y., 1979. 'Density wave theory of spiral structure of galaxies', *Studies in App. Maths.*, **60**, 97.

Toomre, A., 1969. 'Group velocity of spiral waves in galactic disks', *Ap. J.*, **158**, 899.
Toomre, A., 1981. 'What amplifies the spirals?', in *The Structure and Evolution of Normal Galaxies*, S.M. Fall and D. Lynden-Bell (eds.) (Cambridge UP).

Section 63:

Binney, J., 1978. 'On the rotation of elliptical galaxies', *MNRAS*, **183**, 501.
Gerhard, O.E., 1983. 'A quasi-stable stellar system with prolate inner and oblate outer parts', *MNRAS*, **202**, 1159.
Heisler, J., Merritt, D. & Schwarzschild, M., 1982. 'Retrograde closed orbits in a rotating triaxial potential', *Ap. J.*, **258**, 490.
Petrou, M., 1983. 'Models for elliptical galaxies – II. Oblate spheroids with realistic rotation curves', *MNRAS*, **202**, 1209.
Richstone, D.O., 1982. 'Scale-free models of galaxies. II. A complete survey of orbits', *Ap. J.*, **252**, 496.
Schwarzschild, M., 1979. 'A numerical model for a triaxial stellar system in dynamical equilibrium', *Ap. J.*, **232**, 236.
Vandervoort, P.O., 1980. 'The equilibrium of a galactic bar', *Ap. J.*, **240**, 478.

Section 64:

Ostriker, J.P., Spitzer, L., Jr & Chevalier, R.A., 1972. 'On the evolution of globular clusters', *Ap. J. Lett.*, **176**, L51.
Spitzer, L., Jr & Chevalier, R.A., 1973. 'Random gravitational encounters and the evolution of spherical systems. V. Gravitational shocks', *Ap. J.*, **183**, 565.

Section 65:

Alladin, S.M. & Narasimhan, K.S.V.S., 1982. 'Gravitational interactions between galaxies', *Phys. Reports*, **92**, 341.
Alladin, S.M. & Parthasarathy, M., 1978. 'Tidal disruption and tidal coalescence in binary stellar systems', *MNRAS*, **184**, 871.
Gerhard, O.E., 1981. '*N*-body simulations of disk–Halo galaxies: isolated systems, tidal interactions, and merging', *MNRAS*, **197**, 179.
Lin, D.N.C. & Tremaine, S., 1983. 'Numerical simulations of the decay of satellite galaxy orbits', *Ap. J.*, **264**, 364.
Spitzer, L., 1958. 'Disruption of galactic clusters', *Ap. J.*, **127**, 17.
Toomre, A. & Toomre, J., 1972. 'Galactic bridges and tails', *Ap. J.*, **178**, 623.

Index

Index-learning turns no student pale,
Yet holds the eel of science by the tail.

Alexander Pope